Transition Metals in the Synthesis of Complex Organic Molecules

Transition Metals in the Synthesis of Complex Organic Molecules

Third Edition

Louis S. Hegedus
Colorado State University

Björn C. G. Söderberg
West Virginia University

University Science Books
Sausalito, California

University Science Books
www.uscibooks.com

Production Manager: Mark Ong
Manuscript Editor: John Murdzek
Design: Mark Ong
Compositor: Publishers' Design and Production Services, Inc.
Cover Design: Genette Itoko McGrew
Printer & Binder: Maple-Vail Book Manufacturing Group

This book is printed on acid-free paper.

Library of Congress Cataloging-in-Publication Data

Hegedus, Louis S.
    Transition metals in the synthesis of complex organic molecules / Louis S. Hegedus. -- 3rd ed. /
Björn C. G. Söderberg.
                p. cm.
    Includes index.
    ISBN 978-1-891389-59-7 (alk. paper)
    1. Organic compounds--Synthesis. 2. Organometallic compounds. 3. Transition metal
compounds.   I. Söderberg, Björn C. G.   II. Title.
    QD262.H35 2009
    547'.2--dc22
        2009005653
Printed in the United States of America

10 9 8 7 6 5 4 3 2 1

Brief Contents

Contents

Preface

Since the second edition of this book was published in 1999, thousands of papers have appeared in the literature describing all aspects of transition metals in organic synthesis including novel reactions, new catalysts, ligands, and reaction conditions and applications in synthesis of complex organic molecules. With better computational methods to probe the mechanisms of transition metal promoted reactions, a deeper understanding of the often very complex pathways for catalytic reactions has been obtained. This third edition updates the very dynamic and exponentially growing field of transition metal chemistry including literature references up to early 2008. The number of references included has almost doubled to about 1600.

Transition metal catalyzed polymerization, synthesis of compounds of interest for material research, the use of non-conventional "green" solvents such as water, fluorous solvents, supercritical fluids, and ionic liquids, and reactions employing polymer supported reactants or catalysts have all enjoyed enormous attention. However these reactions are not mechanistically different from the more standard transformations and no significant differences in reactivity or selectivity compared to solution phase reactions in more standard organic solvents can be discerned. The focus of this edition is on the synthesis of small well-defined organic target molecules using more standard reaction conditions.

Chapter 1—*Formalisms, Electron Counting, and Bonding*—and Chapter 2—*Organometallic Reaction Mechanisms*—dealing with the basic principles of transition metal chemistry have not been extensively updated. Chapter 3—*Synthetic Applications of Transition Metal Hydrides*—details advances in homogenous hydrogenations in particular asymmetric hydrogenations, transfer hydrogenations, hydrofunctionalizations, and alkene isomerizations. Chapter 4—*Synthetic Applications of Complexes Containing Metal–Carbon σ-Bonds*—updates and expands on the very rich chemistry of σ-metal complexes especially the synthetically important palladium catalyzed coupling reactions. Chapter 5—*Synthetic Applications of Transition Metal Carbonyl Complexes*—remains the most static of all chapters and only minor revisions have been made. Chapter 6—*Synthetic Applications of Transition Metal Carbene Complexes*—treats stoichiometric and catalytic reactions of metal bound carbenes. The section on metathesis processes has been significantly expanded. The formation and reactivity of metal alkene complexes are described in

Chapter 7—*Synthetic Applications of Transition Metal Alkene, Diene, and Dienyl Complexes*. Chapter 8—*Synthetic Applications of Transition Metal Alkyne Complexes*—covers alkyne complexes in organic synthesis and this chapter has been expanded with examples of the often unique reactions catalyzed by gold, silver, and platinum complexes. In Chapter 9—*Synthetic Applications of η^3-Allyl Transition Metal Complexes*—new examples of η^3-allyl chemistry in synthesis using primarily palladium catalysts. Other transition metals have emerged as competitors or complements to palladium. Finally, Chapter 10—*Synthetic Applications of Transition Metal Arene Complexes*—describes the chemistry of mainly chromium, manganese, iron, and ruthenium η^6-complexes of arenes but also η^2-complexes of osmium and arenes.

Björn C. G. Söderberg
West Virginia University

When it became clear that the previous version of this book was badly in need of updating and revision, I enlisted the aid of Professor Björn Söderberg, a former coworker of mine, to take on this task. I felt he was ideally suited for the job, since he had been annually reviewing the field of transition metals in organic synthesis for a number of years, a task I had done for 25 years previous to him. Although strongly based on the previous version, this revision was done almost exclusively by Professor Söderberg, with minimal help from me. Whatever added value this revision has is due to his diligence and hard work.

Louis S. Hegedus, Professor Emeritus
Colorado State University

Formalisms, Electron Counting, and Bonding (How Things Work)

1.1 Introduction

Why should an organic chemist even consider using transition metals in complex syntheses? There are many reasons. Virtually every organic functional group will coordinate to some transition metal, and upon coordination, the reactivity of that functional group is often dramatically altered. Electrophilic species can become nucleophilic and vice versa, stable compounds can become reactive, and highly reactive compounds can become stabilized. Normal reactivity patterns of functional groups can be inverted, and unconventional (*impossible*, under "normal" conditions) transformations can be achieved with facility. Highly reactive, normally unavailable reaction intermediates can be generated, stabilized, and used as efficient reagents in organic synthesis. Most organometallic reactions are highly specific, able to discriminate between structurally similar sites, thus reducing the need for bothersome "protection–deprotection" sequences that plague conventional organic synthesis. Finally, by careful selection of substrate and metal, multistep cascade sequences can be generated to form several bonds in a single process in which the metal "stitches together" the substrate.

However wonderful this sounds, there are a number of disadvantages as well. The biggest one is that the use of transition metals in organic synthesis requires a new way of thinking, as well as a rudimentary knowledge of how transition metals behave. In contrast to carbon, metals have a number of stable, accessible oxidation states, geometries, and coordination numbers, and their reactivity towards organic substrates is directly related to these features. The structural complexity of the transition metal species involved is, at first, disconcerting. Luckily, a few, easily mastered formalisms provide a logical framework upon which to organize and systematize this large amount of information, and this chapter is designed to provide this mastery.

The high specificity, cited as an advantage above, is also a disadvantage, in that specific reactions are not very general, so that small changes in the substrate can turn an efficient reaction into one that does not proceed at all. However, most transition metal systems carry "spectator" ligands—ligands that are coordinated to the metal but are not directly involved in the reaction—in addition to the organic substrate of interest, and these allow for the fine-tuning of reactivity, and an increase in the scope of the reaction.

Finally, organometallic mechanisms are often complex and poorly understood, making both the prediction and the rationalization of the outcome of a reaction difficult. Frequently, there is a manifold of different reaction pathways of similar energy, and seemingly minor changes in the features of a reaction cause it to take an entirely unexpected course. By viewing this as an opportunity to develop new reaction chemistry rather than an obstacle to performing the desired reaction, progress can be made.

1.2 Formalisms

Every reaction presented in this book proceeds in the coordination sphere of a transition metal, and it is the precise electronic nature of the metal that determines the course and the outcome of the reaction. Thus, a very clear view of the nature of the metal is critical to an understanding of its reactivity. The main features of interest are (1) the oxidation state of the metal, (2) the number of d electrons on the metal in the oxidation state under consideration, (3) the coordination number of the metal, and (4) the availability (or lack thereof) of vacant coordination sites on the metal. The simple formalisms presented below permit the easy determination of each of these characteristics. A caveat, however, is in order. These formalisms are just that—formalisms—not reality, not the "truth," and in some cases, they aren't even chemically reasonable. However, by placing the entire organic chemistry of transition metals within a single formalistic framework, an enormous amount of disparate chemistry can be systematized and organized, and a more nearly coherent view of the field is thus available. As long as it remains consistently clear that we are dealing with formalisms, exceptions will not cause problems.

a. Oxidation State

The oxidation state of a metal is defined as the charge left on the metal atom after *all* ligands have been removed *in their normal, closed-shell, configuration*—that is, *with their electron pairs*. The oxidation state is *not* a physical property of the metal, and it cannot be measured. It is a formalism that helps us count electrons, but no more. Typical examples are shown in **Figure 1.1**. [An alternative formalism, common in the older literature, removes each covalent, anionic ligand as a *neutral* species with a single electron (i.e., homolytically), leaving the other electron on the metal. Although this results

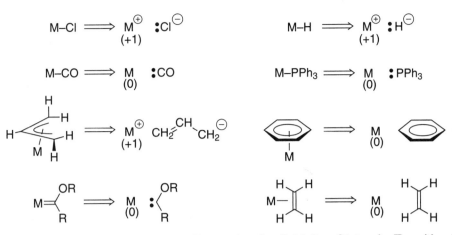

Figure 1.1. Examples of How to Determine the Oxidation State of a Transition Metal

in a different *formal* oxidation state for the metal, it leads to the same conclusions as to the total number of electrons in the bonding shell, and the degree of coordination saturation.]

The *chemical* properties of the ligands are not always consonant with the oxidation state formalism. In metal hydrides, for example, the hydride ligand is *always* formally considered to be H⁻, even though some transition metal "hydrides" are strong acids! Despite this, the formalism is still useful.

b. *d*-Electron Configuration, Coordination Saturation, and the 18-Electron Rule

Having assigned the oxidation state of the metal in a complex, the number of *d* electrons on the metal can easily be assessed by referring to the periodic table. **Figure 1.2** presents the transition elements along with their *d*-electron count. The transition series is formed by the systematic filling of the *d* orbitals. Note that these electron

Group number		4	5	6	7	8	9	10	11
First row	$3d$	Ti	V	Cr	Mn	Fe	Co	Ni	Cu
Second row	$4d$	Zr	Nb	Mo	Tc	Ru	Rh	Pd	Ag
Third row	$5d$	Hf	Ta	W	Re	Os	Ir	Pt	Au

Oxidation state									
0		4	5	6	7	8	9	10	—
I		3	4	5	6	7	8	9	10
II		2	3	4	5	6	7	8	9
III		1	2	3	4	5	6	7	8
IV		0	1	2	3	4	5	6	7

d^n

Figure 1.2. *d*-Electron Configuration for the Transition Metals as a Function of Formal Oxidation State

configurations differ from those presented in most elementary texts in which the $4s$ level is presumed to be lower in energy than the $3d$, and is filled first. Although this is the case for the *free atom* in the elemental state, these two levels are quite close in energy, and, for the *complexes* discussed in this text—which are not free metal atoms, but rather metals surrounded by ligands—the assumption that the outer electrons are d electrons is a good approximation. By referring to the periodic table (or preferably, by remembering the positions of the transition metals), the d-electron count for any transition metal in any oxidation state is easily found.

The d-electron count is critical to an understanding of transition metal organo-metallic chemistry because of the **18-electron rule**, which states "in mononuclear, diamagnetic complexes, the total number of electrons in the bonding shell (the sum of the metal d electrons plus those contributed by the ligands) never exceeds 18" (at least not for very long—see below).[1] This 18-electron rule determines the *maximum* allow-able number of ligands for any transition metal in any oxidation state. Compounds having the maximum allowable number of ligands—that is, having 18 electrons in the bonding shell—are said to be coordinatively saturated—that is, there are *no* remaining coordination sites on the metal. Complexes *not* having the maximum number of ligands allowed by the 18-electron rule are said to be coordinatively unsaturated—that is, they have vacant coordination sites. Since vacant sites are usually required for catalytic processes (the substrate must coordinate before it can react), the degree of coordination is central to many of the reactions presented below.

c. Classes of Ligands

The very large number of ligands that are involved in organotransition metal chem-istry can be classified into three families: (1) *formal* anions, (2) *formal* neutrals, and (3) *formal* cations. These families result from the oxidation state formalism requiring the removal of ligands in their closed-shell [with their pair(s) of electrons] state. Depend-ing on the ligand, it can be either electron donating or electron withdrawing, and its specific nature has a profound effect on the reactivity of the metal center. Ligands with additional unsaturation may coordinate to more than one site, and thus contrib-ute more than two electrons to the total electron count. Examples of each of these are presented below, listed in approximate order of decreasing donor ability.

Formal anionic ligands that act as two electron donors are:

$$R^{\ominus} > Ar^{\ominus} > H^{\ominus} > R\overset{\overset{\textstyle O}{\|}}{C}{}^{\ominus} > halide^{\ominus} \sim CN^{\ominus}$$

These ligands fill one coordination site through one point of attachment and are called "monohapto" ligands, designated as η^1. The allyl group, $C_3H_5^-$, can act as a monohapto (η^1), two-electron donor, or a trihapto (η^3) (π-allyl) four-electron donor (Figure 1.3). In the latter case, *two* coordination sites are filled, and all three carbons are bonded to the metal, but the ligand as a whole is still a formal mono anion. The

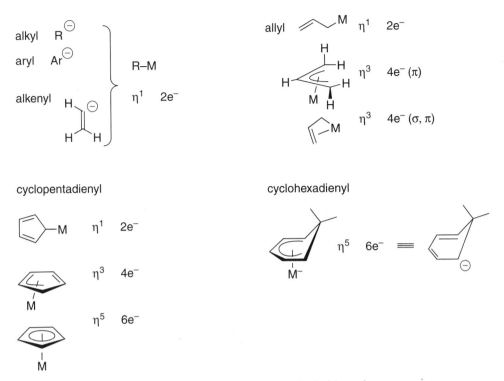

Figure 1.3. Bonding Modes for Mononegative Anionic Ligands

cyclopentadienyl ligand, $C_5H_5^-$, most commonly bonds in an η^5-fashion, filling three coordination sites and acting as a six-electron donor, although η^3 (four electrons, two sites, equivalent to η^3-allyl) and η^4-coordinations are known. The cyclohexadienyl ligand, produced from nucleophilic attack on η^6-arene metal complexes (Chapter 10), is almost invariably a six-electron, mononegative, η^5-ligand that fills three coordination sites. These ligands are illustrated in **Figure 1.3**.

Formal neutral ligands abound, and they encompass not only important classes of "spectator" ligands—ligands such as phosphines and amines introduced to moderate the reactivity of the metal, but which are not *directly* involved in the reaction under consideration—but also ligands such as carbon monoxide, alkenes, alkynes, and arenes, which are often *substrates* (**Figure 1.4**). Ligands such as phosphines and amines are good σ-donors in organometallic reactions, and increase the electron density at the metal, while ligands such as carbon monoxide, isonitriles, and alkenes are π-acceptors, and decrease the electron density at the metal. The reason for this is presented in the next section.

Formally cationic ligands are much less common, since species that bear a full formal positive charge *and* a lone pair of electrons are rare. The nitrosyl group (:NO⁺) is one of these, being a cationic two-electron donor. It is often used as a spectator ligand, or to replace a carbon monoxide, thus converting a neutral carbonyl complex to a cationic nitrosyl complex.

With all of the preceding information in hand, it is now possible to consider virtually any transition metal complex, assign the oxidation state of the metal, assess the

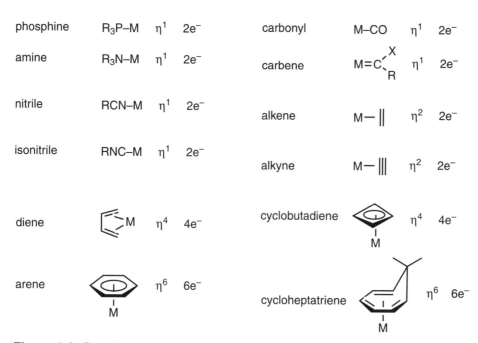

Figure 1.4. Bonding Modes for Neutral Ligands

total number of electrons in the bonding shell, and decide if that complex is coordinatively saturated or unsaturated. For example, the complex $CpFe(CO)_2(C_3H_7)$ is a stable, neutral species, containing two formally mononegative ligands, the propyl (C_3H_7) group, an η^1, $2e^-$ donor ligand, and the Cp ligand, an η^5, $6e^-$ donor ligand, and two neutral carbon monoxide ligands that contribute two electrons each. Since the overall complex is neutral, and has two mononegative ligands, the iron must have a formal 2+ charge, making it Fe(II), d^6. For an overall electron count, there are six electrons from the metal, a total of four from the two CO's, two electrons from the propyl group, and six electrons from the Cp group, for a grand total of $18e^-$. Thus, this complex is an Fe(II), d^6, coordinatively saturated complex. Other examples, with explanatory comments, are given in **Figure 1.5**.

The last example in **Figure 1.5** illustrates one of the difficulties in the strict application of formalisms. Strictly speaking, this treatment of the cyclohexadienyliron tricarbonyl cation is the formally correct one, since the rules state that each ligand shall be removed *in its closed-shell form*, making the cyclohexadienyl ligand a six-electron donor *anion*. However, this particular complex is prepared by a *hydride* abstraction from the neutral cyclohexadieneiron tricarbonyl, making it reasonable to assume that the cyclohexadienyl ligand is a four-electron, donor cation. Furthermore, the cyclohexadienyl group in the complex is quite reactive towards nucleophilic attack (Chapter 7). The true situation is impossible to establish. In reality, the plus charge resides neither exclusively on the metal, as the first, formally correct, treatment would assume, nor completely on the cyclohexadienyl ligand from which the hydride was abstracted, but rather is distributed throughout the entire metal–ligand array. Thus, *neither* treatment represents the true situation, while *both* treatments come to the conclusion that the complex

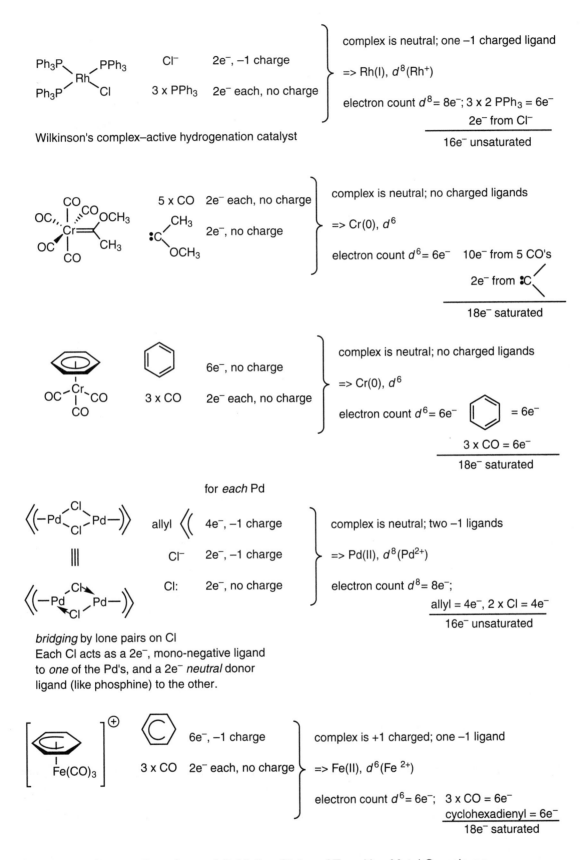

Figure 1.5. Electron Counting and Oxidation States of Transition Metal Complexes

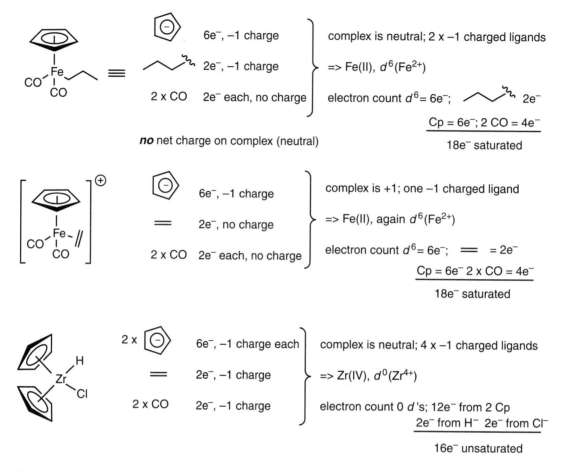

Figure 1.5. *(Continued)*

is an 18e⁻ saturated metal complex. The power of the formalistic treatment is that it is consistent, and does *not* require knowledge of how the complex is synthesized or how it reacts. As long as formalisms are not taken as a serious representation of reality, but rather treated as a convenient way of organizing and unifying a vast amount of information, they will be useful. To quote Roald Hoffmann, "Formalisms are convenient fictions which contain a piece of the truth . . . and it is so sad that people spend a lot of time arguing about the deductions they draw . . . from formalisms without worrying about their underlying assumptions."[2]

1.3 Bonding Considerations[3]

The transition series is formed by the systematic filling of the *d* shell that, in complexes, is of lower energy than the next *s*, *p* level. Thus, transition metals have partially filled *d* orbitals, and vacant *s* and *p* orbitals. In contrast, most of the ligands have filled "*sp*ⁿ" hybrid orbitals and, for unsaturated organic ligands, vacant antibonding π^* orbitals. The *d* orbitals on the metal have the same symmetry *and* similar energy as the antibonding π^* orbitals of the unsaturated ligands (**Figure 1.6**).

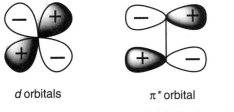

d orbitals π^* orbital

Figure 1.6. Symmetry of *d* Orbitals and π^* Orbitals

Transition metals participate in two types of bonding, often simultaneously. σ-*Donor* bonds are formed by the overlap of filled sp^n hybrid orbitals on the ligand (including π-bonding orbitals of unsaturated hydrocarbons) with vacant "*dsp*" hybrid orbitals on the metal. Ligands that are primarily σ-donors, such as R_3P, R_3N, H^-, and R^-, *increase* the electron density on the metal. Unsaturated organic ligands, such as alkenes, alkynes, arenes, carbon monoxide, and isonitriles, which have π^* antibonding orbitals, bond somewhat differently. They, too, form σ-donor bonds by overlap of their filled π-bonding orbitals with the vacant "*dsp*" hybrid orbitals of the metal. In addition, *filled* d *orbitals* of the metal can overlap with the *vacant* π^* antibonding orbital of the ligand, back-donating electron density from the metal to the ligand (**Figure 1.7**). Thus, unsaturated organic ligands can act as π-acceptors or π-acids, *decreasing* electron density on the metal.

Because of these two modes of bonding—σ-donation and π-accepting "back bonding"—metals can act as an electron sink for ligands, either supplying or accepting electron density. As a consequence, the electron density on the metal, and hence its reactivity, can be modulated by varying the ligands around the metal. This offers a major method for fine-tuning the reactivity of organometallic reagents.

There are two modes of π-back bonding and two types of π-acceptor ligands: (1) longitudinal acceptors, such as carbon monoxide, isonitriles, and linear nitrosyls, and (2) perpendicular acceptors, such as alkenes and alkynes (**Figure 1.8**).

There is ample physical evidence for π-back bonding with good π-acceptor ligands. Carbon monoxide, which always acts as an acceptor, experiences a lengthening of the

organic π-bond as a σ-donor   organic π^* orbital as π-acceptor

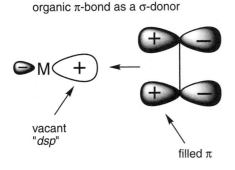

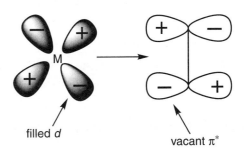

back donation – back bonding

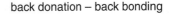

Figure 1.7. Details of π-Bonding

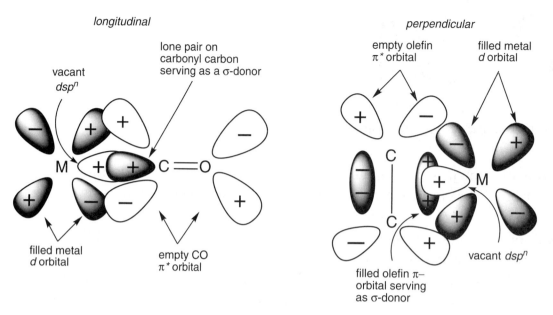

longitudinal   *perpendicular*

Figure 1.8. Types of π-Acceptors

C–O bond and a decrease in C–O stretching frequency in the infrared spectrum upon complexation, both indications of the population of a π^* orbital on CO, decreasing the C–O bond order. The situation is a little more complex with alkenes. With electrophilic metals, such a Pd^{2+} or Pt^{2+}, alkenes are primarily σ-donor ligands, and the C=C bond length in such complexes is virtually the same as for the free alkene. With electron-rich metals, such as Pd(0), substantial back bonding results, the alkene C–C bond is lengthened, and the hybridization at the alkeneic carbons changes towards sp^3. The degree of π-back bonding has a serious effect on the reactivity of π-acceptor ligands. Organometallic processes involving reactions of π-acceptor ligands are the next topic for consideration.

1.4 Structural Considerations

Although the focus of this book is the *reactivity* of metal-bound ligands, it is important to appreciate the geometries adopted by organometallic complexes. For 18-electron (saturated) complexes, these geometries are determined by steric factors, such that four-coordinate complexes such as $Ni(CO)_4$ and $Pd(PPh_3)_4$ are tetrahedral, five-coordinate complexes such as $Fe(CO)_5$ and $(diene)Fe(CO)_3$ are trigonal bipyramidal, and six-coordinate complexes such as $Cr(CO)_6$, Ir(III) complexes, and Rh(III) complexes are octahedral.

In contrast, the d^8 complexes of Pd(II), Pt(II), Ir(I), and Rh(I) find it energetically favorable to form square planar, four-coordinate unsaturated 16e⁻ complexes. The planar geometry makes the $d_{x^2-y^2}$ orbital so high in energy that it remains unoccupied in stable complexes, but enables such complexes to undergo facile ligand substitution by an associative mechanism. This and other mechanistic considerations are the topic of Chapter 2.

References

(1) For a historic account of the 18-electron rule, see Jensen, W. B. *J. Chem. Ed.* **2005**, *82*, 28.

(2) Sailland, J.-V.; Hoffmann, R. *J. Am. Chem. Soc.* **1984**, *106*, 2006.

(3) For a more rigorous treatment of bonding, see Collman, J. P.; Hegedus, L. S.; Finke, R. O.; Norton, J. R. *Principles and Applications of Organotransition Metal Chemistry*, 2nd ed.; University Science Books: Mill Valley, CA, 1987; pp 21–56.

2

Organometallic Reaction Mechanisms

2.1 Introduction

The focus of this entire book is the *organic* chemistry of organometallic complexes, and the effect coordination to a transition metal has on the reactivity of organic substrates. In transition-metal-catalyzed reactions of organic substrates, it is the chemistry of the *metal* that determines the course of the reaction, and a rudimentary understanding of the mechanisms by which transition metal organometallic complexes react is critical to the rational use of transition metals in organic synthesis.

Mechanistic organometallic chemistry is both more complex and less developed than mechanistic organic chemistry, although many parallels exist. The organic chemist's reliance on "arrow pushing" to rationalize the course of a reaction is equally valid with organometallic processes, provided the appropriate rules are followed. Some organometallic mechanisms are understood in exquisite detail, while others are not understood at all. The field is very active, and new insights abound. However, this book treats the topic only at a level required to permit the planning and execution of organometallic processes, and to facilitate the understanding of the literature in the field. More specialized treatises should be consulted for a more thorough treatment.[1]

2.2 Ligand Substitution Processes

Ligand substitution (exchange) processes (**Eq. 2.1**) are central to virtually all organometallic reactions of significance for organic synthesis. For catalytic processes, it is common that a stable catalyst precursor must lose a ligand, coordinate the substrate, promote whatever reaction is being catalyzed, and then release the substrate. Three of the four steps are ligand exchange processes.

Eq. 2.1

$$M-L + L' \rightleftharpoons M-L' + L$$

In other instances, it may be necessary to replace an innocent (spectator) ligand in an organometallic complex to adjust the reactivity of a coordinated substrate (**Eq. 2.2**) or to stabilize an unstable intermediate for isolation or mechanistic studies (**Eq. 2.3**). Thus, a fundamental understanding of ligand exchange processes is central to the utilization of organometallic complexes in synthesis.

Eq. 2.2

Eq. 2.3

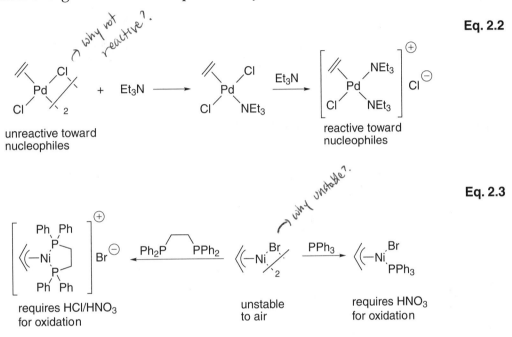

Ligand exchange processes at metal centers share many characteristics with organic nucleophilic displacement reactions at carbon centers, in that the outgoing ligand departs with its pair of electrons, while the incoming ligand attacks the metal with *its* pair of electrons. An obvious difference is that metals can have coordination numbers larger than four, and stable, coordinatively unsaturated metal complexes are common. Thus, the *details* of ligand exchange are somewhat different, and more varied, than typical organic nucleophilic displacement reactions. However, ligand substitution processes can be classified in much the same way as organic nucleophilic displacement reactions, and can involve either two-electron or one-electron processes, in associative (S_N2-like) or dissociative (S_N1-like) reactions (**Figure 2.1**).

By far the most extensively studied ligand exchange reactions are those of coordinatively unsaturated, 16-electron, d^8, square planar complexes of Ni(II), Pd(II), Pt(II), Rh(I), and Ir(I). These typically involve two-electron, associative processes, which resemble S_N2 reactions, but with some differences. The platinum system shown in **Eq. 2.4** is typical.[2] Because the metal is coordinatively unsaturated, apical attack at the vacant site occurs, producing a square pyramidal, saturated intermediate. This

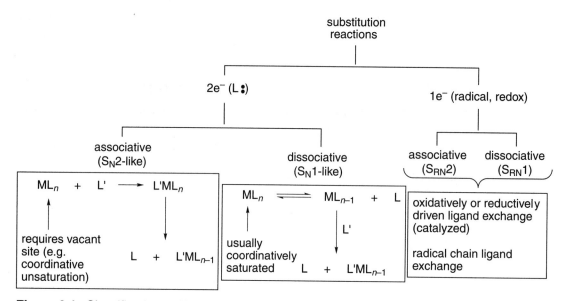

Figure 2.1. Classifications of Ligand Substitution Processes

rearranges via a trigonal bipyramidal intermediate (analogous to the transition state in organic S_N2 reactions) to the square pyramidal intermediate with the leaving group axial; the leaving group then leaves.

Eq. 2.4

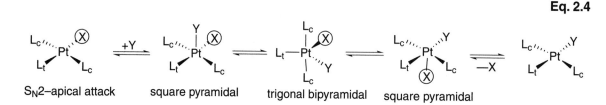

The rate is second order, and depends upon the metal, in the order Ni > Pd >> Pt, with a range of ~10^6. This is thought to reflect the relative abilities of the metals to form five-coordinate, 18-electron intermediates. As a consequence of this rate difference, if a reaction depends on ligand exchange, platinum-catalyzed processes may be too slow to be synthetically useful. The rate also depends on the incoming ligand, Y, in the order R_3P > Py > NH_3, Cl^- > H_2O > OH^-, spanning a range of ~10^5. Furthermore, the rate depends on the leaving group, X, in the order NO_3^- > H_2O > Cl^- > Br^- > I^- > N_3^- > SCN^- > NO_2^- > CN^-. The rate depends on the ligand trans to the ligand being displaced, in the order R_3Si^- > H^- ≈ CH_3^- ≈ CN^- ≈ alkenes ≈ CO > PR_3 ≈ NO_2^- ≈ I^- ≈ SCN^- > Br^- > Cl^- > RNH_2 ≈ NH_3 > OH^- > NO_3^- ≈ H_2O. A practical consequence of this "trans effect" is the ability to activate ligand exchange by replacing the ligand trans to the one to be displaced.[3]

In contrast, 18-electron, *saturated* systems undergo ligand exchange reactions much slower, and then by a dissociative, S_N1-like process. That is, a coordination site must be vacated before ligand exchange can occur. Typical complexes that undergo ligand exchange by this process are the saturated metal carbonyl complexes, such as $Ni(CO)_4$

[M(0), d^{10}, saturated], Fe(CO)$_5$ [M(0), d^8, saturated], and Cr(CO)$_6$ [M(0), d^6, saturated]. Nickel tetracarbonyl is the most labile and most reactive toward ligand exchange (**Eq. 2.5**). The rate of exchange is first order, and proportional to the concentration of nickel carbonyl.

Eq. 2.5

$$Ni(CO)_4 \; \underset{slow}{\rightleftharpoons} \; Ni(CO)_3 \; + \; CO \; \xrightarrow[fast]{L} \; LNi(CO)_3$$

tetrahedral
labile

With ligands other than CO, the rate of ligand exchange can be accelerated by bulky ligands, since loss of one ligand leads to the release of steric strain. This is best illustrated by the nickel(0) (tetrakis)phosphine complexes seen in **Table 2.1**. A measure of the bulk of a phosphine is its "cone angle" (**Figure 2.2**).[4] As the size of the phosphine increases, the equilibrium constant for loss of a ligand changes by more than 10^{10}. A practical consequence is that, in catalytic processes involving loss of a phosphine ligand, reactivity can be "fine-tuned" by alteration of the phosphine ligand.

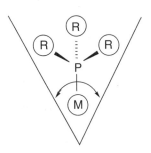

Figure 2.2. Cone Angle

In contrast to nickel tetracarbonyl, iron pentacarbonyl (trigonal bipyramidal) and chromium hexacarbonyl (octahedral) are *not* labile, and are quite inert to ligand substitution, and either heat, sonication, or irradiation (light) is required to affect ligand substitution in these systems. Carbon monoxide displacement in these substitutionally

$$L_4Ni \; \underset{PhH,\,25\,°C}{\overset{K_D}{\rightleftharpoons}} \; L_3Ni{-}L \; + \; L$$

L =	P(OEt)$_3$	P(O-p-tolyl)$_3$	P(O-i-Pr)$_3$	P(O-o-tolyl)$_3$	PPh$_3$
cone angle	109°	128°	130°	141°	145°
K_D	< 10^{-10}	6 x 10^{-10}	2.7 x 10^{-5}	4 x 10^{-2}	no NiL$_4$

Table 2.1. Ligand Dissociation as a Function of Phosphine Cone Angles

inert complexes can also be chemically promoted. Tertiary amine oxides irreversibly oxidize metal-bound carbon monoxide to carbon dioxide, which readily dissociates. Since amines are not strong ligands for metals in low oxidation states, a coordination site is opened (**Eq. 2.6**).[5]

Eq. 2.6

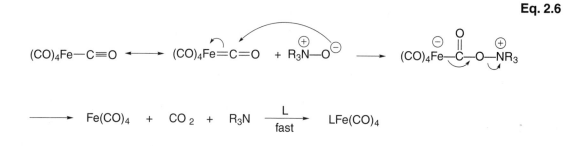

Amine oxides are not particularly powerful oxidants for organic compounds, and have been used to remove metal carbonyl fragments—by "chewing off" the CO's—from organic ligands reluctant to leave on their own. This has been used extensively in organic synthetic applications of cationic dienyliron carbonyl complexes (Chapter 7). Phosphine oxides can *catalyze* CO exchange, presumably by a similar mechanism (**Eq. 2.7**),[6] although in this case the strength of the P–O bond prevents fragmentation and oxidation of CO. Tertiary amine oxides and phosphine oxides are frequently used to accelerated reactions of metal carbonyl complexes, such as in the Pauson–Khand reaction (Chapter 8) and some reactions of Fischer carbenes (Chapter 6).

Eq. 2.7

$$Fe(CO)_5 + R_3PO \longrightarrow (CO)_4Fe-\overset{O}{\underset{}{C}}-O-PR_3 \longrightarrow Fe(CO)_4 + CO + R_3PO \xrightarrow{L} LFe(CO)_4$$

strong P–O bond

Ligand exchanges can also be promoted by one-electron oxidation or reduction of 18e$^-$ systems, transiently generating the much more labile 17e$^-$ or 19e$^-$ systems. However, to date these processes have found little application in the use of transition metals in organic synthesis.

Not all 18e$^-$ complexes undergo ligand substitution by a dissociative mechanism. In these cases, an exchange phenomenon called "slippage" is important. Ligands such as η^3-allyl, η^5-cyclopentadienyl, and η^6-arene normally fill multiple coordination sites. However, even in saturated compounds, they can open coordination sites on the metal by "slipping" to a lower hapticity (coordination number). This is illustrated in **Figure 2.3**.

This "slippage" can account for a number of otherwise difficult to explain observations, including apparent associative S$_N$2-like ligand exchange with coordinatively saturated complexes (e.g., **Eq. 2.8**). With arene complexes in particular, the first step ($\eta^6 \rightarrow \eta^4$) requires the most energy, since aromaticity is disrupted. With fused arene

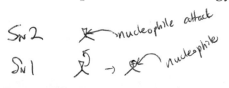

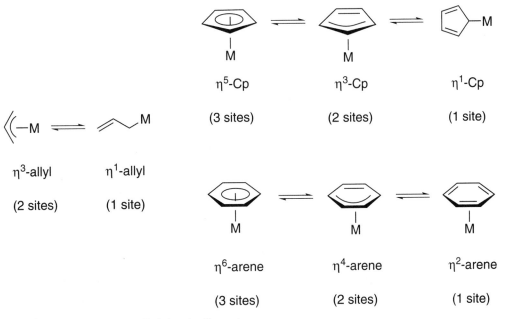

Figure 2.3. Slippage in Polyhapto Complexes

ligands, such as naphthalene or indene, this step is easier, and rate enhancements of 10^3–10^8 have been observed.

Eq. 2.8

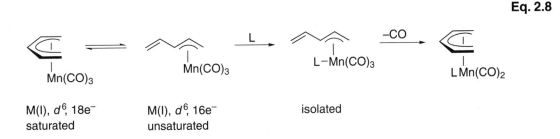

M(I), d^6, 18e⁻   M(I), d^6, 16e⁻   isolated
saturated   unsaturated

2.3 Oxidative Addition–Reductive Elimination[7]

Oxidative addition–reductive elimination processes are central to a vast array of synthetically useful organometallic reactions (e.g., the "Heck" reaction, Chapter 4), and occur because of the ability of transition metals to exist in several different oxidation states, in contrast to nontransition metal compounds, which usually have closed-shell configurations. In fact, it is this facile shuttling between oxidation states that makes transition metals so useful in organic synthesis. The terms "oxidative addition" and "reductive elimination" are generic, describing an overall transformation, but not the specific mechanism by which that transformation occurs. There are many mechanisms for oxidative addition, some of which are presented below. The general transformation is described in **Eq. 2.9**.

Eq. 2.9

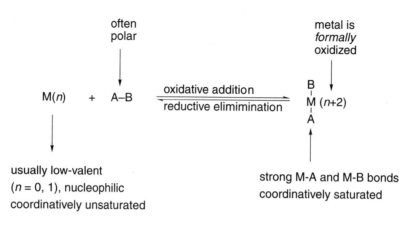

A coordinatively unsaturated transition metal, usually in a low (0, +1) oxidation state and hence relatively electron rich and nucleophilic, undergoes reaction with some substrate, A–B, to form a new complex, A–M–B, in which the metal has formally inserted into the A–B bond. Because of the oxidation state formalism, which demands each ligand be removed with its pair of electrons (although A and B really contribute only one electron each to A–M–B), the metal is formally oxidized in the A–M–B adduct. Since the "oxidant" A–B has added to the metal, the process is called "oxidative addition." In general, ligands such as R_3P, R^-, and H^-, which are good σ-donors and increase the electron density at the metal, facilitate oxidative addition, whereas ligands such as CO, CN^-, and alkenes, which are good π-acceptors and decrease electron density at the metal, suppress oxidative addition. The reverse of oxidative addition is reductive elimination, and both the transformation and terminology are obvious. Oxidative addition–reductive elimination processes are central to the use of transition metals in organic synthesis, because one *need not reductively eliminate the same two groups that were oxidatively added to the metal*. This is the basis of a very large number of cross-coupling reactions (Chapter 4).

The most commonly encountered systems are Ni(0), Pd(0) → Ni(II), Pd(II) (d^{10} → d^8) and Rh(I), Ir(I) → Rh(III), Ir(III) (d^8 → d^6). The most extensively studied system, and the one that contributed heavily to the development of many of the formalisms presented above, is Vaska's compound, the lemon yellow, Ir(I), d^8, 16-electron, coordinatively unsaturated complex that undergoes oxidative addition with a wide array of substrates to form stable Ir(III), d^6, 18-electron, saturated complexes (**Eq. 2.10**).

Eq. 2.10

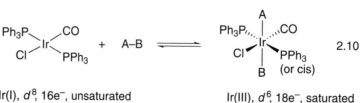

Ir(I), d^8, 16e⁻, unsaturated
Vaska's compound

Ir(III), d^6, 18e⁻, saturated

The range of addenda that participate in oxidative addition reactions to transition metal complexes is large, and contains many members useful in organic synthesis. There are three general classes of addenda: polar electrophiles, nonpolar substrates, and multiple bonds (**Figure 2.4**). Many of these result in the formation of metal–hydrogen and metal–carbon bonds, and in the "activation" of a range of organic substrates.

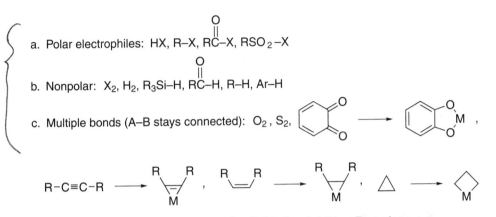

Figure 2.4. Classes of Substrates for Oxidative Addition Reactions

There are several documented mechanisms for oxidative addition, and the one followed depends heavily on the nature of the reacting partners. They include (1) concerted, associative, one-step "insertions" into A–B by M, (2) ionic, associative, two-step S_N2 reactions, and (3) electron transfer–radical chain mechanisms.

Concerted oxidative additions are best known for nonpolar substrates, particularly the oxidative addition of H_2 (central to catalytic hydrogenation), and oxidative addition into C–H bonds (hydrocarbon activation). The process is thought to involve prior coordination of the H–H or C–H bond to the metal in an "agostic" (formally a two-electron, three-center bond) fashion, followed by cis insertion (**Eq. 2.11**).

Eq. 2.11

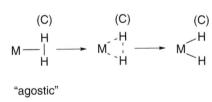

"agostic"

This idea is supported by the isolation of molecular hydrogen complexes of tungsten (**Eq. 2.12**)[8] in which just this sort of bonding is observed. This same process is likely but not yet demonstrated for the oxidative addition of M(0) d^{10} complexes with sp^2 halides, a process that goes with retention of stereochemistry at the alkene (**Eq. 2.13**).

Eq. 2.12

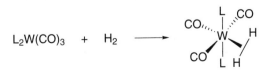

Eq. 2.13

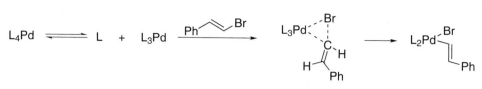

Oxidative addition via S_N2-type processes is most often observed with strongly nucleophilic, low-valent metals and classic S_N2 substrates, such as primary and secondary organic halides and tosylates. An early example is shown in **Eq. 2.14**,[9] and its features are typical of metals that react by S_N2-type processes. The rate law is second order, first order in metal and first order in substrate, and polar solvents accelerate the reaction, as expected for polar intermediates.

Eq. 2.14

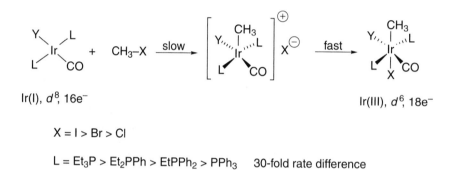

Ir(I), d^8, 16e⁻ → Ir(III), d^6, 18e⁻

X = I > Br > Cl

L = Et_3P > Et_2PPh > $EtPPh_2$ > PPh_3 30-fold rate difference

The order of substrate reactivity is I > Br > Cl, as expected for S_N2-type reactions. The rate also depends on the nature of the spectator phosphine ligands, and, in general, increases with increasing basicity (i.e., ability to donate electrons to the metal) of the phosphine. This is an important observation, since it confirms the claim that the reactivity of transition metal complexes can be "fine-tuned" by changing the nature of the spectator ligands. This feature is exceptionally valuable when using transition metals in complex total syntheses, since the inherent specificity of organometallic reactions can be applied over a wide range of substrates.

A synthetically more useful complex that reacts by an S_N2-type mechanism is Collman's reagent, $Na_2Fe(CO)_4$ (**Eq. 2.15**).[10] This very potent nucleophile is prepared by reducing the Fe(0) complex $Fe(CO)_5$ with sodium benzophenone ketyl to produce the Fe^{2-} complex. This is an unusually low oxidation state for iron, but because carbon monoxide ligands are powerful π-acceptors and strongly electron withdrawing, they stabilize negative charge, and highly reduced metal carbonyl complexes are common. The reaction of this complex with halides has all the features common to S_N2 chemistry (**Eq. 2.15**), and the S_N2 mechanism is the most likely pathway for the reaction of d^8–d^{10} transition metal complexes with 1°, 2°, allylic, and benzylic halides.

Eq. 2.15

$$Na_2Fe(CO)_4 \ + \ R–X \ \longrightarrow \ [RFe(CO)_4]Na \ + \ NaX$$

rate proportional to [RX][Fe(CO)$_4^{2-}$] for RX 1° > 2°   X = I > Br > OTs > Cl

stereochemistry at carbon: clean inversion

However, in many cases there is a competing radical-chain process that can affect the same overall transformation, but by a completely different mechanism. For example, Vaska's complex (**Eq. 2.14**) is unreactive towards organic bromides and even 2° iodides when the reaction is carried out with the strict exclusion of air. However, traces of air or addition of radical initiators promotes a smooth reaction with these substrates, via a radical-chain mechanism (**Eq. 2.16**).[11] In this case, racemization is observed, as expected. The existence of an accessible radical-chain mechanism for oxidative addition explains the empirical observation that traces of air sometimes promote the reaction of complexes that normally require the strict exclusion of air, and serves as a warning that the mechanism by which a reaction proceeds may be a function of the care with which it is carried out.

Eq. 2.16

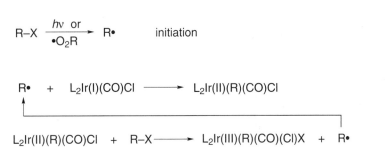

A more synthetically useful example of a radical-chain oxidative addition is the nickel-catalyzed coupling of aryl halides to biaryls, which was originally thought to proceed by two sequential two-electron oxidative additions. However, careful studies showed this was not the case. Rather, a complicated radical-chain process involving Ni(I) → Ni(III) oxidative additions is the true course of the reaction (**Eq. 2.17**).[12]

Eq. 2.17

$$L_nNi \ + \ Ar–X \ \longrightarrow \ Ar–Ar \ + \ NiX_2$$

original (incorrect) mechanism

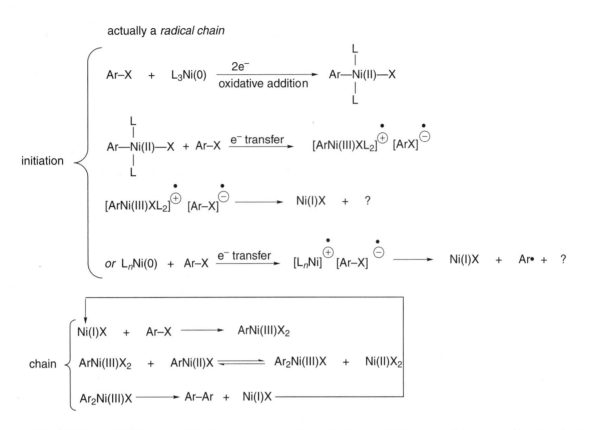

Reductive elimination is the reverse of oxidative addition and is usually the last step of many catalytic processes. It is exceptionally important for organic synthetic applications because *it is the major way in which transition metals are used to make carbon–carbon, carbon–hydrogen, carbon–oxygen, carbon–nitrogen, carbon–sulfur, and carbon–phosphine bonds!* The mechanism of the reductive elimination has been studied to some extent.[13] It is known that the groups to be eliminated must occupy cis positions on the metal, or must rearrange to cis if they are trans (**Eq. 2.18**).[14] Reductive elimination can be promoted. Anything that reduces the electron density at the metal facilitates reductive elimination. This can be as simple as the dissociation of a ligand, either spontaneously or by the application of heat or light. Both chemical and electrochemical oxidation promotes reductive elimination (although oxidatively driven reductive elimination sounds like an oxymoron). Of practical use is the observation that the addition of strong π-acceptor ligands, such as carbon monoxide, maleic anhydride, quinones, or tetracyanoethylene, promotes reductive elimination.

Eq. 2.18

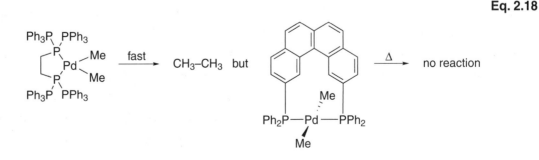

2.4 Migratory Insertion and β-Hydride Elimination

The third organometallic mechanism of interest for organic synthesis is the process of migratory insertion (**Eq. 2.19**), in which an unsaturated ligand—usually CO, RNC, alkenes, or alkynes—*formally* inserts into an adjacent (cis) metal–ligand bond.

Eq. 2.19

$$
\underset{M-Y}{\overset{X}{|}} \rightleftharpoons M-Y-X \xrightarrow{\;L\;} \underset{M-Y-X}{\overset{L}{|}}
$$

Y = CO, RNC, C=C, C≡C   X = H, R, etc.

When this adjacent ligand is hydride, or σ-alkyl, the process forms a new carbon–hydrogen or carbon–carbon bond, and in the case of alkenes, a new metal–carbon bond. The term "insertion" is somewhat misleading, in that the process actually proceeds by *migration* of the adjacent ligand to the metal-bound unsaturated species, generating a vacant site. Insertions are usually reversible. If the migrating group has stereochemistry, it is usually retained in both the insertion step and the reverse.

The insertion of carbon monoxide into a metal–carbon σ-bond is one of the most important processes for organic synthesis because it permits the direct introduction of molecular carbon monoxide into organic substrates, to produce aldehydes, ketones, and carboxylic acid derivatives. The reverse process is also important, because it allows for the decarbonylation of aldehydes to alkanes and acid halides to halides.

The general process of CO insertion, shown in **Eq. 2.20**, involves reversible migration of an R group to a cis-bound CO as the rate-determining step. The vacant site generated by this migration is then filled by some external ligand present.

Eq. 2.20

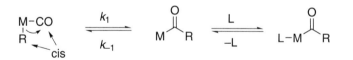

Similarly, the reverse process requires a vacant coordination site cis to the acyl ligand. Both the migratory insertion and the reverse deinsertion occur with retention of configuration of the migrating group and of the metal, if the metal center is chiral. The migratory aptitude for metal carbonyl complexes is Et > Me > PhCH$_2$ > η^1-allyl > alkenyl ≥ aryl, ROCH$_2$ > propargyl > HOCH$_2$. Hydride (H), CH$_3$CO, and CF$_3$ usually do not migrate, and heteroatoms such as RO and R$_2$N rarely migrate. Lewis acids often accelerate CO insertion reactions, as does oxidation, although in this case the metal–acyl species often continues to react, and is oxidatively cleaved from the metal.

Alkenes also readily insert into metal–hydrogen bonds (a key step in catalytic hydrogenations of alkenes) and metal–carbon σ-bonds, resulting in alkylation of the alkene (**Eq. 2.21**). The reverse of hydride insertion is β-hydrogen elimination, a very common pathway for the decomposition of σ-alkylmetal complexes. Alkene insertions share many features with CO insertions. The alkene *must* be coordinated to the metal prior to insertion, and must be cis to the migrating hydride or alkyl group. The migration generates a vacant site on the metal, and for the reverse process, β-hydride elimination (β-alkyl elimination does not occur), a vacant cis site on the metal is required.

Eq. 2.21

Both the metal and the migrating group add to the same face of the alkene. When an alkyl group migrates, its stereochemistry is maintained. A rough order of migrating aptitude for metal–alkene complexes is H >> alkyl, alkenyl, aryl > RCO >> RO, R_2N. Again, heteroatom groups migrate only with difficulty, since the extra lone pair(s) on these ligands can form multiple bonds to the metal (**Eq. 2.22**).

Eq. 2.22

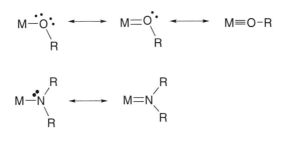

Alkynes also insert into metal–hydrogen and metal–carbon bonds, again with cis stereochemistry and with retention at the migrating center (**Eq. 2.23**). This has been less studied, and is complicated by the fact that alkynes often insert into the product σ-vinyl complex, resulting in oligomerization.

Eq. 2.23

2.5 Nucleophilic Attack on Ligands Coordinated to Transition Metals

The reaction of nucleophiles with coordinated, unsaturated organic ligands is one of the most useful processes for organic synthesis.[15] Unsaturated organic compounds such as carbon monoxide, alkenes, alkynes, and arenes are electron rich, and are usually quite unreactive towards nucleophiles. However, complexation to electron-deficient metals inverts their normal reactivity, making them generally subject to nucleophilic attack, and opening up entirely new synthetic transformations. The more electron deficient a metal center is, the more reactive towards nucleophiles its associated ligands become. Thus, cationic complexes and complexes having strong π-accepting spectator ligands, such as carbon monoxide, are especially reactive in these processes.

Metal-bound carbon monoxide is generally reactive towards nucleophiles, providing a major route for the incorporation of carbonyl groups into organic substrates. Neutral metal carbonyls usually require strong nucleophiles, such as organolithium reagents, and the reaction [labeled (*a*) in **Eq. 2.24**] produces anionic metal–acyl complexes (i.e., acyl "ate" complexes). (For the purposes of electron bookkeeping, the resonance structure denoting both the σ-donor bond and the π-acceptor bond in metal-bound CO's is used. CO is still a formal two-electron donor ligand.) Many of these acyl "ate" complexes are quite stable and can be isolated and manipulated.

Eq. 2.24

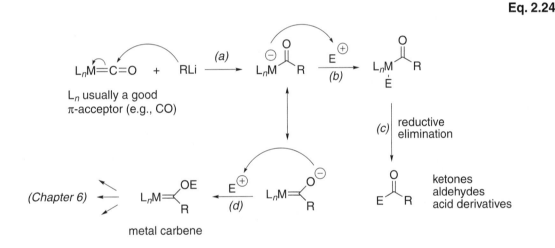

Electrophiles can react at the metal [reaction (*b*) in **Eq. 2.24**] (analogous to C-alkylation of an enolate), giving a neutral acyl electrophile complex, which can undergo reductive elimination [reaction (*c*) in **Eq. 2.24**] to produce ketones, aldehydes, or carboxylic acid derivatives. Alternatively, the reaction of electrophiles at oxygen [reaction (*d*) in **Eq. 2.24**] (analogous to O-alkylation of an enolate) produces heteroatom-stabilized carbene complexes, which have a very rich organic chemistry in their own right (Chapter 6). The site of electrophilic attack depends on both the metal and the electrophile. Hard electrophiles, such as $R_3O^+BF_4^-$, ROTf, and $ROSO_2F$, undergo reaction at oxygen,[16] whereas softer electrophiles undergo reaction at the metal.[17]

Alkoxides and amines attack metal carbonyls, particularly the more reactive cationic ones (**Eq. 2.25**), to produce alkoxycarbonyl or carbamoyl complexes,[18] species that are involved in a variety of metal-assisted carbonylation reactions via insertion into the metal–acyl carbon bond.

Eq. 2.25

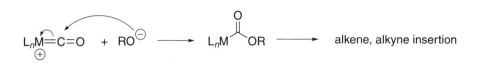

Hydroxide ion attacks metal carbonyls to produce (normally) unstable metal carboxylic acids, which usually decompose to anionic metal hydrides and carbon dioxide[19] (**Eq. 2.26**).

Eq. 2.26

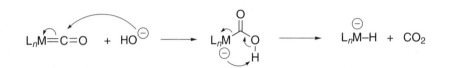

Electrophilic carbene complexes undergo nucleophilic attack at the carbene carbon, resulting in either heteroatom exchange (**Eq. 2.27**)[20] or the formation of stable adducts (**Eq. 2.28**).[21]

Eq. 2.27

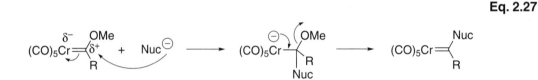

Nuc = RS, R$_2$NH, PhLi, etc.

Eq. 2.28

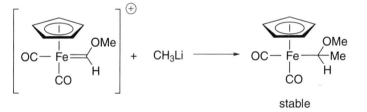

stable

Nucleophilic cleavage of metal–carbon σ-bonds is another process of interest in organic synthesis, since it is frequently involved in freeing an organic substrate from the metal. Many metal–acyl complexes undergo direct cleavage by alcohols, amines, or water, a key step in many metal-catalyzed acylation reactions (**Eq. 2.29**).[22] Others are more stable and require an oxidative cleavage (**Eq. 2.30**).[23] σ-Alkyliron complexes undergo oxidative cleavage, and this process has been studied in some detail (**Eq. 2.31**).[24] Oxidation of the metal makes it a better leaving group, and promotes displacement with clean inversion.

Eq. 2.29

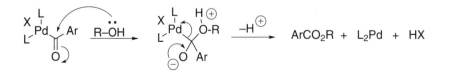

Eq. 2.30

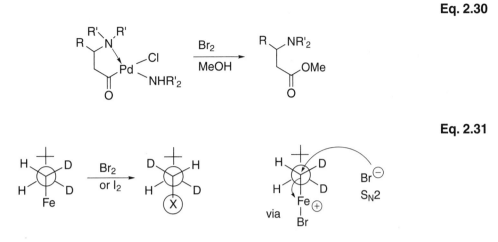

Eq. 2.31

Another nucleophilic reaction of some consequence to organic synthesis is the self-exchange reaction. Oxidative addition of organic halides to low-valent metals is known to go with inversion of configuration at carbon.[25] However, occasionally racemization or partial racemization is observed. This is thought to result from nucleophilic displacement of the metal(II) in the σ-alkyl complex by the nucleophilic metal(0) starting complex (**Eq. 2.32**).[26] Since the reaction is "thermally neutral," this need not be a difficult process, and it should be considered whenever unexpected racemization is observed.

Eq. 2.32

$$L_4Pd \; + \; \underset{Ph}{\overset{HD}{\diagdown}}Br \xrightarrow{\text{inversion}} \underset{L_2BrPd}{\overset{H,D}{\diagup}}Ph \xrightarrow[\text{inversion}]{L_4Pd} \underset{Ph}{\overset{HD}{\diagdown}}PdL_2Br$$

Nucleophilic attack on complexed π-unsaturated hydrocarbons is among the most useful of organometallic processes for organic synthesis, since it is exactly the opposite of "normal" reaction chemistry for these substrates. Nucleophilic attack on 18-electron, cationic complexes has been particularly well studied,[27] and based upon a large number of examples, the following order of reactivity for kinetically controlled reactions of this class of complexes was derived:

Note that the cyclopentadienyl ligand (Cp) is among the least reactive, and as a consequence, is often used as a spectator ligand.

The typical reaction for electrophilic metal complexes of alkenes, particularly those of Pd(II), Pt(II), and Fe(II), is nucleophilic attack on the alkene.[28] In most cases, the nucleophile attacks the more-substituted position of the alkene, from the face opposite the metal (i.e., trans), forming a carbon–nucleophile bond and a carbon–metal bond (**Eq. 2.33**). A few nucleophiles, such as chloride, acetate, and nonstabilized carbanions, can also attack first at the metal. This is followed by insertion of the alkene, resulting in an overall cis addition, with attack at the less-substituted position. This chemistry is normally restricted to mono- and 1,2-disubstituted alkenes, since more highly substituted alkenes coordinate only weakly, and displacement of the alkene by the nucleophile competes.

Eq. 2.33

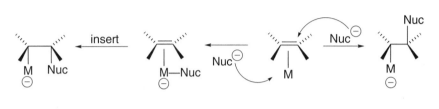

Nuc = Cl, AcO, ROH, RNH$_2$, R

Alkynes complexed to electrophilic metals also undergo nucleophilic attack, although this is a much less-studied process, since stable alkyne complexes of electrophilic metals are rare. (These metals tend to oligomerize alkynes.) Cationic iron complexes of alkynes undergo clean nucleophilic attack from the face opposite the metal, to give stable σ-alkenyl iron complexes (**Eq. 2.34**).[29] This process has not been used in organic synthesis to any extent.

Eq. 2.34

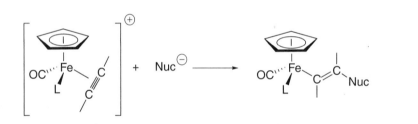

Nuc = PhS, CN, CH(CO$_2$Et)$_2$ [Also: Ph, Me, H$_2$C=CH, MeC≡C from R$_2$Cu(CN)Li$_2$]

L = PPh$_3$, P(OPh)$_3$

Nucleophilic attack on η^3-allyl metal complexes, particularly those of Pd(II), has been extensively developed as an organic synthetic method and is discussed in detail in Chapter 9. The general features are shown in **Eq. 2.35**. Nucleophilic attack occurs in most cases from the face opposite the metal. However, a few nucleophiles, such as carboxylates and nonstabilized carbanions, can first attack the metal, followed by migration to the allylic system. From a synthetic point of view, the most important reaction involving η^3-allyl complexes is the Pd(0)-catalyzed reaction of allylic substrates, a process that proceeds with overall retention (two inversions) of configuration at carbon (**Eq. 2.36**).[30] In certain rare cases, nucleophilic attack can occur at the central carbon of an η^3-allyl complex, generating a metallacyclobutane. In the case of the cationic molybdenum complex in **Eq. 2.37**, the complex was stable and isolated.[31]

Eq. 2.35

Eq. 2.36

Eq. 2.37

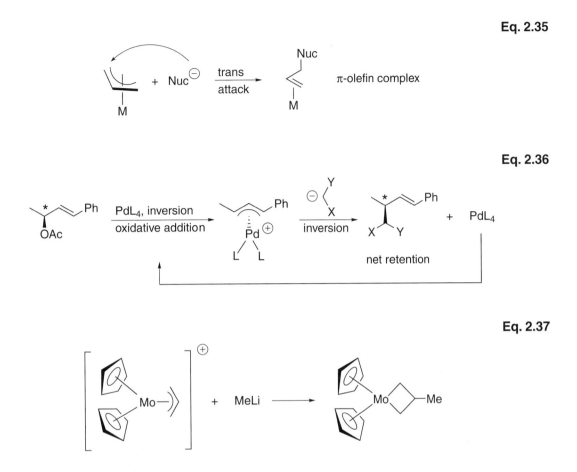

Although extended Hückel calculations[32] suggested that attack at the central carbon of η^3-allylpalladium complexes was unfeasible, under appropriate conditions (i.e., α-branched ester enolates, polar solvent) just such a process occurred smoothly to produce cyclopropanes (**Eq. 2.38**).[33] The proposed palladiacyclobutane intermediate, stabilized by TMEDA, has been isolated and structurally characterized.[34]

Eq. 2.38

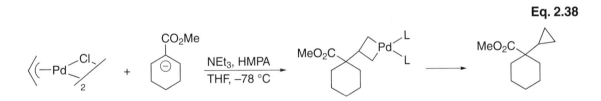

Neutral η^4-dieneiron tricarbonyl complexes undergo reaction with strong nucleophiles (e.g., LiCMe$_2$CN) at both the terminal position, to produce an η^3-allyl complex, and the internal position, to produce a σ-alkyl-η^2-alkene complex (**Eq. 2.39**).[35] In contrast, the much more electrophilic palladium dichloride promotes exclusive terminal attack to produce stable η^3-allylpalladium complexes, which can be further functionalized (Chapter 9) (**Eq. 2.40**).[36] Again, nucleophilic attack occurs in most cases from the face opposite the metal but a few exceptions exist.

Eq. 2.39

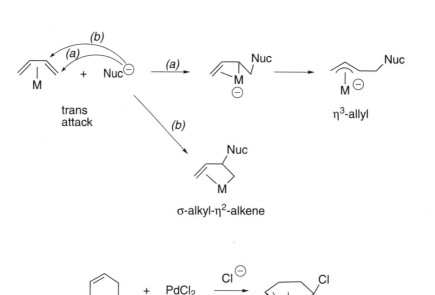

Eq. 2.40

In contrast to neutral η^4-diene complexes, cationic η^5-dienyl complexes are broadly reactive towards nucleophiles, and iron tricarbonyl complexes of the η^5-cyclohexadienyl ligand have found extensive use in organic synthesis. Nucleophiles ranging from electron-rich aromatic compounds through organocopper species readily add to these complexes, attacking, as usual, from the face opposite the metal, to produce stable η^4-diene complexes (**Eq. 2.41**).[37] With substituted cyclohexadienyl ligands, the site of nucleophilic attack is generally governed by electronic factors. By coupling this nucleophilic addition to η^5-dienyl complexes with the generation of η^5-dienyl complexes by hydride abstraction from η^4-diene complexes, regiospecific and stereospecific polyfunctionalizations of this class of ligands have been achieved (Chapter 7).

Eq. 2.41

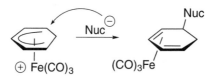

Arenes normally undergo electrophilic attack and, except in special cases, are quite inert to nucleophilic attack. However, by complexation to electron-deficient metal fragments, particularly metal carbonyls (recall that CO groups are strongly electron withdrawing), arenes become generally reactive toward nucleophiles. By far the most extensively studied complexes are the η^6-arenechromium tricarbonyl complexes (**Eq. 2.42**).[38] Again, the nucleophile attacks from the face opposite the metal to give a relatively unstable anionic η^5-cyclohexadienyl complex, which itself has a rich organic chemistry (Chapter 10). Oxidation regenerates the arene, while protolytic cleavage gives the cyclohexadiene.

Eq. 2.42

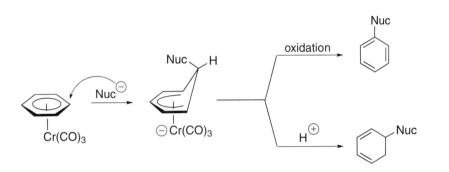

The reactions of nucleophiles with metal-coordinated ligands are summarized in **Table 2.2**, along with the d-electron count and formal oxidation states for reactants and products. Although there is a profound reorganization of electrons about the metal, most of these processes result in no change in the formal oxidation state of the metal, differentiating them from the oxidative addition–reductive elimination family of reactions. This is because in each case a formally neutral ligand is converted to a formally mononegative ligand in the process—that is, nucleophilic attack reduces the ligand, not the metal. The two exceptions are the η^3-allyl and the η^5-dienyl complexes, in which formally negative ligands are converted to formally neutral ligands by the process and the metal is reduced.

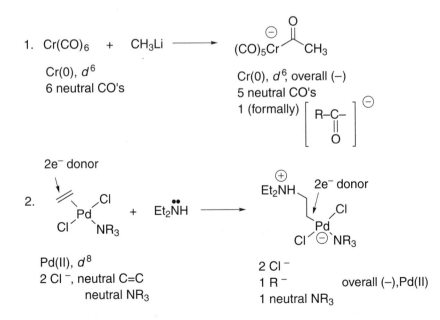

Table 2.2. Summary of Nucleophilic Reactions on Transition Metals

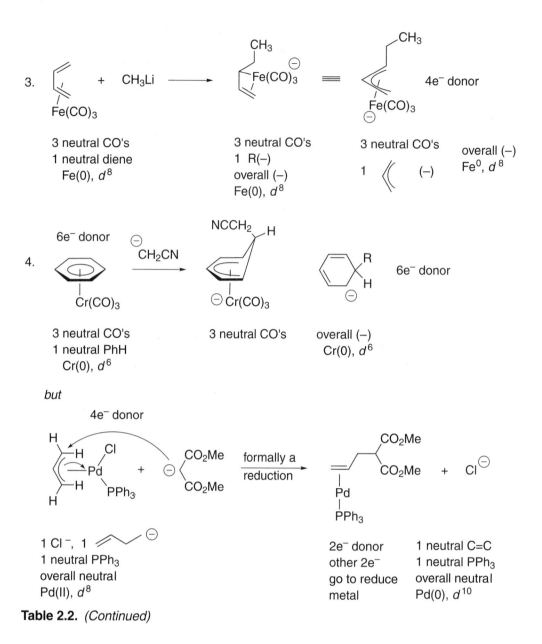

Table 2.2. *(Continued)*

2.6 Transmetallation

Transmetallation is a process of increasing importance in the area of transition metals in organic synthesis, but has been little studied and is not well understood. The general phenomenon involves the transfer of an R group from a main group organometallic compound to a transition metal complex (**Eq. 2.43**).[39] When combined with reactions that introduce an R group into the transition metal complex, such as oxidative addition or nucleophilic attack on alkenes, efficient carbon–carbon bond forming (cross-coupling) reactions ensue. In cases such as these, the transmetallation step is almost always the rate-limiting step, and when catalytic cycles involving a transmetallation step fail, it is usually this step that needs attention.

Eq. 2.43

$$R-M \ + \ M'-X \ \underset{\longleftarrow}{\overset{K}{\longrightarrow}} \ R-M' \ + \ M-X$$

M = Mg, Zn, Zr, B, Hg, Si, Sn, Ge
M' = transition metal

In order for the transmetallation step to proceed, the main group organometallic M must be more electropositive than the transition metal M'. However, since this is an equilibrium, if RM' is irreversibly consumed in a subsequent step, then the process can be used even if it is not favorable (i.e., K is small). This, coupled with the uncertain accuracy of the values for electronegativity, make experiment the wisest course.

An often neglected fact is that **Eq. 2.43** *is* an equilibrium, and that both partners must profit, thermodynamically, from the process. Therefore, the nature of X may well be as important as the nature of M or M', and transmetallation procedures can often be promoted by adding appropriate counterions, X (Chapter 4).

2.7 Electrophilic Attack on Transition Metal Coordinated Ligands

Reactions of coordinated organic ligands with electrophiles have found substantially less use in organic synthesis than have reactions with nucleophiles. Perhaps the most common electrophilic reaction is the electrophilic cleavage of metal–carbon σ-bonds as a method to free an organic substrate from a metal template. If the metal complex has d electrons, electrophilic attack usually occurs at the metal, a formal oxidative addition. This is followed by reductive elimination to result in cleavage with retention of configuration at carbon (**Eq. 2.44**). If a sufficiently good nucleophile is present (e.g., Br⁻ from Br$_2$), cleavage with *inversion* may result.

Eq. 2.44

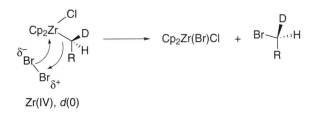

An example of this kind of cleavage has already been discussed (**Eq. 2.31**). Even σ-alkyl complexes of metals having no d electrons undergo facile electrophilic cleavage, usually with retention of configuration. In these cases, direct electrophilic attack at the metal–carbon σ-bond (S$_E$2) is the likely process (**Eq. 2.45**).[40]

Eq. 2.45

Electrophiles also can react at *carbon* in organometallic complexes, and a variety of useful processes involve this type of reaction. These are summarized in **Table 2.3**. Some, particularly reactions of electrophiles with η^1-allyl complexes (i.e., γ-attack) and hydride abstraction from η^4-diene complexes, are quite useful in organic synthesis and are discussed in detail later.

1. Electrophilic cleavage of σ-alkyl metal bonds

$$R\text{–}M \;+\; E^{\oplus} \longrightarrow R\text{–}E \;+\; M^{\oplus} \qquad \textit{retention at R}$$

2. Attack at α-position

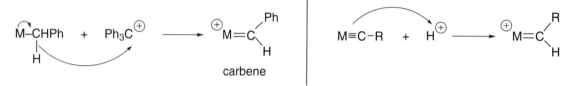

3. Attack at β-position

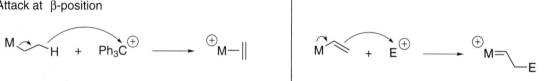

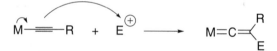

4. Attack at γ-position

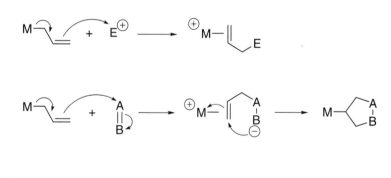

5. Attack on coordinated polyenes

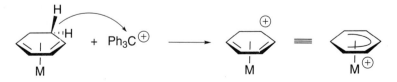

Table 2.3. Electrophilic Attack on Transition Metal Coordinated Organic Ligands

References

(1) a) Collman, J. P.; Hegedus, L. S.; Norton, J. R.; Finke, R. G. *Principles and Applications of Organotransition Metal Chemistry*, 2nd ed.; University Science Books: Mill Valley, CA, 1987; pp 235–458. b) Twigg, M. V., Ed. *Mechanisms of Inorganic and Organometallic Reactions*; Plenum Publishing: New York, annually since 1983; see part 3 in each volume.

(2) Odell, L. A.; Raethel, H. A. *J. Chem. Soc., Chem. Commun.* **1968**, 1323; **1969**, 87.

(3) For a review of this effect in octahedral complexes, see Coe, B. J.; Glenwright, S. J. *Coord. Chem. Rev.* **2000**, *203*, 5.

(4) Tolman, C. A. *Chem. Rev.* **1977**, *77*, 313.

(5) Albers, M. O.; Coville, N. J. *Coord. Chem. Rev.* **1984**, *53*, 227.

(6) a) Darensbourg, D. J.; Walker, N.; Darensbourg, M. *J. Am. Chem. Soc.* **1980**, *102*, 1213. b) Darensbourg, D. J.; Darensbourg, M. Y.; Walker, N. *Inorg. Chem.* **1981**, *20*, 1918. c) Darensbourg, D. J.; Ewen, J. A. *Inorg. Chem.* **1981**, *20*, 4168.

(7) For reviews, see a) Collman, J. P.; Roper, W. R. *Adv. Organomet. Chem.* **1968**, *7*, 53. b) Collman, J. P. *Acc. Chem. Res.* **1968**, *1*, 136. (c) Stille, J. R.; Lau, K. S. Y. *Acc. Chem. Res.* **1977**, *10*, 434. d) Stille, J. K. in *The Chemistry of the Metal-Carbon σ-Bond*; Hartley, F. R., Patai, S., Eds.; Wiley: New York, 1985; Vol. 2, pp 625–787.

(8) a) Sweany, R. L. *J. Am. Chem. Soc.* **1985**, *107*, 2374. b) Upmacis, R. K.; Gadd, G. E.; Poliakoff, M.; Simpson, M. B.; Turner, J. J.; Whyman, R.; Simpson, A. F. *J. Chem. Soc. Chem. Commun.* **1985**, 27.

(9) Chock, P. B.; Halpern, J. *J. Am. Chem. Soc.* **1966**, *88*, 3511.

(10) Collman, J. P.; Finke, R. G.; Cawse, J. N.; Brauman, J. I. *J. Am. Chem. Soc.* **1977**, *99*, 2515.

(11) a) Labinger, J. R.; Osborn, J. A. *Inorg. Chem.* **1980**, *19*, 3230. b) Labinger, J. A.; Osborn, J. A.; Coville, N. J. *Inorg. Chem.* **1980**, *19*, 3236. c) Jenson, F. R.; Knickel, B. *J. Am. Chem. Soc.* **1971**, *93*, 6339.

(12) Kochi, J. K. *Pure Appl. Chem.* **1980**, *52*, 571.

(13) For a review, see Hartwig, J, F. *Inorg. Chem.* **2007**, *46*, 1936.

(14) Gillie, A.; Stille, J. K. *J. Am. Chem. Soc.* **1980**, *102*, 4933.

(15) For a review, see Hegedus, L. S. "Nucleophilic Attack on Transition Metal Organometallic Compounds" in *The Chemistry of the Metal-Carbon σ-Bond*; Hartley, F. R., Patai, S., Eds.; Wiley: New York, 1985; Vol. 2, pp 401–512.

(16) Dötz, K. H. *Angew. Chem., Int. Ed. Engl.* **1984**, *23*, 587.

(17) Collman, J. P. *Acc. Chem. Res.* **1975**, *8*, 342.

(18) Angelici, R. L. *Acc. Chem. Res.* **1972**, *5*, 335.

(19) Ungermann, C.; Landis, V.; Moya, S. A.; Cohen, H.; Walker, H.; Pearson, R. G.; Rinker, R. G.; Ford, P. C. *J. Am. Chem. Soc.* **1979**, *101*, 5922 and references therein.

(20) Duplicate ref. Dötz, K. H. *Angew. Chem. Int. Ed. Engl.* **1984**, *23*, 587.

(21) Casey, C. P.; Miles, W. M. *J. Organomet. Chem.* **1983**, *254*, 333.

(22) Heck, R. F. *Pure Appl. Chem.* **1978**, *50*, 691.

(23) Hegedus, L. S.; Anderson, O. P.; Zetterberg, K.; Allen, G.; Siirala-Hansen, K.; Olsen, D. J.; Packard, A. B. *Inorg. Chem.* **1977**, *16*, 1887.

(24) Whitesides, G. M.; Boschetto, D. J. *J. Am. Chem. Soc.* **1971**, *93*, 1529.

(25) For a recent example, see Rodriguez, N.; Ramirez de Arellano, C.; Asenio, G.; Medio-Simon, M. *Chem.—Eur. J.* **2007**, *13*, 4223.

(26) Lau, K. S. Y.; Fries, R. M.; Stille, J. K. *J. Am. Chem. Soc.* **1974**, *96*, 4983. Similar loss of stereochemistry by Pd self-exchange has been observed in η³-allyl palladium chemistry: Granberg, K. L.; Bäckvall, J.-E. *J. Am. Chem. Soc.* **1992**, *114*, 6858.

(27) Davies, S. G.; Green, M. L. H.; Mingos, D. M. P. *Tetrahedron* **1978**, *34*, 3047.

(28) a) Lennon, P. M.; Rosan, A. M.; Rosenblum, M. *J. Am. Chem. Soc.* **1977**, *99*, 8426. b) Hegedus, L. S.; Åkermark, B.; Zetterberg, K.; Olsson, L. F. *J. Am. Chem. Soc.* **1984**, *106*, 7122. c) Hegedus, L. S.; Williams, R. E.; McGuire, M. A.; Hayashi, T. *J. Am. Chem. Soc.* **1980**, *102*, 4973. d) For a review, see McDaniel, K. Nucleophilic Attack on Alkene Complexes. In *Comprehensive Organometallic Chemistry II*. Abel, E. W., Stone, F. G. A., Wilkinson, G., Eds.; Pergamon Press: Oxford, UK, 1995; Vol. 12, pp 601–622.

(29) Reger, D. L.; Belmore, K. A.; Mintz, E.; McElligott, P. J. *Organometallics* **1984**, *3*, 134; 1759.

(30) a) Hayashi, T.; Hagihara, T.; Konishi, M.; Kumada, M. *J. Am. Chem. Soc.* **1983**, *105*, 7767. b) For a review, see Harrington, P. J. Transition Metal Allyl Complexes. In *Comprehensive Organometallic Chemistry II*; Abel, E. W., Stone, F. G. A., Wilkinson, G., Eds.; Pergamon Press: Oxford, UK, 1995; Vol. 12, pp 797–904.

(31) Periana, R. A.; Bergman, R. G. *J. Am. Chem. Soc.* **1984**, *106*, 7272.

(32) Curtis, M. D.; Eisenstein, O. *Organometallics* **1984**, *3*, 887.

(33) a) Hegedus, L. S.; Darlington, W. H.; Russel, C. E. *J. Org. Chem.* **1980**, *45*, 5193. b) Carfagna, C.; Mariani, L.; Musco, M.; Sallese, G.; Santi, R. *J. Org. Chem.* **1991**, *56*, 3924.

(34) a) Hoffmann, H. M. R.; Otte, A. R.; Wilde, A.; Menzer, S.; Williams, D. J. *Angew. Chem., Int. Ed. Engl.* **1995**, *34*, 100. For a consideration of ligand effects on the site of nucleophilic attack on η³-allylpalladium complexes, see b) Aranyos, A.; Szabo, K. J.; Castaño, A. M.; Bäckvall, J.-E. *Organometallics* **1997**, *16*, 1997; 1058.

(35) a) Semmelhack, M. F.; Le, H. T. M. *J. Am. Chem. Soc.* **1984**, *106*, 2715. b) Semmelhack, M. F.; Herndon, J. W. *Organometallics* **1983**, *2*, 363.

(36) Bäckvall, J.-E.; Nyström, J.-E.; Nordberg, R. E. *J. Am. Chem. Soc.* **1985**, *107*, 3676.

(37) a) Pearson, A. J. *Comp. Org. Synth.* **1991**, *4*, 663. b) Pearson, A. J. Nucleophilic Attack on Diene and Dienyl Complexes. In *Comprehensive Organometallic Chemistry II*; Abel, E. W., Stone, F. G. A., Wilkinson, G., Eds.; Pergamon Press: Oxford, UK, 1995; Vol. 12, pp 637–684.

(38) a) Semmelhack, M. F. *Comp. Org. Synth.* **1991**, *4*, 517. b) Semmelhack, M. F. Transition Metal Arene Complexes. In *Comprehensive Organometallic Chemistry II*; Abel, E. W., Stone, F. G. A., Wilkinson, G., Eds.; Pergamon Press: Oxford, UK, 1995; Vol. 12, pp 979–1038.

(39) Negishi, E.-I. *Organometallics in Organic Synthesis*; Wiley: New York, 1980.

(40) Labinger, J. A.; Hart, D. W.; Seibert, W. E., III; Schwartz, J. *J. Am. Chem. Soc.* **1975**, *97*, 3851.

Synthetic Applications of Transition Metal Hydrides

3.1 Introduction

Transition metal hydrides[1] are an important class of organometallic complexes, primarily because of their role in homogeneous hydrogenation, hydroformylation, and hydrometallation reactions. The reactivity of transition metal hydrides depends very much on the metal and the other ligands, and ranges all the way from hydride donors to strong protic acids![2] However, their major use in organic synthesis relies on neither of these characteristics, but rather their propensity to insert alkenes and alkynes into the metal–hydrogen bond, generating σ-alkylmetal complexes for further transformations.

Transition metal hydrides can be synthesized in a number of different ways (**Figure 3.1**), but the most commonly used method is oxidative addition. Transition metals are unique in their ability to activate hydrogen under very mild conditions, a feature that

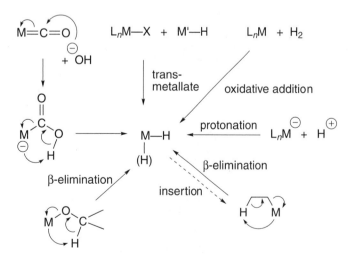

Figure 3.1. Syntheses of Transition Metal Hydrides

accounts for their utility in catalytic hydrogenation. The vast majority of reductions employ rhodium, ruthenium, and iridium complexes, although other transition metals such as platinum, zirconium, and titanium have seen some use.

3.2 Homogeneous Hydrogenation[3]

When synthetic chemists are faced with the reduction of a carbon–carbon double bond, their first choice is, almost invariably, *heterogeneous* catalysis, using, for example, palladium on carbon. Heterogeneous catalysts are convenient to handle, efficient, and easy to remove by filtration after the reaction is complete. In contrast, homogeneous catalysts are sometimes difficult to handle, sensitive to impurities, such as traces of oxygen, have a tendency to cause alkene isomerization, and are difficult to recover, because they are soluble. However, their enormous advantage—namely, selectivity—mitigates these shortcomings.

There are two general classes of homogeneous hydrogenation catalysts—monohydrides and dihydrides—and they react by different mechanisms and have different specificities. The best-studied monohydride catalyst is the lemon yellow, air stable, crystalline solid $Rh(H)(PPh_3)_3(CO)$, and the mechanism by which it hydrogenates alkenes is shown in **Figure 3.2**. It is *completely* specific for terminal alkenes, and will tolerate internal alkenes, aldehydes, nitriles, esters, and chlorides elsewhere in the molecule. Its major limitation is that alkene isomerization competes with hydrogenation, and the isomerized alkene cannot be reduced by the catalyst.

The starting complex is a coordinatively saturated, Rh(I), d^8 complex, and the first step requires loss of a phosphine (ligand dissociation). As a consequence, added phosphine inhibits catalysis. Coordination of the substrate alkene to the unsatu-

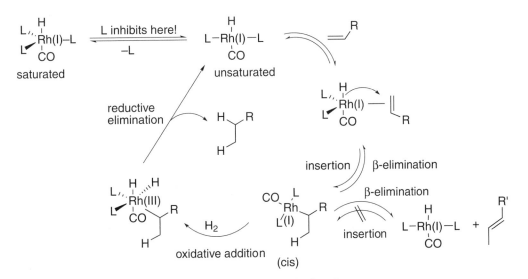

Figure 3.2. Reduction of Alkenes by Monohydride Catalysts

rated metal center is followed by migratory insertion, to produce an unsaturated σ-alkylrhodium(I) complex. The regiochemistry of insertion is unknown, and the rhodium may occupy either the terminal or internal position. Oxidative addition of hydrogen to this σ-alkylrhodium(I) complex, followed by reductive elimination of R–H, results in reduction of the alkene and regeneration of the catalytically active, unsaturated rhodium(I) monohydride. Irreversible alkene isomerization occurs if a 2° σ-alkylrhodium(I) intermediate undergoes β-hydrogen elimination (the reverse of migratory insertion) to give an internal alkene and the catalytically active rhodium(I) monohydride. Because this catalyst cannot reduce internal alkenes, this isomerization is irreversible, and competes with reduction. Although this particular monohydride catalyst is little used in synthesis, the monohydride pathway may be important in the ruthenium(II) asymmetric hydrogenation catalysts discussed below.

Dihydride catalysts are much more versatile, and much better understood. The most widely used catalyst precursor is the burgundy red solid RhCl(PPh₃)₃, called Wilkinson's catalyst. The mechanism of hydrogenation of alkenes by this catalyst has been studied in great detail and is quite complex (**Figure 3.3**).[4] Although Wilkinson's catalyst is already coordinatively unsaturated [Rh(I), d^8, 16e⁻], the kinetically active species is the 14-electron complex RhCl(PPh₃)₂ with, perhaps, solvent loosely associated. In this case, oxidative addition of H₂ is the initial step, followed by coordination of the alkene. Migratory insertion is the rate-limiting step, followed by a fast reductive elimination of the alkane and regeneration of the active catalyst. In addition to the catalytic cycle, several other equilibria are operating, complicating the system.

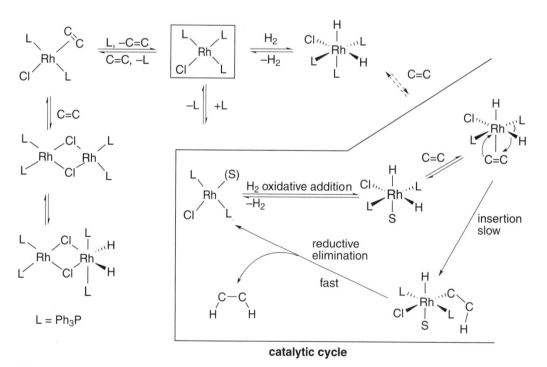

catalytic cycle

Figure 3.3. Reduction of Alkenes by Wilkinson's Catalyst

Wilkinson's catalyst is attractive for synthesis because it is selective and efficient, tolerant of functionality, and, in contrast to monohydride catalysts, does not promote alkene isomerization under neutral conditions. The reactivity towards alkenes parallels their coordination ability:

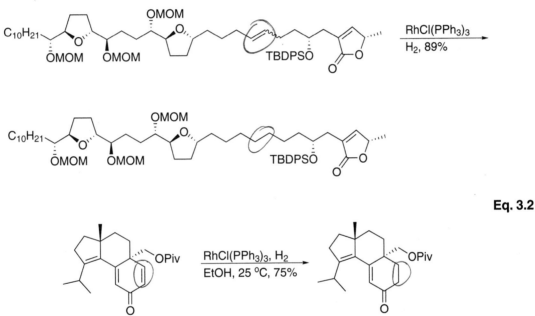

with a 50-fold difference in rate over this range of alkenes. Noncyclic tri- and tetrasubstituted alkenes are not reduced, and ethene, which coordinates too strongly, poisons the catalyst. Examples of the ability of this catalyst system to differentiate between alkenes are seen in **Eqs. 3.1**[5] and **3.2**.[6]

Eq. 3.1

Eq. 3.2

Functional groups can coordinate to the catalyst and direct the reduction. Face-selective reduction of an exocyclic double bond, probably via coordination of the catalyst to a hydroxy group, was used in a synthesis of the A-ring of gambieric acids (**Eq. 3.3**).[7] In contrast, reduction of the exocyclic double bond using palladium on carbon gave a mixture of products. Chelation-controlled hydrogenation of alkenes, wherein the substrate both coordinates to the metal and directs the facial selectivity to give one isomer, is also seen in **Eq. 3.4**.[8]

Eq. 3.3

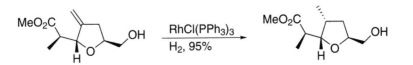

Eq. 3.4

Alkynes are rapidly reduced to alkanes using Wilkinson's catalyst. This rapid reduction of an alkyne to an alkane without cleavage of a benzyl ether was utilized in a synthesis toward macroviracin A (**Eq. 3.5**).[9] Nitro alkenes are reduced to nitro alkanes without competitive reduction of the nitro group.

Eq. 3.5

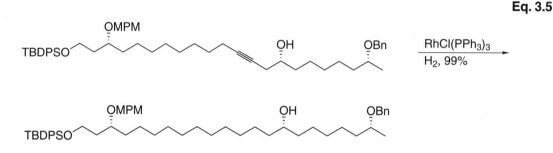

There are several limitations using Wilkinson's catalyst in reductions. In solution, it catalyzes the oxidation of triphenylphosphine, which in turn, leads to destruction of the complex, so air cannot be tolerated. Strong ligands, such as ethene, thiols, and very basic phosphines, poison the system. Carbon monoxide and substrates prone to decarbonylation, such as acid halides, are not tolerated. Notwithstanding these limitations, Wilkinson's catalyst is quite useful in synthesis, and is usually the first choice for routine homogeneous hydrogenations.

Chelation-controlled reduction of homoallylic alcohols was nonselective for Wilkinson's complex, but afforded selective reduction using a binap–Ru(OAc)$_2$ system (**Eq. 3.6**).[10] Wilkinson's complex was used to reduce an α,β-unsaturated ketone in a synthesis of (−)-Bao Gong Teng A (**Eq. 3.7**).[11]

Eq. 3.6

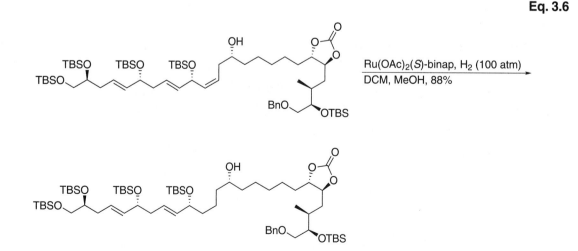

Eq. 3.7

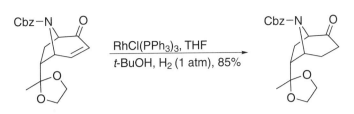

Of more interest to synthetic chemists is asymmetric homogeneous hydrogenation,[12,13] wherein prochiral alkenes are reduced to enantiomerically enriched products. This topic has been studied in excruciating detail, and when it works, it is spectacular! The requirements for successful asymmetric hydrogenation were initially so stringent that the process was limited to a very narrow range of substrates. However, more recently developed catalysts have expanded the substrate scope to some extent.

$Rh(cod)_2BF_4$, $Rh(cod)_2OTf$, and $Rh(nbd)BF_4$ can be used as precursors to the active catalyst. The catalysts used are cationic rhodium(I) complexes of optically active bidentate diphosphines, monodentate[14] phosphines, phosphonites, phosphites, and phosphoramidites generated in situ by reduction of the corresponding diphosdialkene complex (**Eq. 3.8**). A large variety of phophorous ligands (literally hundreds) have been developed (**Figure 3.4**). These catalysts are extremely efficient in the asymmetric hydrogenation of a narrow range of prochiral alkenes that can bind in a bidentate, chelate manner.

Eq. 3.8

S = solvent

By far the best substrates are Z-acetamidopropenoates, which are reduced to optically active α-amino acid derivatives. This system has been studied in minute detail, and several key intermediates in the proposed catalytic cycle have been fully characterized by ^{1}H, ^{13}C, and ^{31}P NMR spectroscopy and by X-ray crystallography.[15]

Most studies indicate that the substrate alkene coordinates in a bidentate manner, with the amide oxygen acting as the second ligand, and that two diastereoisomeric alkene complexes can be formed. The only problem with these early studies was that the diastereoisomer detected would lead to the opposite absolute configuration of the product than was observed! Careful kinetic studies[15b–d] resolved this problem, and led to the following observation that, if you can detect intermediates in a catalytic process, they probably are not involved. That is, anything that accumulates sufficiently to be detected is kinetically inactive; kinetically active species just react, converting starting material to product, and are never present in any appreciable amount. This is not always true since in recent studies, the diastereoisomer observed by NMR and X-ray crystallography was also the one that gave the observed product configuration.[16]

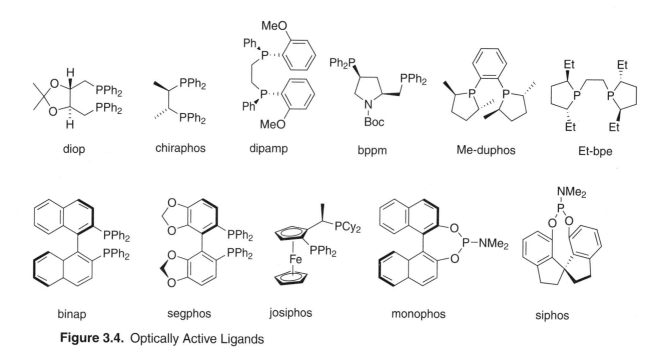

Figure 3.4. Optically Active Ligands

The mechanism for the asymmetric hydrogenation of Z-acetamidopropenoates is shown in **Figure 3.5**. Its importance to synthetic chemists lies not with its details, but rather the clear illustration it provides of how closely balanced a series of complex steps must be to achieve efficient asymmetric hydrogenation. It is for this reason that the range of prochiral alkenes asymmetrically hydrogenated by these catalysts is so narrow, and why extrapolation to different substrates is difficult.

There are two features of the process critical to high asymmetric induction; the two diastereoisomers must undergo reaction at substantially different rates, and they must equilibrate rapidly. The rate-limiting step is the oxidative addition of hydrogen to the rhodium(I) alkene complex, and one (the minor) diastereomer undergoes this reaction at a rate 10^3 greater than the other. Provided the diastereoisomeric alkene complexes can equilibrate rapidly, high asymmetric induction can be observed. Interestingly, higher hydrogen pressure and lower temperatures decrease the enantioselectivity, since both interfere with this equilibration.

The above catalysts are primarily effective with Z-acetamidopropenoates. Other effective carbonyl functionalities on the nitrogen include Bz-, Boc-, and Cbz-groups. The E isomer reduces to give the opposite enantiomer of the amino acid, but E–Z isomerization during reduction often compromises the enantioselectivity of the reduction of E-acetamidopropenoates.

For Z-acetamidobutadienoates, only the double bond closest to the acetamido group is reduced (**Eq. 3.9**).[17] In contrast, related catalysts containing the phospho ligands duphos, dipamp, or bpe reduce both E- and Z-acetamidopropenoates with equally high enantioselection and, for a given catalyst configuration, give the same product absolute configuration regardless of alkene geometry (**Eq. 3.10**).[18] In addition, β,β-disubstituted

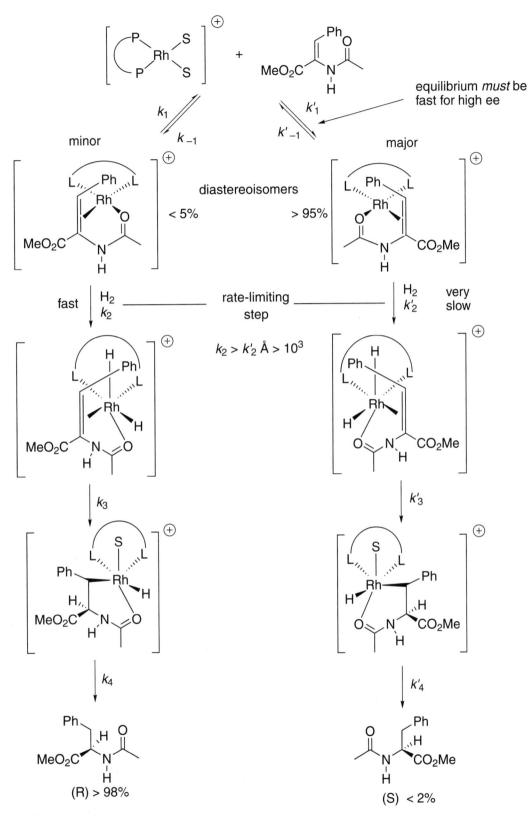

ee *lower* at high H_2 pressure-k'_2 increased

ee *lower* at low temperature-equilibration decreased; Major diastereomer accumulates

Figure 3.5. Mechanism for Asymmetric Catalytic Hydrogenation of *Z*-Acetamidopropenoates

acetamidoalkenoates are reduced to β-branched α-amino acids with high enantiose-lectivity by these catalysts.[19] One example is presented in **Eq. 3.11**.[20]

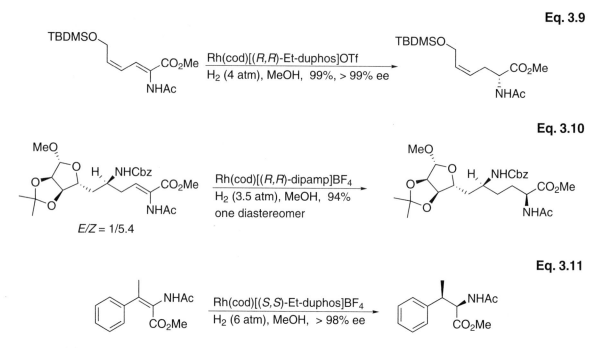

Eq. 3.9

Eq. 3.10

Eq. 3.11

In addition to acetamidopropenoates, itaconic acid esters, enamides, enol esters, α-acetoxy propenoates, dehydro-β-amino acid derivatives (**Eq. 3.12**),[21] and α,β-unsaturated nitriles (**Eq. 3.13**)[22] are also reduced with high enantioselectivity using this type of catalyst system.[23]

Eq. 3.12

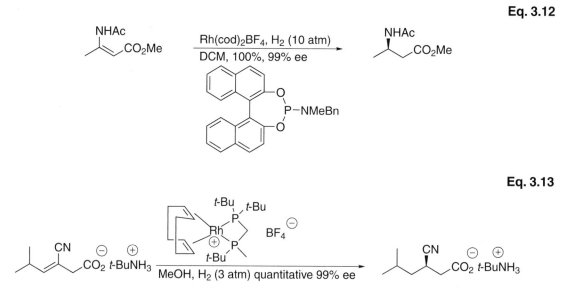

Eq. 3.13

A much more broadly useful class of asymmetric hydrogenation catalysts, which also relies upon bidentate chelate coordination of the alkene substrate, is the ruthenium(II)–binap system.[24] These complexes reduce a wide range of substrates with high asymmetric induction, relying on the rigidity of complexation conferred by chelation to assure high enantioselectivity. For example, α,β-unsaturated carboxylic acids were reduced selectively and efficiently (**Eq. 3.14**),[25] and a very wide range of functionality was tolerated. The anti-inflammatory compound naproxen (**Eq. 3.15**) was produced with 97% ee in a high-pressure reduction in methanol, as was thienamycin (**Eq. 3.16**), albeit with lower enantioselectivity, and a galaxolide (**Eq. 3.17**)[26] precursor.

Eq. 3.14

Eq. 3.15

Eq. 3.16

Eq. 3.17

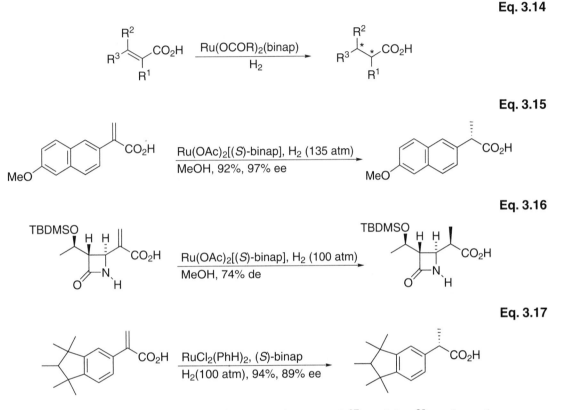

α,β-Unsaturated esters, lactones, ketones (**Eq. 3.18**),[27] amides,[28] and enol acetates were also efficient coordinating groups, and 1,2,3,4-tetrahydroisoquinoline derivatives, including morphine, benzomorphans, and morphinans, were produced in 95–100% ee by reduction under 1–4 atm of hydrogen (**Eq. 3.19**).[29]

Eq. 3.18

Eq. 3.19

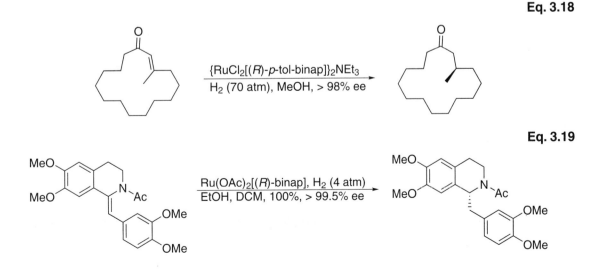

Allylic and homoallylic alcohols also undergo selective reduction with high enantioselectivity (**Eq. 3.20**).[30] Geraniol was hydrogenated to citronellol with high ee (> 96% ee) with no observed reduction of the remote double bond (**Eq. 3.21**). Either geometrical isomer could be converted to either enantiomer by appropriate choice of binap ligand. Homogeraniol was also reduced selectively, but bis-homogeraniol was inert (**Eq. 3.21**). This indicates that bidentate, chelate coordination of the alkene substrates is required for *reactivity* as well as enantioselectivity. This feature was used in a synthesis toward chatancin, wherein only the allylic alcohol alkene was reduced (**Eq. 3.22**[31]).[32]

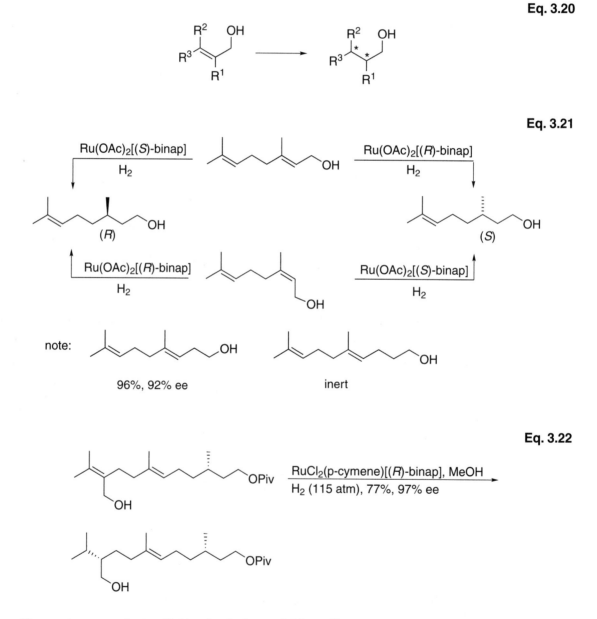

Eq. 3.20

Eq. 3.21

note:

96%, 92% ee   inert

Eq. 3.22

RuCl$_2$(p-cymene)[(R)-binap], MeOH
H$_2$ (115 atm), 77%, 97% ee

Racemic secondary allylic alcohols could be efficiently resolved by carrying out a partial hydrogenation over a ruthenium–binap catalyst (**Eq. 3.23**).[33] A wide variety of allylic alcohols undergo this reaction efficiently.

Eq. 3.23

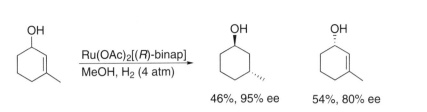

46%, 95% ee   54%, 80% ee

Although the overall transformation using these ruthenium–binap catalysts is the same as that using rhodium–chiral ligand catalyst systems, the mechanism is completely different,[34] in that ruthenium remains in the same (+2) oxidation state throughout the catalytic cycle.[35] A simplified catalytic cycle is depicted in **Figure 3.6**. Several competing catalytic cycles have been indicated.[36] A monohydride-chelated alkene intermediate has been fully characterized by ^{1}H, ^{13}C, and ^{31}P NMR spectroscopy.[37]

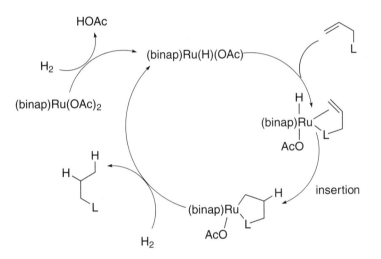

Figure 3.6. Hydrogenation Mechanism for Ru–binap Catalysts

Another synthetically useful class of hydrogenation catalyst is the "super unsaturated" iridium or rhodium complex catalyst system known as Crabtree's catalysts.[38] These are generated in situ by reduction of for example cationic cyclooctadiene iridium(I) complexes in the "noncoordinating" solvent dichloromethane (**Eq. 3.24**).

Eq. 3.24

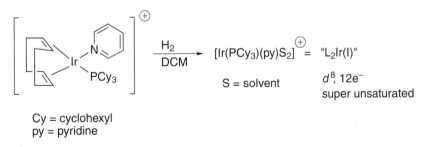

Cy = cyclohexyl
py = pyridine

The solvent is apparently so loosely coordinated that it is replaced by virtually any alkene substrate. These catalysts are exceptionally efficient for the reduction of tetrasubstituted alkenes, substrates most hydrogenation catalysts will not reduce. A comparison of Crabtree's catalyst with Wilkinson's catalyst is shown in **Figure 3.7**.

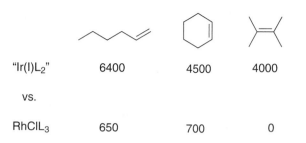

"Ir(I)L$_2$"	6400	4500	4000
vs.			
RhClL$_3$	650	700	0

Figure 3.7. Comparison of Wilkinson's and Crabtree's Catalysts

Nonpolar solvents must be used with Crabtree's catalysts. Acetone and ethanol act as competitive inhibitors, and greatly reduce catalytic activity.

The observation that Crabtree's catalyst coordinates to alcohols has been used to an advantage in synthesis, by using adjacent "hard" ligands in the substrate to coordinate to the catalyst and deliver it from a single face of the alkene. **Eq. 3.25** compares Crabtree's catalyst to the normally used palladium on carbon. The iridium catalyst was delivered exclusively from the same face of the alkene as that occupied by the OH group, while palladium on carbon was nonselective.[39] This same effect was seen with much more complex substrates (**Eqs. 3.26,**[40] **3.27,**[41] **3.28**[42] and **3.29**[43]). In all cases, the rate of hydrogenation was slower than that for substrate without coordinating functional groups.

Eq. 3.25

Eq. 3.26

Eq. 3.27

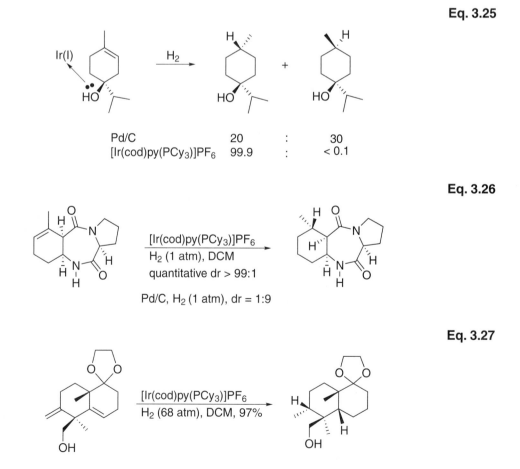

Pd/C   20   :   30
[Ir(cod)py(PCy$_3$)]PF$_6$ 99.9 : < 0.1

[Ir(cod)py(PCy$_3$)]PF$_6$
H$_2$ (1 atm), DCM
quantitative dr > 99:1

Pd/C, H$_2$ (1 atm), dr = 1:9

[Ir(cod)py(PCy$_3$)]PF$_6$
H$_2$ (68 atm), DCM, 97%

Eq. 3.28

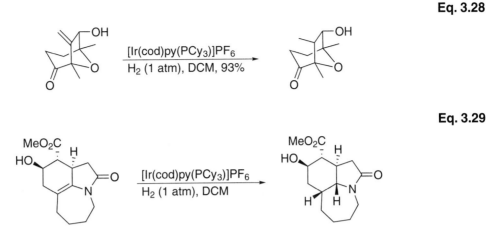

Eq. 3.29

Relatively unactivated alkenes can also be reduced with high enantiomeric induction using a cationic Ir(cod)–chiral ligand catalyst system.[44] An aryl group is required for the reaction to give high ee's. Example of alkenes and their corresponding enantiomeric excesses are shown in **Figure 3.8**.[45] Asymmetric reductions of allylic alcohols have also been achieved using iridium catalysts.[46]

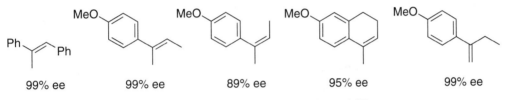

Figure 3.8. Asymmetric Reduction of Relatively Unactivated Alkenes

Most homogeneous hydrogenation catalysts are specific for carbon–carbon double bonds, and other multiply bonded species are inert. In marked contrast, halogen- or methallyl-containing ruthenium(II) complexes[47] catalyze the efficient asymmetric reduction of the carbonyl group, provided there is a heteroatom in the α-, β-, or γ-position to provide the requisite second point of attachment to the catalyst (**Eq. 3.30**).

Eq. 3.30

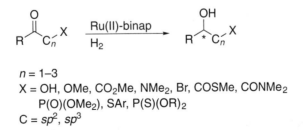

$n = 1–3$
$X = OH, OMe, CO_2Me, NMe_2, Br, COSMe, CONMe_2$
     $P(O)(OMe_2), SAr, P(S)(OR)_2$
$C = sp^2, sp^3$

β-Keto ester compounds are efficiently reduced to β-hydroxy esters by this system (**Eq. 3.31**).[48,49] Originally, high pressures of hydrogen (50–100 atm) were required, but simply adding an acid co-catalyst allows efficient catalysis under much milder conditions.[50] A larger scale application can be seen in **Eq. 3.32**.[51] This reduction has found extensive application in total synthesis (**Eqs. 3.33**[52] and **3.34**[53]).

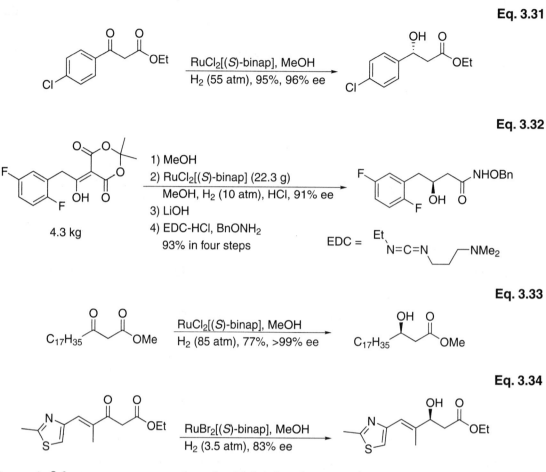

Eq. 3.31

Eq. 3.32

EDC = Et$\diagdown$N=C=N$\diagdown$—NMe$_2$

Eq. 3.33

Eq. 3.34

Racemic β-keto esters were reduced with high selectivity for each diastereoisomer (**Eq. 3.35**). However, when the reaction was carried out under conditions that ensured rapid equilibration of the α-position, high yields of a single diastereoisomer were obtained (**Eq. 3.36**).[54] The reduction is usually syn-selective, although the substrate, reaction conditions, and ruthenium catalyst all affect the syn–anti selectivity of the reaction.

Eq. 3.35

Eq. 3.36

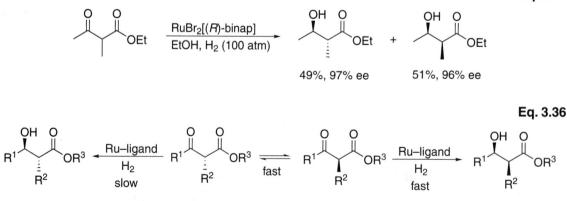

A few examples of this very powerful reaction can be seen in **Eqs. 3.37**[55] and **3.38**.[56] Ruthenium-catalyzed anti-selective reductions have also been reported (**Eqs. 3.39**[57] and **3.40**[58]).

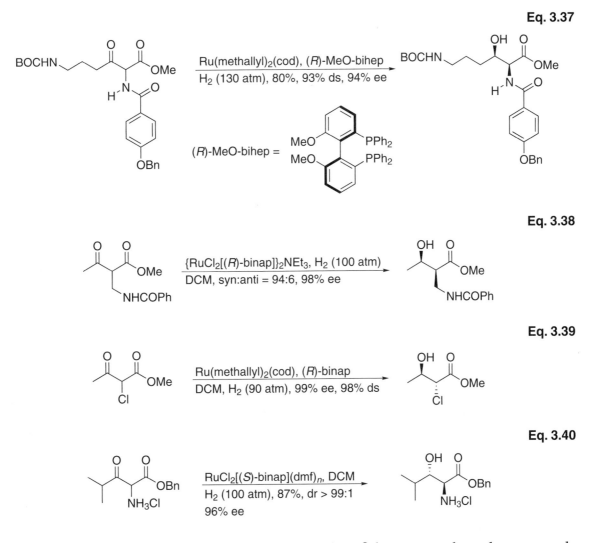

Eq. 3.37

Eq. 3.38

Eq. 3.39

Eq. 3.40

This asymmetric reduction is not restricted to β-ketoesters, but also proceeds efficiently with a wide range of carbonyl compounds having appropriately disposed directing groups.[59]

The ruthenium–bpe dibromide catalyst was also efficient in the reduction of β-ketoesters.[60] The reduction proceeded under mild conditions (35°C, 4 atm H_2, MeOH/ H_2O). The normally high enantiomeric excess observed was significantly reduced with halogen-containing substrates, and t-butyl or aryl substituents resulted in low conversions. These catalysts were also efficient when used for the dynamic resolution of chiral racemic β-ketoesters (**Eq. 3.36**), for the reduction of enamides to amines,[61] and the reduction of enol esters to α-hydroxy esters.[62]

Of substantial synthetic importance is the asymmetric reduction of ketones. Chiral rhodium and iridium complexes can be utilized, affording products with high enantiomeric excesses. However, the development of ruthenium(II) diamine phosphine complexes in the presence of an alkaline base offers a general methodology for this reduction. Both aromatic and aliphatic ketones can be reduced. However, aromatic ketones usually afford a higher enantioselectivity. Two examples of this type of reduction are given in **Eqs. 3.41**[63] and **Eq. 3.42.**[64] Dynamic kinetic resolution has been achieved using this type of catalyst system (**Eq. 3.43**).[65]

Eq. 3.41

Eq. 3.42

Eq. 3.43

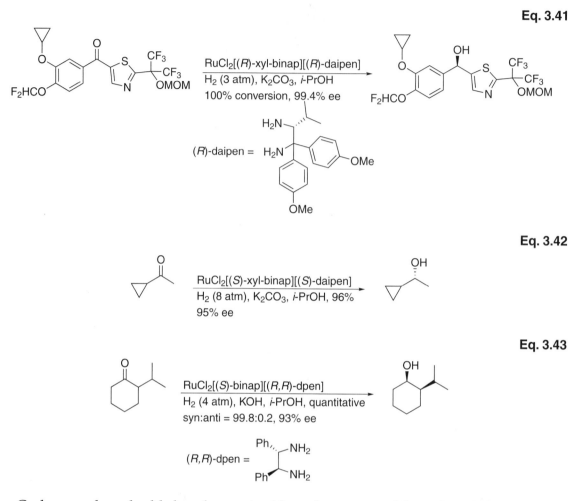

Carbon–carbon double bonds are significantly more readily reduced than are carbon–oxygen double bonds. This represents one of the more difficult and longstanding problems in catalytic reductions. The use of a Lewis base solved this problem. For example, reduction of a carbonyl group in the presence of two alkenes was achieved using a ruthenium complex in an alkaline solution (**Eq. 3.44**).[66]

Eq. 3.44

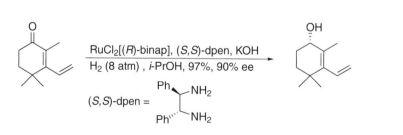

Imines and other carbon–nitrogen double bonds are reduced to amines using this system (**Eq. 3.45**).[67] Alkyl aryl ketone-derived imines are also reduced to optically active amines using iridium(I) complexes, a chiral ligand, and hydrogen gas.[68,69,70]

Eq. 3.45

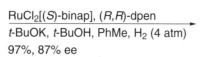

RuCl$_2$[(S)-binap], (R,R)-dpen
t-BuOK, t-BuOH, PhMe, H$_2$ (4 atm)
97%, 87% ee

The mechanism for the ruthenium(II)-catalyzed reduction of ketones described above has been examined in some detail (**Figure 3.9**).[71] With this catalyst system, the hydrogen is activated *heterolytically*, with the coordinated NH group acting as a *base* to remove H$^+$ from H$_2$ and deliver H$^-$ to the metal.

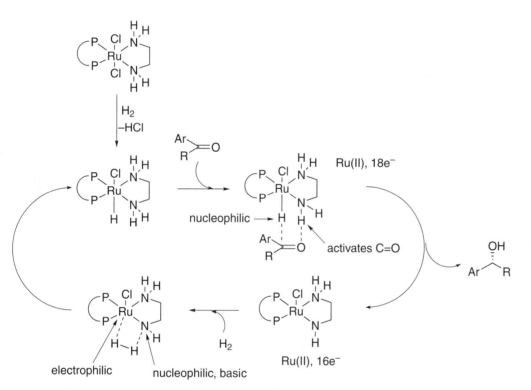

Figure 3.9. Mechanism for the Ruthenium(II)-Catalyzed Reduction of Ketones

3.3 Transfer Hydrogenations[72]

In all of the reactions presented above, molecular hydrogen was the source of hydrogen for the reduction, and the ability of transition metals to break the hydrogen–hydrogen bond in an oxidative addition reaction was central to the process. Metal hydrides are produced by several other pathways, however, and some of these have proven synthetically useful.

Alcohols can be used as the source of hydride via coordination of the oxygen to the metal followed by a β-hydride elimination. Two general mechanisms for the generation of metal hydrides from alcohols have been proposed, corresponding to

Monohydride

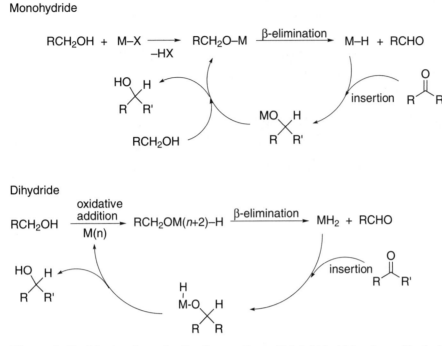

Dihydride

Figure 3.10. Mechanisms for the Generation of Metal Hydrides from Alcohols

the monohydride and dihydride reduction processes (**Figure 3.10**). [73] In this context, ruthenium-catalyzed asymmetric reductions of ketones using Noyori's catalyst[74] have found extensive use in organic synthesis. For example, chemoselective ruthenium-catalyzed enantioselective reduction of an α,β-unsaturated ketone able to achieve an S-cis conformation was used in a synthesis of (+)-tricycloclavulone (**Eq. 3.46**).[75] Alkynones are reduced in a similar fashion (**Eqs. 3.47**[76] and **3.48**[77])

Eq. 3.46

Eq. 3.47

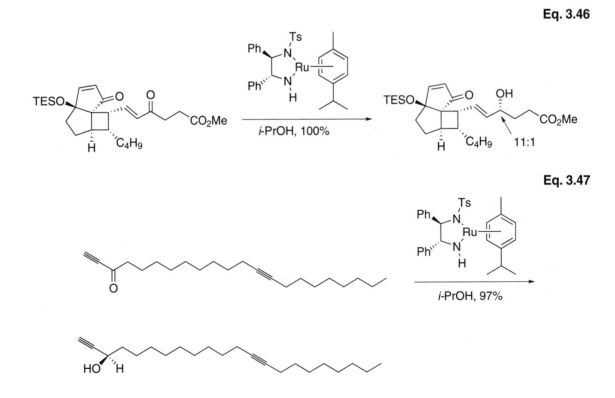

Eq. 3.48

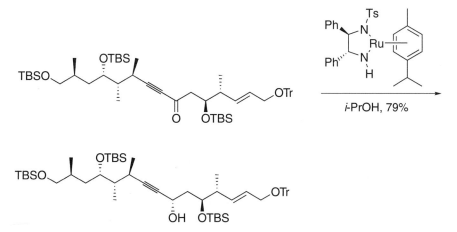

Formic acid or formates also serve as convenient sources of hydrides in the ruthenium-catalyzed dynamic kinetic resolution of 1,3-diketones[78] using Noyori's catalyst (**Eq. 3.49**),[79] a process that can also be used to reduce imines (**Eq. 3.50**).[80]

Eq. 3.49

Eq. 3.50

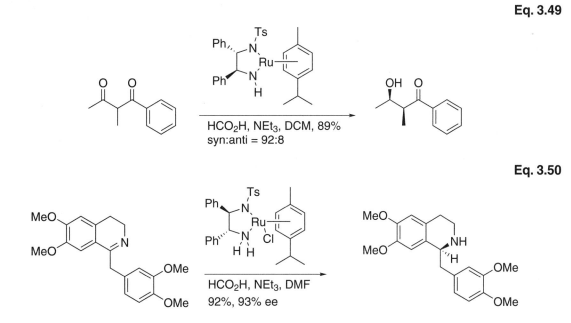

Ruthenium-catalyzed asymmetric transfer hydrogenations of ketones having a second functionality for coordination can be achieved with excellent enantiomeric excess (**Eq. 3.51**).[81,82] Rhodium hydrides formed from formates can also be used in asymmetric reductions of aromatic ketones (**Eq. 3.52**).[83]

Eq. 3.51

$$\text{Ph}\overset{\text{O}}{\underset{}{\|}}\text{R} \xrightarrow[\substack{\text{R = NMeBOC, 82\%, 99\% ee} \\ \text{R = CN, 100\%, 98\% ee} \\ \text{R = N}_3\text{, 65\%, 92\% ee} \\ \text{R = NO}_2\text{, 90\%, 98\% ee}}]{(p\text{-cymene})\text{RuClL*, NEt}_3\text{, HCO}_2\text{H}} \text{Ph}\overset{\text{OH}}{\underset{*}{}}\text{R}$$

Eq. 3.52

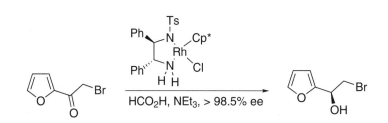

The hexamer $[(Ph_3P)CuH]_6$ ("Stryker's reagent") reduces carbon–carbon double bonds of α,β-unsaturated ketones, aldehydes, esters, nitriles, and nitro compounds stoichiometrically, presumably via a conjugate addition to produce a copper enolate intermediate.[84,85] Ynones are reduced to enones or saturated ketones (**Eq. 3.53**).[86]

Stryker's reagent does not reduce unactivated alkenes and carbon–oxygen double bonds, and *E*-enones are reduced significantly faster compared to *Z*-enones (**Eq. 3.54**).[87] Selective 1,4- versus 1,6-reduction was observed in a synthesis of the cripowellin skeleton (**Eq. 3.55**).[88]

Eq. 3.53

Eq. 3.54

Eq. 3.55

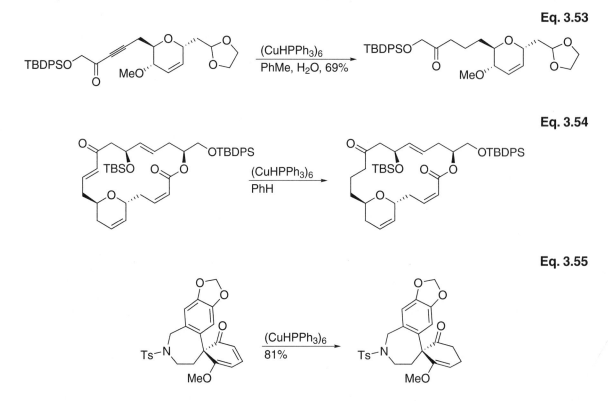

Interestingly, under modest hydrogen pressure in the presence of excess phosphine, or in the presence of polymethylhydrosiloxane (PMHS), this hexamer catalyzes this reduction.[89] Strykers's reagent or copper hydride prepared in situ from a variety of copper salts, such as $Cu(OAc)_2$, $CuCl$, $CuCl_2$, and CuF_2, by reaction with silanes, such as phenylsilane, diphenylsilane, and polymethylhydrosiloxane, are useful catalysts for conjugate reduction of α,β-unsaturated ketones, nitro compounds, esters, lactones, lactams, and nitriles. In conjunction with a chiral ligand, excellent asymmetric induction in the reduction of α,β-unsaturated compounds can be achieved[90] (**Eq. 3.56**).[91]

Eq. 3.56

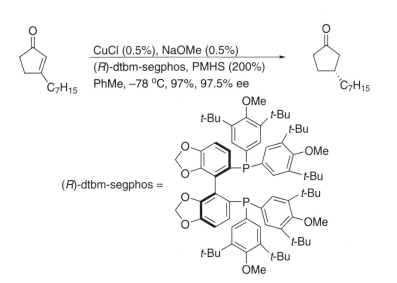

Yet another efficient system for the reduction of conjugated enals and enones is the combination of a palladium(0) catalyst and tributyltin hydride (**Eqs. 3.57**[92] and **3.58**).[93] The mechanism of the process is unknown. A possibility is shown in **Figure 3.11**.

Eq. 3.57

Eq. 3.58

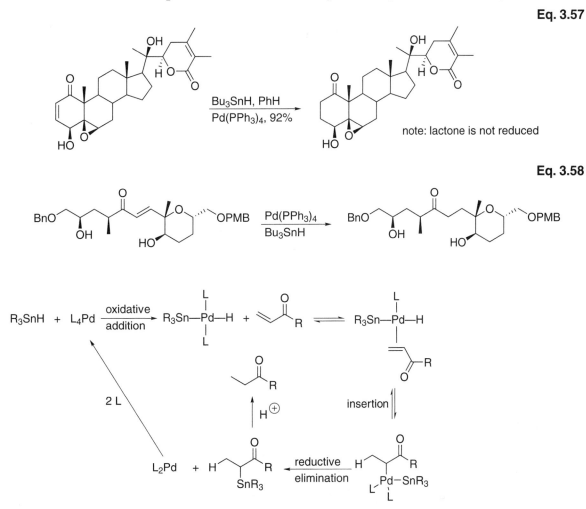

Figure 3.11. Mechanism for Palladium-Catalyzed Trialkyltin Hydride Reduction of Conjugated Enals and Enones

3.4 Hydrofunctionalizations

A number of interesting and synthetically useful transition-metal-catalyzed additions of a hydrogen and one more functional group to alkenes or alkynes have been developed.[94] Mechanistically, the reactions can be viewed as an oxidative addition of a low-valent transition metal to give a metal hydride, alkene/alkyne coordination–insertion, with reductive elimination completing the catalytic cycle (**Figure 3.12**).[95]

Synthetically important transition-metal-functionalized compounds can be obtained by metal-catalyzed hydrometallation of alkenes and alkynes. Hydrostannation of alkynes has been mediated by a number of transition metals, including palladium, molybdenum, rhodium, nickel, cobalt, copper, platinum, and ruthenium. By far the most utilized and synthetically important of these reactions is the palladium(0)-catalyzed hydrostannation of terminal alkynes, leading to stable E-alkenylstannanes with high regio- and stereoselectivity (**Eq. 3.59**).[96]

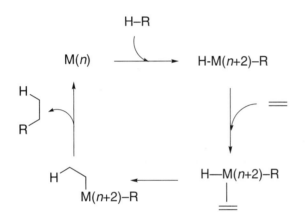

H–R = "H–Sn," "H–B," "H–Si," "H–Zr," "H–C=O"

Figure 3.12. Mechanism for "Hydrofunctionalization" of Alkenes

Eq. 3.59

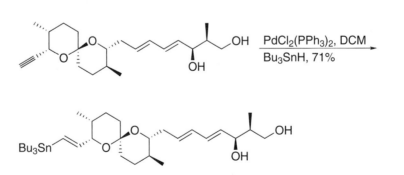

Alkynes react faster in hydrostannations compared to alkenes. Regioselective hydrostannations of internal alkynes have been reported. The tin moiety usually occupies the less-hindered position in the product. These reactions are probably aided by both electronic and steric factors (**Eq. 3.60**).[97] Probably for electronic reasons, alkynoates are preferentially or exclusively stannylated in the α-position regardless of the type of alkyne (**Eq. 3.61**).[98]

Eq. 3.60

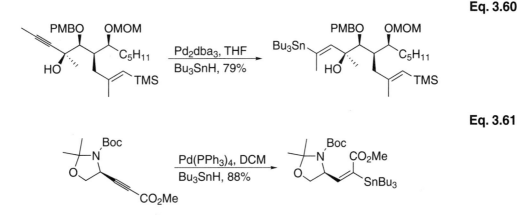

Eq. 3.61

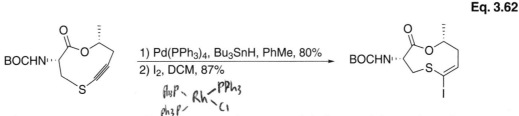

Alkenylstannanes are important intermediates in Stille-type coupling reactions (see Chapter 4). In addition, oxidative cleavage of the alkenylstannanes using iodine affords alkenyl iodides with retention of regio- and stereochemistry (**Eq. 3.62**).[99]

Eq. 3.62

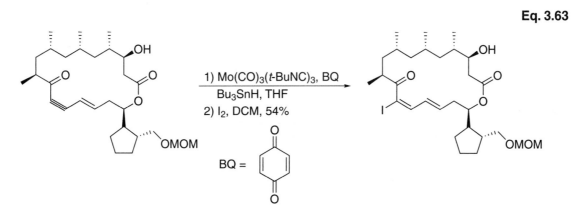

Related reactions using Wilkinson's catalyst or molybdenum(0)-catalyzed reactions have also been reported (**Eq. 3.63**).[100] In the example shown in **Eq. 3.63**, the palladium-catalyzed reaction gave mixtures of regioisomers.

Eq. 3.63

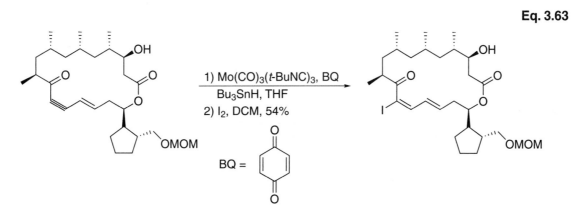

A number of transition-metal-catalyzed hydroborations and hydrosilylations have also been developed and used in organic synthesis. Hydroborations of alkenes using Wilkinson's catalyst are mild and very regio- and chemoselective (**Eq. 3.64**).[101] An example of an intramolecular platinum-catalyzed hydrosilylation is seen in **Eq. 3.65**.[102] Enone hydrosilylation using Wilkinson's complex and phenyldimethylsilanes can be used to produce silylenol ethers. An example of chemoselective hydrosilylation of a cross-conjugated enone was used in a synthesis of guanacastepene A (**Eq. 3.66**).[103]

Eq. 3.64

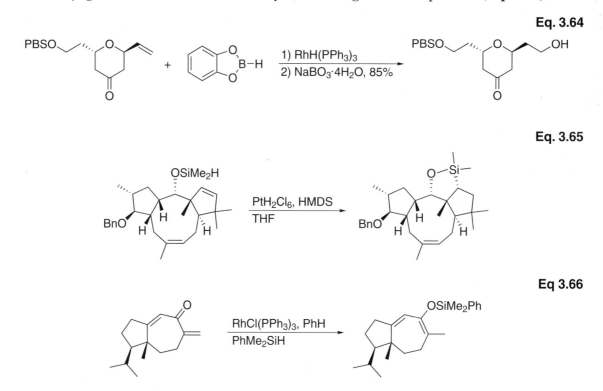

Eq. 3.65

Eq 3.66

Another hydrofunctionalization reaction involves the hydroesterification of alkenes using rhodium or ruthenium catalysts. An oxygen-directed ruthenium-catalyzed hydroesterification is presented in **Eq. 3.67**.[104]

Eq. 3.67

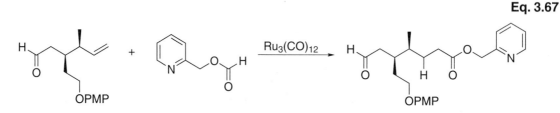

The commercially available, electron-poor, d^0, Zr(IV) complex, $Cp_2Zr(H)Cl$ ("Schwartz's reagent"), reacts under mild conditions with a variety of alkenes to give stable, isolable σ-alkyl complexes of the type $Cp_2Zr(R)Cl$ (**Eq. 3.68**).[105]

Eq. 3.68

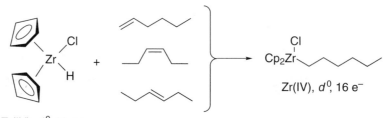

Zr(IV), d^0, 16 e⁻

This reaction is remarkable in several respects. The addition occurs under mild conditions, and the resulting alkyl complexes are quite stable. Regardless of the initial position of the double bond in the substrate, the zirconium ends up at the sterically least-hindered accessible position of the alkene chain. This rearrangement of the zirconium to the terminal position occurs by a Zr–H β-elimination, followed by a readdition that places zirconium at the less-hindered position of the alkyl chain in each instance (**Figure 3.13**). This migration proceeds because of the steric congestion about Zr caused by the two cyclopentadienyl rings. The order of reactivity of various alkenes toward hydrozirconation is terminal alkenes > cis internal alkenes > trans internal alkenes > exocyclic alkenes > cyclic alkenes and terminal disubstituted alkenes > trisubstituted alkenes. Tetrasubstituted alkenes and trisubstituted cyclic alkenes fail to react. Finally, 1,3-dienes add Zr–H to the less-hindered double bond to give γ,δ-unsaturated alkylzirconium complexes.

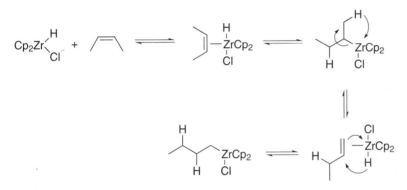

Figure 3.13. Mechanism of Zirconium Migration

Alkynes also react readily with Cp$_2$Zr(H)Cl, adding Zr–H in a cis manner. With unsymmetric alkynes, the addition of Zr–H gives mixtures of alkenyl Zr complexes with the less-hindered complex predominating, regardless of electronic factors. Equilibration of this mixture with a slight excess of Cp$_2$Zr(H)Cl leads to mixtures greatly enriched in less-hindered complex.

Zirconium(IV) alkyls are readily cleaved by electrophiles, such as bromine, iodine, or peroxides, with retention of configuration, to give straight-chain alkyl halides or alcohols, respectively,[106] even from mixtures of alkene isomers. However, this transformation has not been used extensively in organic synthesis. Zirconium(IV) alkenyls undergo electrophilic cleavage with retention of alkene geometry (**Eq. 3.69**)[107] and they can be added to aldehydes in the presence of a Lewis acid (**Eq. 3.70**).[108] In addition, they insert carbon monoxide, giving stable acylzirconium(IV) complexes that can be cleaved by acid to give aldehydes.[109] However, it was not until the development of efficient transmetallation processes that hydrozirconation achieved real utility in organic synthesis (see Chapter 4).

Eq. 3.69

Eq. 3.70

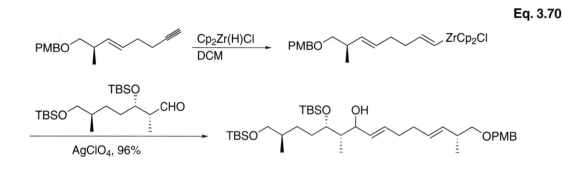

3.5 Metal Hydride-Catalyzed Alkene Isomerizations

The isomerization of carbon–carbon multiple bonds was initially viewed as a nuisance and a dead end in hydrogenation reactions of terminal alkenes. However, isomerizations have more recently found a place in organic synthesis. The most utilized alkene isomerizations in organic synthesis of complex molecules are alkenes to more-substituted alkenes, allylic amines/amides to enamines/amides, and allylic ethers to enol ethers. Not surprisingly, the transition metals usually employed in alkene reductions are also used in isomerization reactions. These particularly include palladium, rhodium, ruthenium, and iridium catalysts. Examples of the formation of a more-substituted alkene are shown in **Eqs. 3.71**[110] and **3.72**.[111]

Eq. 3.71

Eq. 3.72

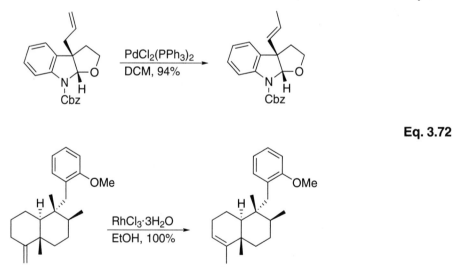

Related reactions of allylic ethers and allylic amines to give enol ethers or enamines have found use in organic synthesis.[112] An example of a reaction using Wilkinson's catalyst can be seen in **Eq. 3.73**.[113] Of substantial commercial importance is the asymmetric isomerization of *N,N*-diethylgeranylamine to the corresponding citronellal enamine used in the synthesis of (–)-menthol in >1000 ton per year (**Eq. 3.74**).[114]

Eq. 3.73

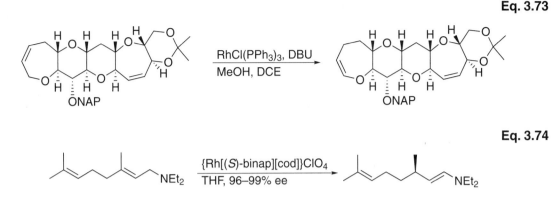

Eq. 3.74

Isomerizations of primary and secondary allylic alcohols afford saturated aldehydes and ketones, respectively. In related reactions, propargylic alcohols are isomerized to α,β-unsaturated aldehydes or ketones. This constitutes a reduction–oxidation reaction within the same molecule (**Eq. 3.75**).[115] Asymmetric variations of this type of isomerization have been achieved using a variety of chiral ligands and transition metals.

Eq. 3.75

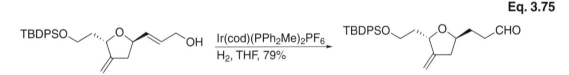

The mechanism of the isomerizations has been examined in some detail.[116] In general, a few different mechanistic routes have been proposed. For transition metal hydride complexes, alkene coordination, migratory insertion, followed by β-hydride elimination is the most likely mechanistic pathway (**Figure 3.14**). The metal remains in the same oxidation state throughout the reaction.

For all substrates discussed above, the isomerization may occur via coordination of a low-valent metal to the alkene, oxidative formation of a η^3-allylmetal hydride complex, followed by readdition of the hydride to the opposite end of the η^3-allyl system (**Figure 3.15**). Alternatively, the metal may coordinate to a heteroatom, undergo a β-hydride elimination to give a metal hydride complex that undergoes a migratory insertion. Isomerizations of allylic alcohols using ruthenium(II) complexes probably

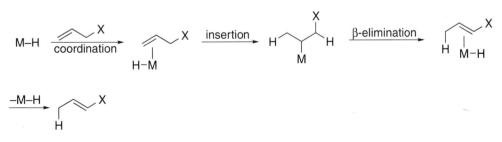

Figure 3.14. Alkene Isomerization via Migratory Insertion-β-Elimination

proceed via the formation of a ruthenium alkoxide and β-hydride elimination (**Figure 3.16**).[117] Migratory insertion and alkoxide exchange completes the catalytic cycle.

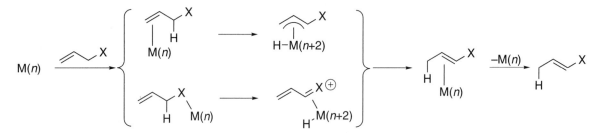

Figure 3.15. Alkene Isomerization via Oxidative Addition

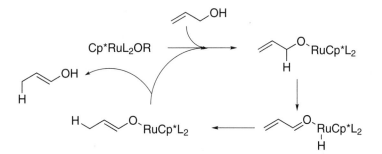

Figure 3.16. Ruthenium-Catalyzed Isomerization of Allylic Alcohols

3.6 Other Reactions of Transition Metal Hydrides

Transition metal hydrides are involved in a host of other synthetically useful reactions, most of which proceed by "hydrometallation" (i.e., the insertion into a metal–hydrogen bond) of an alkene or alkyne to produce σ-alkylmetal complexes. The subsequent reactions of these σ-alkylmetal species provide a variety of synthtically useful transformations, and these are discussed in Chapter 4.

References

(1) For a review, see McGrady, G. S.; Guilera, G, *Chem. Soc. Rev.* **2003**, *32*, 383.

(2) Moore, E. J.; Sullivan, J. M.; Norton, J. R. *J. Am. Chem. Soc.* **1986**, *108*, 2257.

(3) For reviews of asymmetric hydrogenations, see a) Tang, W.; Zhang, X. *Chem. Rev.* **2003**, 3029. b) Noyori, R. *Angew. Chem. Int. Ed.* **2002**, *41*, 2008. c) Noyori, R.; Ohkushima, T. *Angew. Chem. Int. Ed.* **2001**, *41*, 1999.

(4) Halpern, J.; Okamoto, T.; Zakhariev, A. *J. Mol. Catal.* **1976**, *2*, 65.

(5) Zhu, L; Mootoo, D. R. *J. Org. Chem.* **2004**, *69*, 3154.

(6) Drege, E.; Morgant, G.; Desmaële, D. *Tetrahedron Lett.* **2005**, *46*, 7263.

(7) Clark, J. S.; Fessard, T. C.; Wilson, C. *Org. Lett.* **2004**, *6*, 1773.

(8) Nagamitsu, T.; Takano, D.; Fukuda, T.; Otoguro, K.; Kuwajima, I.; Harigaya, Y.; Omura, S. *Org. Lett.* **2004**, *6*, 1865.

(9) Takahashi, S.; Souma, K.; Hashimoto, R.; Koshino, H.; Nakata, T. *J. Org. Chem.* **2004**, *69*, 4509.

(10) In a synthesis toward amphidinol 3. Flamme, E. M.; Roush, W. R. *Org. Lett.* **2005**, *7*, 1411.

(11) Zhang, Y.; Liebeskind, L. S. *J. Am. Chem. Soc.* **2006**, *128*, 465.

(12) For reviews, see a) Noyori, R. *Asymmetric Catalysis in Organic Synthesis*; Wiley: New York, 1994; pp 16–94. b) Ojima, I. Transition Metal Hydrides: Hydrocarboxylation Hydroformylation and Asymmetric Hydrogenation. In *Comprehensive Organometallic Chemistry II*; Abel, E. W., Stone, F. G. A., Wilkinson, G., Eds.; Pergamon Press: Oxford, UK, 1995; Vol. 12, pp 9–38.

(13) For reviews, see a) Komarov, I. V.; Börner, A. *Angew. Chem. Int. Ed.* **2001**, *40*, 1197. b) Berthod, M.; Mignani, G.; Woodward, G.; Lemaire, M. *Chem. Rev.* **2005**, *105*, 1801.

(14) For a review of monodentate ligands in asymmetric hydrogenations, see Jerphagnon, T.; Renaud, J.-L.; Bruneau, C. *Tetrahedron: Asymmetry* **2004**, *15*, 2101.

(15) a) Brown, J. M.; Chaloner, P. A. *J. Am. Chem. Soc.* **1980**, *102*, 3040. b) Brown, J. M.; Parker, D. *Organometallics* **1982**, *1*, 950. c) Halpern, J. *Science* **1982**, *217*, 401. d) Halpern, J. *Acc. Chem. Res.* **1982**, *15*, 332. e) Gridnev, I. D.; Higashi, N.; Asakura, K.; Imamoto, T. *J. Am. Chem. Soc.* **2000**, *122*, 7183.

(16) a) Drexler, H.-J.; Bauman, W.; Schmidt, T.; Zhang, S.; Sun, A.; Spannenberg, A.; Fischer, C.; Buschmann, H.; Heller, D. *Angew. Chem. Int. Ed.* **2005**, *44*, 1184. b) Evans, D.; Michael, F. E.; Tedrow, J. S.; Campos, K. R. *J. Am. Chem. Soc.* **2003**, *125*, 3534.

(17) In a synthesis of (+)-bulgecinine. Burk, M. J.; Allen, J. G.; Kiesman, W. F. *J. Am. Chem. Soc.* **1998**, *120*, 657.

(18) In a synthesis of (+)-sinefungin. Ghosh, A. K.; Wang, Y. *J. Chem. Soc., Perkin Trans. 1* **1999**, 3597.

(19) a) Burke, M. J.; Gross, M. F.; Harper, T. G. P.; Kalberg, C. S.; Lee, J. R.; Martinez, J. P. *Pure Appl. Chem.* **1996**, *68*, 37. b) Burk, M. J.; Wang, Y. M.; Lee, J. R. *J. Am. Chem. Soc.* **1996**, *118*, 5142. c) Burk, M. J.;

(20) Gross, M. F.; Martinez, J. P. *J. Am. Chem. Soc.* **1995**, *117*, 9375.

(20) Roff, G. J.; Lloyd, R. C.; Turner, N. J. *J. Am. Chem. Soc.* **2004**, *126*, 4098.

(21) Pena, D.; Minnaard, A. J.; de Vries, J. G.; Feringa, B. L. *J. Am. Chem. Soc.* **2002**, *124*, 14552.

(22) Hoge, G.; Wu, H.-P.; Kissel, W. S.; Pflum, D.; Greene, D. J.; Bao, J. *J. Am. Chem. Soc.* **2004**, *126*, 5966.

(23) Lotz, M.; Polborn, K.; Knochel, P. *Angew. Chem. Int. Ed.* **2002**, *41*, 4708.

(24) For reviews, see a) Noyori, R.; Takaya, H. *Acc. Chem. Res.* **1990**, *23*, 345. b) Noyori, R. *Acta Chem. Scand.* **1996**, *50*, 380. c) For a synthesis of the catalyst, see Takaya, H.; Akatagawa, S.; Noroyi, R. *Org. Synth.* **1988**, *67*, 20.

(25) Ohta, T.; Takaya, H.; Kitamura, M.; Nagai, K.; Noyori, R. *J. Org. Chem.* **1987**, *52*, 3174.

(26) Ciappa, A.; Matteoli, U.; Scivanti, A. *Tetrahedron: Asymmetry* **2002**, *13*, 2193.

(27) Yamamoto, T.; Ogura, M.; Kanisawa, T. *Tetrahedron* **2002**, *58*, 9209.

(28) For a review, see van den Berg, M.; Haak, R. M.; Minnaard, A. J.; Vries, A. H. M.; de Vries, J. G.; Feringa, B. L. *Adv. Synth. Catal.* **2002**, *344*, 1003.

(29) a) Noyori, R.; Ohta, M.; Hsiao, Y.; Kitamura, M.; Ohta, T.; Takaya, H. *J. Am. Chem. Soc.* **1986**, *108*, 7117. b) Kitamura, M.; Hsiao, Y.; Ohta, M.; Tsukamoto, M.; Ohta, T.; Takaya, H.; Noyori, R. *J. Org. Chem.* **1994**, *59*, 297. c) Uematsu, N.; Fujii, A.; Hishiguchi, S.; Ikoriza, T.; Noyori, R. *J. Am. Chem. Soc.* **1996**, *118*, 4916.

(30) Takaya, H.; Ohta, T.; Sayo, N.; Kumobayashi, H.; Akutagawa, S.; Inoue, S.-I.; Kasahara, I.; Noyori, R. *J. Am. Chem. Soc.* **1987**, *109*, 1596.

(31) Toro, A.; L'Heureux, A.; Deslongchamps, P. *Org. Lett.* **2000**, *2*, 2737.

(32) Imperiali, B.; Zimmerman, J.W. *Tetrahedron Lett.* **1988**, *29*, 5343.

(33) Kitamura, M.; Kasahara, I.; Manabe, K.; Noyori, R.; Takaya, H. *J. Org. Chem.* **1988**, *53*, 708.

(34) a) Ohta, T.; Takaya, H.; Noroyi, R. *Tetrahedron Lett.* **1990**, *31*, 7189. b) Ashby, M. T.; Halpern, J. *J. Am. Chem. Soc.* **1991**, *113*, 589.

(35) Kitamura, M.; Tsukamoto, M.; Bessho, Y.; Yoshimura, M.; Kobs, U.; Widhalm, M.; Noyori, R. *J. Am. Chem. Soc.* **2002**, *124*, 6649.

(36) Yoshimura, M.; Ishibashi, Y.; Miyata, K.; Bessho, Y.; Tsukamoto, M.; Kitamura, M. *Tetrahedron* **2007**, *63*, 11399.

(37) Wiles, J. A.; Bergens, S. H.; *Organometallics* **1999**, *18*, 3709.

(38) Crabtree, R. *Acc. Chem. Res.* **1979**, *12*, 331.

(39) Crabtree, R. H.; Davis, M. W. *Organometallics* **1983**, *2*, 681.

(40) Schultz, A. G.; McCloskey, P. J. *J. Org. Chem.* **1985**, *50*, 5905.

(41) Watson, A. T.; Park, K.; Wiemer, D. F.; Scott, W. J. *J. Org. Chem.* **1995**, *60*, 5102.

(42) Hodgson, D. M.; Le Strat, F.; Avery, T. D.; Donohue, A. C.; Brückl, T. *J. Org. Chem.* **2004**, *69*, 8796.

(43) Padwa, A.; Ginn, J. D. *J. Org. Chem.* **2005**, *70*, 5197.

(44) For a review, see Xiuhua, C.; Burgess, K. *Chem. Rev.* **2005**, *105*, 3272.

(45) Smidt, S. P.; Menges, F.; Pfaltz, A. *Org. Lett.* **2004**, *6*, 2023.

(46) Lightfoot, A.; Schneider, P.; Pfaltz, A. *Angew. Chem. Int. Ed.* **1998**, *37*, 2897.

(47) a) Mashima, K.; Kusano, K.; Sato, N.; Matsumura, Y.; Nozaki, K.; Kumbayashi, H.; Sayo, N.; Hori, Y.; Ishizaki, T.; Akutagawa, S.; Takaya, H. *J. Org. Chem.* **1994**, *59*, 3064. b) Kitamura, M.; Tokamages, M.; Ohkuma, T.; Noyori, Y. *Org. Synth.* **1992**, *71*, 1.

(48) Noyori, R.; Ohkuma, T.; Kitamura, M.; Takaya, H.; Sayo, N.; Kumobayashi, H.; Akutagawa, S. *J. Am. Chem. Soc.* **1987**, *109*, 5856.

(49) In a synthesis of (*R*)-baclofen. Thakur, V. V.; Nikalje, M. D.; Sudalai, A. *Tetrahedron: Asymmetry* **2003**, *14*, 581.

(50) a) King, S. A.; Thompson, A. S.; King, A. O.; Verhoeven, T. R. *J. Org. Chem.* **1992**, *57*, 6689. b) Hartman, R.; Chen. P. *Angew. Chem. Int. Ed.* **2001**, *40*, 3581.

(51) In a synthesis of a β-amino acid pharmacophore. Angelaud, R.; Zhong, Y.-L.; Maligres, P.; Lee, J.; Askin, D. *J. Org. Chem.* **2005**, *70*, 1949.

(52) In a synthesis of (*R*)-3-hydroxyicosonoic acid. Zamyatina, A.; Sekljic, H.; Brade, H.; Kosma, P. *Tetrahedron* **2004**, *60*, 12113.

(53) In a synthesis of the C12–C21 fragment of epothilone. Reiff, E. A.; Nair, S. K.; Reddy, B. S. N.; Inagaki, J.; Henri, J. T.; Greiner, J. F.; Georg, G. I. *Tetrahedron Lett.* **2004**, *45*, 5845.

(54) a) Noyori, R.; Ikeda, T.; Ohkuma, T.; Widhalm, M.; Kitamura, M.; Takaya, H.; Akutagawa, S.; Sayo, N.; Saito, T.; Taketomi, T.; Kumobayashi, H. *J. Am. Chem. Soc.* **1989**, *111*, 9134. b) Noyori, R.; Tokunaga, M.; Kitamura, M. *Bull. Chem. Soc. Jpn.* **1995**, *68*, 36.

(55) In a synthesis of (–)-balanol. Phasavath, P.; Duprat de Paul, S.; Ratovelomana-Vidal, V.; Genet, J. P. *Eur. J. Org. Chem.* **2000**, 3903.

(56) Noyori, R.; Ikeda, Y.; Ohkuma, T.; Widhalm, M.; Kitamura, M.; Takaya, H.; Akutagawa, S.; Sayo, N.; Saito, T.; Taketomi, T.; Kumobayashi, H. *J. Am. Chem. Soc.* **1989**, *111*, 9134.

(57) Genet, J. P.; Cano de Andrade, M. C.; Ratovelomana-Vidal, V. *Tetrahedron Lett.* **1995**, *36*, 2063.

(58) Makino, K.; Takayuki, G.; Hiroki, Y.; Hamada, Y. *Angew. Chem. Int. Ed.* **2004**, *43*, 882.

(59) Kitamura, M.; Ohnkuma, T.; Inoue, S.; Sayo, N.; Kumobayashi, H.; Akutagawa, S.; Ohta, T.; Takaya, H.; Noyori, R. *J. Am. Chem. Soc.* **1988**, *110*, 629.

(60) Burk, M. J.; Harper, T. G. P.; Kalberg, C. S. *J. Am. Chem. Soc.* **1995**, *117*, 4423.

(61) a) Burk, M. J.; Wang, Y. M.; Lee, J. R. *J. Am. Chem. Soc.* **1996**, *118*, 5143. b) Burk, M. J.; Allen, J. G.; Kiesman, W. F. *J. Am. Chem. Soc.* **1998**, *120*, 657.

(62) a) Burk, M. J. *J. Am. Chem. Soc.* **1991**, *113*, 8518. b) Burk, M. J.; Allen, J. G.; Kiesman, W. F. *J. Am. Chem. Soc.* **1998**, *120*, 4345.

(63) In a synthesis of a PDE4 inhibitor. O'Shea, P.; Chen, C.-i.; Chen, W.; Dagneau, P.; Frey, L. F.; Grabowski, E. J. J.; Marcantonio, K. M.; Reamer, R. A.; Tan, L.; Tillyer, R. D.; Roy, A.; Wang, X.; Zhao, D. *J. Org. Chem.* **2005**, *70*, 3021.

(64) Ohkuma, T.; Koizumi, M.; Doucet, H.; Pham, T.; Kozawa, M.; Murata, K.; Katayama, E.; Yokozawa, T.; Ikariya, T.; Noyori, R. *J. Am. Chem. Soc.* **1998**, *120*, 13529.

(65) Ohkuma, T.; Ooka, H.; Yamakawa, M.; Ikariya, T.; Noyori, R. *J. Org. Chem.* **1996**, *61*, 4872.

(66) In a synthesis of (+)-arisugacins A and B. Sunazuka, T.; Handa, M.; Nagai, K.; Shirahata, T.; Harigaya, Y.; Otogurao, K.; Kuwajima, I.; Omura, S. *Tetrahedron* **2004**, *60*, 7845.

(67) In a synthesis of S18986. Cobley, C. J.; Foucher, E.; Lecouve, J.-P.; Lennon, I. C.; Ramsden, J. A.; Thominot, G. *Tetrahedron: Asymmetry* **2003**, *14*, 3431.

(68) Jiang, X.-b.; Minnaard, A. J.; Hessen, B.; Feringa, B. L.; Duchateau, A. L. L.; Andrien, J. G. O.; Boogers, J. A. F.; de Vries, J. G. *Org. Lett.* **2003**, *5*, 1503.

(69) Trifonova, A.; Diesen, J. S.; Chapman, C. J.; Andersson, P. G. *Org. Lett.* **2004**, *6*, 3825.

(70) Moessner, C.; Bolm, C. *Angew. Chem. Int. Ed.* **2005**, *44*, 7564.

(71) a) Sandoval, C. A.; Ohkuma, T.; Muniz, K.; Noyori, R. *J. Am. Chem. Soc.* **2003**, *125*, 13490. b) Clapham, S. E.; Hadovic, A.; Morris, R. H.; *Coord. Chem. Rev.* **2004**, *248*, 2201.

(72) For a review of asymmetric transfer hydrogenations, see Gladiali, S.; Alberico, E. *Chem. Soc. Rev.* **2006**, *35*, 226.

(73) Bäckvall, J.-E. *J. Organomet. Chem.* **2002**, *652*, 105.

(74) Matsumura, K.; Hashiguchi, S.; Ikariya, T.; Noyori, R. *J. Am. Chem. Soc.* **1997**, *119*, 8738.

(75) Ito, H.; Hasegawa, M.; Takenaka, Y.; Kobayashi, T.; Iguchi, K. *J. Am. Chem. Soc.* **2004**, *126*, 4520.

(76) In a synthesis of (*R*)-strongylodiol A and B. Kirkham, J. E. D.; Courtney, T. D. L.; Lee, V.; Baldwin, J. E. *Tetrahedron* **2005**, *61*, 7219.

(77) In a synthesis of (–)-dictyostatin. Shin, Y.; Fournier, J.-H.; Fukui, Y.; Brückner, A. M.; Curran. D. P. *Angew. Chem. Int. Ed.* **2004**, *43*, 4634.

(78) Eustache, F.; Dalko, P. I.; Cossy, J. *Org. Lett.* **2002**, *4*, 1263.

(79) In a synthesis of (+)-conagenine. Matsukawa, Y.; Isobe, M.; Katsuki, H.; Ichikawa, Y. *J. Org. Chem.* **2005**, *70*, 5339.

(80) In a synthesis of (+)-laudanosine and (–)-xylopine. Mujahidin, D.; Doye, S. *Eur. J. Org. Chem.* **2005**, 2689.

(81) Kawamoto, A. M.; Wills, M. *J. Chem. Soc., Perkin Trans. I* **2001**, 1916.

(82) Watanabe, M.; Murata, K.; Ikariya, T. *J. Org. Chem.* **2002**, *67*, 1712.

(83) In a synthesis of eriolanin and eriolangin. Merten, J. Fröhlich, R.; Metz, P. *Angew. Chem. Int. Ed.* **2004**, *43*, 5991.

(84) Mahoney, W. S.; Stryker, J. M. *J. Am. Chem. Soc.* **1989**, *111*, 8818.

(85) For examples of sequential reduction and intramolecular aldol or Henry reactions of the intermediate enolate, see a) Chiu, P.; Leung, S. K. *Chem. Commun.* **2004**, 2308. b) Chung, W. K.; Chiu, P. *Synlett* **2005**, 55.

(86) In a synthesis of a novel laulimalide analog. Gallagher, B. M., Jr.; Zhao, H.; Pesant, M.; Fang, F. G. *Tetrahedron Lett.* **2005**, *46*, 923.

(87) Paterson, I; Bergmann, H.; Menche, D.; Berkessel, A. *Org. Lett.* **2004**, *6*, 1293.

(88) Moon, B.; Han, S.; Yoon, Y.; Kwon, H. *Org. Lett.* **2005**, *7*, 1031.

(89) Mahoney, W. S.; Brestensky, D. M.; Stryker, J. M. *J. Am. Chem. Soc.* **1988**, *110*, 291.

(90) Appella, D. H.; Moritani, Y.; Shintani, R.; Ferreira, E. M.; Buchwald, S. L. *J. Am. Chem. Soc.* **1999**, *121*, 9473.

(91) Lipshutz, B. H.; Servesko, J. M.; Petersen, T. B.; Papa, P. P.; Lover, A. A. *Org. Lett.* **2004**, *6*, 1273.

(92) Keinan, E.; Gleize, P. A. *Tetrahedron Lett.* **1982**, *23*, 477.

(93) In the synthesis of a C7–C19 fragment of lituarine. Smith, A. B., III; Frohn, M. *Org. Lett.* **2001**, *3*, 3979.

(94) For a review of additions to alkynes, see Trost, B. M.; Ball, Z. T. *Synthesis* **2005**, 853.

(95) For a study of the mechanism of rhodium-catalyzed hydroboration, see Evans, D. A.; Fu, G. C.; Anderson, B. A. *J. Am. Chem. Soc.* **1992**, *1143*, 6679.

(96) In a synthesis of spirofungin B. Zanatta, S. D.; White, J. M.; Rizzacasa, M. A. *Org. Lett.* **2004**, *6*, 1041.

(97) In a synthesis of formamicinone. Savall, B. M.; Blanchard, N.; Roush, W. R. *Org. Lett.* **2003**, *5*, 377.

(98) In a synthesis of asperazine. Govek, S. P.; Overman, L. E. *J. Am. Chem. Soc.* **2001**, *123*, 9468.

(99) In a synthesis toward griseoviridin. Kuligowski, C.; Bezzenine-Lafollee, S.; Chaume, G.; Mahuteau, J.; Barriere, J.-C.; Bacque, E.; Pancrazi, A.; Ardisson, J. *J. Org. Chem.* **2002**, *67*, 4565.

(100) In a synthesis of (–)-borrelidin. Vong, B. G.; Kim, S. H.; Abraham, S.; Theodorakis, E. A. *Angew. Chem. Int. Ed.* **2004**, *43*, 3947.

(101) In a synthesis of (+)-phorboxazole. Smith, A. B., III; Minbiole, K. P.; Verhoest, P. R.; Schelhaas, M. *J. Am. Chem. Soc.* **2001**, *123*, 10942.

(102) In a synthesis of jatrophatrione and citlalitrione. Yang, J.; Long, Y. O.; Paquette, L. A. *J. Am. Chem. Soc.* **2003**, *125*, 1567.

(103) Tan, D. S.; Dudley, G. B.; Danishefsky, S. J. *Angew. Chem. Int. Ed.* **2002**, *41*, 2185.

(104) In a synthesis of the C16–C35 fragment of integramycin. Wang, L.; Floreancig, P. E. *Org. Lett.* **2004**, *6*, 569.

(105) Schwartz, J.; Labinger, J. A. *Angew. Chem., Int. Ed. Engl.* **1976**, *15*, 333.

(106) Gibson, T. *Tetrahedron Lett.* **1982**, *23*, 157.

(107) In a synthesis of (+)-scyphostatin. Inoue, M.; Yokota, W.; Murugesh, M. G.; Izuhara, T.; Katoh, T. *Angew. Chem. Int. Ed.* **2004**, *43*, 4207.

(108) In a synthesis of antascomicin B. Brittain, D. E. A.; Griffiths-Jones, C. M.; Linder, M. R.; Smith, M. D.; McCusker, C.; Barlow, J. S.; Akiyama, R.; Yasuda, K.; Ley, S. V. *Angew. Chem. Int. Ed.* **2005**, *44*, 2732.

(109) Bertelo, C. A.; Schwartz, J. *J. Am. Chem. Soc.* **1976**, *98*, 262.

(110) In a synthesis of (–)-physovenine. Sunazuka, T.; Yoshida, K.; Kojima, N.; Shirata, T.; Hirose, T.; Handa, M.; Yamamoto, D.; Harigaya, Y.; Kuwajima, I.; Omura, S. *Tetrahedron Lett.* **2005**, *46*, 1459.

(111) In a synthesis toward stachyflin. Nakatani, M.; Nakamura, M.; Suzuki, A.; Inoue, M.; Katoh, T. *Org. Lett.* **2002**, *4*, 4483.

(112) Corey, E. J.; Suggs, J. W. *J. Org. Chem.* **1978**, *38*, 3224.

(113) In a synthesis toward ciguatoxin. Kobayashi, S.; Alizadeh, B. H.; Sasaki, S.-y.; Oguri, H.; Hirama, M. *Org. Lett.* **2004**, *6*, 751.

(114) Tani, K.; Yamagata, T.; Akutagawa, S.; Kumobayashi, H.; Taketomi, T.; Takaya, H.; Miyashita, A.; Noyori, R.; Ohtsuka, S. *J. Am. Chem. Soc.* **1984**, *106*, 5208.

(115) In a synthesis of the F-ring of halichondrin. Jiang, L.; Burke, S. D. *Org. Lett.* **2002**, *4*, 3411.

(116) a) Tanaka, K.; Fu, G. C. *J. Org. Chem.* **2001**, *66*, 8177. b) Uma, R.; Crevisy, C.; Gree, R. *Chem. Rev.* **2003**, *103*, 27. c) Ito, M.; Kitahara, S.; Ikariya, T. *J. Am. Chem. Soc.* **2005**, *127*, 6172.

(117) Martin-Matute, B.; Bogar, K.; Edin, M.; Kaynak, F. B.; Bäckvall, J.-E. *Chem.—Eur. J.* **2005**, *11*, 5832.

4

Synthetic Applications of Complexes Containing Metal–Carbon σ-Bonds

4.1 Introduction

Transition metal complexes containing metal–carbon σ-bonds are central to a majority of transformations in which transition metals are used to form carbon–carbon, carbon–heteroatom, and carbon–hydrogen bonds, and thus are of supreme importance for organic synthesis. The transition-metal-to-carbon σ-bond is usually covalent rather than ionic, and this feature strongly moderates the reactivity of the bound organic group, restricting its reactions to those accessible to the transition metal (e.g., oxidative addition, insertion, reductive elimination, β-hydride elimination, or transmetallation). That is, the transition metal is much more than a sophisticated counterion for the organic group; it is the major determinant of the chemical behavior of that organic group.

σ-Carbon–metal complexes can be prepared by the methods summarized in **Figure 4.1**. This variety makes virtually every class of organic compound a potential source of the organic group in these complexes, and thus subject to all of the carbon–carbon bond-forming reactions of σ-carbon–metal complexes. The use of these processes in organic synthesis is the subject of this chapter.

4.2 σ-Carbon–Metal Complexes from the Reaction of Carbanions and Metal Halides: Organocopper Chemistry

Starting with the early observation that small amounts of copper(I) salts catalyzed the 1,4-addition of Grignard reagents to conjugated enones,[1] organocopper chemistry has been very actively investigated and widely applied by synthetic chemists. As a

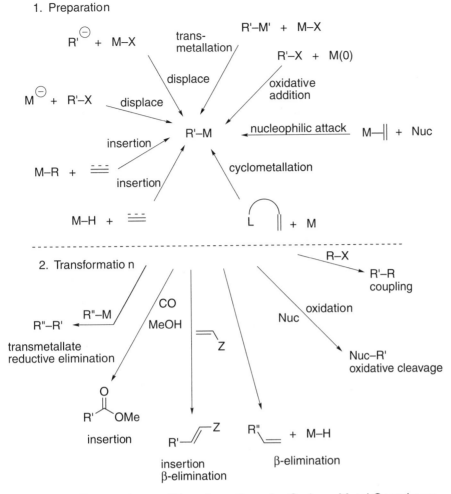

Figure 4.1. Preparation and Transformation of σ-Carbon–Metal Complexes

broad class, they are among the most extensively used organometallic reagents, and the variety of species and the transformations they effect is staggering.[2,3] The simplest and most extensively used complexes are the lithium diorganocuprates,"R_2CuLi," which are soluble, thermally unstable complexes generated in situ by the reaction of copper(I) iodide with two equivalents of an organolithium reagent (**Eq. 4.1**). These diorganocuprates are actually a complex aggregation of several "R_2CuLi" units and the structure depends on the solvent, concentration, and other factors. (The *monoalkyl* copper complexes, RCu, are yellow, insoluble, oligomeric complexes little used in synthesis.) Diorganocuprates efficiently alkylate a variety of organic halides (**Eq. 4.2**). Their low basicity favors displacement over competing elimination processes, and allows hydroxy-groups to be present in the substrate.

Eq. 4.1

$$2RLi + CuI \longrightarrow R_2CuLi + LiI$$

Eq. 4.2

$$R_2CuLi \ + \ R'X \ \longrightarrow \ R-R'$$

The range of useful halide substrates is very broad. The order of reactivity of alkyl (sp^3) halides is primary > secondary >> tertiary, with iodides more reactive than bromides and chlorides. Alkyl tosylates and triflates (**Eq. 4.3**)[4] also undergo substitution by diorganocuprates at a rate comparable to iodides. Alkenyl halides and triflates (enol triflates derived from ketones[5]) are also very reactive toward lithium diorganocuprates, and react with essentially complete retention of geometry of the double bond, a feature that has proved useful in organic synthesis. Aryl iodides and bromides are alkylated by lithium organocuprates, although halogen–metal exchange is sometimes a problem. (RCu reagents are sometimes more efficient in this reaction.) Benzylic, allylic, and propargylic halides also react cleanly. Propargylic halides produce primarily allenes (via S_N2'-type reactions), but allylic chlorides and bromides react without allylic transposition. In contrast, allylic esters and oxiranes (**Eq. 4.4**)[6] react mainly with allylic transposition. Acid halides are converted to ketones by lithium dialkylcuprates, although alkyl hetero (mixed) cuprates are more efficient in this process. Finally, oxiranes undergo ring opening, resulting from alkylation at the less-substituted carbon.

Consideration of a great deal of experimental data leads to the following order of reactivity of lithium diorganocuprates with organic substrates: acid chlorides > aldehydes > tosylates ≈ oxiranes > iodides > ketones > esters > nitriles. This is only a generalization; many exceptions are caused by unusual features in either the organocopper reagent or the substrate.

Eq. 4.3

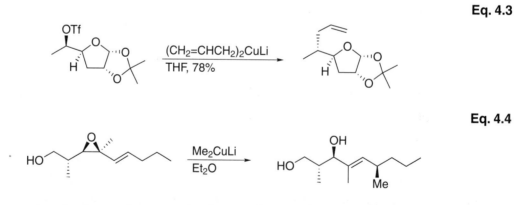

Eq. 4.4

Another extensively used process in the synthesis of complex organic molecules is the conjugate addition of lithium diorganocuprates to enones (**Eq. 4.5**).[7,8] The range of R groups available for this useful reaction is similar to that for the alkylation reaction. Virtually all types of sp^3-hybridized dialkylcuprate reagents react cleanly, including straight-chain and branched primary, secondary, and tertiary alkyl complexes. Phenyl and substituted arylcuprates react in a similar fashion, as do alkenyl-, benzyl-, and allylcuprates. Alkynyl copper reagents do *not* transfer an alkyne group to conjugated enones, a feature used to advantage in the chemistry of mixed cuprates discussed later. The addition occurs preferentially from the less-hindered side of cyclic systems (**Eq. 4.6**).[9]

Eq. 4.5

Eq. 4.6

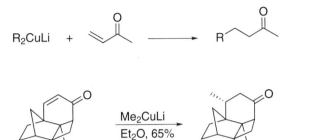

The range of α,β-unsaturated carbonyl substrates that undergo 1,4-addition with diorganocuprates is indeed broad. Although the reactivity of any particular system depends on many factors, including the structure of the copper complex and the substrate, sufficient data are available to allow some generalizations. Conjugated ketones are among the most reactive substrates, combining very rapidly with diorganocuprates (the reaction is complete in less than 0.1 s at 25°C) to produce the 1,4-adduct in excellent yield. Alkyl substitution at the α-, α'-, or β-positions of the parent enone causes only slight alterations in the rate and course of this conjugate addition with acyclic enones, although this same substitution, as well as remote substitution, does have stereochemical consequences with cyclic enone systems.

Conjugated esters are somewhat less reactive than conjugated ketones. With such esters, α,β- and β,β-disubstitution decreases reactivity drastically. Conjugated carboxylic acids do not react with lithium diorganocuprates, and conjugated aldehydes suffer competitive 1,2-additions. Conjugated anhydrides and amides have not been studied to any extent.

Conjugate additions of organocuprates to enones initially produce an enolate, and further reaction with an electrophile is possible. An example of a conjugate addition to a α,β-unsaturated sulfone followed by intramolecular acylation is seen in **Eq. 4.7**[10] The enolate alkylation is at times slow and inefficient. This problem has been overcome, in the context of prostaglandin synthesis, by use of a tributylphosphine-stabilized copper reagent, and by conversion of the unreactive, initially formed copper (or lithium) enolate into a more reactive tin enolate by addition of triphenyl tin chloride (**Eq. 4.8**).[11]

Eq. 4.7

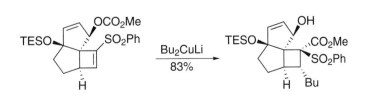

Eq. 4.8

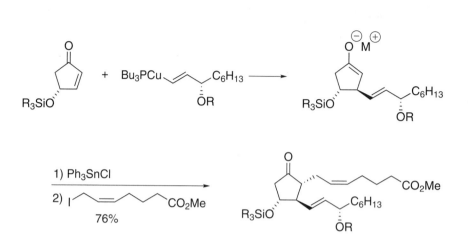

Although lithium dialkylcuprates are very useful reagents, they suffer several serious limitations, the major one being that only one of the two alkyl groups is transferred. Because of the instability of the reagent, a three-fold to five-fold excess (a six-fold to ten-fold excess of R) is often required for complete reaction. When the R group is large or complex, as it often is in prostaglandin chemistry, this lack of efficiency is intolerable. Cuprous alkoxides, mercaptides, cyanides, and acetylides are considerably more stable than cuprous alkyls, and, furthermore, are relatively unreactive. By making mixed alkyl heterocuprates with one of these nontransferable, stabilizing groups, this problem can be avoided.[12] Among the heterocuprates, lithium phenylthio(alkyl) cuprates, lithium *t*-butoxy(alkyl)cuprates, and lithium 2-thienylcuprates[13] are the most useful, and are stable at about –20°C to 0°C.

Mixed alkyl heterocuprates are the most efficient reagents for the conversion of acid chlorides to ketones. Only a 10% excess of these reagents is required to provide high yields, even in the presence of remote halogen, keto, and ester groups. However, large excesses of lithium dialkylcuprates are required for this reaction, and secondary and tertiary organocuprates never work well. In addition, although aldehydes are unstable toward lithium dialkylcuprates, even at –90°C, equal portions of benzaldehyde and benzoyl chloride react with one equivalent of lithium phenylthio(*t*-butyl)cuprate to produce pivalophenone in 90% yield with 73% recovery of benzaldehyde. The lithium phenylthio(*t*-alkyl)cuprate is also the reagent of choice for the alkylation of primary alkyl halides with *t*-alkyl groups. Secondary and tertiary halides fail to react with this reagent, however.

Perhaps most importantly, these mixed cuprate reagents, especially the acetylide and 2-thienyl complexes, are extremely efficient in alkylation reactions and additions to α,β-unsaturated ketones. Since these reactions are the most extensively used in the synthesis of complex molecules, the development of mixed cuprates was an important synthetic advance. An alkylation using a mixed 2-thienyl cuprate is shown in **Eq. 4.9**.[14]

Eq. 4.9

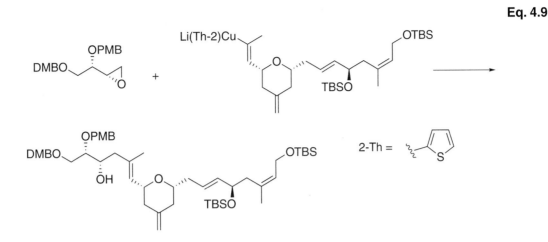

Organocopper chemistry has been developed far beyond these original boundaries. Thermally stable cuprates have been synthesized by using sterically hindered nontransferable ligands, such as diphenylphosphide (Ph_2P) and dicyclohexylamide (Cy_2N), to prepare RCu(L)Li.[15] These reagents undergo typical organocuprate reactions, and are stable at 25°C for over an hour. Even more impressive is the very hindered reagent RCu[P(t-Bu)$_2$]Li, which is stable in THF at reflux for several hours, but still enjoys organocuprate reactivity (**Eq. 4.10**),[16] and the more easily prepared β-silyl organocuprates $RCuCH_2SiR_3$.[17]

Eq. 4.10

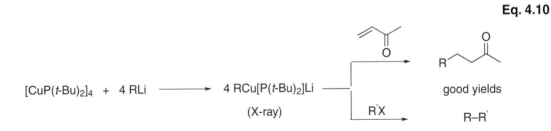

Treatment of copper(I) cyanide with two equivalents of an organolithium reagent produces the "higher-order" cuprate "$R_2Cu(CN)Li_2$," which is more stable than R_2CuLi, and is particularly effective for the reaction with normally unreactive secondary bromides or iodides, as well as for 1,4-additions to conjugated ketones (**Eq. 4.11**)[18] and esters.[19] Spectroscopic evidence suggests that "higher-order cuprates" are monomeric, not higher order, with the general structure R_2CuLi–$LiCN$.[20] Regardless of the true structure, its reactivity is dramatically different from R_2CuLi.

Eq. 4.11

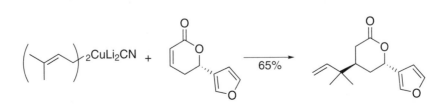

By taking advantage of the observation that mixed diaryl "higher-order" diaryl-cuprates do not scramble aryl groups at low (< –100°C) temperatures, an efficient synthesis of unsymmetrical biaryls was developed (**Eq. 4.12**).[21] This chemistry was used in efficient syntheses of axially chiral natural products (**Eqs. 4.13**[22] and **4.14**[23]).

Eq. 4.12

$$\text{ArLi} + \text{CuCN} \xrightarrow{-78\ ^\circ\text{C}} \text{ArCu(CN)Li} \xrightarrow[-125\ ^\circ\text{C}]{\text{Ar'Li}} [\text{ArCuAr'}]\text{CNLi}_2 \xrightarrow[-125\ ^\circ\text{C}]{\text{oxidation}} \text{Ar–Ar'}$$

78–90%, > 96% cross coupling

Eq. 4.13

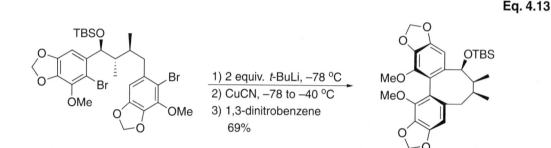

1) 2 equiv. *t*-BuLi, –78 °C
2) CuCN, –78 to –40 °C
3) 1,3-dinitrobenzene
69%

The addition of Lewis acids such as BF_3–OEt_2 or TMSCl to organocopper compounds dramatically increases their reactivity and broadens their synthetic utility. The addition of BF_3 to the normally unreactive RCu reagents produces species that are highly reactive for the alkylation of allylic substrates,[24] as well as conjugate additions to enones and α,β-unsaturated carboxylic acids, themselves unreactive towards R_2CuLi.[25] The reagent RCu–BF_3 monoalkylates ketals with displacement of an alkoxy group, a very unconventional transformation (**Eq. 4.14**).[26] Optically active allylic acetals undergo S_N2' cleavage with very high diastereoselectivity[27] (**Eq. 4.15**[28]).

Eq. 4.14

R'Cu–BF₃ +

Eq. 4.15

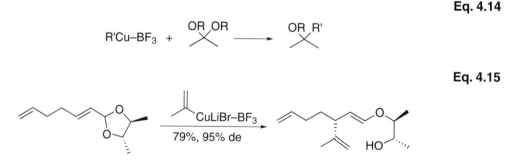

79%, 95% de

Addition of BF_3–OEt_2 to diorganocuprates, R_2CuLi, also dramatically enhances their reactivity, and permits the alkylation of very sterically hindered unsaturated ketones,[29] the ring opening of aziridines,[30] and the diastereoselective alkylation of chiral ketals. These Lewis acid modified cuprate reagents also efficiently add 1,4- to conjugated enones, and effective diastereoselection in this process has been achieved using optically active α,β-unsaturated amides, esters, and sultam amides.[31,32] In most cases, both high yields and high diastereomeric excesses are observed (**Eq. 4.16**).[33]

Eq. 4.16

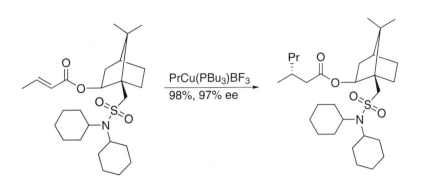

Stoichiometric reactions involving preformed alkylcopper reagents have in many cases been replaced by copper-catalyzed reactions. Copper has proven to be a versatile catalyst for S_N2-type reactions and conjugate additions employing Grignards and organolithium reagents. Common catalyst precursors are CuCN, CuI, CuBr–DMS, $Cu(OTf)_2$, and Li_2CuCl_4. This reaction has been extensively utilized in organic synthesis and an example of an alkylation is shown in **Eq. 4.17**.[34]

Eq. 4.17

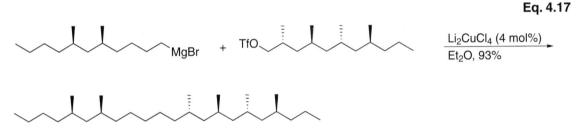

Copper-catalyzed additions of organolithium (**Eq. 4.18**),[35] Grignard, organoaluminum, and in particular organozinc reagents[36,37] (**Eq. 4.19**)[38] to conjugated enones in the presence of optically active ligands offers a more general approach to asymmetric induction.[39] Although this methodology was initially restricted to simple cyclopentenones and cyclohexenones, high asymmetric induction has been observed using a variety of both cyclic and acyclic enones and α,β-unsaturated esters and nitroalkenes. A staggering number of ligands (hundreds) have been prepared and examined in this context.

Eq. 4.18

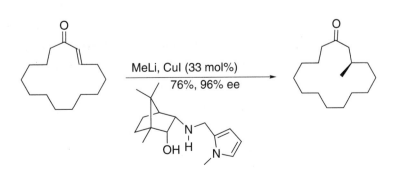

Eq. 4.19

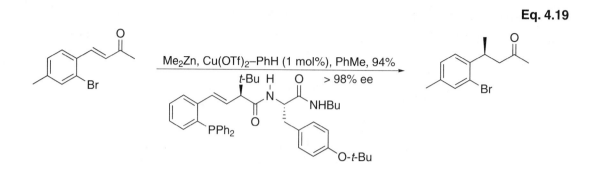

Most of the organocopper chemistry discussed above relies upon the generation of the reactive organocopper species from either a Grignard reagent or an organolithium compound. This seriously limits the nature of the functional groups contained in the organocopper species, since they must be stable to organolithium or Grignard reagents. Several solutions to this problem have recently been developed.

Due to the relatively low pK_a values of terminal alkynes, copper acetylides are readily formed from alkynes using amine bases and a copper(I) salt. The copper acetylides alkylate bromoalkynes to give unsymmetrical 1,3-diynes ("Cadiot–Chodkiewicz coupling").[40] Efficient catalytic versions of this reaction using copper(I) salts and excess amines have more recently been developed[41] and found use in organic synthesis (**Eq. 4.20**).[42] The copper acetylides can also be oxidatively coupled to form symmetrical 1,3-diynes ("Glaser coupling"), a reaction reported as early as 1869.[43]

Eq. 4.20

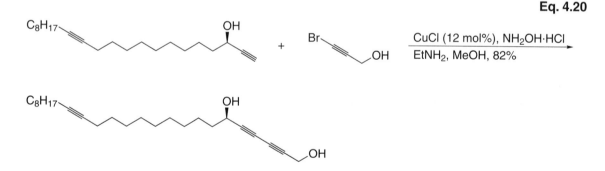

Reduction of CuCN–2LiBr or CuI-PR$_3$ with lithium naphthalenide at –100°C produces a highly active copper that can react with *functionalized* organic halides (I > Br >> Cl) to generate *functionalized* organocopper species in solution. The addition of electrophiles directly to these solutions produces alkylation products in excellent yield (**Figure 4.2**).[44] By this process, ester groups, nitriles, and both aryl and alkyl chlorides are inert, and cuprates containing these functional groups can be generated and transferred.

This class of reagent efficiently cyclized ω-haloepoxides (**Eq. 4.21**), alkylated imines with functionalized allylic halides (**Eq. 4.22**),[45] and even coupled 2-haloarylcopper reagents (prone to benzyne formation when lithiation is attempted) to reactive electrophiles (**Eq. 4.23**).[46] Reaction of the arylcopper reagents with carbon dioxide and alkylation of the intermediate copper carboxylate affords aromatic esters (**Eq. 4.24**).[47]

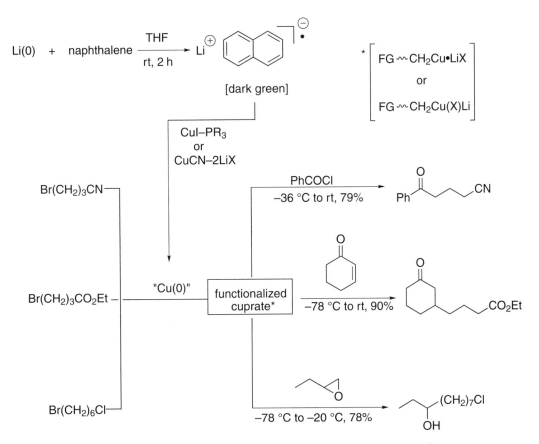

Figure 4.2. Preparation and Reactions of Functionalized Organocopper Species

Eq. 4.21

Eq. 4.22

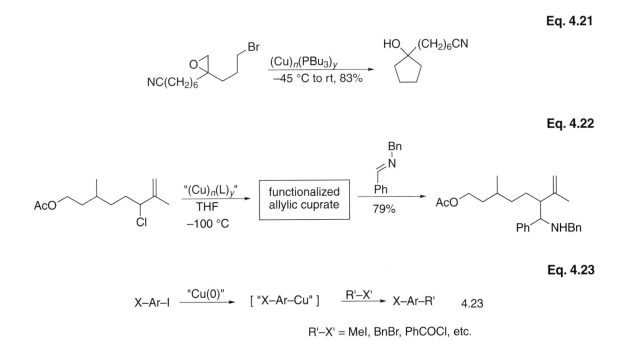

Eq. 4.23

$$X{-}Ar{-}I \xrightarrow{\text{"Cu(0)"}} [\text{ "X–Ar–Cu" }] \xrightarrow{R'{-}X'} X{-}Ar{-}R' \qquad 4.23$$

R'–X' = MeI, BnBr, PhCOCl, etc.

Eq. 4.24

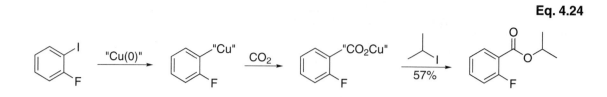

Another route to functionalized organocopper reagents involves transmetallation to copper from less-reactive organometallic reagents, particularly organozinc compounds.[48] Organozinc halides can be made directly from organic halides and zinc metal or diethyl zinc, or from the treatment of organolithium species with zinc chloride. Many functional groups are quite stable to these organometallic reagents (**Figure 4.3**). The treatment of CuCN–2LiCl with functionalized organozinc halides produces functionalized organocuprates, which then undergo normal cuprate coupling processes, introducing *functionalized* R groups into the substrate. An example of an allylic alkylation using transmetallation from zinc to copper is shown in **Eq. 4.25**.[49] Note that the CH$_2$TMS group is not transferred to the product. The process has been made catalytic in copper[50] (**Eq. 4.26**).[51]

Eq. 4.25

Eq. 4.26

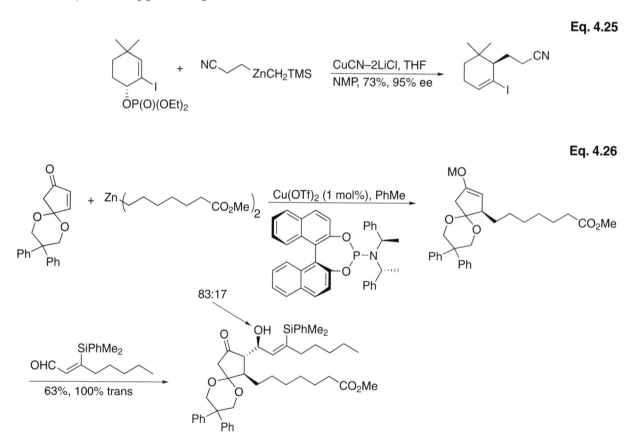

A particularly nice application of this type of zinc–copper chemistry is in the synthesis of cyclopentenones from zinc homoenolates and ynones, which are catalyzed

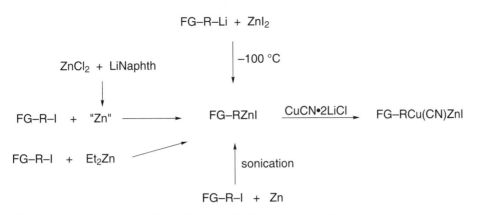

Figure 4.3. Synthesis of Functionalized Organocopper Reagents

by copper salts (**Eq. 4.27**).[52] This reaction has been used in the synthesis of ginkolides (**Eq. 4.28**).[53]

Eq. 4.27

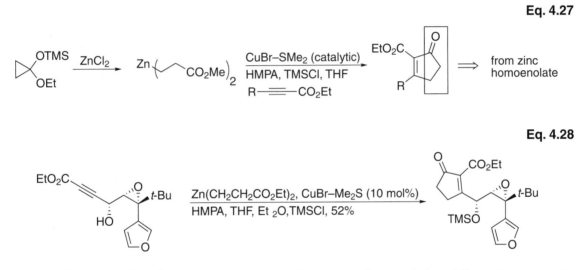

Eq. 4.28

Other functionalized organocopper complexes are also useful. α-Alkoxytin reagents are easily prepared by the reaction of trialkyl stannyllithium or bis(trialkylstannyl) zinc reagents with aldehydes. The α-alkoxyalkyl group is readily cleaved from tin by butyllithium, producing an α-alkoxyalkyllithium reagent.[54] The addition of copper(I) cyanide, followed by the typical array of electrophiles, results in the introduction of an α-alkoxyalkyl group (**Eq. 4.29**). More synthetically useful is the direct transmetallation to copper from α-alkoxytin reagents having coordinating functionalities.[55] Retention of the stereochemistry at the tin-substituted carbon is observed (**Eq. 4.30**).[56]

Eq. 4.29

$$Bu_3SnSnBu_3 + 2Li \longrightarrow 2Bu_3SnLi \xrightarrow[\substack{2)\ R'^{\oplus}}]{1)\ RCHO} \underset{R}{\overset{OR'}{\diagup}} SnBu_3 \xrightarrow[\substack{2)\ CuCN \\ 3)\ E^{\oplus}}]{1)\ BuLi} \underset{R}{\overset{OR'}{\diagup}} E$$

Eq. 4.30

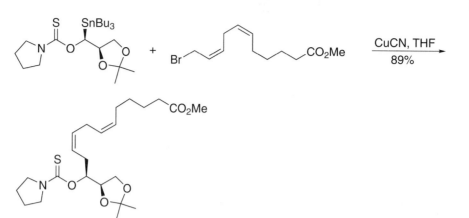

Although dialkyl or alkyl heterocuprates are by far the most extensively used organocopper reagents, the simple organocopper complexes RCuMX perform much of the same chemistry, and are clearly superior for certain types of reactions, particularly the addition of RCuMgX$_2$ to terminal alkynes. R$_2$CuLi complexes often abstract the acidic acetylenic proton from terminal alkynes, and pure RCu or RCuLiX do not add at all. However, RCuMgX$_2$ complexes generated from Grignard reagents and copper(I) iodide cleanly add syn to simple terminal alkynes to produce alkenylcuprates both regiospecifically and stereospecifically ("carbocupration"). Note that the acidic acetylenic C–H bond is left intact. These alkenyl complexes enjoy the reactivity common to organocopper species and react with a variety of organic substrates (**Figure 4.4**).[57]

An elegant application of carbocupration in organic synthesis is the double ethyne carbocupration alkylation used in a synthesis toward ajudazole A (**Eq. 4.31**).[58] For electronic reasons, reversed regioselectivity is observed using alkynyl ethers (**Eq. 4.32**).[59]

$$R-CuMgX_2 \ + \ R'-C\equiv CH \ \longrightarrow \ \underset{R \quad\quad CuMgX_2}{\overset{R' \quad\quad H}{\diagdown\diagup}} \ \longrightarrow \ \underset{R \quad\quad E}{\overset{R' \quad\quad H}{\diagdown\diagup}}$$

E = R''–X, D$^+$, I$_2$, SO$_2$, CO$_2$, CH$_2$=CHCOMe, etc.

Figure 4.4. Carbocupration of Alkynes

Eq. 4.31

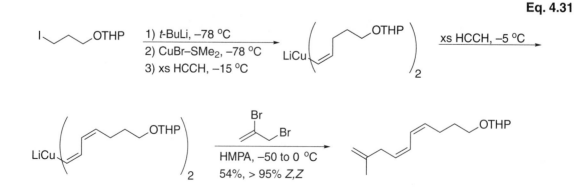

Eq. 4.32

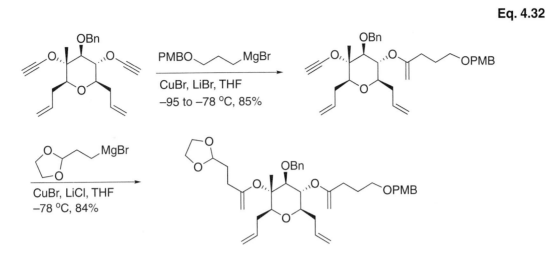

Functionalized cuprates generated from organozinc reagents as in **Figure 4.3** also "carbocuprate" alkynes, producing functionalized alkenylcuprates for use in synthesis.[60] These reagents exhibit lower reactivity compared to the lithium and magnesium reagents and do not react with unactivated terminal alkynes apart from ethyne. When carried out intramolecularly, exocyclic alkenes are produced (**Eq. 4.33**).

Eq. 4.33

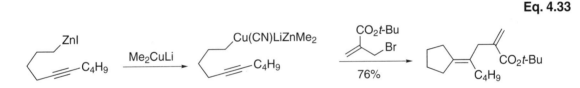

4.3 σ-Carbon–Metal Complexes from Insertion of Alkenes and Alkynes into Metal–Hydrogen Bonds

The insertion of an alkene or alkyne into a metal–hydrogen bond to form a σ-carbon–metal complex is a key step in a number of important processes, including catalytic hydrogenation and hydroformylation. This insertion is an equilibrium reaction, with β-hydride elimination being the reverse step. With electron-rich metals in low oxidation states [e.g., Rh(I) or Pd(II)], the equilibrium lies far to the left, presumably because π-acceptor (electron-withdrawing) alkene ligands stabilize these electron-rich metals more than the corresponding σ-donor alkyl ligands resulting from insertion (**Eq. 4.34**).

Eq. 4.34

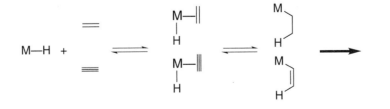

The situation is the reverse with electron-poor metals in high oxidation states [e.g., d^0 Zr(IV)]. In these cases, σ-donor alkyl ligands are stabilizing (and π-accepting alkene ligands are destabilizing) and the equilibrium lies to the right. From a synthetic point of view, the position of the equilibrium is unimportant, as long as the σ-carbon–metal complex formed by insertion is more reactive than the alkene–hydride complex. σ-Carbon–metal complexes of both electron-rich and electron-poor metals, formed from metal hydrides and alkenes, have been used in the synthesis of complex molecules.

Palladium(II) complexes catalyze the cycloisomerization of 1,n-enynes, 1,n-diynes, 1,n-dienes, and 1,2,n-dienynes (n = 6 and 7) to form five- or six-membered rings (**Eq. 4.35**).[61] Although the mechanism of this reaction is not yet known in detail, deuterium labeling supports an in situ generation of a palladium(II) hydride species.[62] Reaction of Pd(PPh$_3$)$_4$ with acetic acid in DMF generates HPd(II)(DMF)$_2$OAc.[63] "Insertion" (hydrometallation) of the alkyne affords a σ-alkenylpalladium(II) complex with an adjacent alkene group. Coordination of this alkene, followed by insertion into the σ-alkenylpalladium complex, forms a carbon–carbon bond (the ring) and a σ-*alkyl*palladium(II) complex having two sets of β-hydrogens. β-Hydride elimination can occur in either direction, generating the cyclic diene and regenerating the palladium(II) hydride species to re-enter the catalytic cycle. The direction of β-elimination can be influenced by the addition of ligands (**Eq. 4.36**), although the reason for this is not entirely clear. Note that palladium remains in the same oxidation state throughout the catalytic cycle.

Eq. 4.35

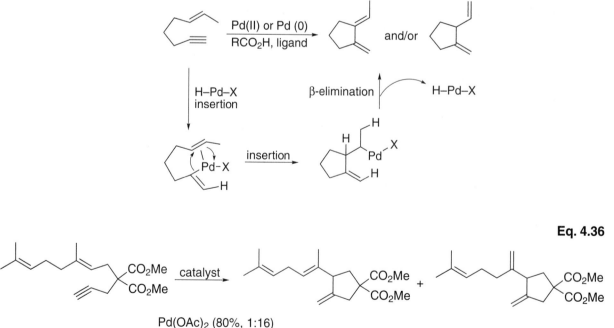

Eq. 4.36

Pd(OAc)$_2$ (80%, 1:16)
Pd(OAc)$_2$, 2 PPh$_3$ (73%, 1:2.9)

The reaction tolerates a range of functional groups as well as structural complexity in the substrate (**Eq. 4.37**).[64] When carried out in the presence of an appropriate reducing agent, such as a silane, the ultimate σ-alkylpalladium(II) complex can be reduced prior to β-elimination, giving monoenes rather than dienes (**Eq. 4.38**).[65] In the presence of organotin reagents, transmetallation/reductive elimination alkylates the final product (**Eq. 4.39**).[66]

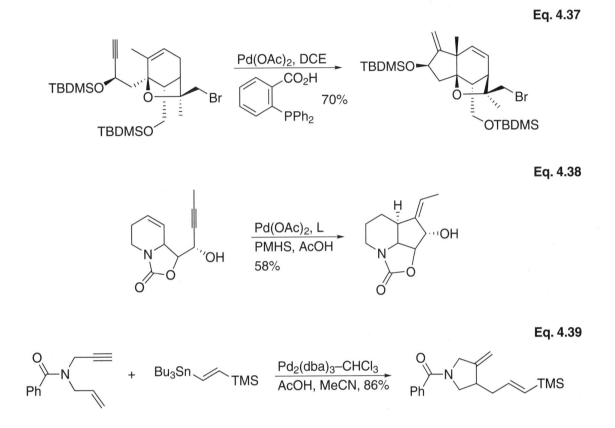

Eq. 4.37

Eq. 4.38

Eq. 4.39

Multiple insertions are possible, provided β-hydrogen elimination is slower than the next insertion, a feature achieved by careful selection of substrate.[67] The example in **Eq. 4.40** is illustrative. Insertion is a cis process, and β-elimination requires that the metal and hydrogen achieve a syn coplanar relationship. In **Eq. 4.40**, a cis insertion leads to a σ-alkylpalladium(II) complex that has no β-hydrogen syn coplanar, making β-elimination disfavored. Instead, a suitably situated alkene undergoes insertion, and the resulting σ-alkylpalladium(II) complex can then undergo β-elimination. The ultimate example of this is shown in **Eq. 4.41** in which all but the final center formed are quaternary and have no β-hydrogens.[68]

Eq. 4.40

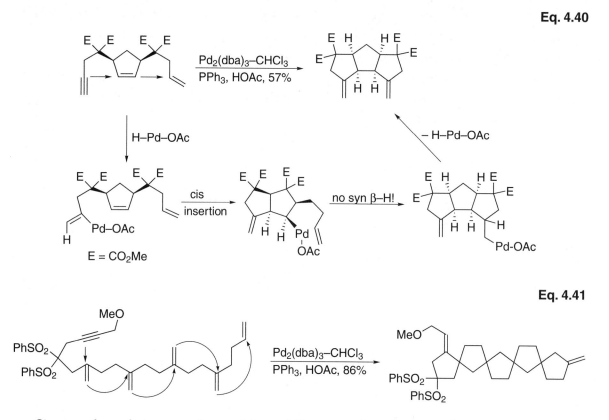

Eq. 4.41

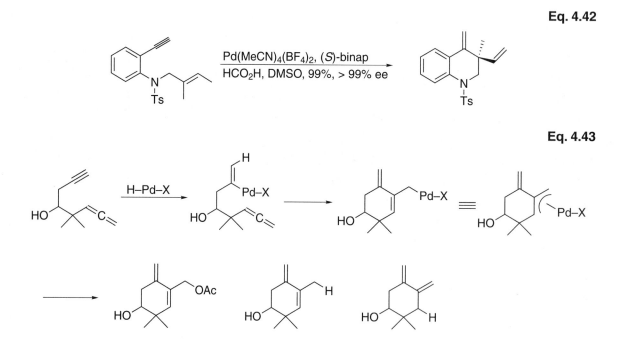

Six-membered rings are formed from 1,7-enynes, but the rate of cycloisomerization is slower compared to the corresponding 1,6-enynes. Asymmetric cycloisomerizations[69] can be realized in the presence of a chiral ligand (**Eq. 4.42**).[70] 1,2-Diene-7-ynes also afford six-membered rings, probably via an η^3-palladium intermediate. This intermediate can undergo a variety of termination reactions depending on the reagents used (**Eq. 4.43**).[71]

Eq. 4.42

Eq. 4.43

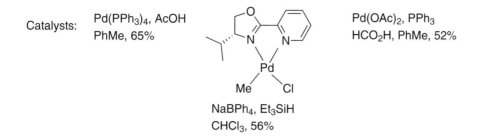

Catalysts: Pd(PPh₃)₄, AcOH   Pd(OAc)₂, PPh₃
            PhMe, 65%   HCO₂H, PhMe, 52%

            NaBPh₄, Et₃SiH
            CHCl₃, 56%

Substrates having tethered unsaturations, such as 1,6-diynes, 1,6-dienes, and 1,2-diene-7-ynes, also undergo palladium-catalyzed cycloisomerizations to form synthetically interesting compounds. Reactions of 1,6-diynes are terminated by reduction of the intermediately formed σ-alkenylpalladium complex, thus affording five-membered rings (**Eq. 4.44**).[72] This process has been used to generate 1,3,5-hexatrienes that undergo a facile electrocyclic reaction to produce tricyclic material (**Eq. 4.45**).[73] In none of these cases could σ-carbon–metal complexes be detected, since they were present only in low concentrations because the insertion equilibria lay far to the left.

Eq. 4.44

Eq. 4.45

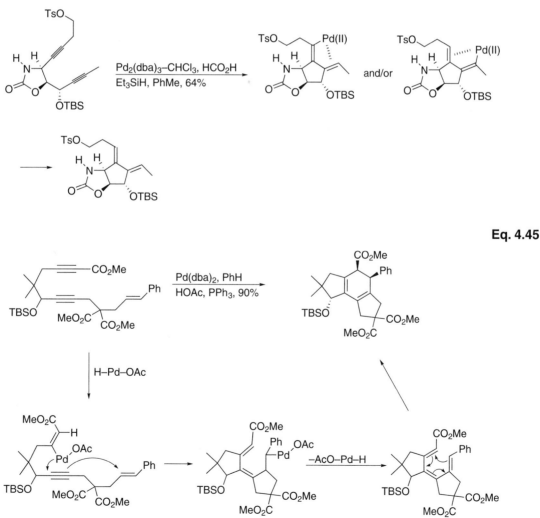

Finally, ruthenium also catalyzes cycloisomerization of 1,6-enynes, probably via a hydrometallation mechanism. An example of a ruthenium–hydride-catalyzed reaction is seen in **Eq. 4.46**.[74]

Eq. 4.46

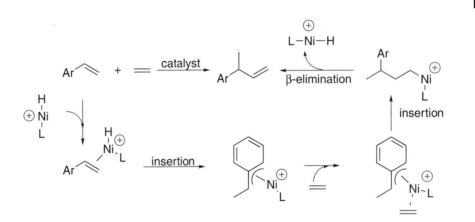

Transition metals catalyze hydroalkenylations of alkenylarenes and 1,3-dienes using ethene.[75] Commonly used metals include palladium, nickel, iridium, and ruthenium. The mechanism is not known in detail, but probably involves the formation of a metal hydride that coordinates to the alkenylarene, alkene insertion to form a η³-allyl complex, coordination and insertion of ethene, followed by β-hydride elimination (**Eq. 4.47**).

Eq. 4.47

A chiral center is produced in the reaction of alkenylarenes and asymmetric reactions are possible using nickel catalysts together with a chiral ligand (**Eq. 4.48**).[76] The NaBAr₄ abstracts the bromide from the precatalyst, thereby generating a cationic complex. Asymmetric reactions forming quarternary centers are also possible.[77] Reactions of 1,3-dienes produce 1,4-dienes (**Eq. 4.49**).[78]

Eq. 4.48

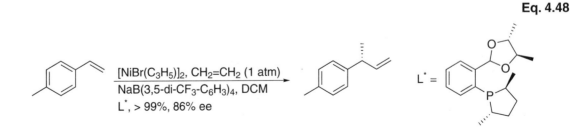

Eq. 4.49

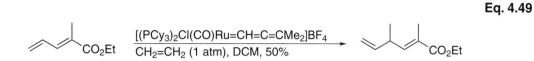

The formation of a η^3-allyl-metal intermediate has also been proposed for palladium-catalyzed hydroamination of alkenylarenes (**Eq. 4.50**)[79] and 1,3-dienes[80] (For a different mechanism for alkene hydroamination, see Chapter 7.) Addition of the amine to the cationic η^3-allyl metal intermediate can occur from either the same side or the opposite side of the metal and both products have been observed. In stoichiometric reactions of isolated cationic η^3-allyl complexes, products with inverted stereochemistry predominate as result of an external attack. In contrast, catalytic reactions produce compounds with retention of stereochemistry.

Eq. 4.50

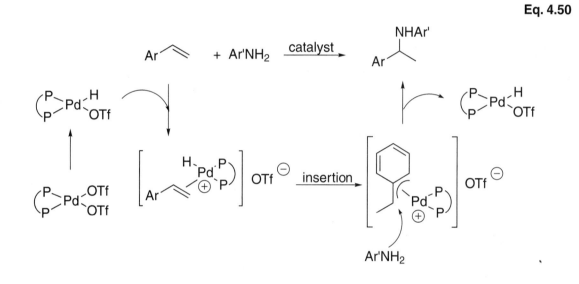

4.4 σ-Carbon–Metal Complexes from Transmetallation–Insertion Processes

Transmetallation (the transfer of an R group) from main group metals to transition metals is extensively exploited in organic synthesis (Chapter 2). Despite the utility of transmetallation, the process is poorly understood. Its utility lies in the ability to use main group organometallic chemistry to "activate" organic substrates (e.g., hydroboration of alkenes or direct mercuration of aromatics), then transfer the organic to transition metals to take advantage of their unique reactivity (e.g., oxidative addition or insertion). One such very useful class of reaction is transmetallation–insertion (**Eq. 4.51**). As expected, the most common termination steps of the σ-alkylmetal complex formed are β-hydride elimination and protonolysis (electrophilic cleavage).

Eq. 4.51

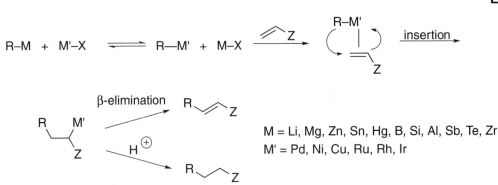

M = Li, Mg, Zn, Sn, Hg, B, Si, Al, Sb, Te, Zr
M' = Pd, Ni, Cu, Ru, Rh, Ir

As mentioned in Chapter 3, alkenes and alkynes are readily hydrozirconated, but the resulting σ-alkylzirconium(IV) species have only minimal reactivity. However, they will undergo transmetallation to nickel(II) complexes, and the resulting organonickel complexes undergo 1,4-addition to conjugated enones (**Eq. 4.52**).[81] σ-Alkenylzirconium(IV) species also transmetallate to rhodium(I) and, in the presence of a chiral ligand, enantioselective 1,4-additions can be realized (**Eq. 4.53**).[82,83]

Eq. 4.52

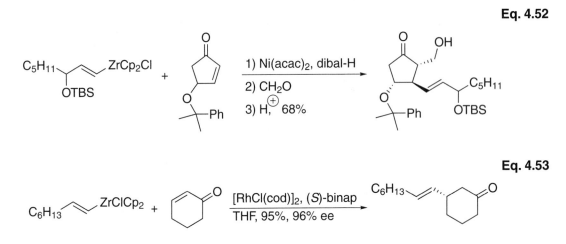

Eq. 4.53

Multiple transmetallations increase the utility of the process. In **Eq. 4.54**,[84] for example, hydrozirconation is used to activate the alkyne, but the σ-alkenylzirconium(IV) species does not directly transmetallate to copper in this case. However, organolithium reagents will alkylate the alkenylzirconium species which then transmetallates to copper, which then does 1,4-additions to an enone. This process can be made catalytic in the copper salt by using appropriate conditions (**Eq. 4.55**).[85] A copper iodide DMS complex catalyst has recently been shown to alleviate the need for organolithium reagents in 1,4-additions of alkenylzirconium species.[86]

Eq. 4.54

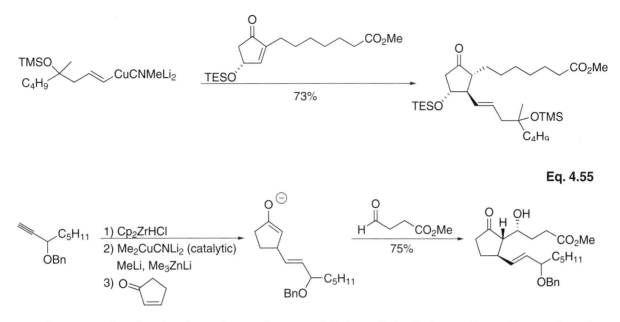

Eq. 4.55

Copper salts also catalyze the conjugate addition of vinyl alanes (from Zr-catalyzed carboalumination of alkynes) to enones (**Eq. 4.56**),[87] and the addition of *alkyl*zirconium species (from hydrozirconation of alkenes rather than alkynes) to enones.[88]

Eq. 4.56

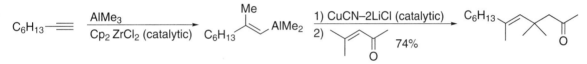

Transmetallation from mercury, thallium, and tin to palladium(II) is also a powerful synthetic technique.[89] Aromatic compounds undergo direct, electrophilic mercuration or thallation and, although these main group organometallics have limited reactivity, they will transfer their aryl groups to palladium(II), which has a very rich organic chemistry. When transmetallation–insertion chemistry is used, the reaction is, in most cases, stoichiometric in palladium, since Pd(II) is required for insertion, and Pd(0) is produced in the last step (**Eq. 4.57**). Efficient reoxidation has only been achieved in a few cases.[90] However, this can still be a useful process, since Pd(0) can be recovered and reoxidized in a separate step[91] (**Eq. 4.58[0]**). In the special case of an allylic chloride as the alkene (**Eq. 4.59**),[92] the reaction *is* catalytic in Pd(II) since, in this case, β-Cl elimination to give PdCl$_2$, rather than β-hydride elimination to give Pd(0) and HCl, is more facile.

Eq. 4.57

Ar–H + HgX$_2$ ⟶ Ar–Hg–X $\xrightarrow{\text{PdX}_2}$ Ar–Pd–X + HgX$_2$ (alkene-Z) →

Eq. 4.58

Eq. 4.59

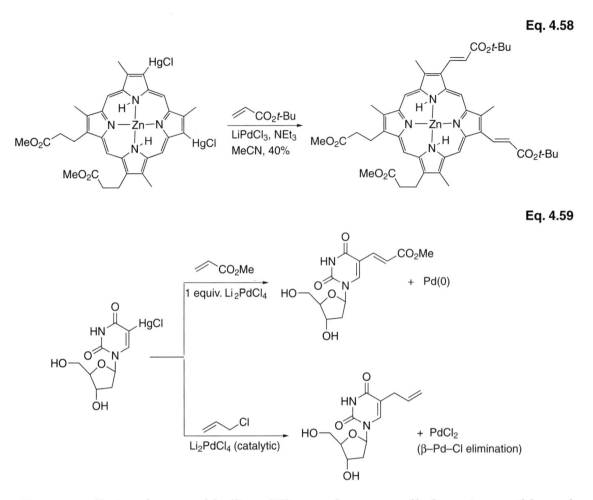

Transmetallation from arylthallium(III) complexes to palladium is possible and the process can be made moderately catalytic (**Eq. 4.60**).[93] In this case, Tl(III) oxidizes the formed Pd(0) to Pd(II). Transmetallation from tin has been used to synthesize arylglycosides (**Eq. 4.61**).[94] Even arylmanganese(I) complexes transmetallate to Pd(II), permitting alkenylation (**Eq. 4.62**).[95]

Eq. 4.60

Eq. 4.61

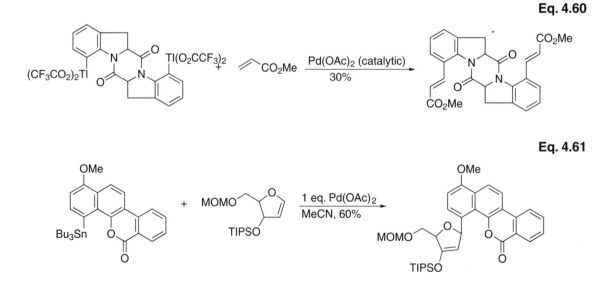

Eq. 4.62

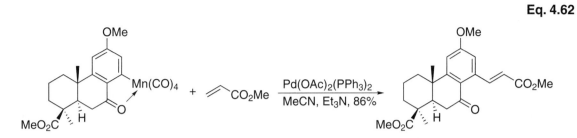

Organoboron reagents also readily undergo transmetallation to palladium and rhodium,[96] followed by insertion of an alkene. Rhodium in particular catalyzes very general asymmetric 1,4-additions of organoboron compounds to electron-deficient alkenes. A variety of alkenyl- and arylboronic acids can be used in reactions with cyclic and acyclic enones, α,β-unsaturated esters and sulfones, as well as nitroalkenes. Even some unactivated alkenes participate in this reaction (**Eq. 4.63**).[97] The reaction is catalytic in palladium since O_2 oxidizes the palladium(0) formed back to palladium(II).

Eq. 4.63

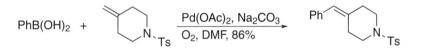

4.5 σ-Carbon–Metal Complexes from Oxidative Addition–Transmetallation Sequences[98]

One of the most general approaches to the formation of σ-carbon–metal complexes is the oxidative addition (Chapter 2) of organic halides and triflates to low-valent transition metals, most commonly palladium(0) and nickel(0) [see path (a) in **Figure 4.5**]. The reaction was initially restricted to aryl and alkenyl halides, and sulfonates, or substrates lacking β-hydrogens, since β-hydride elimination is very fast at temperatures above –20°C. New catalyst systems have recently been developed wherein transmetallation–reductive elimination is faster compared to β-hydride elimination, allowing for a variety of *alkyl* halides to be used in coupling reactions.[99] With alkenyl halides, retention of stereochemistry is observed. The order of reactivity is I > OTf > Br >> Cl.[100] Alkenyl triflates undergo oxidative addition much faster then do aryl triflates.[101] Chlorides are by far the least reactive substrates for this type of reaction due to a very slow oxidative addition process. However, bulky ligands (e.g., aryldialkylphosphines, trialkylphosphines, and carbenes) accelerated the oxidative addition to chlorides and a variety of coupling reactions can now be performed under relatively mild conditions.[102]

The mechanism of the oxidative addition–transmetallation–reductive elimination process is very complex and the exact details depend on solvents, ligands, the transition metals, and additives. A simplified mechanism for the overall reaction is shown in **Figure 4.5**, starting with an oxidative addition [path (a)]. From the viewpoint of the substrate, the metal is oxidized and the substrate is reduced. Thus, electron-deficient substrates are more reactive than electron-rich substrates. The resulting (relatively unstable) σ-carbon–metal halide complex can undergo a variety of useful reactions. Treatment with a main group organometallic results in transmetallation [path (b) in **Figure 4.5**], producing a diorganometal complex, which undergoes rapid reductive elimination [path (c)] to give coupling (i.e., carbon–carbon bond formation) with regeneration of the low-valent metal catalyst. Because reductive elimination is *faster* than β-hydride elimination, the main group organometallic *may* contain β-hydrogens, and is not restricted to *sp²* carbon groups. Transmetallation and reductive elimination usually occur with retention of stereochemistry.

Because CO insertion into metal–carbon σ-bonds is facile at temperatures as low as –20°C, carrying out the oxidative addition under an atmosphere of CO results in the capture of the σ-carbon–metal complex by CO to produce a σ-acyl species [path (d) in **Figure 4.5**]. Transmetallation [path (e)] to this complex is followed by carbonylative coupling [path (f)]. Both of these processes have been extensively used in synthesis.

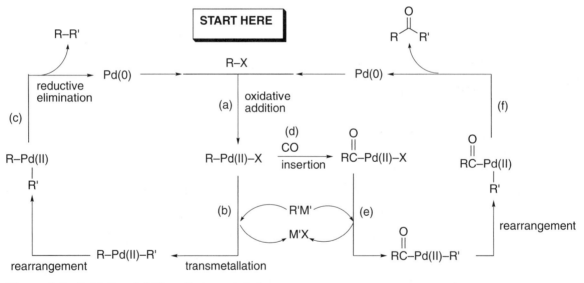

Figure 4.5. Oxidative Addition–Transmetallation

The classical mechanistic picture of the oxidative addition step involves the reaction of a neutral palladium(0) complex with the organohalide/triflate. However, more recent detailed experimental and theoretical studies indicate the formation of trivalent anionic palladium(0) complexes in the presence of added anions, such as chloride or acetate, prior to oxidative addition (**Eq. 4.64**).[103]

Eq. 4.64

$$Pd(0)L_n + X^{\ominus} \rightleftharpoons Pd(0)L_nX^{\ominus}$$

By far the most extensively developed oxidative addition–transmetallation chemistry involves palladium(0) catalysis, and transmetallation from Li, Mg, Zn, Zr, B, Al, Sn, Si, Ge, Hg, Tl, Cu, In, Bi, Ga, Ti, Ag, Mn, Sb, Te, and Ni. There is an overwhelming amount of literature on the subject, but all examples have common features. A very wide range of palladium catalysts or catalyst precursors have been used (**Figure 4.6**) and, in many cases, catalyst choice is not critical. These catalysts include preformed, stable palladium(0) complexes, such as $Pd(PPh_3)_4$ ("tetrakis") or $Pd(dba)_2/Pd_2(dba)_3CHCl_3$ plus a phosphine[104]; in situ generated palladium(0) phosphine complexes, made by reducing palladium(II) complexes in the presence of phosphines; and "naked" or ligandless palladium(0) species, including, in some cases, palladium on carbon. Palladium(II) is very readily reduced to palladium(0) by alcohols, amines, CO, alkenes, phosphines, and main group organometallics.[105]

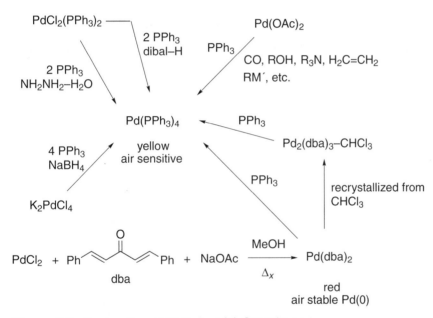

Figure 4.6. Preparation of Palladium(0) Complexes

Phosphine ligands are normally required for the oxidative addition of bromides, but they suppress the reaction with aryl iodides, so discrimination between halides is efficient. In these coupling processes, almost invariably it is the transmetallation step that is rate limiting and is the least understood. It is important that *both* metals ultimately benefit energetically from the transmetallation step and, depending on the metal, additives are sometimes required.

One of the earliest examples of oxidative addition–transmetallation is the nickel- or palladium-catalyzed reaction with Grignard reagents ("Kumada coupling").[106] [Although Ni(II) or Pd(II) complexes are the most common catalyst precursors, the active catalyst species is the metal(0) formed by reduction of the metal(II) complex by the Grignard reagent.] Nickel phosphine complexes catalyze the selective cross coupling of Grignard reagents with aryl and alkenyl halides and triflates (**Eq. 4.65**).[107] The reaction works well with primary and secondary alkyl, aryl, alkenyl, and allyl Grignard reagents. A wide variety of simple, fused, or substituted aryl and alkenyl halides and triflates are reactive substrates. In addition, conditions have been developed for reactions of simple alkyl halides with Grignard reagents.[108]

Eq. 4.65

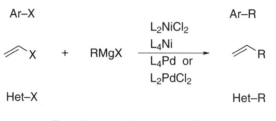

R = alkyl, aryl, heteroaryl, alkenyl

This process is particularly useful for the coupling of heterocyclic substrates.[109] In this case the Grignard reagent, the halide, or both can be heterocycles. Some applications of coupling of Grignard reagents in the context of total synthesis are seen in **Eqs. 4.66**[110] and **4.67**.[111]

Eq. 4.66

Eq. 4.67

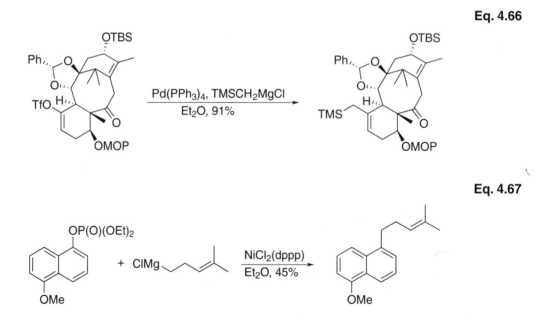

Configurationally unstable secondary alkyl Grignard reagents can be coupled to a variety of aryl and alkenyl halides with high asymmetric induction using chiral bidentate P,N-ligands (**Eq. 4.68**).[112,113] This is a *dynamic kinetic resolution* of the racemic (but rapidly equilibrating) Grignard reagent—that is, one enantiomer of the Grignard reagent reacts with the optically active phosphine metal(II) σ-carbon complex faster than does the other enantiomer, and the Grignard reagent equilibrates (racemizes) faster than it couples. Aryl triflates also couple efficiently to Grignard reagents. With prochiral bis triflates, desymmetrization takes place and optically active biaryls can be produced with high enantioselectivity (**Eq. 4.69**).[114]

Eq. 4.68

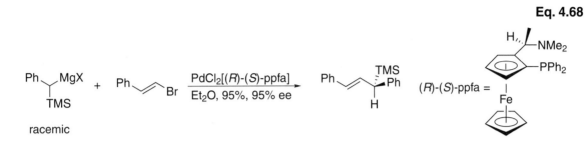

Eq. 4.69

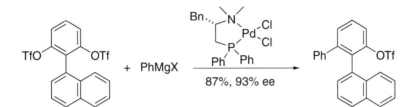

Aryl and alkenyl ethers[115] (**Eq. 4.70**)[116] and sulfides (**Eq. 4.71**)[117] can also be used in oxidative addition–transmetallations using Grignard reagents in the presence of a nickel catalyst.

Eq. 4.70

Eq. 4.71

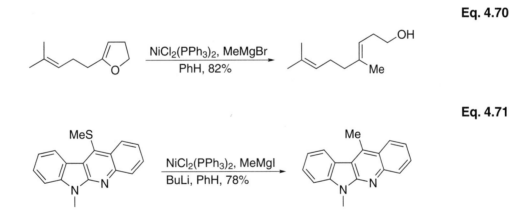

Iron and cobalt are more recent catalysts for the coupling of Grignard reagents with organic halides (I, Br, or Cl) and triflates.[118] These catalysts offer some advantages compared to palladium and nickel, such as they are easy to handle and inexpensive pre-catalysts, their reactions are easily scaled, and they have low toxicity. Catalyst precursors include iron trichloride, iron tris(acetonylacetonate), or cobalt dichloride. Superior yields are obtained using NMP as solvent or co-solvent. β-Hydride elimination is minimized and often not observed in iron- and cobalt-catalyzed coupling reactions of primary and secondary alkyl halides.[119] Tertiary halides are inert under iron catalysis, but couple using cobalt complexes (**Eq. 4.72**).[120] Remarkable functional group tolerance is observed wherein ketones, esters, nitriles, and R–N=C=O remain intact (**Eqs. 4.73**[121] and **4.74**[122]).

Eq. 4.72

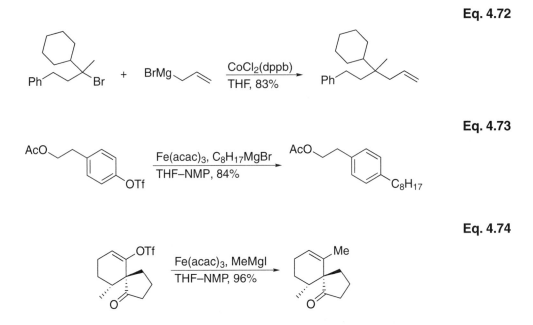

Eq. 4.73

Eq. 4.74

The mechanism for these reactions is not known, but experiments indicate a very complex radical process.[123] A simplified mechanism is shown in **Figure 4.7**. This involves reduction of the metal catalyst precursor to an M(n) species, halogen atom abstraction from the organohalide to give a radical and M(n+1)X. Transmetallation from magnesium to iron, thus yielding a σ-metal complex, followed by coupling with the radical, generates the product and the active catalyst.

Organozinc reagents are very versatile partners in palladium-catalyzed cross-coupling reactions with organic halides and triflates ("Negishi coupling"). Organozinc halides are among the most efficient main group organometallics in transmetallation reactions to palladium(II) (**Eq. 4.75**),[124] while organolithium reagents are among the least.[125] However, treatment of an organolithium reagent with one equivalent of anhydrous zinc chloride in THF generates, in situ, an organozinc halide that couples effectively (**Eq. 4.76**).[126]

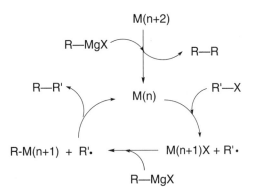

Figure 4.7. Mechanism for the Coupling of Gignard Reagents Using Cobolt and Iron Catalysts

Eq. 4.75

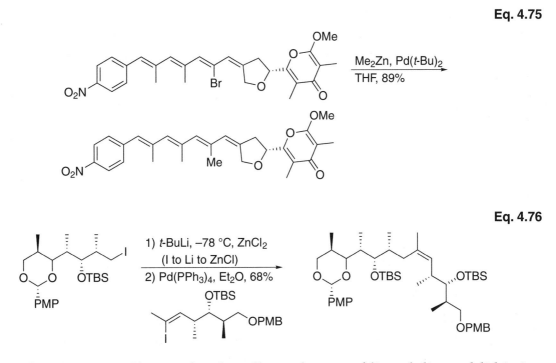

Eq. 4.76

This facile Li to $ZnCl_2$ transfer also allows the use of ligand-directed lithiation to provide coupling partners in this process. Perhaps the most important feature of transmetallation to zinc followed by cross coupling is the facile coupling of alkyl zinc reagents with little or no β-hydride elimination. With the development of new ligands, even alkyl bromides, chlorides, triflates, and tosylates can be coupled with alkyl zinc reagents.[127] Another attractive feature of organozinc halides is that a wide variety of *functionalized* reagents are available directly from functionalized halides and activated zinc (see **Figure 4.3**), permitting the introduction of carbonyl groups, nitriles, halides, and other species into these coupling reactions. Recent applications of this type of coupling in synthesis are seen in **Eqs. 4.77**[128] and **4.78**[129].

Eq. 4.77

Eq. 4.78

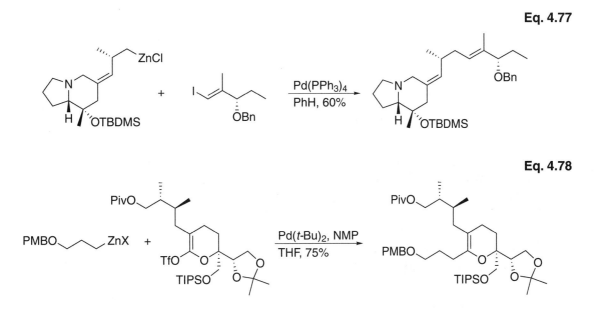

σ-Alkenylzirconium(IV) complexes formed by hydrozirconation of *terminal* alkynes also transmetallate efficiently to palladium(II)–carbon complexes and are effective coupling partners in Pd(0)- and Ni(0)-catalyzed oxidative addition–transmetallation processes (**Eq. 4.79**[130]).[131] The hydrozirconation is a cis-addition process and very high stereoselectivity is observed in the *E*-alkenylzirconium intermediate and coupling products.

Eq. 4.79

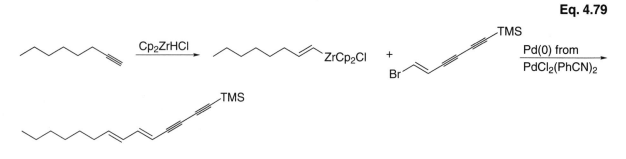

However, transmetallation from the corresponding σ-alkenylzirconium(IV) complexes from hydrozirconation of *internal* alkynes to palladium(II) fail entirely! Addition of zinc chloride to these reactions promotes this process, probably via a Zr → Zn → Pd/Ni transmetallation sequence (**Eqs. 4.80**[132] and **4.81**[133]). This, again, illustrates the efficacy of zinc organometallics in these coupling processes.

Eq. 4.80

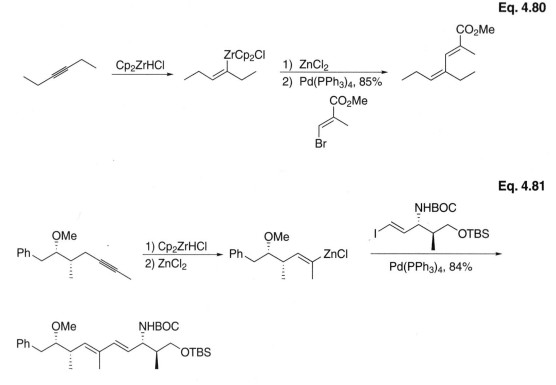

Eq. 4.81

Although hydrozirconation has been extensively studied, alkylzirconocenes do not "carbozirconate" alkynes. However, zirconium catalyzes carboaluminations of alkynes, again with very high syn stereoselectivity. Particularly useful is the reaction of alkynes with trimethylaluminum in the presence of zirconocene dichloride. The readily formed σ-alkenylaluminum species undergo palladium-catalyzed coupling reactions (**Eq. 4.82**).[134]

Eq. 4.82

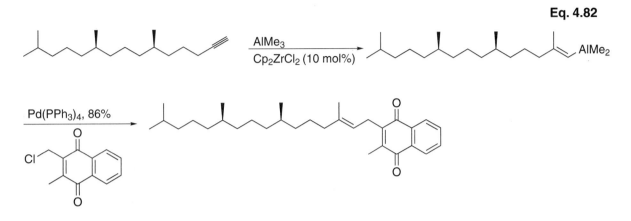

Hydroboration is potentially a very attractive way to bring alkynes and alkenes into oxidative addition–transmetallation cycles. This process has been the subject of long term, intense development, and a truly staggering number of cases have been studied. However, boron is electrophilic, and attached alkyl groups lack sufficient nucleophilicity to transmetallate to palladium(II). The addition of anionic bases, such as hydroxide, to neutral organoboranes dramatically increases the nucleophilicity of the organic groups, now attached to a *borate*, and transmetallation from boron to palladium can easily be achieved. This is the basis of the "Suzuki" or "Suzuki–Miyaura"[135,136] reaction, which has, to a great extent, supplanted transmetallation from zirconium or aluminum. Thus, alkylboranes and alkenylboranes couple efficiently to a wide range of highly functionalized alkenyl halides and triflates. Even alkyl halides can be used as the substrate. Two spectacular examples of the Suzuki reaction are seen in **Eqs. 4.83**[137] and **4.84**.[138] With complex substrates, the use of TlOH or TlOEt as the base often results in an increase of both the rate and the yield of the process.[139] With base-sensitive substrates, fluoride can be used in many cases.[140,141] Nickel complexes also catalyze Suzuki reactions. For example, a more recently developed catalyst system was used to couple unactivated alkyl halides with alkyl boron reagents at room temperature![142]

Eq. 4.83

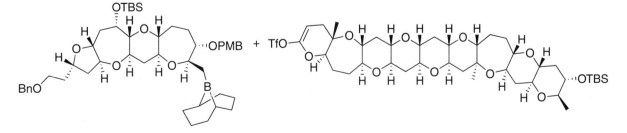

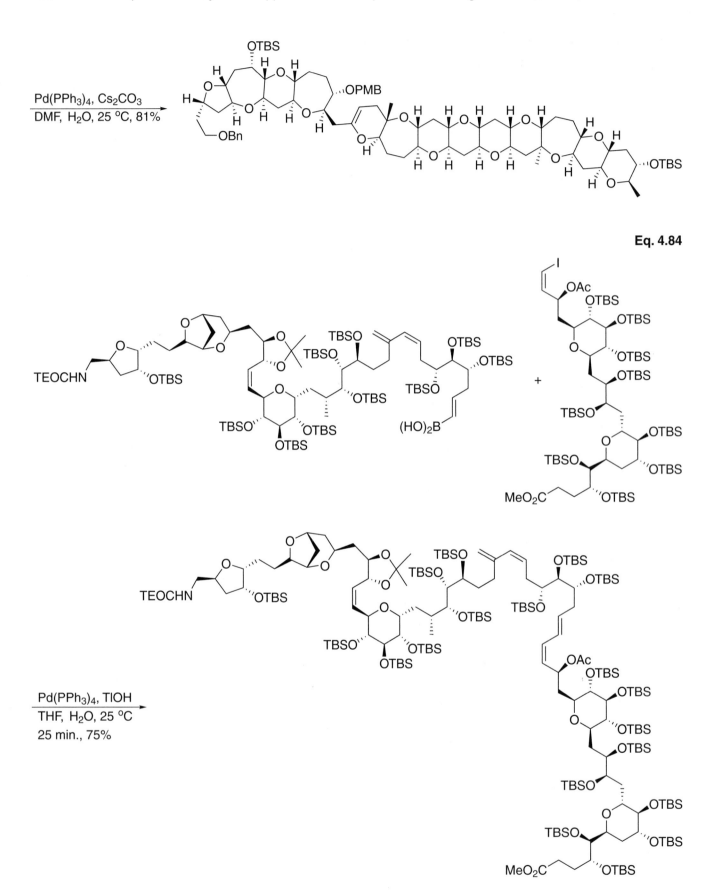

$$\frac{\text{Pd(PPh}_3)_4,\ \text{Cs}_2\text{CO}_3}{\text{DMF, H}_2\text{O, 25 }^{\circ}\text{C, 81\%}}$$

Eq. 4.84

$$\frac{\text{Pd(PPh}_3)_4,\ \text{TlOH}}{\text{THF, H}_2\text{O, 25 }^{\circ}\text{C}}$$
25 min., 75%

Suzuki coupling is generally the method of choice for making biaryls,[143] because arylboronic acids are easy to synthesize, relatively stable to handle, and the boron-containing byproducts are water soluble and nontoxic (**Eq. 4.85**[144]). Highly hindered, functionalized systems couple efficiently (**Eq. 4.86**),[145] and the reaction works well even on a very large scale (**Eq. 4.87**).[146]

Eq. 4.85

Eq. 4.86

Eq. 4.87

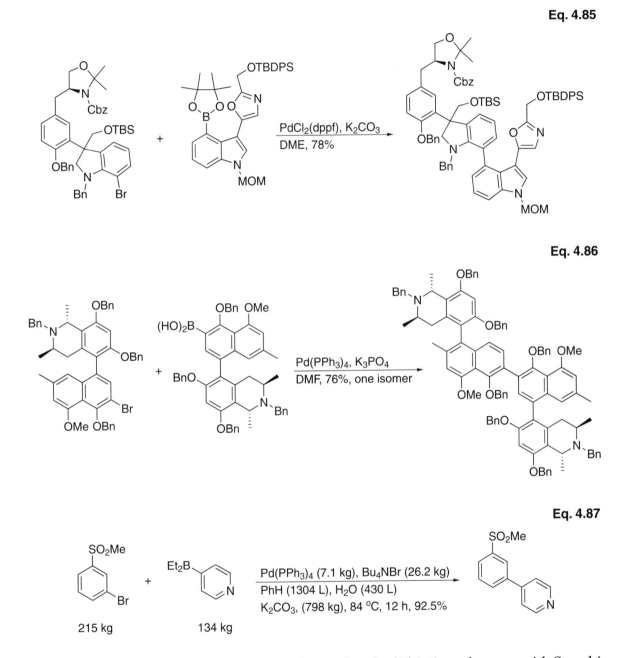

By combining the functional-group-directed ortho lithiation of arenes with Suzuki coupling[147] (**Eq. 4.88**[148]), a wide range of functionalized arenes, even quite hindered ones, can be synthesized (**Eq. 4.89**).[149] Asymmetric Suzuki couplings can be achieved using either a chiral catalyst system (**Eq. 4.90**)[150] or a chiral coupling partner (**Eq. 4.91**).[151]

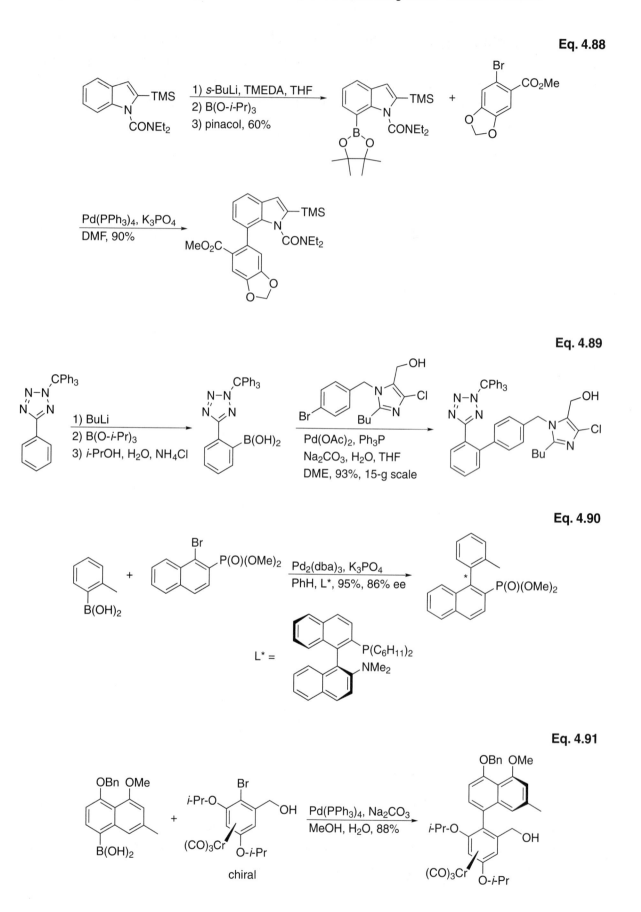

Eq. 4.88

Eq. 4.89

Eq. 4.90

Eq. 4.91

A more recent development is the formation of boronic esters by the reaction of organic triflates and halides with pinacol borane or bis(pinacolato)diboron in the presence of a palladium catalyst (**Eq. 4.92**).[152] The addition of organolithium reagents to B-OMe-9-BBN, or related compounds, furnish boron derivatives that readily transfer to palladium (**Eq. 4.93**)[153].

Eq. 4.92

Eq. 4.93

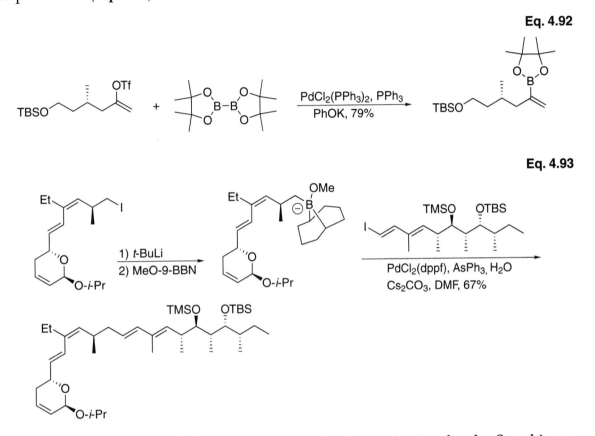

Macrocyclic compounds have been prepared using an intramolecular Suzuki coupling (**Eqs. 4.94**[154] and **4.95**[155]). Organoboronic esters and acids are sometimes unstable and hard to purify. Trifluoroborates, on the other hand, are easily handled crystalline solids that are indefinitely stable in air, and are useful in oxidative addition–transmetallation reactions.[156] An intramolecular cross coupling is shown in **Eq. 4.96**.[157]

Eq. 4.94

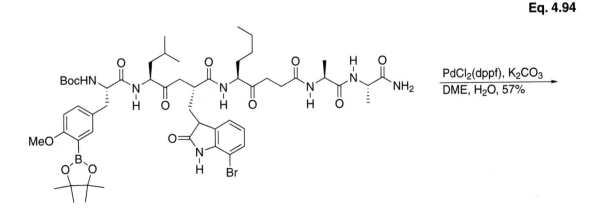

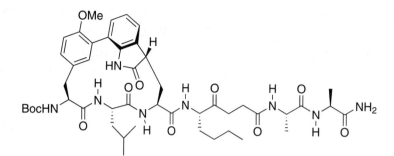

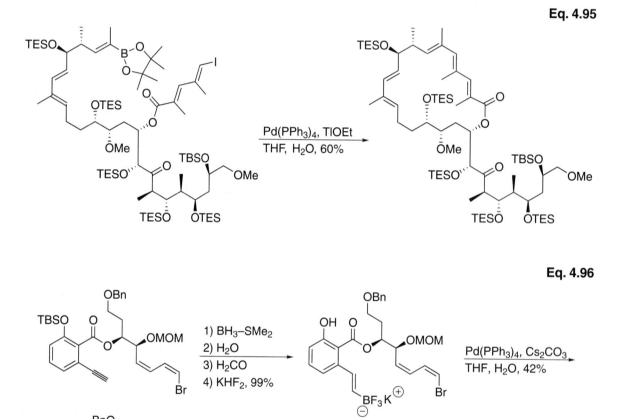

Eq. 4.95

Eq. 4.96

Transmetallation from tin to palladium (the "Stille" or "Kosugi–Migita–Stille" reaction[158]) is one of the most extensively utilized and highly developed palladium-catalyzed processes in organic synthesis.[159,160] A wide variety of structurally elaborate organotin compounds are easily synthesized, and they are relatively unreactive toward a wide range of functional groups, yet they readily transmetallate to palladium, making them ideal reagents for complex oxidative addition–transmetallation sequences. The inherent toxicity of tin compounds and the difficulty in removing tin byproducts from reactions have done little to dampen this enthusiasm, and have been somewhat mitigated by the development of tin reagents more readily removed by extraction.[161]

The rate and ease of transfer from tin to palladium is alkynyl > alkenyl > aryl > allyl >> alkyl. Alkenyl groups transfer with retention of alkene geometry, and allyl groups with allylic transposition. Because simple alkyl groups transfer very slowly, trimethyl- or tributylstannyl compounds can be used to transfer a single R group. A methyl group can be transferred if it is the only group on tin and conditions are sufficiently vigorous.[162] Transmetallation is almost invariably the rate-limiting step in these processes, and relatively unreactive tin reagents often require high (> 100°C) reaction temperatures and long reaction times. The use of lower donicity ligands such as AsPh₃ or P(2-furyl)₃ can often enhance transmetallation rates and thus permit the use of milder conditions,[163] as can the use of copper(I) co-catalysts.[164]

In principle, any substrate that will undergo oxidative addition with palladium(0) can be coupled to any organotin group that will transfer. Among the earliest studied substrates were acid chlorides, which are readily converted to ketones (**Eq. 4.97**[165]).[166] A carbon monoxide atmosphere is required to prevent competitive decarbonylation (Chapter 5) of the acid chloride.

Eq. 4.97

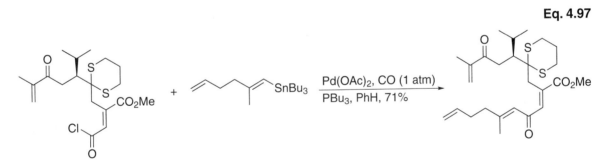

Alkenyl halides couple efficiently with alkynyl-[167] (**Eq. 4.98**)[168] and simple alkenyl-tin compounds,[169] with retention of alkene geometry for both partners (**Eq. 4.99**).

Eq. 4.98

Eq. 4.99

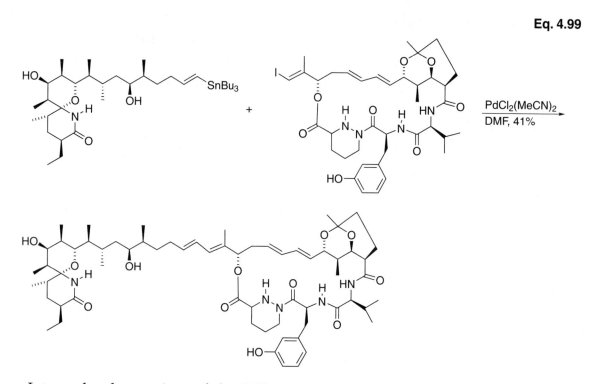

Intramolecular versions of the Stille reaction are efficient, and a variety of very complex macrocycles have been synthesized using this methodology (**Eqs. 4.100**[170] **4.101**,[171] **4.102**,[172] and **4.103**[173]). Although rarely used, alkynyl halides also couple efficiently to organostannanes under palladium catalysis (**Eq. 4.104**).[174]

Eq. 4.100

Eq. 4.101

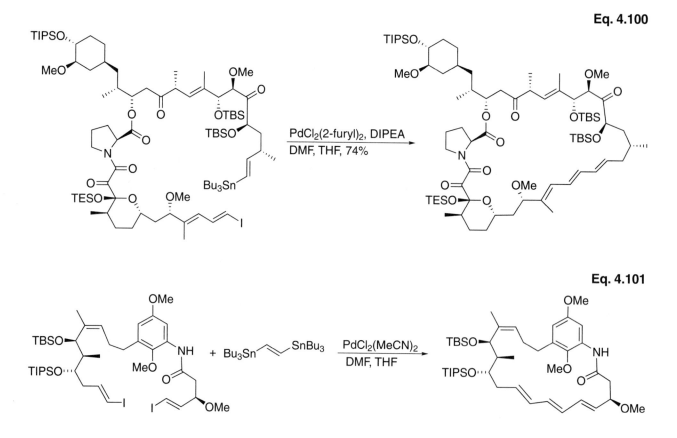

Eq. 4.102

Eq. 4.103

Eq. 4.104

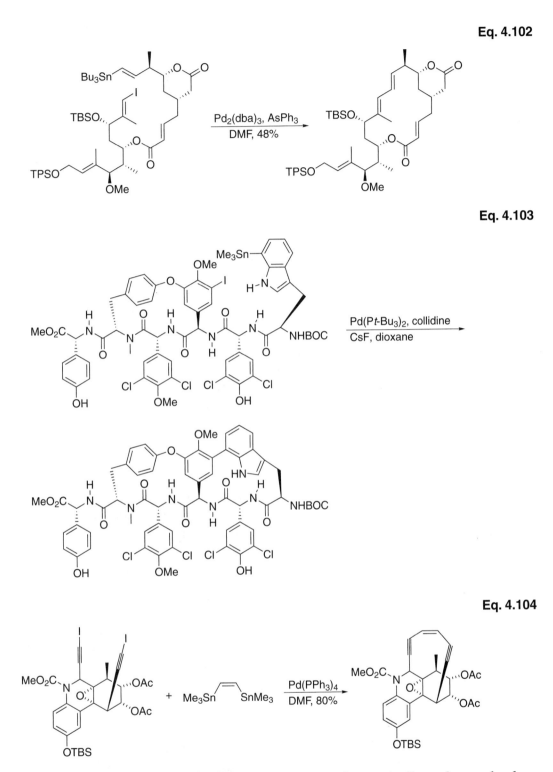

The synthetic utility of this coupling process was dramatically enhanced when conditions were found to make aryl and alkenyl triflates,[175] fluorosulfonates,[176] and phosphonates[177] participate, since this allowed phenols, ketones, and esters (via their enol triflates or fluorosulfonates) to serve as coupling substrates. As with other substrates, extensive functionality is tolerated in both coupling partners (**Eqs. 4.105**[178] and **4.106**[179]).

Eq. 4.105

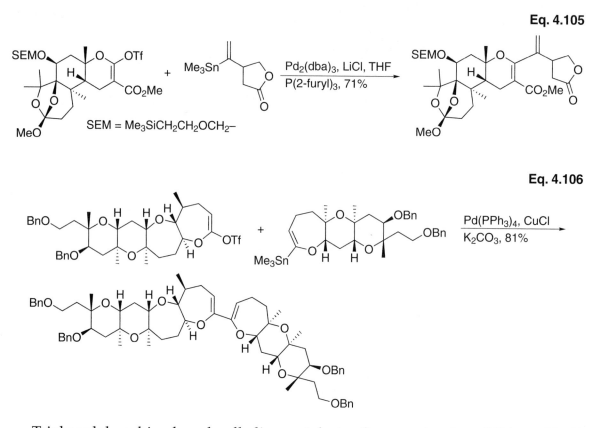

Eq. 4.106

Eq. 4.107

Triphenylphosphine-based palladium catalysts often require the addition of LiCl or ZnCl₂ to promote cross coupling, but the use of AsPh₃ in polar solvents such as DMF obviates the need for additives. Even unreactive aryl triflates participate under these conditions.[180] A mechanism for the coupling reaction of triflates with organotin reagents in the absence or presence of added chloride is presented in **Eq. 4.107**.

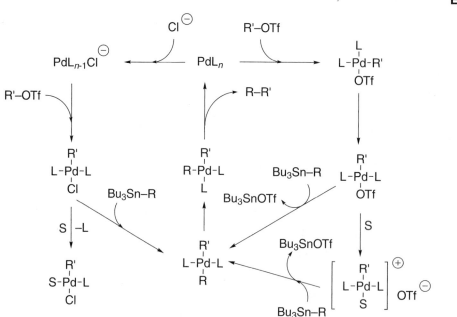

Aryl halides were the first substrates to be coupled with tin reagents using palladium catalysts.[181] Although Suzuki coupling has largely supplanted Stille coupling for the synthesis of biaryls, Stille coupling continues to be extensively used for the synthesis of heteroaromatic systems. An example of a large-scale reaction is shown in **Eq. 4.108**.[182]

Eq. 4.108

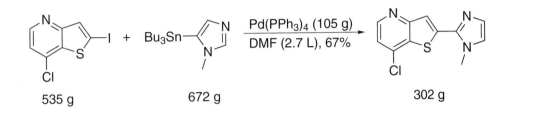

Hexaalkylditin compounds undergo facile transmetallation to σ-palladium complexes, ultimately forming alkenyl- or aryltin reagents (**Eq. 4.109**[183]). The formation of aryltins from hexaalkylditin and an aryl halide was probably the first palladium-catalyzed coupling reaction described using tin reagents.[184] When a second halide/triflate is present in the same molecule, both formation of a tin reagent and an intramolecular Stille reaction can occur (**Eq. 4.110**).[185]

Eq. 4.109

Eq. 4.110

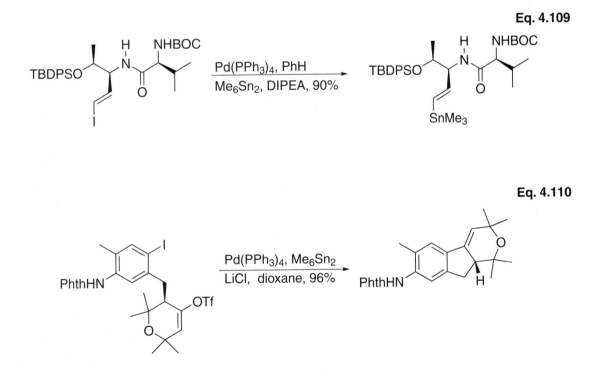

A significant improvement in the Stille reaction has recently been achieved by the use of a catalytic amount of an organotin reagent (**Eq. 4.111**), although the scope of this reaction is at present limited.[186] The reaction involves two palladium-catalyzed steps, a hydrostannation and a cross coupling. In addition, reactions of primary alkyl halides with organotin reagents have been developed.[187]

Eq. 4.111

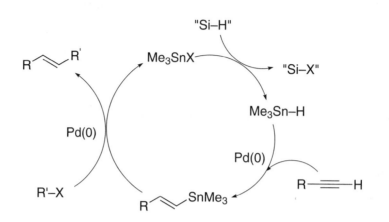

Copper(I) salts mediate Stille-type reactions of alkenyl iodides with alkenyl- and aryltin reagents (**Eq. 4.112**).[188] A particularly useful reagent is copper(I) thiophene carboxylate [Cu(TC)], which mediates[189] the coupling at room temperature using NMP as the solvent (**Eq. 4.113**).[190] A mechanism involving transmetallation from tin to copper to give an alkyl copper reagent, followed by oxidative addition to the alkenyl iodide, and reductive elimination has been suggested. Nickel has more recently been reported as a catalyst for "Stille" reactions of secondary and primary halides with monoorganotin reagents.[191] This reaction probably occurs via the formation of an alkyl radical.

Eq. 4.112

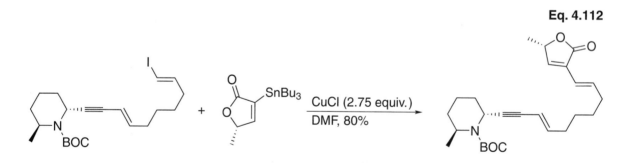

In general, transmetallation from silicon to palladium is not a facile process, and alkenyl silanes can often be carried along as unreactive functional groups in other transmetallation-coupling schemes (e.g., **Eq. 4.114**[192]). However, in the presence of a fluoride source, such as tetrabutylammonium fluoride (TBAF) and tris(diethylamino) sulfonium difluorotrimethylsilicate (TASF),[193] or sodium hydroxide,[194] a pentacoordinated silicate forms that aids in Si–C bond cleavage and provides stable silicon products. Transmetallation from these silicon compounds proceeds smoothly with a wide variety of substrates (**Eq. 4.115**[195]).[196]

Eq. 4.113

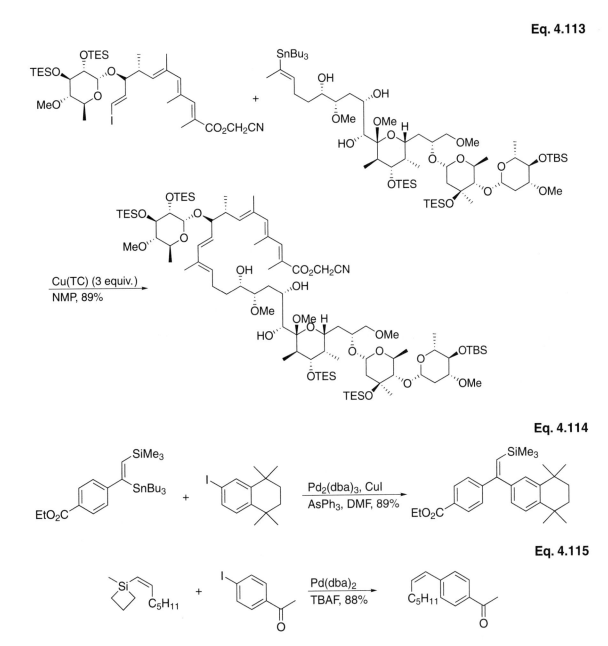

Eq. 4.114

Eq. 4.115

The reaction is commonly referred to as the "Hiyama" reaction. In general, conservation of alkene geometry is observed. Organosilanols and organosilyl ethers[197] also undergo transmetallation to palladium in the presence of silver(I) oxide or TBAF (**Eqs. 4.116**[198] and **4.117**[199]). An important feature of this reaction is the observation that iodides reacts exclusively, even in the presence of bromides and triflates.

Eq 4.116

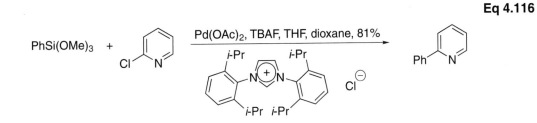

Eq. 4.117

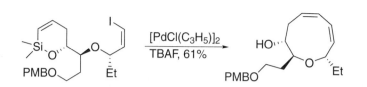

Preformed organocuprates, R₂CuLi, transmetallate to palladium only slowly, and most decompose at temperatures well below that required for transmetallation. However, terminal alkynes couple to aryl and alkenyl halides and triflates (the "Sonogashira" or "Sonogashira–Hagihara" reaction) upon treatment with palladium catalyst, an amine, and copper(I) iodide. This process almost certainly involves oxidative addition to palladium(0) followed by transmetallation from copper.[200] Although the mechanism has not been studied in detail, a likely one is shown in **Eq. 4.118**. This reaction has been known for a long time, but only became appreciated with the advent of the ene-diyne antibiotic, antitumor agents such as calicheamycin or esperamycin.[201] It turns out that this is an excellent reaction, which retains the stereochemistry of the alkenyl halide and tolerates functionality (**Eqs. 4.119**[202] and **4.120**[203]). Some additional spectacular coupling reactions are shown in **Eqs. 4.121**[204] and **4.122**.[205]

Eq. 4.118

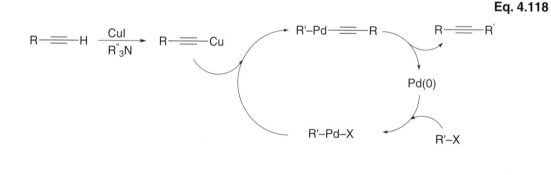

Eq. 4.119

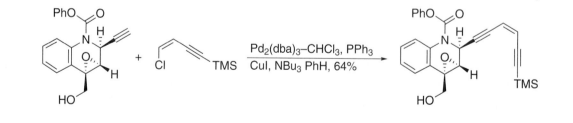

Eq. 4.120

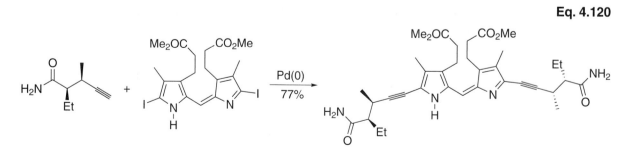

Eq. 4.121

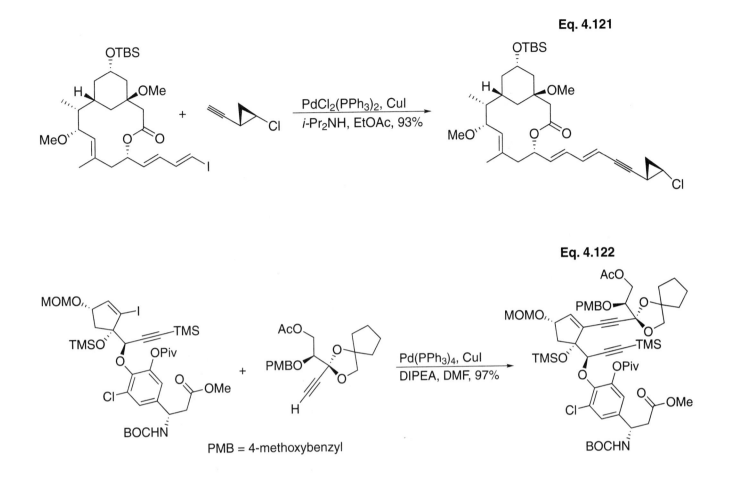

Eq. 4.122

PMB = 4-methoxybenzyl

As this process has become more important in the synthesis of highly function-alized organic compounds, modified conditions have been developed, including the use of Pd/C as the palladium catalyst,[206] amine and/or copper free conditions (**Eq. 4.123**),[207,208] and the use of water as the reaction solvent.[209] By using tetrabutylam-monium iodide as a promoter, aryl triflates can be made to react efficiently.[210] Even primary alkyl bromides and iodides[211] and secondary alkyl bromides[212] can be coupled with terminal alkynes.

Eq. 4.123

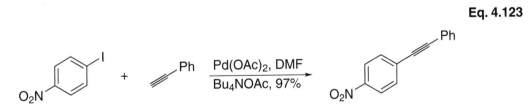

Finally, coupling reactions of allyl, alkenyl, and alkynyl halides with aldehydes mediated by chromium(II) (the "Nozaki–Hiayama–Kishi" reaction) have found exten-sive use in organic total synthesis (**Eq. 4.124**).[213] The mechanism of these reactions is still unclear, but evidence for radical intermediates has been presented.[214]

Eq. 4.124

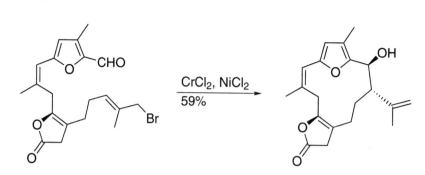

Of particular interest are the nickel–chromium (i.e., NiCl$_2$ and CrCl$_2$) mediated reactions of alkenyl iodides with aldehydes. The reaction can be made catalytic in nickel. This reaction has been employed in a large number of synthetic applications leading to complex organic molecules. Two examples are depicted in **Eqs. 4.125**[215] and **4.126**.[216]

Eq. 4.125

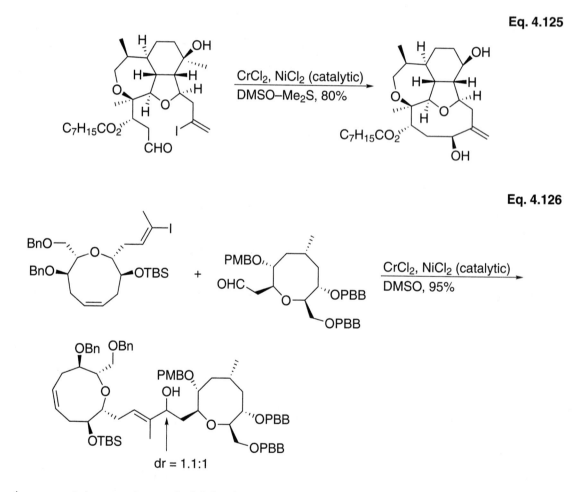

Eq. 4.126

Asymmetric reactions of aldehydes with allylic or propargylic halides have also been reported, affording homoallylic, homopropargylic,[217] and allenylic alcohols[218] in high enantiomeric excess (**Eq. 4.127**).[219]

Eq. 4.127

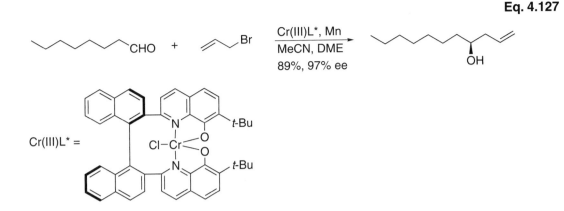

4.6 σ-Carbon–Metal Complexes from Oxidative Addition–Nucleophilic Substitution Sequences

Palladium-catalyzed amination of aryl halides by stannyl amines ($R_3SnNR'_2$) was reported in 1983,[220] but was later shown to not require tin reagents.[221] Rather, treatment of aryl bromides, iodides, or triflates with primary or secondary amines or amides in the presence of palladium–phosphine catalysts and sodium *tert*-butoxide or cesium carbonate[222] results in efficient aryl amination or amidation (a "Hartwig–Buchwald coupling"). By using activated palladium[223] or nickel[224] catalysts, even aryl chlorides are aminated or amidated. The key to the success of this reaction is the choice of ligand and numerous have been developed. The scope of this reaction is truly remarkable and virtually any primary or secondary amine can be N-arylated. In addition, N–H containing compounds, including heterocycles (indoles, pyrroles, etc.), hydrazines, hydrazones, sulfamides, imines, and sulfoximines, also participate in carbon–nitrogen coupling reactions. Some examples of both intra- and intermolecular reactions can be seen in **Eqs. 4.128,**[225] **4.129,**[226] and **4.130.**[227]

Enamines and enamides can readily be prepared via amination or amidation of alkenyl halides and triflates.[228] Copper has also been used as the catalyst, thus extending the scope of the reaction.[229]

Eq. 4.128

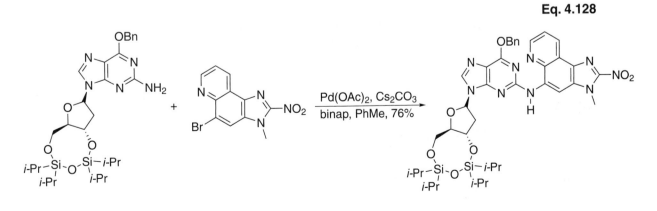

Eq. 4.129

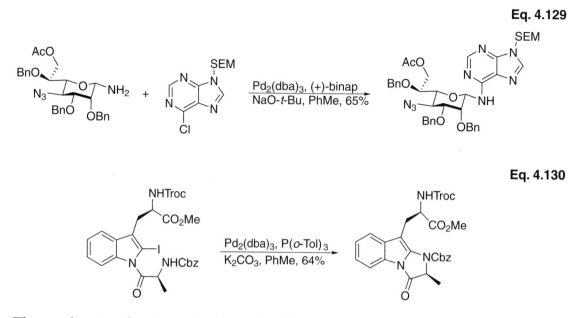

Eq. 4.130

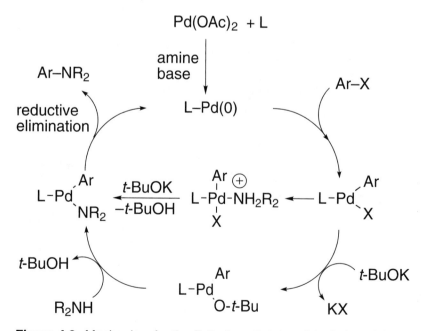

The mechanism for the amination of aryl halides with amines is quite complex.[230] However, a possible mechanism for the coupling is illustrated in **Figure 4.8**.

A further proof of the significance of ligand development is the very recent direct introduction of NH_2 or OH groups on aromatic rings using ammonia[231] or potassium hydroxide,[232] respectively. Alkoxides and thiolates react in a similar fashion in the presence of a palladium catalyst to give ethers (**Eq. 4.131**)[233] and thioethers (**Eq. 4.132**).[234]

Figure 4.8. Mechanism for the Palladium-Catalyzed Arylation of Amines

Eq. 4.131

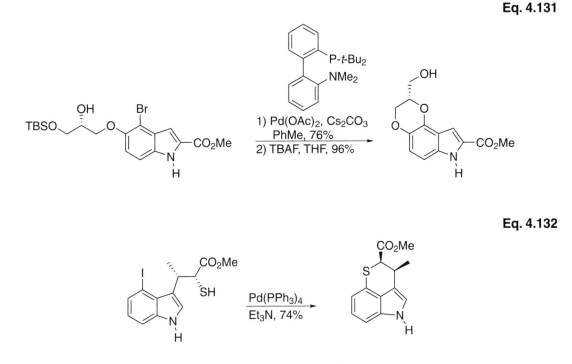

Eq. 4.132

Carbon nucleophiles also participate in oxidative addition–nucleophilic substitution reactions. For example, malonate ester and ethyl acetoacetate ester anions, cyanide, and enolates can be used as the nucleophile (**Eqs. 4.133**[235] and **4.134**[236]).

Eq. 4.133

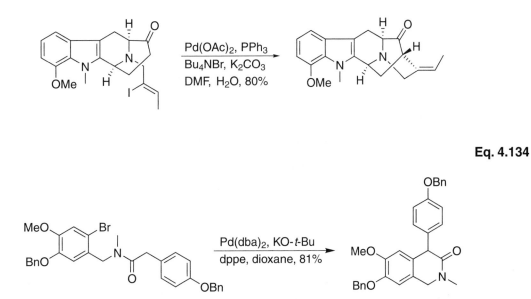

Eq. 4.134

σ-Palladium complexes derived from organic halides and triflates are readily reduced to alkanes or alkenes, depending on the substrate, using a variety of hydride sources (**Eq. 4.135**). Common reducing agents are tributyltin hydride, ammonium formates, and trialkylsilanes. One example from a total synthesis is shown in **Eq. 4.136.**[237]

Eq. 4.135

R–Pd–X $\xrightarrow{H^{\ominus}}$ R–Pd–H + X$^{\ominus}$ $\xrightarrow[\text{elimination}]{\text{reductive}}$ R–H + Pd(0)

R–X

Eq. 4.136

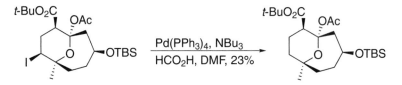

In the case of alkyl halides, the hydride transfer and reductive elimination steps must be faster than β-hydride elimination (**Eq. 4.137**).[238] Carboxylic acid chlorides are reduced to aldehydes when tributyltin hydride is used.[239]

Eq. 4.137

Two equivalents of tributyltin hydride, together with palladium(0) catalyst, reduce bromoalkynes to *E*-tributylstannyl alkenes in a highly regioselective fashion.[240] The reaction probably proceeds via initial hydrostannation of the bromoalkyne, followed by reduction of the intermediate bromotin-substituted alkene, and not by reduction to the alkyne followed by a hydrostannation. A higher selectivity for the formation of the terminal tributylstannylalkene was observed starting from a bromoalkyne compared to the corresponding alkyne (**Eq. 4.138**).[241]

Eq. 4.138

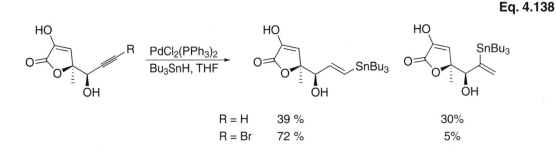

R = H	39 %		30%
R = Br	72 %		5%

4.7 σ-Carbon–Metal Complexes from Oxidative Addition– Insertion Processes (the Heck Reaction)

Metal–carbon σ-bonds are readily formed by oxidative addition processes, and carbon monoxide and alkenes readily insert into these bonds, forming new metal–carbon σ-complexes. These two processes form the basis of a number of synthetically useful transformations. Because of the very wide range of organic substrates that undergo oxidative addition, carbon monoxide insertion offers one of the best ways to introduce carbonyl groups into organic substrates. Depending on the nucleophile present, carboxylic acids, esters, or amides can be prepared using water, alcohols, or amines, respectively. The general process is shown in **Figure 4.9**.

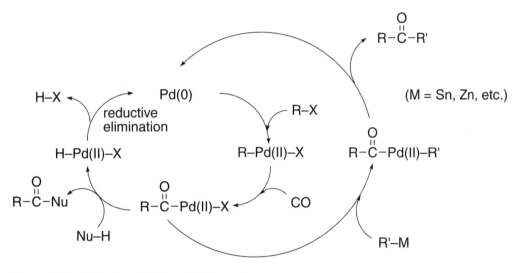

Figure 4.9. Oxidative Addition–CO Insertion

As usual, this process is limited to substrates lacking β-hydrogens, since β-hydride elimination often occurs under the conditions required for oxidative addition. Thus, alkenyl, aryl, benzyl, and allyl halides, triflates, and fluorosulfonates[242] are readily carbonylated. Pressure equipment is usually not required, since insertion occurs readily at one atmosphere pressure. A base is required to neutralize the acid produced from the nucleophilic cleavage of the σ-acyl complex. When palladium(II) catalyst precursors are used, rapid reduction to palladium(0) by carbon monoxide ensues. Some ligand, usually triphenylphosphine, is required both to facilitate oxidative addition with bromides and to prevent precipitation of metallic palladium. Alkenyl triflates have been extensively used (**Eq. 4.139**),[243] and the reaction is efficient even on large scales (**Eq. 4.140**[244]).[245] Because they are directly available from ketones, this provides a convenient method for the conversion of this common functional group to conjugated esters. Intramolecular versions are efficient and provide lactones[246] (**Eq. 4.141**),[247] lactams[248] (**Eq. 4.142**)[249] and even β-lactams (**Eq. 4.143**).[250]

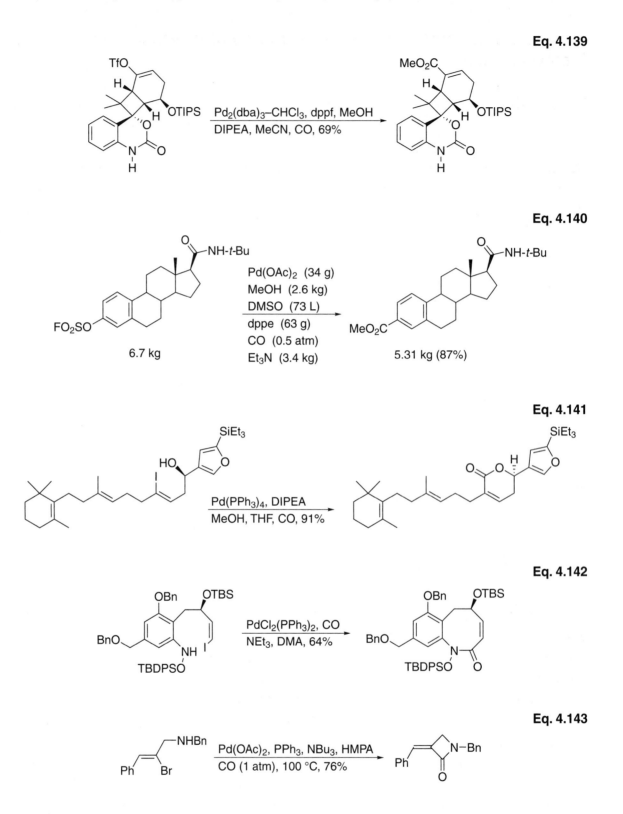

Eq. 4.139

Eq. 4.140

Eq. 4.141

Eq. 4.142

Eq. 4.143

The combination of oxidative addition–CO insertion with transmetallation results in the efficient conversion of organic halides and triflates into aldehydes[251,252] or ketones. The requirement that CO insertion *must* precede transmetallation can usually be met by adjusting the CO pressure (3–6 atm is usually required, rather than atmospheric pressure). Some examples are shown in **Eqs. 4.144,**[253] **4.145,**[254] **4.146,**[255] and **4.147.**[256]

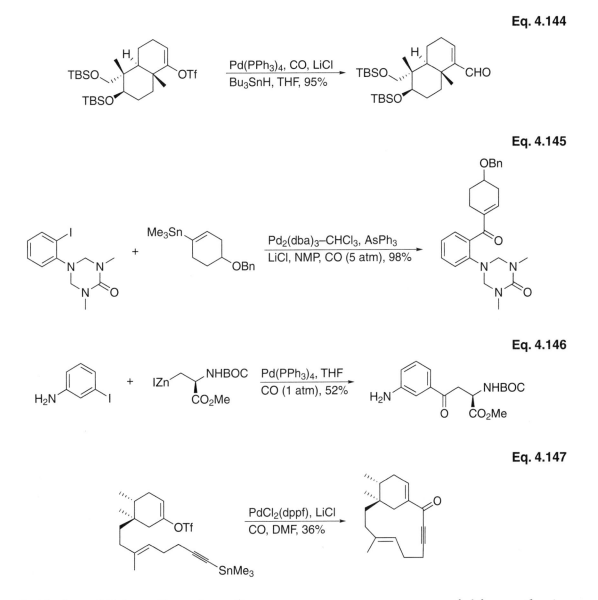

Eq. 4.144

Eq. 4.145

Eq. 4.146

Eq. 4.147

Oxidative addition–alkene insertion sequences are even more useful for synthesis, because a greater degree of elaboration is introduced. The general process is known as the "Heck" or "Mizoroki–Heck"[257] reaction.[258] Two general mechanistic rationales (i.e., neutral vs. cationic) are shown in **Figure 4.10.**[259] These are generalizations of more complex mechanisms. In this case, it is normally the insertion step that is most difficult and it is the structural features on the alkene that impose the most limitations. *All* substrates subject to oxidative addition can, in principle, participate in this process, although, again, β-hydrogens are not usually tolerated because β-elimination competes with insertion.

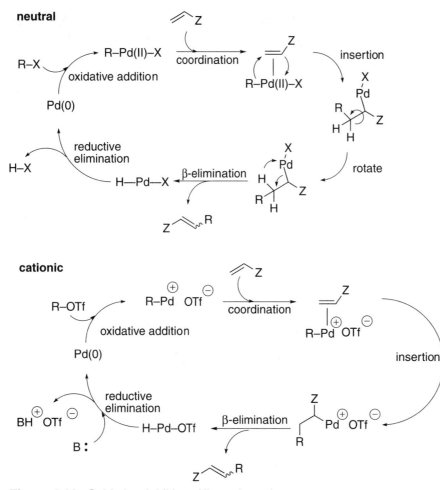

Figure 4.10. Oxidative Addition–Alkene Insertion

Typically, the Heck reaction is carried out by heating the substrate halide or triflate, the alkene, a catalytic amount of palladium(II) acetate, and an excess of a tertiary amine in acetonitrile. The tertiary amine rapidly reduces Pd(II) to Pd(0) by coordination, β-elimination, and reductive elimination (**Eq. 4.148**).[260]

Eq. 4.148

$$R_2NCH_2R' \;+\; PdX_2 \;\rightleftharpoons\; \left[\begin{array}{c} H \\ | \\ R_2N-C-R' \\ | \;\;| \\ Pd\;\;H \\ | \\ X \end{array} \right]^{\oplus} X^{\ominus} \xrightarrow{\text{β-elimination}} \left[R_2\overset{\oplus}{N}=CHR' \right] X^{\ominus} + \text{H–Pd–X} \longrightarrow Pd(0) + HX$$

With aryl iodides as substrates, the addition of phosphine is not required, and in fact, inhibits the reaction. Even palladium on carbon will catalyze Heck reactions of aryl iodides.[261] In contrast, with aryl bromides as substrates, a hindered tertiary phosphine like tri-*o*-tolylphosphine is required to suppress quaternization of the phosphine to give Ar'Ph₃PX. Recently, considerably milder, more efficient conditions for the Heck reaction have been developed. These involve using polar solvents such as DMF with added quaternary ammonium chlorides to promote the reaction.[262] The soluble quaternary ammonium chloride keeps the concentration of soluble chloride high, and halide ions stabilize and activate palladium(0) complexes during these reactions.[263] Quaternary ammonium sulfates allow the reaction to be carried out under aqueous conditions.[264] For industrial applications, related conditions result in very high catalyst turnover numbers and frequencies (TON = 10^6 and TOF 4×10^4 per hour).[265,266]

A very wide range of alkenes will undergo this reaction, including electron-deficient alkenes, such as α,β-unsaturated esters, amides, ketones, and nitriles; unactivated alkenes, such as ethene and styrene; and electron-rich alkenes, such as enol ethers and enamides. With electron-poor and unactivated alkenes, the insertion step is sterically controlled and insertion occurs to place the R group at the less-substituted position. The R group is delivered from the same face as the metal (cis insertion). Substituents on the β-position of the alkene inhibit insertion and, in intermolecular reactions, β-disubstituted alkenes are unreactive. The β-hydride elimination is also a cis process (Pd and H must be syn coplanar).

With electron-rich alkenes, the regioselectivity of insertion is more complex.[267,268] With cyclic enol ethers and enamides, arylation α to the heteroatom is always favored, but mixtures are often obtained with acyclic systems. Bidentate phosphine ligands favor α-arylation, as do electron-rich aryl halides, while electron-poor aryl halides tend to result in β-arylation. The regioselectivity of *intra*molecular processes is also difficult to predict, and seems to depend more on ring size than alkene substitution. Even "aromatic" double bonds can insert (see below).

The Heck reaction has become very popular with synthetic chemists and a very broad range of applications to complex molecule syntheses has been reported. The process enjoys a wide tolerance to functionality and normally proceeds with high regio- and stereoselectivity (**Eqs. 4.149**[269] and **4.150**[270]). Heck reactions using allylic alcohols afford either aldehydes or ketones via β-elimination toward the alcohol and tautomerization of the corresponding enol (**Eq. 4.151**).[271]

Eq. 4.149

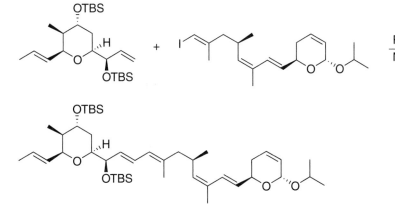

Eq 4.150

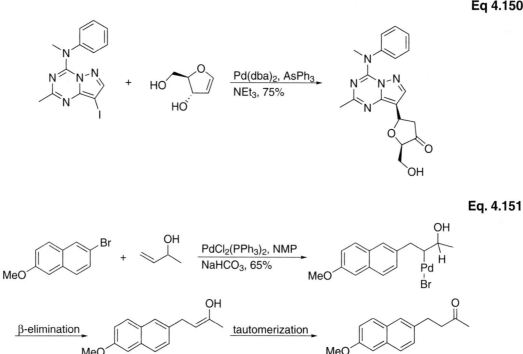

Eq. 4.151

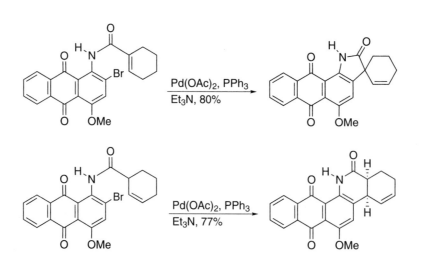

As is often the case with organometallic reactions, intramolecular versions are more efficient than intermolecular versions.[272] Although the regioselectivity of intramolecular Heck reactions is usually high, the mode of addition (exo vs. endo) is less predictable and is often dictated by ring size (**Eq. 4.152**).[273] In general, 5-exo or 6-exo cyclizations are strongly favored. In some cases, the regioselectivity can be altered by the choice of conditions (**Eq. 4.153**).[274] (The *stereochemistry* of cyclization depends on the orientation of the alkene relative to the palladium–carbon σ-bond[275]). By use of high dilution or site isolation by polymer supports, macrocycles are readily available by the Heck reaction.[276] Examples of some more spectacular intramolecular Heck reactions are shown in **Eqs. 4.154,**[277] **4.155,**[278] **4.156,**[279] and **4.157.**[280] Even intramolecular Heck reactions of *alkyl* bromides have recently been reported.[281]

Eq. 4.152

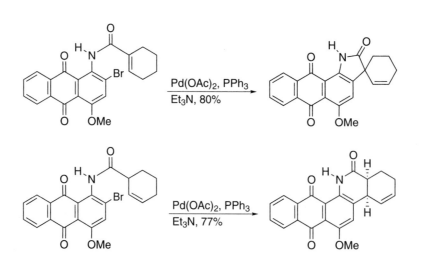

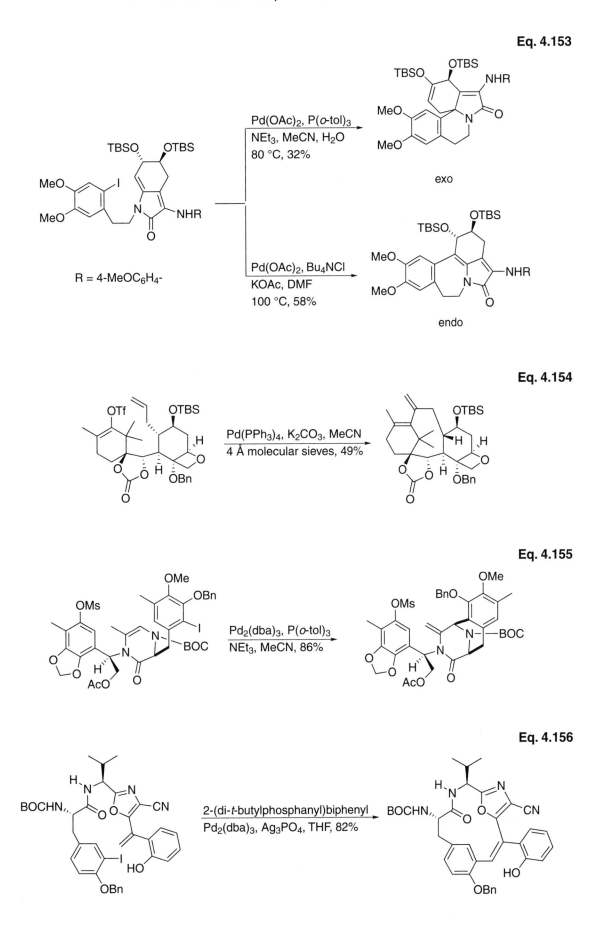

Eq. 4.153

Pd(OAc)₂, P(o-tol)₃
NEt₃, MeCN, H₂O
80 °C, 32%

exo

Pd(OAc)₂, Bu₄NCl
KOAc, DMF
100 °C, 58%

endo

R = 4-MeOC₆H₄-

Eq. 4.154

Pd(PPh₃)₄, K₂CO₃, MeCN
4 Å molecular sieves, 49%

Eq. 4.155

Pd₂(dba)₃, P(o-tol)₃
NEt₃, MeCN, 86%

Eq. 4.156

2-(di-t-butylphosphanyl)biphenyl
Pd₂(dba)₃, Ag₃PO₄, THF, 82%

Eq. 4.157

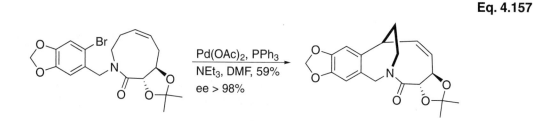

A major advance came with the ability to induce asymmetry during the Heck reaction.[282] One requirement for asymmetric induction is that the newly formed stereogenic center must be an sp^3 center (as in **Eq. 4.151**) rather than the more usual sp^2 centers (**Eq. 4.149**). For this to be the case, β-hydride elimination must occur *away* from the newly formed center (β'), rather than toward it (β) (**Eq. 4.158**). This restricts asymmetric induction to cyclic systems or to the formation of quaternary centers since it is only in these cases that β-elimination towards the newly formed center is prevented, either by the inability of that β-hydrogen to achieve the syn orientation required for β-elimination, or by the lack of that β-hydrogen. When these conditions are met, excellent enantioselectivity can be achieved (**Eqs. 4.159,**[283] **4.160,**[284] **4.161,**[285] and **4.162**[286]).

Eq. 4.158

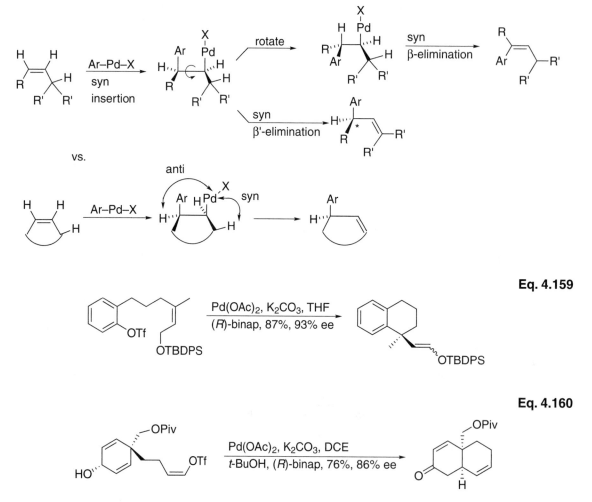

Eq. 4.159

Eq. 4.160

Eq. 4.161

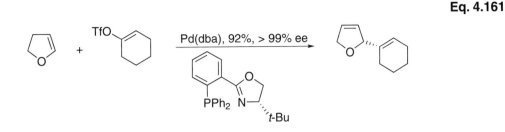

Eq. 4.162

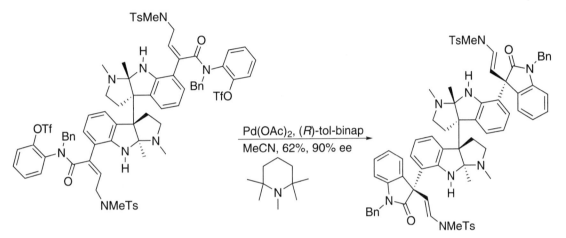

Heck-type coupling of aromatic halides and triflates is not limited to alkenes. Even aromatic π-systems, in particular electron-rich π-systems, can "insert" (**Eq. 4.163**[287]).[288] Two somewhat different mechanisms, a metathesis mechanism[289,290] that forms a palladium biaryl intermediate (**Eq. 4.164**)[291] or an electrophilic addition (**Eq. 4.165**) followed by reductive elimination, have been proposed based on experimental and computational evidence. Even benzene undergoes coupling reactions with aromatic halides (**Eq. 4.166**).[292] This type of Heck reaction has been utilized in a number of synthetic applications toward complex polycyclic ring systems (**Eqs. 4.167**[293] and **4.168**).[294]

Eq. 4.163

Eq. 4.164

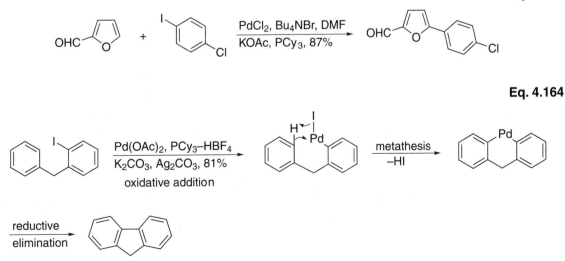

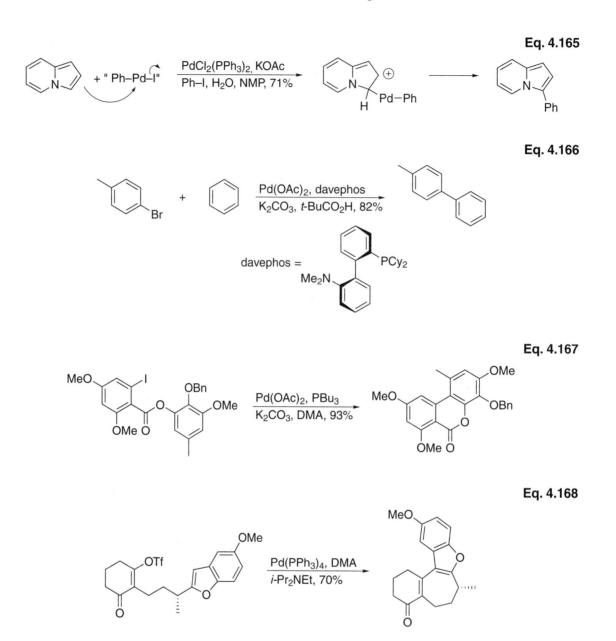

Eq. 4.165

Eq. 4.166

Eq. 4.167

Eq. 4.168

The alkene insertion step in the Heck reaction generates a new metal–carbon σ-bond which, in principle, could undergo any of the reactions of metal–carbon σ-bonds (**Figure 4.1**), if only β-hydride elimination did not occur so readily. If β-elimination could be suppressed or if a process even more favored than β-elimination (such as CO insertion) could be arranged, then several new bonds could be formed, in sequence, in a single catalytic process. This is, indeed, possible, and these "cascade" reactions are becoming powerful tools in the synthesis of polycyclic compounds.[295]

The least elaborate application of this concept is "the reductive Heck reaction" wherein a hydride is transferred from a formate, usually ammonium formate, to palladium, followed by a reductive elimination of the resulting organopalladium hydride to form an alkane (**Eq. 4.169**). An example of this reaction is shown in **Eq. 4.170**.[296]

Eq. 4.169

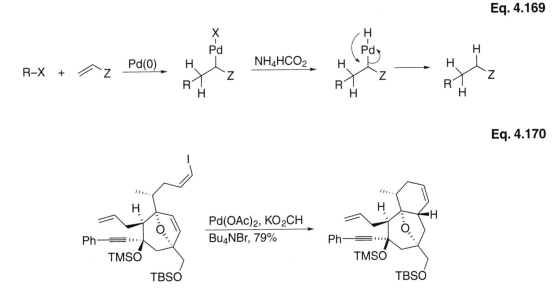

Eq. 4.170

Formation of a new carbon–carbon bond from the σ-palladium intermediate is synthetically much more interesting. For example, the reaction in **Eq. 4.171**[297] involves an oxidative addition–CO insertion–alkene insertion–CO insertion–cleavage process that forms three carbon–carbon bonds per turn of catalyst. This process works because the initial oxidative adduct has no accessible β-hydrogens, and CO insertion is faster than alkene insertion for σ-alkylpalladium complexes, but slower for σ-acyl complexes. The major challenge in the system was to make the final CO insertion into a σ-alkylpalladium(II) complex, which does have β-hydrogens, competitive with β-elimination. This was achieved by using a relatively high pressure of CO. Indeed, under one atmosphere of CO, β-elimination was the main course of the reaction.

Eq. 4.171

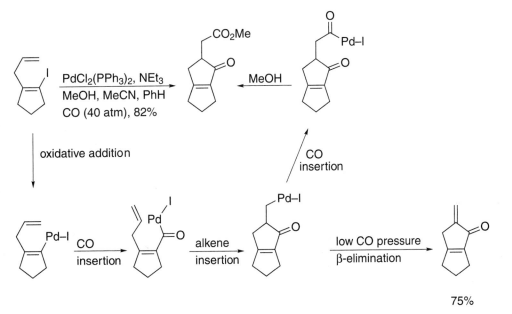

This downhill sequence of events is critical in these multiple-step cascade reactions, and when met, an impressive number of bonds per catalyst cycle can be achieved[298] (**Eqs. 4.172**[299] and **4.173**).[300]

Eq. 4.172

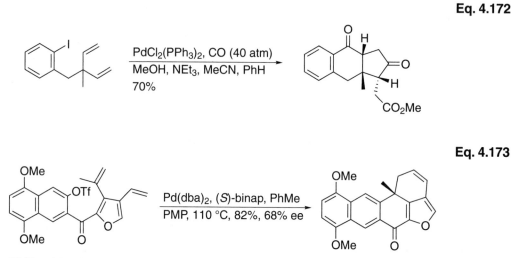

Eq. 4.173

PMP = 1,2,2,6,6-pentamethylpiperidine

Alkynes insert more readily than alkenes, and a variety of "carbocyclization" procedures[301] involving this functional group have been developed (**Eqs. 4.174**,[302] **4.175**,[303] and **4.176**).[304]

Eq. 4.174

Eq. 4.175

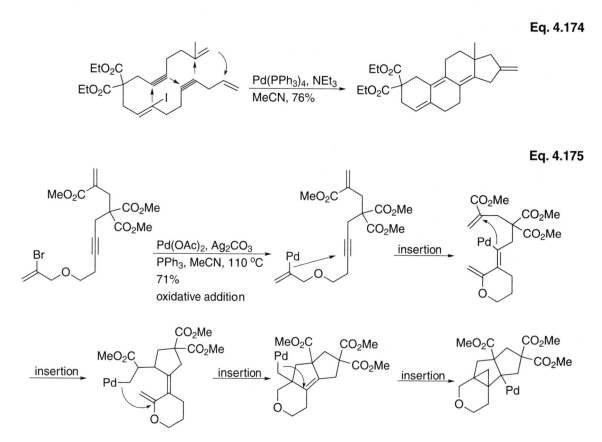

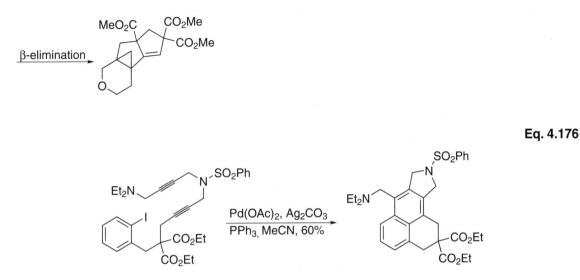

Eq. 4.176

In most of the above cases, the last step of the cascade was a β-hydride elimination. However, truncation by transmetallation (**Eq. 4.177**)[305] or nucleophilic attack (**Eqs. 4.178**[306] and **4.179**)[307] is also possible. This methodology offers an attractive route to a large variety of 2,3-disubstituted indoles and related pyrrolo-fused heterocycles. Almost endless variations of these themes are possible, and most are likely to be attempted.

Eq. 4.177

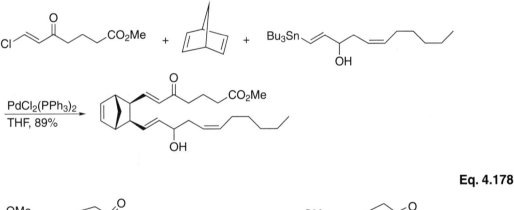

Eq. 4.178

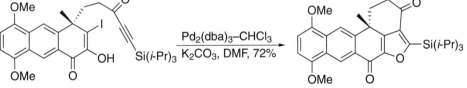

Eq. 4.179

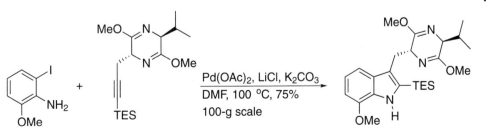

4.8 σ-Carbon–Metal Complexes from Carbon–Hydrogen Bond Insertion Processes

Palladium(II) salts are reasonably electrophilic, and under appropriate conditions [i.e., electron-rich arenes, Pd(OAc)$_2$, and boiling acetic acid] can directly palladate arenes via an electrophilic aromatic substitution process. Although this process is neither general nor efficient at the present, this chemistry is being developed and some applications in synthesis have recently appeared. Much more general is the ligand-directed ortho palladation shown in **Eq. 4.180**.[308,309] This process is very general, the only requirement being a lone pair of electrons in a benzylic position to precoordinate to the metal, and conditions conducive to electrophilic aromatic substitution (usually HOAc solvent). Nitrogen, oxygen, sulfur, and phosphorous are the most common donor atoms. The site of palladation is always ortho to the directing ligand and, when the two ortho positions are inequivalent, palladation results at the sterically less-congested position, regardless of the electronic bias. (This is in direct contrast to o-lithiation, which is electronically controlled and will usually occur at the more-hindered position on rings having several electron-donating groups.) The resulting chelate-stabilized, σ-arylpalladium(II) complexes are almost invariably quite stable, are easily isolated, purified, and stored, and are quite unreactive.

Eq. 4.180

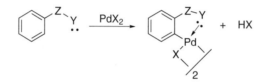

These ortho-palladated species *can* be made to react like "normal" σ-arylmetal complexes by decomplexing the ortho ligand, usually accomplished by treating the complex with the substrate and a large excess of triethylamine, to compete with the ortho ligand. Under these conditions, alkene insertion (**Eq. 4.181**)[310] and CO insertion (**Eq. 4.182**)[311] are feasible.

Eq. 4.181

Eq. 4.182

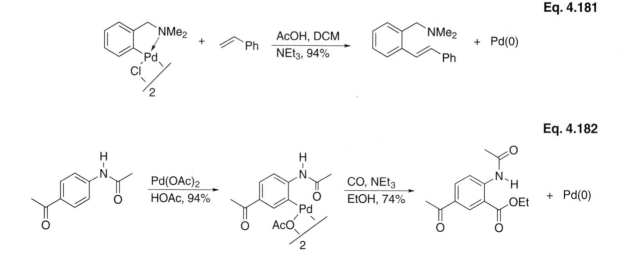

The potential power of ligand-directed metallation is illustrated in **Eq. 4.183**,[312] in which an unactivated methyl group was palladated solely because it had the misfortune of being within the range of a ligand-directed palladation. Sequential ligand-directed palladations were used in a synthesis of the core of teleocidin B4 (**Eq. 4.184**).[313]

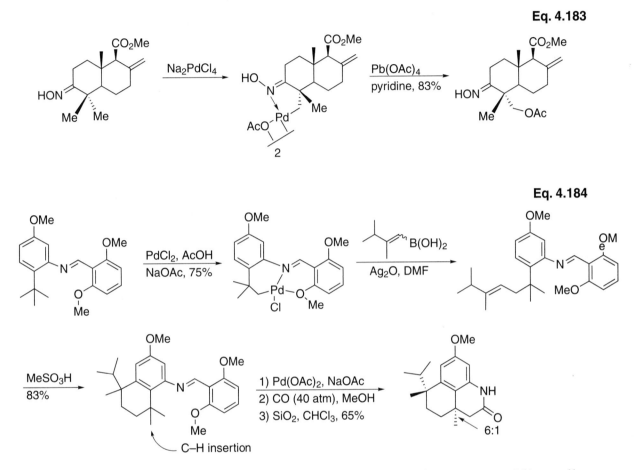

Eq. 4.183

Eq. 4.184

Catalytic reactions have been developed that use copper diacetate to oxidize palladium(0) to palladium(II). Examples include carbonylations to give lactams (**Eq. 4.185**)[314] and chlorinations (**Eq. 4.186**).[315] The activation of sp^3-hybridized carbon–hydrogen bonds has also been achieved, as in the β-arylation of carboxylic acids (**Eq. 4.187**).[316]

Eq. 4.185

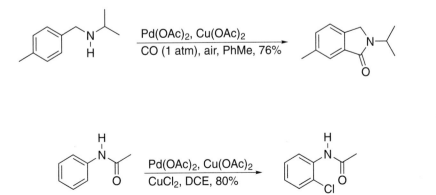

Eq. 4.186

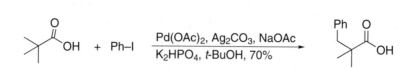

Eq. 4.187

Catalytic reactions that probably involve the direct palladation of the 2-position of indoles, followed by the insertion of a pendant alkene, have been developed and used in synthesis (**Eqs. 4.188**[317] and **4.189**[318]). Benzoquinone (BQ) or oxygen is used to oxidize formed palladium(0) to palladium (II).

Eq. 4.188

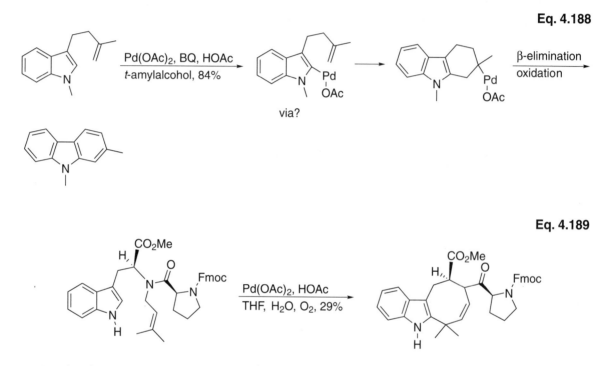

Eq. 4.189

Multiple aromatic carbon–hydrogen bond activations can be achieved using palladium as the catalyst. This type of reaction is initiated by an oxidative addition followed by an alkene insertion (**Eq. 4.190**.[319] Palladium subsequently migrates to an adjacent aromatic ring via a cyclometallated complex. Both palladium(IV) and palladium(II) intermediates have been proposed.[320] A second cyclometallation, followed by reductive elimination, affords a polyaromatic compound. Alkynes undergo related reactions probably via cyclometallated intermediates (**Eq. 4.191**).[321] Related cyclometallated intermediates were proposed in the formation of benzocyclobutene (**Eq. 4.192**)[322] and benzofuran derivatives (**Eq. 4.193**).[323]

Eq. 4.190

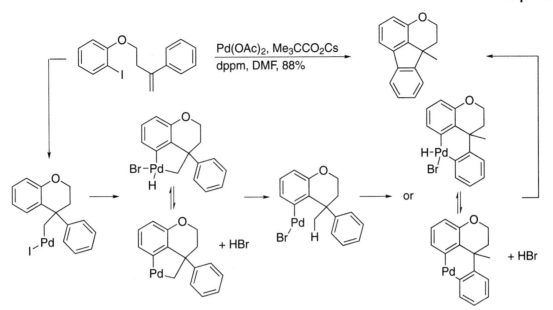

Eq. 4.191

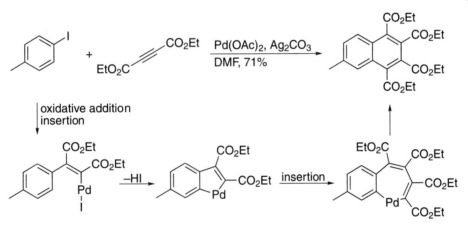

oxidative addition
insertion

Eq. 4.192

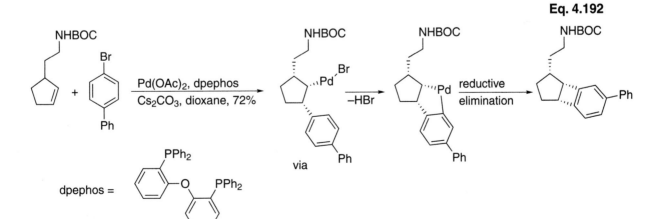

dpephos =

Eq. 4.193

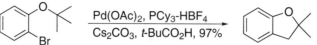

Catalytic aromatic carbon–hydrogen bond activation is very nicely achieved in the presence of norbornene.[324] A plausible mechanism involves oxidative addition of the aromatic iodide followed by insertion of norbornene (**Eq. 4.194**).[325] Since no syn β-hydrogens are available in the σ-palladium complex that forms, palladium inserts into one of the adjacent carbon–hydrogen bonds, thus forming a metallacycle. Oxidative addition of an alkyl or aryl halide affords a palladium(IV) complex that undergo reductive elimination. This cyclopalladation–oxidative addition–reductive elimination sequence can occur at the other ortho position if available (not shown). If not, palladium de-inserts norbornene, giving a σ-aryl palladium complex wherein palladium is bound to the position of the original iodide. This intermediate can now react with the usual array of reagents used in palladium-catalyzed coupling reactions (e.g., cyanide).[326] The final σ-palladium complex reacts with alkenes (**Eq. 4.195**),[327] alkynes,[328] and organoboronic acids[329] to produce Heck-, Sonogashira-, and Suzuki-type products, respectively. Reactions of aromatic iodides in the presence of isopropylboronic acid as a hydride source furnish 1,3-disubstituted aromatic compounds (**Eq. 4.196**).[330]

Eq. 4.194

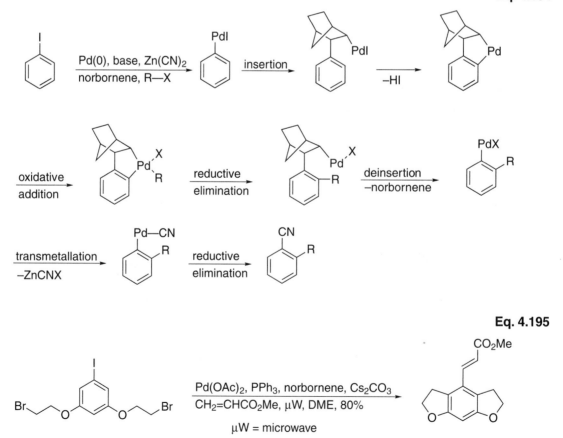

Eq. 4.195

μW = microwave

Eq. 4.196

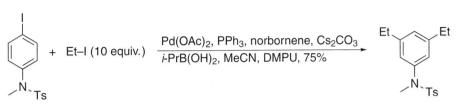

Carbon–hydrogen activation of aromatic compounds has been significantly expanded by the use of rhodium, ruthenium, and iridium catalysts.[331] Sequential alkene and carbon monoxide insertion can be achieved (**Eq. 4.197**).[332] A plausible mechanism for the reaction involves the coordination of ruthenium to the nitrogen atom, oxidative addition (forming a ruthenacycle), and insertion of carbon monoxide and ethene followed by reductive elimination. Related intramolecular alkene insertions of cyclometallated intermediates afford bicyclic compounds in excellent yield and in some cases in high enantiomeric excess (**Eq. 4.198**).[333] The metallacycle that forms can also be intercepted by organoboron compounds, yielding unsymmetrically substituted ketones (**Eq. 4.199**).[334]

Eq. 4.197

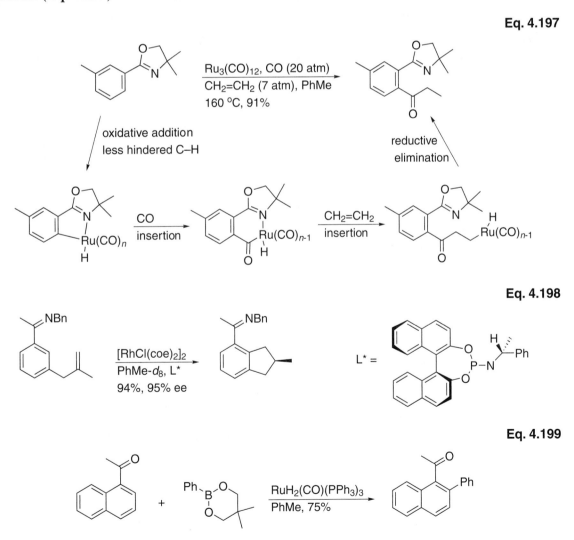

Eq. 4.198

Eq. 4.199

4.9 σ-Carbon–Metal Complexes from the Reductive Coupling of Alkenes and Alkynes

Low-valent metals from the two extremes of the transition series (i.e., Ni, Ti, and Zr) undergo reactions with alkenes and alkynes to form five-membered metallacycles in which the unsaturated substrates are formally reductively coupled, and the metal is formally oxidized (**Eq. 4.200**). These metallacycles can be converted to a number of interesting organic compounds.

Eq. 4.200

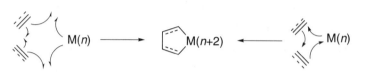

The most extensively studied complex to effect this reaction is "zirconocene," "Cp_2Zr," an unstable, unsaturated, d^2 Zr(II) complex.[335] It can be generated in many ways, but the most convenient is to treat Cp_2ZrCl_2 with butyllithium and to allow the unstable dibutyl complex to warm to above –20 °C in the presence of substrate (**Figure 4.11**). This generates a zirconium–butene complex that can exchange with substrate alkene or alkyne and promote the cyclodimerization. The zirconium–butene complex can be considered either as a Zr(II)–alkene complex or a Zr(IV) metallacyclopropane complex. These are the two extreme bonding forms for metal alkene complexes (resonance forms), with the metallacyclopropane form denoting complete two-electron transfer from the metal to the alkene. For the zirconium–butene complex, the metallacyclopropane form is probably the more accurate representation, based on its reaction chemistry and the isolation of a stable PMe_3 complex. Given that "Cp_2Zr" in fact reductively dimerizes alkenes and alkynes (**Eq. 4.200**) (2 e⁻ transfer from Zr(II) to

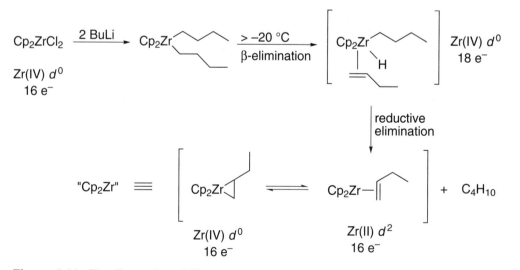

Figure 4.11. The Formation of Zirconocene

alkenes), it is not surprising that the zirconium–butene complex also involves extensive electron transfer to the alkene.

The five-membered zirconacycles formed by the cyclodimerization of alkenes can be transformed into a variety of interesting products. Protolytic cleavage produces the alkane, iodine cleavage the diiodide, reaction with carbon monoxide and protolytic cleavage a cyclopentanol, and treatment with carbon monoxide and iodine a cyclopentanone (**Eq. 4.201**).[336] The cyclodimerization of alkynes produces zirconacyclopentadienes, which undergo electrophilic cleavage to produce, for example, a number of unusual heterocycles (**Eq. 4.202**).[337]

Eq. 4.201

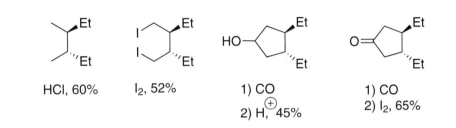

HCl, 60%   I$_2$, 52%   1) CO 2) H,$^{\oplus}$ 45%   1) CO 2) I$_2$, 65%

Eq. 4.202

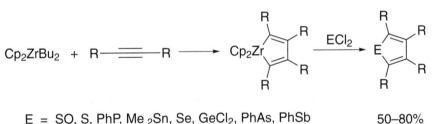

E = SO, S, PhP, Me$_2$Sn, Se, GeCl$_2$, PhAs, PhSb   50–80%

Intramolecular enyne, diyne, and diene cyclizations are also effective, forming mono- or bicyclic compounds, depending on the method of cleavage (**Eq. 4.203**).[338] A range of functional groups is tolerated, and the reaction is highly stereoselective (**Eqs. 4.204,**[339] **4.205,**[340] and **4.206**).[341] Note the insertion of an isonitrile, which is isoelectronic and isostructural with CO (**Eq. 4.205**).

Eq. 4.203

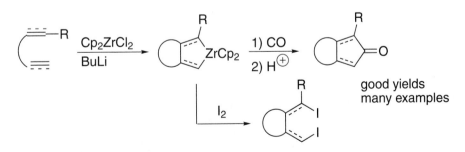

good yields
many examples

Eq. 4.204

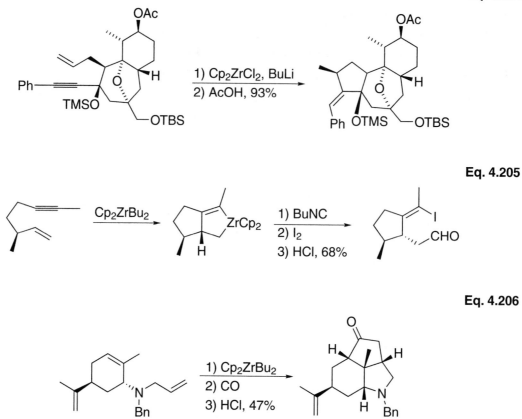

Eq. 4.205

Eq. 4.206

When cleaved with anions, such as lithiated allylic chlorides (and related carbenoids), η¹-alkyl-η³-allylzirconium complexes are produced, which undergo a number of useful transformations with iminium ions, aldehydes, acetals, and TMS cyanide.[342] For example, a reaction with an aldehyde was used in a synthesis of acetoxyodontoschismenol (**Eq. 4.207**).[343]

Eq. 4.207

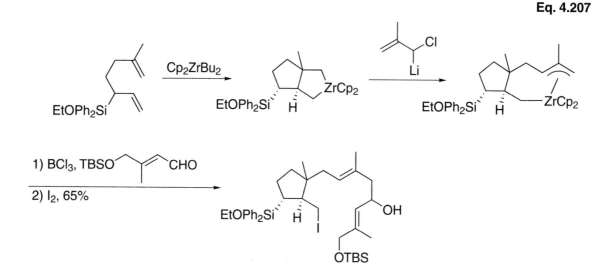

Imines (**Eq. 4.208**)[344] and hydrazones[345] derived from enals undergo cyclometalla-tion with zirconocene to give aminated products after hydrolytic workup.

Eq. 4.208

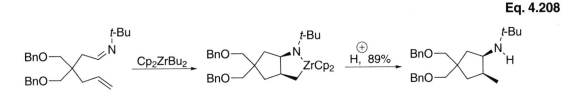

Related low-valent titanium complexes similarly "reductively couple" enynes or dienes, giving metallacyclopentenes and -pentanes that are readily cleaved by electro-philes to produce interesting compounds[346,347] (**Eqs. 4.209**[348] and **4.210**[349]).

Eq. 4.209

Eq. 4.210

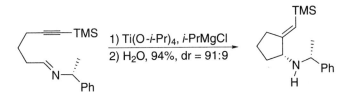

CpTiMe$_2$Cl and related titanium(IV) complexes form transient imidotitanium spe-cies that undergo cycloadditions to alkynes, again making metallacycles that undergo useful electrophilic cleavages (**Eq. 4.211**).[350] This reaction is catalytic in titanium and has been used to some extent in total synthesis (**Eq. 4.212**).[351]

Eq. 4.211

Eq. 4.212

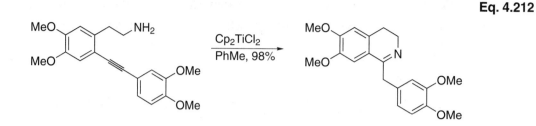

Treatment of Cp$_2$Zr(Me)Cl with lithiated amines produces unstable zirconium–imine complexes, probably by a sequence of Cl$^-$ displacement, β-hydride elimination, and reductive elimination (**Eq. 4.213**). These insert alkenes and alkynes to give five-membered azazirconacycles that can be cleaved using electrophilic reagents.[352]

Eq. 4.213

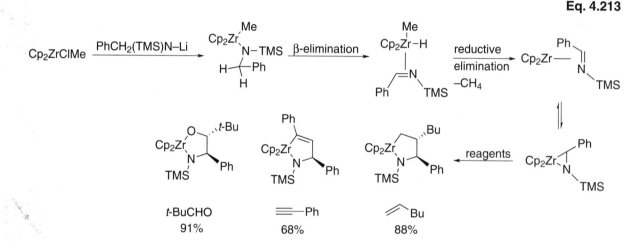

Titanium(IV) metallacyclopropane [or, alternatively, titanium(II) alkene] complexes react with esters and tertiary amides to afford hydroxy- and amino-substituted cyclopropanes (the "Kulinkovich–de Meijere reaction").[353] In the reaction, a titanacyclopropane is formed by the reaction of a catalytic amount of titanium tetraisopropoxide with an excess of a Grignard reagent—usually ethylmagnesium chloride or cyclohexylmagnesium chloride (**Eq. 4.214**). The titanacyclopropane then acts as a 1,2-dinucleophile, which alkylates the ester twice (**Eq. 4.215**). Reaction with two equivalents of Grignard reagent regenerates the catalyst and forms a magnesium cyclopropoxide. For simplicity, two isopropoxides have been shown throughout the mechanism. It is possible and likely that the ester alkoxide (R'O$^-$) also will be found on some of the catalyst.

Eq. 4.214

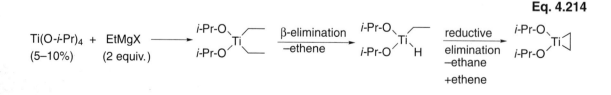

Eq. 4.215

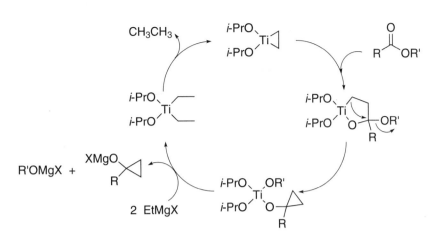

This reaction has been used in a number of syntheses and an example can be seen in **Eq. 4.216**.[354] The titanacylopropane formed from the Grignard reagent can be exchanged using a more-substituted alkene, thus opening an entry to substituted cyclopropanes (**Eq. 4.217**).[355] This also works very nicely in intramolecular reactions (**Eq. 4.218**).[356]

Eq. 4.216

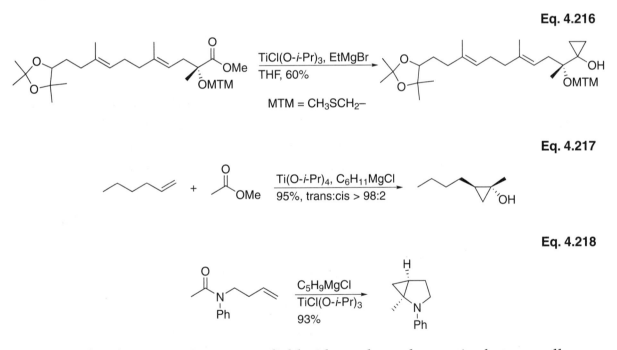

In a related process, zirconocene dichloride catalyzes the reaction between alkenes and Grignard reagents to produce new Grignard reagents. The intramolecular version is the easiest to follow (**Eq. 4.219**).[357] Treatment of 1,6- and 1,7-dienes with butylmagnesium chloride in the presence of catalytic amounts of Cp_2ZrCl_2 produces cyclized, bis-Grignard reagents, which can be further functionalized. The reaction is thought to start as in **Figure 4.11**. Note that the cyclization is not stereospecific and both cis and trans products are formed.

Treatment of zirconocene dichloride with butyl Grignard reagent produces the reactive zirconacyclopropane/zirconocene butene complex (a). This reacts with the diene to produce the zironacyclopentane (b). Once this occurs, the zirconacyclopentane is cleaved by the Grignard reagent to generate the bis-Grignard and regenerate the $Cp_2Zr(Bu)_2$.

Eq. 4.219

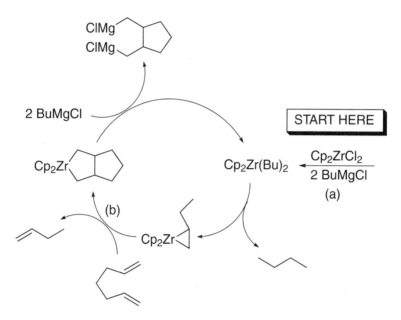

With monoalkenes, Cp_2ZrCl_2 catalyzes the addition of EtMgCl to alkenes to give homologated Grignard reagents (**Eq. 4.220**).[358] The process is similar to that with dienes, except in this case the Grignard provides one "alkene", and the cleavage gives a mono- rather than a bis-Grignard reagent. Note the selective cleavage of the more-substituted carbon–zirconium bond.

Eq. 4.220

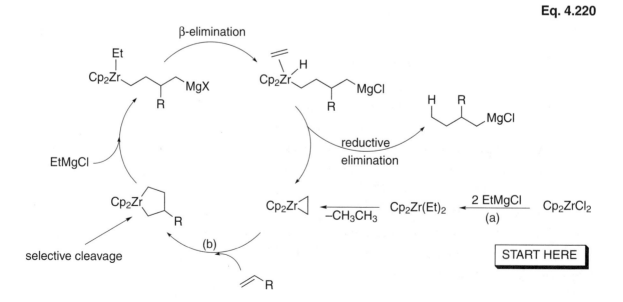

Perhaps most synthetically interesting is the reaction with allylic alcohols and ethers (**Eq. 4.221**).[359] Here the allylic oxygen has both stereo- and regiochemical consequences, indicating complexation to zirconium along the course of the reaction. With allylic alcohols, the zirconium reagent is delivered to the *same* face as the OH group, while with allylic ethers it comes from the opposite face, implying that the alkoxide that is deprotonated using one equivalent of the Grignard reagent complexes to Zr, but OCH_3 does not. The reaction is also sensitive to steric hindrance (**Eq. 4.222**). Given a choice between an allylic methyl ether and the substantially larger *tert*-butyldimethylsilyl ether, reaction occurs exclusively at the less-hindered alkene.

Eq. 4.221

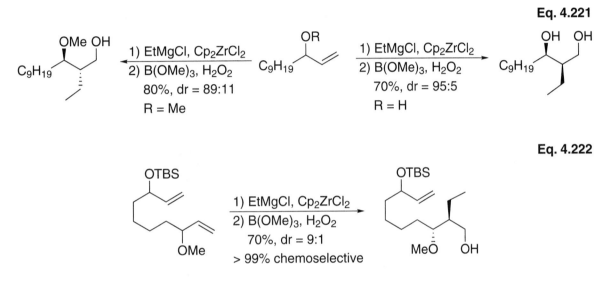

Eq. 4.222

By using optically active zirconocene complexes, this ethylmagnesation of alkenes has been made asymmetric.[360] The most synthetically useful reactions involve oxygen or nitrogen heterocycles, which are ring-opened in the process, to produce optically active, functionalized terminal alkenes (**Eq. 4.223**). The mechanism by which this is thought to proceed is shown in **Eq. 4.224** and involves the same steps seen above.

Eq. 4.223

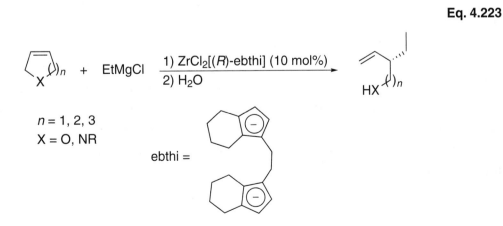

Eq. 4.224

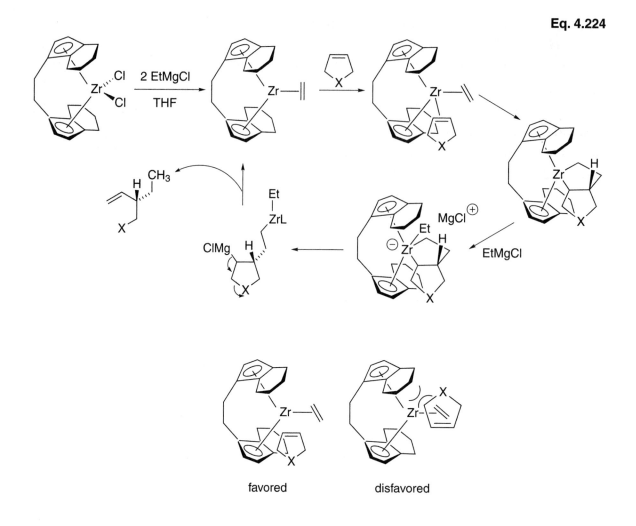

favored   disfavored

Asymmetric induction is thought to result from complexation of the substrate alkene in such a way as to minimize the steric interactions between the chiral ligand and the ring atoms of the substrate. The presence of a β-heteroatom in the final organomagnesium complex leads to production of a terminal alkene by facile elimination. An elegant example of the use of both the allylic alcohol-directed carbomagnesation and the diastereoselective ring opening of dihydrofurans is shown in the total synthesis of fluvivicine (**Eq. 4.225**).[361]

Because the chiral zirconium catalyst so efficiently discriminates in its reactions with cyclic ethers, it can be used to kinetically resolve racemic dihydrofurans and dihydropyrans.[362] The process takes advantage of the fact that one enantiomer of the dihydropyran undergoes ring opening substantially faster than the other, and is consumed, while the less reactive enantiomer can be recovered unchanged and with very good enantiomeric excess (**Eq. 4.226**).[363] With dihydrofurans, both react, but give different products (regioisomers) which are separable.

Eq. 4.225

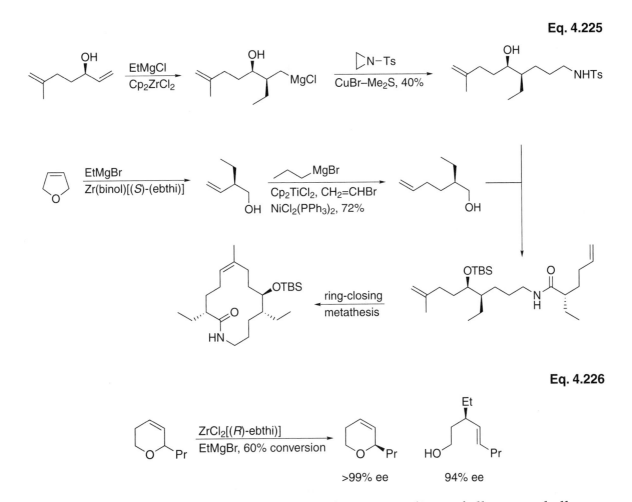

Eq. 4.226

>99% ee　　94% ee

The other class of synthetically useful reductive couplings of alkenes and alkynes comes from the opposite end of the periodic table—that is, they are catalyzed by Ni(0) d^{10} complexes.[364] In this case the metallacycle is unstable, and has not been observed, but is inferred from the products. In the absence of added phosphine, simple reductive coupling, followed by β-elimination–reductive elimination occurs, giving dienes in reasonable yield. With added tributylphosphine, insertion of an isonitrile (see also **Eq. 4.205**) occurs, giving the bicyclic imine which can be hydrolyzed to the ketone (**Eq. 4.227**). Diynes undergo similar cyclizations to give cyclopentadienone imines.

Eq. 4.227

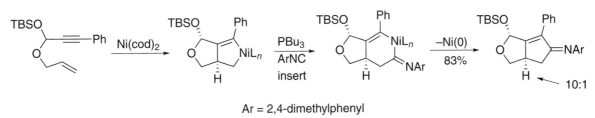

Ar = 2,4-dimethylphenyl

Perhaps more synthetically useful are nickel-catalyzed reductive enone–enone, alkyne–enone (**Eq. 4.228**),[365] alkyne–aldehyde, alkyne–imine, 1,3-diene–aldehyde (**Eq. 4.229**),[366] 1,2-diene–aldehyde, and alkyne–epoxide cyclizations to yield five-membered rings.[367] For example, alkyne–enone cyclization probably occurs via initial cyclometallation to form a five-membered nickelacycle (**Eq. 4.228**). The five-membered complex is in equilibrium with a seven-membered metallacycle. The metallacycles undergo reactions with a variety of organometallic reagents. For reactions of epoxides, an initial oxidative addition of nickel(0) to the epoxide, followed by insertion of an alkyne, has been suggested (**Eq. 4.230**).[368]

Eq. 4.228

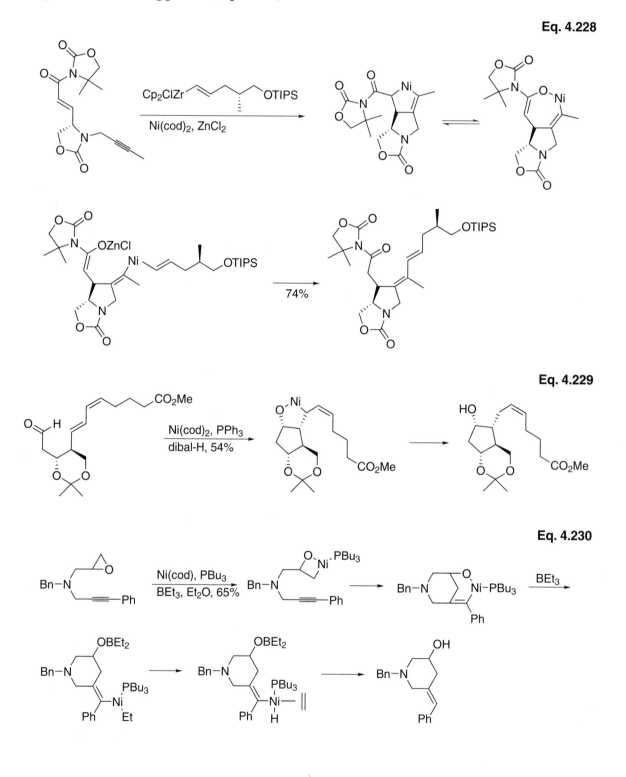

Eq. 4.229

Eq. 4.230

Palladium and platinum also cycloisomerize enynes. However, some ambiguity and overlap with the "hydrometallation" mechanism for cycloisomerization exists. The general mechanism is shown in **Eq. 4.231**. Examples of palladium- and platinum-catalyzed reactions are shown in **Eqs. 4.232**[369] and **4.233**,[370] respectively. Unusual platinum(IV) intermediates were proposed.

Eq. 4.231

Eq. 4.232

Eq. 4.233

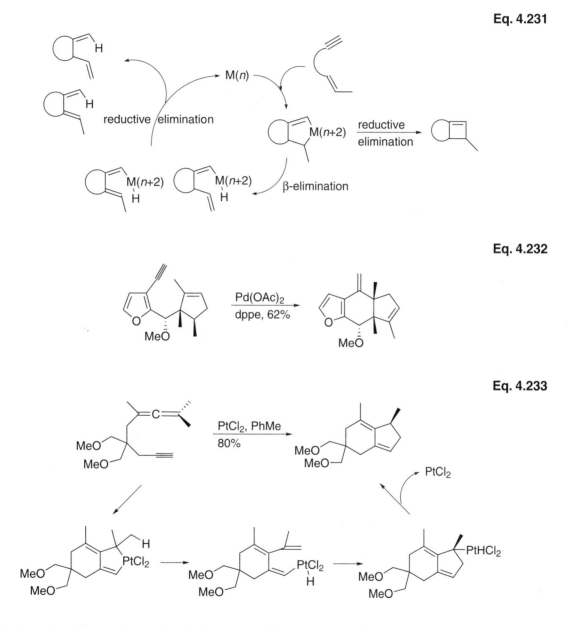

Ruthenium(I) catalyzes both intra- and intermolecular ene–yne coupling reactions. The mechanism may involve the formation of ruthenacyclopentene intermediate, followed by sequential β-elimination–reductive elimination (**Eq. 4.234**).[371] This type of reaction has been used in a number of synthetic applications (**Eqs. 4.235**[372] and **4.236**[373]).

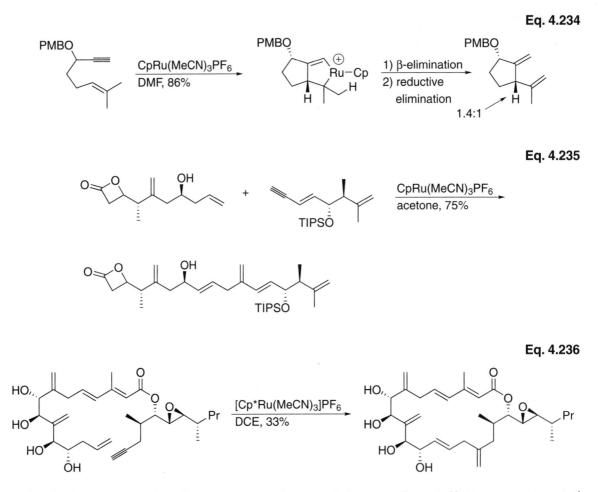

Eq. 4.234

Eq. 4.235

Eq. 4.236

Both rhodium and ruthenium complexes catalyze cyclometallation reactions of cyclopropylalkenes or -imines with alkynes or 1,2-dienes to give seven-membered cyclic compounds. Of particular synthetic interest is the intramolecular variation forming bicyclo[5.3.0]ring systems. Two mechanisms have been proposed, as depicted in **Eq. 4.237**. A computational study has suggested that the rhodium-catalyzed reactions proceed via a metallacyclohexene,[374] whereas experimental results indicate that ruthenium probably forms a metallacyclopentene.[375] An example of this type of reaction is shown in **Eq. 4.238**.[376]

Eq. 4.237

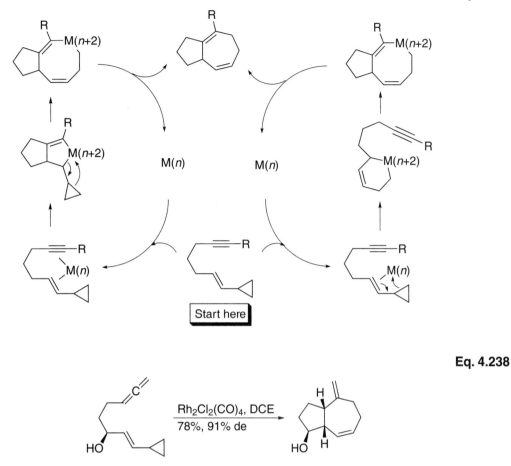

Eq. 4.238

There are a number of closely related cyclizations of unsaturated species using late transition metals that may proceed via carbeniod or η^3-allylmetal intermediates. These are discussed in Chapters 6, 8, and 9, respectively.

References

(1) Kharasch, M.S.; Tawney, P.O. *J. Am. Chem. Soc.* **1941**, *63*, 2308.

(2) For early reviews, see a) Posner, G. H. *Org. React.* **1972**, *19*, 1. b) **1979**, *22*, 253. c) *An Introduction to Synthesis Using Organocopper Reagents*; Wiley: New York, 1980.

(3) For reviews, see a) Lipshutz, B. H.; Sengupta, S. *Org. React.* **1992**, *41*, 135. b) Lipshutz, B. H. Transition Metal Alkyl Complexes from RLi and CuX. In *Comprehensive Organometallic Chemistry II*; Abel, E. W., Stone, F. G. A., Wilkinson, G., Eds.; Pergamon: Oxford, U.K., 1995; Vol. 12, pp 59–130. c) Lipshutz, B. H. Organocopper Reagents and Procedures. In *Organometallics in Synthesis. A Manual II*; Schlosser, M., Ed.; Wiley: Chichester, U.K., 1998.

(4) In a synthesis of the core of salicylhalamide. George, G. I.; Ahn, Y. M.; Blackman, B.; Farokhi, F.; Flaherty, P. T.; Mossman, C. J.; Roy, S.; Yang, K. *Chem. Commun.* **2001**, 255.

(5) McMurry, J. E.; Scott, W. J. *Tetrahedron Lett.* **1980**, *21*, 4313.

(6) In a synthesis of zincoporin. Komatsu, K.; Tanino, K.; Miyashita, M. *Angew. Chem. Int. Ed.* **2004**, *43*, 4341.

(7) McMurry, J. E.; Scott, W. J. *Tetrahedron Lett.* **1980**, *21*, 4313.

(8) Feringa, B. L.; Naasz, R.; Imbos, R.; Arnold, L. A. In *Modern Organocopper Chemistry*; Krause, N., Ed.; Wiley-VCH: New York, 2002.

(9) In a synthesis of (+)-solanascone. Srikrishna, A. Ramasastry, S. S. V. *Tetrahedron Lett.* **2005**, *46*, 7373.

(10) In a synthesis of (+)-tricycloclavulone. Ito, H.; Hasegawa, M.; Takenaka, Y.; Kobayashi, T.; Iguchi, K. *J. Am. Chem. Soc.* **2004**, *126*, 4520.

(11) For reviews, see a) Taylor, J. K. *Synthesis* **1985**, 365. b) Suzuki, M; Yanagisawa, A.; Noyori, R. *J. Am. Chem. Soc.* **1988**, *110*, 4718.

(12) Posner, G. H.; Whitten, C. E.; Sterling, J. J. *J. Am. Chem. Soc.* **1973**, *95*, 7788.

(13) Lindstedt, E.-L.; Nilsson, M.; Olsson, T. *J. Organomet. Chem.* **1987**, *334*, 255.

(14) In a synthesis of (+)-zampanolide. Smith, A. B., III; Safonov, I. G.; Corbett, R. M. *J. Am. Chem. Soc.* **2002**, *124*, 11102.

(15) a) Bertz, S. H.; Dabbagh, G.; Villacorta, G. M. *J. Am. Chem. Soc.* **1982**, *104*, 5824. b) Bertz, S. H.; Dabbagh, G. *J. Org. Chem.* **1984**, *49*, 1119.

(16) Martin, S. F.; Fishpaugh, J. R.; Power, J. M.; Giolando, D. M.; Jones, R. A.; Nunn, C. M.; Cowley, A. H. *J. Am. Chem. Soc.* **1988**, *110*, 7226.

(17) Bertz, S. H.; Eriksson, M.; Miao, G.; Snyder, J. P. *J. Am. Chem. Soc.* **1996**, *118*, 10906.

(18) In a synthesis of ricciocarpins A and B. Held, C.; Fröhlich, R.; Metz, P. *Angew. Chem. Int. Ed.* **2001**, *40*, 1058.

(19) For reviews, see a) Lipshutz, B. H.; Wilhelm, R. S.; Kozlowski, T. J. *Tetrahedron* **1984**, *40*, 5005. b) Lipshutz, B. H. *Syn. Lett.* **1990**, 119. For a different opinion, see c) Bertz, S. H. *J. Am. Chem. Soc.* **1990**, *112*, 4031. d) Snyder, J. P.; Tipsword, G. E.; Spangler, D. P. *J. Am. Chem. Soc.* **1992**, *114*, 1507. For the latest (but not necessarily the last) word, see e) Huang,

H.; Alvarez, K.; Lui, Q.; Barnhart, T. M.; Snyder, J. P.; Penner-Hahn, J. E. *J. Am. Chem. Soc.* **1996**, *118*, 8808. f) Bertz, S. H.; Miao, G.; Eriksson, M. *Chem. Commun.* **1996**, 815. g) Boche, G.; Bosold, F.; Marsch, M.; Harms, K. *Angew. Chem. Int. Ed.* **1998**, *37*, 1684.

(20) Bertz, S. H.; Nilsson, K.; Davidsson, Ö.; Snyder, J. P. *Angew. Chem. Int. Ed.* **1998**, *37*, 314.

(21) a) Lipshutz, B. H.; Siegmann, K.; Garcia, E. *J. Am. Chem. Soc.* **1991**, *113*, 8161. b) For a review, see Lipshutz, B. H.; Siegmann, K.; Garcia, E. *Tetrahedron* **1992**, *48*, 2579. c) Lipshutz, B. H.; Siegmann, K.; Garcia, E.; Kayser, P. *J. Am. Chem. Soc.* **1993**, *115*, 9276.

(22) In a synthesis of calphostins. Coleman, R. S.; Grant, E. B. *J. Am. Chem. Soc.* **1995**, *117*, 10889.

(23) In synthesis of dibenzocyclooctadiene lignans. Coleman, R. S.; Gurrala, S. R.; Mitra, S.; Raao, A. *J. Org. Chem.* **2005**, *70*, 8932.

(24) Maruyama, K.; Yamamoto, Y. *J. Am. Chem. Soc.* **1977**, *99*, 8068.

(25) Yamamoto, Y.; Maruyama, K. *J. Am. Chem. Soc.* **1978**, *100*, 3241.

(26) Ghribi, A.; Alexakis, A.; Normant, J. F. *Tetrahedron Lett.* **1984**, *25*, 3075.

(27) a) For a review, see Alexakis, A.; Mangeney, P.; Ghribi, A.; Marek, I.; Sedrani, R.; Guir, C.; Normant, J. F. *Pure Appl. Chem.* **1988**, *60*, 49. b) In a synthesis of California red scale pheromone. Mangeney, P.; Alexakis, A.; Normant, J. F. *Tetrahedron Lett.* **1987**, *28*, 2363.

(28) Ghribi, A.; Alexakis, A.; Normant, J. F. *Tetrahedron Lett.* **1984**, *25*, 3083.

(29) Smith, A. B., III; Jerris, P. J. *J. Org. Chem.* **1982**, *47*, 1845.

(30) a) Eis, M. J.; Ganem, B. *Tetrahedron Lett.* **1985**, *26*, 1153. b) Tian, X.; Hudlicky, T.; Konigsberger, K. *J. Am. Chem. Soc.* **1995**, *117*, 3643.

(31) For reviews, see a) Lopez, F.; Minnaard, A. J.; Feringa, B. L. *Acc. Chem. Res.* **2007**, *40*, 179. b) Rossiter, B. E.; Swingle, N. M. *Chem. Rev.* **1992**, *92*, 771. c) Oppolzer, W.; Kingma, A. J. *Helv. Chim Acta* **1989**, *72*, 1337. d) Oppolzer, W. *Tetrahedron* **1987**, *43*, 1969, 4057.

(32) a) Bergdahl, M.; Nilsson, M.; Olsson, T. *J. Organomet. Chem.* **1990**, *391*, C19. b) Bergdahl, M.; Nilsson, M.; Olsson, T.; Stern, K. *Tetrahedron* **1991**, *47*, 9691. c) Eriksson, M.; Johansson, A.; Nilsson, M.; Olsson, T. *J. Am. Chem. Soc.* **1996**, *118*, 10904. d) Urbom, E.; Knühl, G.; Helmchen, G. *Tetrahedron* **1996**, *52*, 971

(33) In a synthesis of the southern corn rootworm pheromone. Oppolzer, W.; Dudfield, P.; Stevenson, T.; Godel, T.; *Helv. Chim. Acta.* **1985**, *68*, 212.

(34) In a synthesis of 4,6,8,10,16,18-hexamethyldocosane. Herber, C.; Breit, B. *Angew. Chem. Int. Ed.* **2005**, *44*, 5267.

(35) a) In syntheses of (*R*)- and (*S*)-muscone. Tanaka, K.; Matsui, J.; Suzuki, H.; Watanabe, A. *J. Chem. Soc., Perkin Trans. 1* **1992**, 1193. b) For a review, see Rossiter, B. E.; Swingle, N. M. *Chem. Rev.* **1992**, *92*, 771.

(36) Alexakis, A.; Vastra, J.; Burton, J.; Benhaim, C.; Mangeney, P. *Tetrahedron Lett.* **1998**, *39*, 7869 and references therein.

(37) For reviews, see a) Alexakis, A.; Benhaim, C. *Eur. J. Org. Chem.* **2002**, 3221. b) Feringa, B. L. *Acc. Chem. Res.* **2000**, *33*, 346.

(38) In a synthesis of erogorgiaene. Cesati, R. R., III; de Armas, J.; Hoveyda, A. H. *J. Am. Chem. Soc.* **2004**, *126*, 96.

(39) For reviews, see a) Krause, N.; Hoffman-Röder, A. *Synthesis* **2001**, 171. b) Alexakis, A.; Benhaim, C. *Eur. J. Org. Chem.* **2002**, 3221.

(40) For a review, see Siemsen, P.; Livingston, R. C.; Diedrich, F. *Angew. Chem. Int. Ed.* **2000**, *39*, 2632.

(41) Alami, M.; Ferri, F. *Tetrahedron Lett.* **1996**, *37*, 2763.

(42) In a synthesis of (*R*)-strongylodiol B. Kirkham, J. E. D.; Courtney, T. D. L.; Lee, V.; Baldwin, J. E. *Tetrahedron* **2005**, *61*, 7219.

(43) Glaser, C. *Ber. Dtsch. Chem. Ges.* **1869**, *2*, 422.

(44) Rieke, R. D.; Stack, D. E.; Dawson, B. T.; Wu, T.-C. *J. Org. Chem.* **1993**, *58*, 2483.

(45) Stack, D. E.; Klein, W. R.; Rieke, R. D. *Tetrahedron Lett.* **1993**, *34*, 3063.

(46) Ekert, G. W.; Pfennig, D. R.; Suchan, S. D.; Donovan, T. A., Jr.; Aouad, E.; Tehrani, S. F.; Gunnersen, J. N.; Dong, L. *J. Org. Chem.* **1995**, *60*, 2361.

(47) Ebert, G. W.; Juda, W. L.; Kosakowski, R. H.; Ma, B.; Dong, L.; Cummings, K. E.; Phelps, M. V. B.; Mostafa, A. E.; Luo, Jianyuan. *J. Org. Chem.* **2005**, *70*, 4314.

(48) For reviews, see a) Knochel, P. Zinc and Cadmium. In *Comprehensive Organometallic Chemistry II*; Abel, E. W., Stone, F. G. A., Wilkinson, G., Eds.; Pergamon: Oxford, U.K., 1995; Vol. 11, pp 159–183. b) Knochel, P. *Synlett* **1995**, 393. c) Knochel, P.; Singer, R. *Chem. Rev.* **1993**, *93*, 2117.

(49) Soorukram, D.; Knochel, P. *Org. Lett.* **2004**, *6*, 2409.

(50) Lipshutz, B. H.; Wood, M. R.; Tirado, R. *J. Am. Chem. Soc.* **1995**, *117*, 6126.

(51) In a synthesis of (–)-prostaglandin E1 methyl ester. Arnold, L. A.; Naasz, R.; Minnaard, A. J.; Feringa, B. L. *J. Org. Chem.* **2002**, *67*, 7244.

(52) Crimmins, M. T.; Nantermet, P. G.; Trotter, B. W.; Vallin, I. M.; Watson, P. S.; McKerlie, L. A.; Reinhold, T. L.; Cheung, A. W.-H.; Stetson, K. A.; Dedopoulow, D.; Gray, J. L. *J. Org. Chem.* **1993**, *58*, 1038. Based on Nakamura, E.; Aoki, S.; Sekiya, K.; Oshino, H.; Kuwajima, I. *J. Am. Chem. Soc.* **1987**, *109*, 8056.

(53) Crimmins, M. J.; Jung, D. K.; Grail, J. L. *J. Am. Chem. Soc.* **1993**, *115*, 3146.

(54) a) Linderman, R. J.; Griedel, B. D. *J. Org. Chem.* **1990**, *55*, 5428. b) Linderman, R. J.; McKenzie, J. R. *J. Organomet. Chem.* **1989**, *361*, 31. c) Linderman, R. J.; Godfrey, A. J. *J. Am. Chem. Soc.* **1988**, *110*, 6249.

(55) Mohapatra, S.; Bandyopadhyay, A.; Barma, D. K.; Capdevila, J. H.; Falck, J. R. *Org. Lett.* **2003**, *5*, 4759.

(56) In a synthesis of 11,12,15(*S*)-trihydroxyeicosa-5(*Z*),8(*Z*),13(*E*)-trienoic acid. Falck, J. R.; Barma, D.; Mohapatra, S.; Bandyopadhyay, A.; Reddy, K. M.; Qi, J.; Campbell, W. *Bioorg. Med. Chem. Lett.* **2004**, *14*, 4987.

(57) a) Alexakis, A.; Commercon, A.; Coulentianos, C.; Normant, J. F. *Pure Appl. Chem.* **1983**, *55*, 1759. b) Normant, J .F.; Alexakis, A. *Synthesis* **1981**, 841. c)

(58) Gardette, M.; Alexakis, A.; Normant, J. F. *Tetrahedron* **1985**, *41*, 5887.

(58) Krebs, O.; Taylor, R. J. K. *Org. Lett.* **2005**, *7*, 1063.

(59) In a synthesis toward F–J fragment of gambieric acids. Clark, J. S.; Kimber, M. C.; Robertson, J. McErlean, C. S. P.; Wilson, C. *Angew. Chem. Int. Ed.* **2005**, *44*, 6157.

(60) Rao, S.A.; Knochel, P. *J. Am. Chem. Soc.* **1991**, *113*, 5735.

(61) For reviews, see a) Aubert, C.; Buisine, O.; Malacria, M. *Chem. Rev.* **2002**, *102*, 813. b) Trost, B. M.; Krische, M. J. *Synlett* **1998**, 1. c) Trost, B. M. *Acc. Chem. Res.* **1990**, *23*, 34.

(62) Bray, K. L.; Charmant, J. P. H.; Fairlamb, I. J. S.; Lloyd-Jones, G. C. *Chem.—Eur. J.* **2001**, *7*, 4205.

(63) Amatore, C.; Jutand, A.; Meyer, G.; Carelli, I.; Chiaretto, I. *Eur. J. Inorg. Chem.* **2000**, 1855.

(64) In a synthesis of picrotoxanes. Trost, B. M.; Krische, M. J. *J. Am. Chem Soc.* **1996**, *118*, 233.

(65) In a synthesis of (–)-4a,5-dihydrostreptazolin. Yamada, H.; Aoyagi, S.; Kibayashi, C. *Tetrahedron Lett.* **1996**, *37*, 8787.

(66) Yamada, H.; Aoyagi, S.; Kibayashi, C. *Tetrahedron Lett.* **1997**, *38*, 3027.

(67) For an example, see Holzapfel, C. W.; Marcus, L. *Tetrahedron Lett.* **1997**, *38*, 8585.

(68) a) Trost, B. M.; Shi, Y. *J. Am. Chem. Soc.* **1991**, *113*, 701. For full papers, see b) Trost, B. M.; Tanoury, G. J.; Lautens, M.; Chan, C.; MacPherson, D. T. *J. Am. Chem. Soc.* **1994**, *116*, 4255. c) Trost, B. M.; Romero, D. L.; Rise, F. *J. Am. Chem. Soc.* **1994**, *116*, 4268.

(69) For a review, see Tietze, L. F.; Ila, H.; Bell, H. P. *Chem. Rev.* **2004**, *104*, 3453.

(70) Hatano, M.; Mikami, K. *J. Am. Chem. Soc.* **2003**, *125*, 4704.

(71) Oh, C. H.; Jung, S. H.; Rhim, C. Y. *Tetrahedron Lett.* **2001**, *42*, 8669.

(72) In a synthesis of (+)-strepazoline. Trost, B. M.; Chung, C. K.; Pinkerton, A. B. *Angew. Chem. Int. Ed.* **2004**, *43*, 4327.

(73) Trost, B. M.; Shi, Y. *J. Am. Chem. Soc.* **1992**, *114*, 791.

(74) Mori, M.; Kozawa, Y.; Nishida, M.; Kanamaru, M.; Onozuka, K.; Takimoto, M. *Org. Lett.* **2000**, *2*, 3435.

(75) For a review, see RajanBabu, T. V. *Chem. Rev.* **2003**, *103*, 2845.

(76) In a synthesis of (–)-curcumene. Zhang, A.; RajanBabu, T. V. *Org. Lett.* **2004**, *6*, 3159.

(77) a) Zhang, A.; RajanBabu, T. V. *J. Am. Chem. Soc.* **2006**, *128*, 5620. b) Shi, W.-J.; Zhang, Q.; Xie, J.-H.; Zhu, S.-F.; Hou, G.-H.; Zhou, Q.-L. *J. Am. Chem. Soc.* **2006**, *128*, 2780.

(78) In a synthesis of the C10–C18 segment of ambruticin. He, Z.; Yi, C. S.; Donaldson, W. A. *Synlett* **2004**, 1312.

(79) For reviews, see a) Hartwig, J. F. *Pure Appl. Chem.* **2004**, *76*, 507. b) Roesky, P. W.; Mueller, T. E. *Angew. Chem. Int. Ed.* **2003**, *42*, 2708. c) Pohlki, F.; Doye, S. *Chem. Soc. Rev.* **2003**, *32*, 104.

(80) Palladium catalyzed: Johns, A. M.; Utsunomiya, M.; Incarvito, C. D.; Hartwig, J. F. *J. Am. Chem. Soc.* **2006**, *128*, 1828. Nickel catalyzed: Pawlas, J.; Nakao, Y.; Kawatsura, M.; Hartwig, J. F. *J. Am. Chem. Soc.* **2002**, *124*, 3669.

(81) Loots, M. J.; Schwartz, J. *Tetrahedron Lett.* **1978**, *19*, 4381 and *J. Am. Chem. Soc.* **1977**, *99*, 8045.

(82) Oi, S.; Sato, T.; Inoue, Y. *Tetrahedron Lett.* **2004**, *45*, 5051.

(83) For a related reaction using a chiral Michael acceptor, see Kakuuchi, A.; Taguchi, T.; Hanzawa. Y. *Tetrahedron* **2004**, *60*, 1293.

(84) a) In a synthesis of misoprostol. Babiak, K. A.; Behling, J. R.; Dygos, J. H.; McLaughlin, K. T.; Ng, J. S.; Kalish, V. J.; Kramer, S. W.; Shone, R. L. *J. Am. Chem. Soc.* **1990**, *112*, 7441. See also b) Lipshutz, B. H.; Ellsworth, E. L. *J. Am. Chem. Soc.* **1990**, *112*, 7440. c) Dygos, J. H.; Adamek, J. P.; Babiak, K. A.; Behling, J. R.; Medich, J. R.; Ng, J. S.; Wieczoerek, J. J. *J. Org. Chem.* **1991**, *56*, 2549. d) Lipshutz, B. H.; Keil, R. *J. Am. Chem. Soc.* **1992**, *114*, 7919.

(85) a) Lipshutz, B. H.; Wood, M. R. *J. Am. Chem. Soc.* **1993**, *115*, 12625. b) Lipshutz, B. H.; Wood, M. R. *J. Am. Chem. Soc.* **1994**, *116*, 11689.

(86) El-Batta, A.; Hage, T. R.; Plotkin, S.; Bergdahl, M. *Org. Lett.* **2004**, *6*, 107.

(87) a) Van Horn, D. E.; Negishi, E.-I. *J. Am. Chem. Soc.* **1978**, *100*, 2252. b) Lipshutz, B. H.; Dimock, S. H. *J. Org. Chem.* **1991**, *56*, 5761.

(88) Wipf, P.; Xu, W. J.; Smitrovich, J. H.; Lehman, R.; Venanzi, L. M. *Tetrahedron* **1994**, *50*, 1935.

(89) For a review, see Leong, W. W.; Larock, R. C., Transition Metal Alkyl Complexes: Main Group Transmetallation/Insertion Chemistry. In *Comprehensive Organometallic Chemistry II*; Abel, E. W., Stone, F. G. A., Wilkinson, G., Eds.; Pergamon: Oxford, U.K., 1995; Vol. 12, pp 131–160.

(90) Morris, I. K.; Snow, K. M.; Smith, N. W.; Smith, K. M. *J. Org. Chem.* **1990**, *55*, 1231.

(91) Takemoto, Y.; Kuraoka, S.; Ohra, T.; Yonetoku, Y.; Iwata, C. *Tetrahedron* **1997**, *53*, 603.

(92) a) Ruth, J. L.; Bergstrom, D. E. *J. Org. Chem.* **1978**, *43*, 2870. b) Bergstrom, D. E.; Inoue, H.; Reddy, P. A. *J. Org. Chem.* **1982**, *47*, 2174 c) Hacksell, U.; Daves, G. D., Jr. *J. Org. Chem.* **1983**, *48*, 2870.

(93) Somei, M.; Kawesaki, T. *Chem. Pharm. Bull.* **1989**, *37*, 3426.

(94) Outten, R. A.; Daves, G. D., Jr. *J. Org. Chem.* **1987**, *52*, 5064.

(95) Cambie, R. C.; Metzler, M. R.; Rutledge, P. S.; Woodgate, P. D. *J. Organomet. Chem.* **1992**, *429*, 59.

(96) For a review, see Hayashi, T.; Yamasaki. K. *Chem. Rev.* **2003**, *103*, 2829.

(97) Furman, B.; Dziedzic, M. *Tetrahedron Lett.* **2003**, *44*, 8249.

(98) For a review, see Farina, V. Transition Metal Alkyl Complexes: Oxidative Additions and Transmetallations. In *Comprehensive Organometallic Chemistry II*; Abel, E. W., Stone, F. G. A., Wilkinson, G., Eds.; Pergamon: Oxford, U.K., 1995; Vol. 12, pp 161–240.

(99) For a review, see Frisch, A. C.; Beller, M. *Angew. Chem. Int. Ed.* **2005**, *44*, 674.

(100) Jutand, A.; Mosleh, A. *Organometallics* **1995**, *14*, 1810.

(101) Jutand, A.; Negri, S. *Organometallics* **2003**, *22*, 4229.

(102) Herrmann, W.A.; Brossmer, C.; Reisinger, C-P.; Priermeier, T.; Beller, M.; Fischer, H. *Angew. Chem., Int. Ed. Engl.* **1995**, *34*, 1844.

(103) For a discussion and references, see Kozuch, S.; Amatore, C.; Jutand, A.; Shaik, S. *Organometallics* **2005**, *24*, 2319.

(104) a) The order of reactivity for phosphine/Pd(dba)$_2$-generated catalysts is Pd(dba)$_2$ + 2 Ph$_3$P > 1 diop > 1 dppf > 1 binap; addition of 2 equivalents of a diphosphine gives an inactive catalyst: Amatore, C.; Broeka, G.; Jutand, A.; Khalil, F. *J. Am. Chem. Soc.* **1997**, *119*, 5176. b) For a review, see Amatore, C.; Jutand, A. *Coord. Chem. Rev.* **1998**, *178–180*, 511.

(105) Amatore, C.; Carre, E.; Jutand, A.; M'Barke, M. A. *Organometallics* **1995**, *14*, 1818.

(106) For reviews, see a) Shinokubo, H.; Oshima, K. *Eur. J. Org. Chem.* **2004**, 2081. b) Tietze, L. F.; Ila, H.; Bell, H. P. *Chem. Rev.* **2004**, *104*, 3453.

(107) Tamao, K.; Sumitani, K.; Kiso, Y.; Zembayashi, M.; Fujioka, H.; Kodama, S-I.; Nakajima, I.; Minato, A.; Kumada, M. *Bull. Chem. Soc. Jpn.* **1976**, *49*, 1958.

(108) Frisch, A. C.; Rataboul, F.; Zapf, A.; Beller, M. *J. Organomet. Chem.* **2003**, *687*, 403.

(109) For a review, see Kalinin, V. N. *Synthesis* **1992**, 413.

(110) In a synthesis of (–)-taxol. Kusama, H, Hara, R.; Kawahara, S.; Nishimori, T.; Kashima, H.; Nakamura, N.; Morihara, K.; Kuwajima, I. *J. Am. Chem. Soc.* **2000**, *122*, 3811.

(111) In a synthesis of cryptotanshinone. Jiang, Y.-Y.; Li, Q.; Lu, W.; Cai, J.-C. *Tetrahedron Lett.* **2003**, *44*, 2073.

(112) Saluzzo, C.; Breuzard, J.; Pellet-Rostaing, S.; Vallet, M.; Le Guyader, F.; Lemaire, M. *J. Organomet. Chem.* **2002**, *643–644*, 98.

(113) a) Hayashi, T.; Konishi, M.; Ito, H.; Kumada, M. *J. Am. Chem. Soc.* **1982**, *104*, 4962. b) Hayashi, T.; Konishi, M.; Fukushima, M.; Kanehira, K.; Hioki, T.; Kumada, M. *J. Am. Chem. Soc.* **1983**, *48*, 2195. c) For a review, see Sawamura, M.; Ito, Y. *Chem. Rev.* **1992**, *92*, 857.

(114) a) Hayashi, T.; Niizuma, S.; Kamikawa, T.; Suzuki, N.; Uozumi, Y. *J. Am. Chem. Soc.* **1995**, *117*, 9101. b) Kamikawa, T.; Nozumi, Y.; Hayashi, T. *Tetrahedron Lett.* **1996**, *37*, 3161. c) Kamikawa, T.; Hayashi, T. *Synlett* **1997**, 163.

(115) a) Dankwardt, J. W. *Angew. Chem. Int. Ed.* **2004**, *43*, 2428. b) Kocienski, P. J.; Pritchard, M.; Wadman, S. N.; Whitby, R. J.; Yeates, C. L. *J. Chem. Soc., Perkin Trans. 1* **1992**, 3419.

(116) In a synthesis of dihydrorhipocephalin. Commeiras, L.; Valls, R.; Santelli. M.; Parrain, J.-L. *Synlett* **2003**, 1719.

(117) In a synthesis of cryptotackiene. Sundaram, G. S. M.; Venkatesh, C.; Kumar, U. K. S.; Ila, H.; Junjappa, H. *J. Org. Chem.* **2004**, *69*, 5760.

(118) For reviews, see a) Fürstner, A.; Martin, R. *Chem. Lett.* **2005**, *34*, 624. b) Blom, C.; Legros, J.; Paih, J. L.; Zani, L. *Chem. Rev.* **2004**, *104*, 6217. c) Shinokubo, H.; Oshima, K. *Eur. J. Org. Chem.* **2004**, 2081.

(119) For an iron-catalyzed coupling of primary and secondary alkyl bromides with alkyl Grignard reagents, see Dongol, C. G.; Koh, S.; Sau, M.; Chai, C. L. L. *Adv. Synth. Catal.* **2007**, *349*, 1015.

(120) Tsuji, T.; Yorimitsu, H.; Oshima, K. *Angew. Chem. Int. Ed.* **2002**, *41*, 4137

(121) In a synthesis of FTY720. Seidel, G.; Laurich, D.; Fürstner, A. *J. Org. Chem.* **2004**, *69*, 3950.

(122) In a synthesis of spirovetivanes. Maulide, N.; Vanherck, J.-C.; Marko, I. E. *Eur. J. Org. Chem.* **2004**, 3962.

(123) Bedford, R. B.; Betham, M.; Bruce, D. W.; Danopoulos, A. A.; Frost, R. M.; Hird, M. *J. Org. Chem.* **2006**, *71*, 1104 and references therein.

(124) In a synthesis of spectinabilin. Jacobsen, M. F.; Moses, J. E.; Adlington, R. M.; Baldwin, J. E. *Org. Lett.* **2005**, *7*, 2473.

(125) a) Kobayashi, M.; Negishi, E.-I. *J. Org. Chem.* **1980**, *45*, 5223. b) Negishi, E.-I.; Owczarczyk, Z. *Tetrahedron Lett.* **1991**, *32*, 6683.

(126) In a synthesis of (−)-discodermolide. Smith, A. B., III; Qiu, Y.; Jones, D. R.; Kobayashi, K. *J. Am. Chem. Soc.* **1995**, *117*, 12011.

(127) a) Organ, M. G.; Avola, S.; Dubovyk, I.; Hadei, N.; Kantchev, E. A. B.; O'Brien, C. J.; Valente, C. *Chem.—Eur. J.* **2006**, *12*, 4749. b) Zhou, J.; Fu, G. C. *J. Am. Chem. Soc.* **2003**, *125*, 12527.

(128) In a synthesis of pumilotoxins. Aoyagi, S.; Hirashima, S.; Saito, K.; Kibayashi, C. *J. Org. Chem.* **2002**, *67*, 5517.

(129) In a synthesis toward pinnatoxins. Pelc, M. J.; Zakarian, A. *Org. Lett.* **2005**, *7*, 1629.

(130) In a synthesis of bupleurynol. Ghasemi, H.; Antunes, L. M.; Organ, M. G. *Org. Lett.* **2004**, *6*, 2913.

(131) For reviews, see a) Negishi-E.-i. *J. Chem. Soc., Dalton Trans.* **2005**, 827. b) Wipf, P.; Nunes, R. L. *Tetrahedron* **2004**, *60*, 1269.

(132) a) For a review, see Negishi, E.-i. *Acc. Chem. Res.* **1982**, *15*, 340. b) Negishi, E.-i.; Okukado, N.; King, A. O.; Van Horn, D. E.; Spiegel, B. I. *J. Am. Chem. Soc.* **1978**, *100*, 2254.

(133) In a synthesis of (−)-motuporin. Hu, T.; Panek, J. S. *J. Am. Chem. Soc.* **2002**, *124*, 11368.

(134) In a synthesis of vitamin K. Huo, S.; Negishi, E.-i. *Org. Lett.* **2001**, *3*, 3253.

(135) For mechanistic studies that show that transmetallation in the Suzuki coupling goes with retention, see a) Ridgeway, B. H.; Woerpel, K. D. *J. Org. Chem.* **1998**, *63*, 458. b) Matos, K.; Soderquist, J. A. *J. Org. Chem.* **1998**, *63*, 461.

(136) For reviews, see a) Suzuki, A. *Chem. Commun.* **2005**, 4759. b) Bellina, F.; Carpita, A.; Rossi, R. *Synthesis* **2004**, 2419. c) Kotha, S.; Lahiri, K.; Kashinath, D. *Tetrahedron* **2002**, 58, 9633. d) Chemler, S. R.; Trauner, D.; Danishefsky, S. J. *Angew. Chem. Int. Ed.* **2001**, *40*, 4544. e) Miyaura, N.; Suzuki, A. *Chem. Rev.* **1995**, *95*, 2457.

(137) In a synthesis of gymnocin A. Tsukano, C.; Ebine, M.; Sasaki, M. *J. Am. Chem. Soc.* **2005**, *127*, 4326.

(138) Uenishi, J.-i.; Bean, J. M.; Armstrong, R. W.; Kishi, Y. *J. Am. Chem. Soc.* **1987**, *109*, 4756.

(139) Humphrey, J. M.; Aggen, J. B.; Chamberlin, A. R. *J. Am. Chem. Soc.* **1996**, *118*, 11759.

(140) Shen, W. *Tetrahedron Lett.* **1997**, *38*, 5575.

(141) Wright, S. W.; Hageman, D. L.; McClure, L. D. *J. Org. Chem.* **1994**, *59*, 6095.

(142) Saito, B.; Fu, G. C. *J. Am. Chem. Soc.* **2007**, *129*, 9602.

(143) For a review on its use in biaryl synthesis, see Stanforth, S. P. *Tetrahedron* **1998**, *54*, 263.

(144) In a synthesis of diazonamide A. Nicolaou, K. C.; Rao, P. B.; Hao, J.; Reddy, M. V.; Rassias, G.; Huang, X.; Chen, D. Y.-K.; Snyder, S. A. *Angew. Chem. Int. Ed.* **2003**, *42*, 1753.

(145) In a synthesis of michellamine B. a) Hobbs, P. D.; Upender, V.; Liu, J.; Pollart, D. J.; Thomas, P. W.; Dawson, M. I. *Chem. Commun.* **1996**, 923. b) Hobbs, P. D.; Upender, V.; Dawson, M. I. *Synlett* **1997**, 965.

(146) In a synthesis of OSU 6162. Lipton, M. F.; Mauragis, M. A.; Maloney, M. T.; Veley, M. F.; VanderBor, D. W.; Newby, J. J.; Appell, R. B.; Daugs, E. D. *Org. Process Res. Dev.* **2003**, *7*, 385.

(147) For reviews, see a) Snieckus, V. *Chem. Rev.* **1990**, *90*, 879. b) Snieckus, V. *Pure Appl. Chem.* **1994**, *66*, 2155.

(148) In a synthesis of hippadine. Hartung, C. G.; Fecher, A.; Chapell, B.; Snieckus, V. *Org. Lett.* **2003**, *5*, 1889.

(149) a) In a synthesis of losartan. Larsen, R. D.; King, A. O.; Chen, C. Y.; Corley, E. G.; Foster, B. S.; Roberts, F. E.; Yang, C.; Lieberman, D. R.; Reamer, R. A.; Tschaen, D. M.; Verhoeven, T. R.; Reider, P. J.; Lo, Y. S.; Rossano, L. T.; Brooks, A. S.; Meloni, D.; Moore, J. R.; Arnett, J. F. *J. Org. Chem.* **1994**, *59*, 6391. b) For a mechanistic study of this process, see Smith, G. B.; Dezeny, G. C.; Hughes, D. L.; King, A. O.; Verhoeven, T. R. *J. Org. Chem.* **1994**, *59*, 8151.

(150) Yin, J.; Buchwald, S. L. *J. Am. Chem. Soc.* **2000**, *122*, 12051.

(151) In a synthesis of korupensamine A. Watanabe, T.; Shakadou, M.; Uemura, M. *Synlett* **2000**, 1141.

(152) In a synthesis toward amphidinolide B. Gopalarathnam, A.; Nelson, S. G. *Org. Lett.* **2006**, *8*, 7.

(153) In a synthesis of (−)-callystatin A. Diaz, L. C.; Meira, P. R. R. *J. Org. Chem.* **2005**, *70*, 4762.

(154) In a synthesis of a TMC-95A analog. Kaiser, M.; Siciliano, C.; Assfalg-Machleidt, I.; Groll, M.; Milbradt, A. G.; Moroder, L. *Org. Lett.* **2003**, *5*, 3435.

(155) In a synthesis of apoptolidinone. Wu, B.; Liu, Q.; Sulikowski, G. A. *Angew. Chem. Int. Ed.* **2004**, *43*, 6673.

(156) a) Puentener, K.; Scalone, M. Eur. Pat. App. 2000 EP 1,057,831 A2, 2000 (b) Molander, G. A.; Ito, T. *Org. Lett.* **2001**, *3*, 393. c) Molander, G. A.; Rivero, M. R. *Org. Lett.* **2002**, *4*, 107. d) Molander, G. A.; Bernardi, C. *J. Org. Chem.* **2002**, *67*, 8424. e) Darses, S.; Genet, J.-P. *Eur. J. Org. Chem.* **2003**, 4313.

(157) In a synthesis of oxamidine II. Molander, G. A.; Dehmel, F. *J. Am. Chem. Soc.* **2004**, *126*, 10313.

(158) For a historical note on the Stille reaction, see Kosugi, M.; Fugami, K. *J. Organomet. Chem.* **2002**, *653*, 50.

(159) For a review, see Farina, V.; Krishnamurthy, V.; Scott, W. J. *Org. React.* **1997**, *50*, 1.

(160) For a review of the mechanism of the Stille reaction, see Espinet, P.; Echavarren, A. M. *Angew. Chem. Int. Ed.* **2004**, *43*, 4704.

(161) Hoshino, M.; Degenkolb, P.; Curran, D. P. *J. Org. Chem.* **1997**, *62*, 8341.

(162) Kende, A. S.; Roth, B.; Sanfillipo, P. J.; Blacklock, T. J. *J. Am. Chem. Soc.* **1982**, *104*, 5808.

(163) Farina, V.; Krishnan, B. *J. Am. Chem. Soc.* **1991**, *113*, 9585.

(164) a) Liebeskind, L. S.; Fengl, R. W. *J. Org. Chem.* **1990**, *55*, 5359. b) Farina, V.; Kapadia, S.; Krishnan, B.; Wang, C.; Liebeskind, L. S. *J. Org. Chem.* **1994**, *59*, 5905.

(165) In a synthesis of methyl sarcoate. Ichige, T.; Kamimura, S.; Mayumi, K.; Sakamoto, Y.; Terashita, S.; Ohteki, E.; Kanoh, N.; Nakata, M. *Tetrahedron Lett.* **2005**, *46*, 1263.

(166) a) Labadie, J. W.; Teuting, D.; Stille, J. K. *J. Org. Chem.* **1983**, *48*, 4634. b) Labadie, J. W.; Stille, J. K. *J. Am. Chem. Soc.* **1983**, *109*, 669.

(167) Stille, J. K.; Simpson, J. H. *J. Am. Chem. Soc.* **1987**, *109*, 2138.

(168) In a synthesis of callipeltoside A. Trost, B. M.; Gunzner, J. L.; Dirat, O.; Rhee, Y. H. *J. Am. Chem. Soc.* **2002**, *124*, 10396.

(169) Stille, J.K.; Groh, B.L. *J. Am. Chem. Soc.* **1987**, *109*, 813.

(170) In a synthesis of (–)-rapamycin. Smith, A. B., III; Condon, S. M.; McCauley, J. A.; Leazer, J. L., Jr.; Leahy, J. W.; Maleczka, R. E., Jr. *J. Am. Chem. Soc.* **1997**, *119*, 962.

(171) In a synthesis of (+)-mycotrienol. Panek, J. S.; Masse, C. E. *J. Org. Chem.* **1997**, *62*, 8290.

(172) In a synthesis of rhizoxin D. Mitchell, I. S.; Pattenden, G.; Stonehouse, J. *Org. Biomolec. Chem.* **2005**, *3*, 4412.

(173) In a synthesis of chloropeptin I. Deng, H.; Jung, J.-K.; Liu, T.; Kuntz, K. W.; Snapper, M. L.; Hoveyda, A. H. *J. Am. Chem. Soc.* **2003**, *125*, 9032.

(174) Shair, M.D.; Yoon, T.; Danishefsky, S.J. *J. Org. Chem.* **1994**, *59*, 3755.

(175) a) For a review on enol triflates, see Scott, W. J.; McMurray, J. E. *Acc. Chem. Res.* **1988**, *21*, 47. b) Echavarren, A. M.; Stille, J. K. *J. Am. Chem. Soc.* **1987**, *109*, 5478.

(176) Roth, G. P.; Fuller, C. E. *J. Org. Chem.* **1991**, *56*, 3493.

(177) Nicolaou, K. C.; Shi, G.-Q.; Gunzner, J. L.; Gärtner, P.; Yang, Z. *J. Am. Chem. Soc.* **1997**, *119*, 5467.

(178) In a synthesis of (–)-austalide B. Paquette, L. A.; Wang, T.-Z.; Sivik, M. R. *J. Am. Chem. Soc.* **1994**, *116*, 11323.

(179) Nicolaou, K. C.; Sato, M.; Miller, N. D.; Gunzner, J. L.; Renaud, J.; Untersteller, E. *Angew. Chem., Int. Ed. Engl.* **1996**, *39*, 889.

(180) Farina, V.; Krishnan, B.; Marshall, D. R.; Roth, G. P. *J. Org. Chem.* **1993**, *59*, 5434.

(181) Kosugi, M.; Shimizu, Y.; Migita, T. *Chem. Lett.* **1977**, 1423.

(182) Ragan, J. A.; Raggon, J. W.; Hill, P. D.; Jones, B. P.; McDermott, R. E.; Munchhof, M. J.; Marx, M. A.; Casavant, J. M.; Cooper, B. A.; Doty, J. L.; Lu, Y. *Org. Process Res. Dev.* **2003**, *7*, 676.

(183) In a synthesis of (–)-motuporin. Hu, T.; Panek, J. S. *J. Am. Chem. Soc.* **2002**, *124*, 11368.

(184) Azarian, D.; Dua, S. S.; Eaborn, C.; Walton, D. R. M. *J. Organomet. Chem.* **1976**, *117*, C55.

(185) In a synthesis of the heptacyclic core of (–)-nodulisporic acid D. Smith, A. B., III; Davulcu, A. H.; Kürti, L. *Org. Lett.* **2006**, *8*, 1669.

(186) a) Gallagher, W. P.; Terstiege, I.; Maleczka, R. E., Jr. *J. Am. Chem. Soc.* **2001**, *123*, 3194 b) Gallagher, W. P.; Maleczka, R. E., Jr. *J. Org. Chem.* **2005**, *70*, 841.

(187) a) Tang, H.; Menzel, K.; Fu, G. C. *Angew. Chem. Int. Ed.* **2003**, *42*, 5079. b) Menzel, K.; Fu, Gregory C. *J. Am. Chem. Soc.* **2003**, *125*, 3718.

(188) In a synthesis of (+)-himbacin. Hofman, S.; Gao, L.-J.; Van Dingenen, H.; Hosten N. G. C.; Van Haver, D.; De Clercq, P. J.; Milanesio, M. Viterbo, D. *Eur. J. Org. Chem.* **2001**, 2851.

(189) Allred, G. D.; Liebeskind, L. S. *J. Am. Chem. Soc.* **1996**, *118*, 2748.

(190) In a synthesis of apoptolidin. Wehlan, H.; Dauber, M.; Fernaud, M.-T. M.; Schuppan, J.; Mahrwald, R.; Ziemer, B.; Garcia, M.-E. J.; Koert, U. *Angew. Chem. Int Ed.* **2004**, *43*, 4597.

(191) Powell, D. A.; Maki, T.; Fu, G. C. *J. Am. Chem. Soc.* **2005**, *127*, 510.

(192) Belema, M.; Nguyen, V. N.; Zusi, F. C. *Tetrahedron Lett.* **2004**, *45*, 1693.

(193) Hiyama, T. *Syn. Lett.* **1991**, 845.

(194) Hagiwara, E.; Gowda, K.-i.; Hatanaka, Y.; Hiyama, T. *Tetrahedron Lett.* **1997**, *38*, 439.

(195) Denmark, S. E.; Choi, J. Y. *J. Am. Chem. Soc.* **1999**, *121*, 5821.

(196) For reviews, see a) Handy, C. J.; Manoso, A. S.; McElroy, W. T.; Seganish, W. M.; DeShong, P. *Tetrahedron* **2005**, *61*, 12201. b) Denmark, S. E.; Sweis, R. F. *Acc. Chem. Res.* **2002**, *35*, 835. c) Denmark, S. E.; Sweis, R. F. *Chem. Pharm. Bull.* **2002**, *50*, 1531. (d) Hiyama, T.; Shirakawa, E. *Top. Curr. Chem.* **2002**, *219*, 61. For a historical review, see e) Hiyama, T. *J. Organomet. Chem.* **2002**, *653*, 58.

(197) Tamao, K.; Kobayashi, K.; Ito, Y. *Tetrahedron Lett.* **1989**, *30*, 6051.

(198) Lee, H. M.; Nolan, S. P. *Org. Lett.* **2000**, *2*, 2053.

(199) In a synthesis of (+)-brasilenyne. Denmark, S. E.; Yang, S.-M. *J. Am. Chem. Soc.* **2002**, *124*, 15196.

(200) For reviews, see a) Negishi, E.-i.; Anastasia, L. *Chem. Rev.* **2003**, *103*, 1979. b) Chinchilla, R. Najera, C. *Chem. Rev.* **2007**, *107*, 874.

(201) a) For the total syntheses of these compounds, see Magnus, P.; Carter, P.; Elliot, J.; Lewis, R.; Harling, J.; Pitterns, T.; Batwa, W. E.; Fort, S. *J. Am. Chem. Soc.* **1992**, *114*, 2544. b) For a review, see Maier, M. E. *Synlett* **1995**, 13.

(202) a) Wender, P. A.; Beckham, S.; O'Leary, J.-G. *Synthesis* **1995**, 1279. b) Nishikawa, T.; Yoshikai, M.; Kawai, T.; Unno, R.; Jomori, T.; Isobe, M. *Tetrahedron* **1995**, *51*, 9339.

(203) Jacobi, P. A.; Guo, J. *Tetrahedron Lett.* **1995**, *36*, 2717.

(204) In a synthesis of callipeltoside aglycon. Nagata, H.; Miyazawa, N.; Ogasawara, K. *Org. Lett.* **2001**, *3*, 1737.

(205) In a synthesis toward C-1027. Inoue, M.; Sasaki, T.; Hatano, S.; Hirama, M. *Angew. Chem. Int. Ed.* **2004**, *43*, 6500.

(206) a) De la Rosa, M. A.; Velardi, E.; Grizman, A. *Synth. Comm.* **1990**, *20*, 2059. b) Bleicher, L.; Cosford, N. P. P. *Synlett* **1995**, 1115.

(207) Urgaonkar, S.; Verkade, J. G. *J. Org. Chem.* **2004**, *69*, 5752.

(208) For a discussion of the mechanism of the copper free reaction, see Tiugerti, A.; Negri, S.; Jutand, A. *Chem.—Eur. J.* **2007**, *13*, 666.

(209) Bumagin, N.; Sukholminova, L. I.; Luzckova, E. V.; Tolstaya, T. P.; Beletskaya, I. P. *Tetrahedron Lett.* **1996**, *37*, 897.

(210) Powell, N. A.; Rychnovsky, S. D. *Tetrahedron Lett.* **1996**, *37*, 7901.

(211) Eckhardt, M.; Fu, G. C. *J. Am. Chem. Soc.* **2003**, *125*, 13642.

(212) Altenhoff, G.; Würtz, S.; Glorius, F. *Tetrahedron Lett.* **2006**, *47*, 2925.

(213) In a synthesis of bipinnatin J. Roethle, P. A.; Trauner, D. *Org. Lett.* **2006**, *8*, 345.

(214) Mulzer, J.; Strecker, A. R.; Kattner, L. *Tetrahedron Lett.* **2004**, *45*, 8867.

(215) In a synthesis of briarellins E and F. Corminboef, O.; Overman, L. E.; Pennington, L. D. *J. Am. Chem. Soc.* **2003**, *125*, 6650.

(216) In a synthesis of the FGHI-ring part of ciguatoxins. Takizawa, A.; Fujiwara, K.; Doi, E.; Murai, A.; Kawai, H.; Suzuki, T. *Tetrahedron Lett.* **2006**, *47*, 747.

(217) Inoue, M.; Nakada, M. *Org. Lett.* **2004**, *6*, 2977.

(218) Inoue, M.; Nakada, M. *Agew. Che. Int. Ed.* **2006**, *45*, 252.

(219) Xia, G.; Yamamoto, H. *J. Am. Chem. Soc.* **2006**, *128*, 2554.

(220) Kosugi, M.; Kamezawa, M.; Migita, T. *Chem. Lett.* **1983**, 927.

(221) a) Wolfe, J. P.; Wagaw, S.; Buchwald, S. L. *J. Am. Chem. Soc.* **1996**, *118*, 7215. b) Wolfe, J. P.; Buchwald, S. L. *J. Org. Chem.* **1996**, *61*, 1133. c) For triflates, see Wolfe, J. P.; Buchwald, S. L. *J. Org. Chem.* **1997**, *62*, 1264. d) Driver, M. J.; Hartwig, J. F. *J. Am. Chem. Soc.* **1996**, *118*, 7217. e) Louis, J.; Driver, M. S.; Hamann, B. C.; Hartwig, J. F. *J. Org. Chem.* **1997**, *62*, 1268. f) Hartwig, J. F. *Synlett* **1997**, 329. g) Wolfe, J. P.; Buchwald, S. L. *J. Org. Chem.* **1997**, *62*, 6066. h) Wolfe, J. P.; Wagaw, S.; Marcoux, J.-F.; Buchwald, S. L. *Acc. Chem. Res.*, **1998**, *31*, 805.

(222) Åhman, J; Buchwald, S. L. *Tetrahedron Lett.* **1997**, *38*, 6363.

(223) a) Reddy, N. P.; Tanaka, M. *Tetrahedron Lett.* **1997**, *38*, 4807. b) Beller, M.; Riermeier, T. H.; Reisinger, C.-P.; Hermann, W. A. *Tetrahedron Lett.* **1997**, *38*, 2073.

(224) Wolfe, J. P.; Buchwald, S. L. *J. Am. Chem. Soc.* **1997**, *119*, 6054.

(225) Stover, J. S.; Rizzo, C. J. *Org. Lett.* **2004**, *6*, 4985.

(226) In a synthesis of spicamycin. Suzuki, T.; Tanaka, S.; Yamada, I.; Koashi, Y.; K. Yamada, N. Chida, *Org. Lett.* **2000**, *2*, 1137.

(227) In synthesis of fumiquinazolines. Snider, B. B.; Zheng, H. *J. Org. Chem.* **2003**, *68*, 545.

(228) For a review, see Barluenga, J.; Valdes, C. *Chem. Commun.* **2005**, 4891.

(229) For a review, see Beletskaya, I. P.; Cheprakov, A. V. *Coord. Chem. Rev.* **2004**, *248*, 2337.

(230) Shekar, S.; Ryberg, P.; Hartwig, J. F.; Mathew, J. S.; Blackmond, D. G.; Streiter, E. R.; Buchwald, S. L. *J. Am. Chem. Soc.* **2006**, *128*, 3584.

(231) a) Shen, Q.; Hartwig, J. F. *J. Org. Chem.* **2006**, *128*, 10028. b) Surry, D. S.; Buchwald, S. L. *J. Am. Chem. Soc.* **2007**, *129*, 10354.

(232) a) Anderson, K. W.; Ikawa, T.; Tundel, R. E.; Buchwald, S. L. *J. Am. Chem. Soc.* **2006**, *128*, 10694. b) For reactions using K_3PO_4-H_2O, see Chen, G.; Chan, A. S. C.; Kwong, F. Y. *Tetrahedron Lett.* **2007**, *48*, 473.

(233) In a synthesis of U86192A. Chae, J.; Buchwald, S. L. *J. Org. Chem.* **2004**, *69*, 3336.

(234) In a synthesis of chuangxinmycins. Kato, K.; Ono, M.; Akita, H. *Tetrahedron* **2001**, *57*, 10055.

(235) In a synthesis of sarpagine alkaloids. Zhou, H.; Liao, X.; Yin, W.; Cook, J. M. *J. Org. Chem.* **2006**, *71*, 251.

(236) In a synthesis of cherylline. Honda, T.; Namiki, H.; Satoh, F. *Org. Lett.* **2001**, *3*, 631.

(237) In a synthesis toward azadirachtin. Nicolaou, K. C.; Sasmal, P. K.; Koftis, T. V.; Converso, A.; Loiz-idou, E.; Kaiser, F.; Roecker, A. J.; Dellios, C. C.; Sun, X.-W.; Petrovic, G. *Angew. Chem. Int. Ed.* **2005**, *44*, 3447.

(238) In a synthesis toward FR182877. Clark, P. A.; Grist, M.; Ebden, M. *Tetrahedron Lett.* **2004**, *45*, 927.

(239) Four, P.; Guibe, F. *J. Org. Chem.* **1981**, *46*, 4439.

(240) Zhang, H. X.; Guibe, F.; Balavoine, G. J. *J. Org. Chem.* **1990**, *55*, 1857.

(241) In a synthesis of nafuredin γ. Nagamitsu, T.; Takano, D.; Shiomi, K.; Ui, H.; Yamaguchi, Y.; Masuma, R.; Harigaya, Y.; Kuwajima, I.; Omura, S. *Tetrahedron Lett.* **2003**, *44*, 6441.

(242) a) Roth, G. P.; Thomas, J. A. *Tetrahedron Lett.* **1992**, *33*, 1959. b) Cacchi, S.; Ciattini, P. G.; Moreta, E.; Ortar, G. *Tetrahedron* **1986**, *27*, 3931. c) Dolle, R. E.; Schmidt, S. J.; Kruse, L. I. *J. Chem. Soc., Chem. Commun.* **1987**, 904.

(243) In a synthesis of welwitindolinone A isonitrile. Reisman, S. E.; Ready, J. M.; Hasuoka, A.; Smith, C. J.; Wood, J. L. *J. Am. Chem. Soc.* **2006**, *128*, 1448.

(244) McGuire, M. A.; Sorenson, E.; Owings, F. W.; Resnick, T. M.; Fox, M.; Baine, N. H. *J. Org. Chem.* **1994**, *59*, 6683.

(245) Tilley, J. W.; Coffen, D. L.; Schaer, B. H.; Lind, J. *J. Org. Chem.* **1987**, *52*, 2469.

(246) a) Cowell, A.; Stille, J. K. *J. Am. Chem. Soc.* **1980**, *102*, 4193. b) Martin, L. D.; Stille, J. K. *J. Org. Chem.* **1982**, *47*, 3630.

(247) In a synthesis of (+)-manoalide. Pommier, A.; Stepanenko, V.; Jarowicki, K.; Kocienski, P. J. *J. Org., Chem.* **2003**, *68*, 4008.

(248) Mori, M.; Chiba, K.; Ban, Y. *J. Org. Chem.* **1978**, *43*, 1684.

(249) In a synthesis toward FR900482. Trost, B. M.; Amerkis, M. K. *Org. Lett.* **2004**, *6*, 1745.

(250) a) Mori, M.; Chiba, K.; Okita, M.; Ban, Y. *J. Chem. Soc., Chem. Commun.* **1979**, 698. b) *Tetrahedron* **1985**, *41*, 387.

(251) Kotsuki, H.; Datta, P. K.; Suenaga, H. *Synthesis* **1996**, 470.

(252) Baillargeon, V. P.; Stille, J. K. *J. Am. Chem. Soc.* **1986**, *108*, 452.

(253) In a synthesis toward nodulisporic acid. Smith, A. B., III; Cho, Y. S.; Ishiyama, H. *Org. Lett.* **2001**, *3*, 3971.

(254) In a synthesis of lapidilectine B. Pearson, W. H.; Lee, I. Y.; Stoy, P. *J. Org. Chem.* **2004**, *69*, 9109.

(255) Jackson, R. F. W.; Turner, D.; Block, M. H. *J. Chem. Soc., Perkin Trans. 1* **1997**, 865.

(256) In a synthesis of the phomactin core. Houghton, T. J.; Choi, S.; Rawal, V. H. *Org. Lett.* **2001**, *3*, 3615.

(257) The reaction was discovered independently. a) Mizoroki, T.; Mori, K.; Ozaki, A. *Bull. Chem. Soc. Jpn.* **1971**, *44*, 581. b) Heck, R. F.; Nolley, J. P., Jr. *J. Org. Chem.* **1972**, *37*, 2320.

(258) For reviews, see a) Heck, R. F. *Org. React.* **1982**, *27*, 345. b) Heck, R. F. *Palladium Reagents in Organic Synthesis*; Academic Press: London, U.K., 1985. c) Söderberg, B. C. Transition Metal Alkyl Complexes: Oxidative Addition and Insertion. In *Comprehensive Organometallic Chemistry II*; Abel, E. W., Stone, F. G. A., Wilkinson, G., Eds.; Pergamon: Oxford, U.K., 1995; Vol. 12, pp 241–297. d) Overman, L. E. *Pure Appl. Chem.* **1994**, *66*, 1423. e) de Meijere, E.; Meyer, F. E. *Angew. Chem., Int. Ed. Engl.*

1994, *33*, 2379. f) Cabri, W.; Candiani, I. *Acc. Chem. Res.* **1995**, *28*, 2.

(259) For a mechanistic review, see Knowles. J. P.; Whiting, A. *Org. Biomolec. Chem.* **2007**, *5*, 31.

(260) McCrindle, R.; Ferguson, G.; Arsenault, G. J.; McAlees, A. J.; Stephenson, D. K. *J. Chem. Res., Synop.* **1984**, 360.

(261) Andersson, C. M.; Karabelas, K.; Hallberg, A.; Andersson, C. *J. Org. Chem.* **1985**, *50*, 3891.

(262) a) Jeffry, T. *Tetrahedron Lett.* **1985**, *26*, 2667. b) Larock, R. C.; Baker, B. E. *Tetrahedron Lett.* **1988**, *29*, 905. c) Jeffrey, T. *J. Chem. Soc., Chem. Comm.* **1984**, 1287. (d) Jeffry, T. *Synthesis* **1987**, 70.

(263) Amatore, C.; Azzabi, M.; Jutand, A. *J. Am. Chem. Soc.* **1991**, *113*, 8375.

(264) a) Jeffery, T. *Tetrahedron* **1996**, *52*, 10113. b) Casalnuovo, W. L.; Calabrese, J. C. *J. Am. Chem. Soc.* **1990**, *112*, 4324. c) Genet, J. P.; Blast, E.; Savignac, M. *Synlett* **1992**, 1715. d) Jeffry, T. *Tetrahedron Lett.* **1994**, *35*, 3501. e) Dibowski, H.; Schmidchen, F. P. *Tetrahedron* **1995**, *51*, 2325.

(265) Herrmann, W. A.; Brossmer, C.; Reisinger, C.-P.; Riermeier, T. H.; Ofele, K.; Beller, M. *Chem.—Eur. J.* **1997**, *3*, 1357.

(266) a) Spencer, A. *J. Organomet. Chem.* **1983**, *258*, 101; **1984**, *265*, 323; **1984**, *270*, 115. See also b) Ohff, M.; Ohff, A.; van der Boom, M. E.; Milstein, D. *J. Am. Chem. Soc.* **1997**, *119*, 11687.

(267) a) Daves, G. D., Jr.; Hallberg, A. *Chem. Rev.* **1989**, *89*, 1433. b) Daves, G. D. *Acc. Chem. Res.* **1990**, *23*, 201.

(268) a) Cabri, W.; Candiani, I.; Bedeschi, A.; Penco, S. *J. Org. Chem.* **1992**, *57*, 1481. b) Cabri, W.; Candiani, I.; Bedeschi, A.; Santi, R. *J. Org. Chem.* **1992**, *57*, 3558.

(269) In a synthesis of ratjadone. Bhatt, U.; Christmann, M.; Qitschalle, M.; Claus, E.; Kalesse, M. *J. Org. Chem.* **2001**, *66*, 1885.

(270) Tietze, L. F.; Nöbel, T.; Speacha, M. *Angew. Chem., Int. Ed. Engl.* **1996**, *35*, 2259.

(271) In a synthesis of nabumeton. Aslam, M.; Elango, V.; Davenport, K. G. *Synthesis* **1989**, 869.

(272) For a review, see Link, J. T. *Org React.* **2002**, *60*, 157.

(273) Okita, T.; Isobe, M. *Tetrahedron* **1994**, *50*, 11143.

(274) a) Rigby, J. H.; Hughes, R. J.; Heeg, M. J. *J. Am. Chem. Soc.* **1995**, *117*, 7834. b) Bombrun, A.; Sageot, O. *Tetrahedron Lett.* **1997**, *38*, 1057.

(275) Overman, L. E.; Abelman, M. M.; Kucera, D. J.; Tran, V. D.; Ricca, D. J. *Pure Appl. Chem.* **1992**, *64*, 813.

(276) Hiroshige, M.; Hauske, J. R.; Zhou, R. *J. Am. Chem. Soc.* **1995**, *117*, 11590.

(277) a) In a synthesis of taxol. Masters, J. J.; Link, J. T.; Snyder, L. B.; Young, W. B.; Danishefsky, S. J. *Angew. Chem., Int. Ed. Engl.* **1995**, *34*, 1723. b) Young, W. B.; Masters, J. J.; Danishefsky, S. J. *J. Am. Chem. Soc.* **1995**, *117*, 5228.

(278) In a synthesis of ecteinascidin 743. Endo, A.; Yanagisawa, A.; Abe, M.; Tohma, S.; Kan, T.; Fukuyama, T. *J. Am. Chem. Soc.* **2002**, *124*, 6552.

(279) In a synthesis of diazonamide A derivatives. Li, J.; Chen, X.; Burgett, A. W. G.; Harran, P. G. *Angew. Chem. Int. Ed.* **2001**, *40*, 2682.

(280) In a synthesis toward cripowellins A and B. Enders, D.; Lenzen, A.; Backes, M.; Janeck, C.; Catlin, K.; Lannou, M.-I.; Runsik, J.; Raabe, G. *J. Org. Chem.* **2005**, *70*, 10538.

(281) Firmansjah, L.; Fu, G. C. *J. Am. Chem. Soc.* **2007**, *129*, 11340.

(282) For reviews, see a) Shibasaki, M.; Boden, C.D.J.; Kojima, A. *Tetrahedron* **1997**, *53*, 7371. b) Dounay, A. B.; Overman, L. E. *Chem Rev.* **2003**, *103*, 2945.

(283) In a synthesis of (–)-eptazocine. Takemoto, T.; Sodeoka, M.; Sasai, H.; Shibasaki, M. *J. Am. Chem. Soc.* **1993**, *115*, 8477.

(284) In a synthesis of (+)-vernolepin. Ohrai, K.; Kondo, K.; Sodeoka, M.; Shibasaki, M. *J. Am. Chem. Soc.* **1994**, *116*, 11737.

(285) a) Loiseleur, O.; Meier, P.; Pfaltz, A. *Angew. Chem., Int. Ed. Engl.* **1996**, *35*, 200. b) Loiseleur, O.; Hayashi, M.; Sames, N.; Pfaltz, A. *Synthesis* **1997**, 1338. See also c) Trabesinger, G.; Albinati, A.; Feiken, N.; Kunz, R. W.; Pregosin, P. S.; Tschoerner, M. *J. Am. Chem. Soc.* **1997**, *119*, 6315.

(286) In a synthesis of (–)-quadrigemin C. Lebsack, A. D.; Link, J. T.; Overman, L. E.; Stearns, B. A. *J. Am. Chem. Soc.* **2002**, *124*, 9008.

(287) McClure, M. S.; Glover, B.; McSorley, E.; Millar, A.; Ostenhout, M. H.; Roschagar, F. *Org. Lett.* **2001**, *3*, 1677.

(288) For reviews, see a) Seregin, I. V.; Gevorgyan, V. *Chem. Soc. Rev.* **2007**, *36*, 1173. b) Campeau, L.-S.; Fagnou, K. *Chem. Commun.* **2006**, 1253. c) Echavarren, A. M.; Gomez-Lor, B.; Gonzales, J. J.; de Frutos, O. *Synlett* **2003**, 585.

(289) Garcia-Cuadrado, D.; Braga, A. A. C.; Maseras, F.; Echavarren, A. M. *J. Am. Chem. Soc.* **2006**, *128*, 1066.

(290) Mota, A. J.; Didieu, A.; Bour, C.; Suffert, J. *J. Am. Chem. Soc.* **2005**, *127*, 7171.

(291) Campeau, L.-C.; Parisien, M.; Jean, A.; Fagnou, K. *J. Am. Chem. Soc.* **2006**, *128*, 581.

(292) Lafrance, M.; Fagnou, K. *J. Am. Chem. Soc.* **2006**, *128*, 16496.

(293) In a synthesis of graphislactone B. Abe, H.; Nishioka, K.; Takeda, S.; Arai, M.; Takeuchi, Y.; Harayama, T. *Tetrahedron Lett.* **2005**, *46*, 3197.

(294) In a synthesis of (–)-frodosin. Hughes, C. C.; Trauner, D. *Angew. Chem. Int. Ed.* **2002**, *41*, 1569.

(295) For reviews, see a) Heumann, A.; Reglier, M. *Tetrahedron* **1996**, *52*, 9289. b) Grigg, R.; Sridharan, V. Transition Metal Alkyl Complexes: Multiple Insertion Cascades, In *Comprehensive Organometallic Chemistry II*; Abel, E.W., Stone, F. G. A., Wilkinson, G., Eds.. Pergamon: Oxford, U.K., 1995; Vol. 12, pp 299–321.

(296) In a synthesis of (+)-phorbol. Lee, K.; Cha, J. K. *J. Am. Chem. Soc.* **2001**, *123*, 5590.

(297) Negishi, E.-i.; Sawada, H.; Tour, J. M.; Wei, Y. *J. Org. Chem.* **1988**, *53*, 913.

(298) For a review, see Catellani, M. *Synlett* **2003**, 298.

(299) a) Coperet, C.; Ma, S.; Negishi, E.-i. *Angew. Chem., Int. Ed. Engl.* **1996**, *35*, 2125. For full papers on the subject, see b) Coperet, C.; Ma, S.; Sugihara, T.; Negishi, E.-i. *Tetrahedron* **1996**, *52*, 11529. c) Negishi, E.-i.; Ma, S.; Amanfu, J.; Coperet, C.; Miller, J. A.; Tour, J. M. *J. Am. Chem. Soc.* **1996**, *118*, 5919.

(300) In a synthesis of (+)-xestoquinone. Maddaford, S. P.; Andersen, N. G.; Cristofoli, W. A.; Keays, B. A. *J. Am. Chem. Soc.* **1996**, *118*, 10766.

(301) For reviews, see a) de Meijere, A; von Zezschwitz, P.; Braese, S. *Acc. Chem. Res.* **2005**, *38*, 413. b) Poli, G.; Giambastiani, G.; Heumann, A. *Tetrahedron*

2000, *56*, 5959. c) Malacria, M. *Chem. Rev.* **1996**, *96*, 289. d) Ojima, I.; Tzamarioudalsi, M.; Li, Z.; Donovan, R.J. *Chem. Rev.* **1996**, *96*, 635. e) Negishi, E.-i.; Copret, C.; Ma, S.; Liou, S.-Y.; Liu, F. *Chem. Rev.* **1996**, *96*, 365.

(302) Zhang, Y.; Wu, G. Z.; Agnel, G.; Negishi, E.-i. *J. Am. Chem. Soc.* **1990**, *112*, 8590.

(303) Meyer, F. E.; Parsons, P. J.; de Meijere, A. *J. Org. Chem.* **1991**, *56*, 6487.

(304) Grigg, R.; Logananthan, V.; Sridharan, V. *Tetrahedron Lett.* **1996**, *37*, 3399.

(305) Oda, H.; Kobayashi, T.; Kosugi, M.; Migita, T. *Tetrahedron* **1995**, *51*, 695.

(306) In a synthesis of halenaquinol. Kojima, A.; Takemoto, T.; Sodeoka, M.; Shibasaki, M. *J. Org. Chem.* **1996**, *61*, 4876.

(307) a) In syntheses of 12-methoxy-substituted sarpagine alkaloids. Zhou, H.; Liao, X.; Yin, W.; Ma, J.; Cook, J. M. *J. Org. Chem.* **2006**, *71*, 251. See also b) Larock, R. C.; Yum, E. K. *J. Am. Chem. Soc.* **1991**, *113*, 6689. c) Park, S. S.; Chgoi, J.-K.; Yum, E. K. *Tetrahedron Lett.* **1998**, *39*, 627.

(308) For reviews, see a) Dupont, J.; Consorti, C. S.; Spencer, J. *Chem. Rev.* **2005**, *105*, 2527. b) Ryabov, A. D. *Synthesis* **1985**, 233. c) Omae, I. *Chem. Rev.* **1979**, *79*, 287. d) Pfeffer, M.; Dehand, P. *Coord. Chem. Rev.* **1976**, *18*, 327.

(309) For reviews on C–H activation, see a) Ritleng, V.; Sirlin, C.; Pfeffer, M. *Chem. Rev.* **2002**, *102*, 1731. b) Dyker, G. *Angew. Chem. Int. Ed.* **1999**, *38*, 1699. (c) Jun, C. H.; Moon, C. W.; Lee, D.-Y. *Chem.—Eur. J.* **2002**, *8*, 2423.

(310) Brisdon, B. J.; Nair, P.; Dyke, S. F. *Tetrahedron* **1981**, *37*, 173.

(311) Horino, H.; Inoue, N. *J. Org. Chem.* **1981**, *46*, 4416.

(312) In a synthesis of aphidicolin. Justicia, J.; Oltra, J. E.; Cuerva, J. M. *J. Org. Chem.* **2005**, *70*, 8265.

(313) Dangel, B. D.; Godula, K.; Youn, S. W.; Sezen, B.; Sames, D. *J. Am. Chem. Soc.* **2002**, *124*, 11856.

(314) Orito, K.; Miyazawa, M.; Nakamura, T.; Horibata, A.; Ushito, H.; Nagasaki, H.; Yuguchi, M.; Yamashita, S.; Yamazaki, T.; Tokuda, M. *J. Org. Chem.* **2006**, *71*, 5951.

(315) Wan, X.; Ma, Z.; Li, B.; Zhang, K.; Cao, S.; Zhang, S.; Shi, Z. *J. Am. Chem. Soc.* **2006**, *128*, 7416.

(316) Giri, R.; Maugel, N.; Li, J.-J.; Wang, D.-H.; Breazzano, S. P.; Saunders, L. B.; Yu, J.-Q. *J. Am. Chem. Soc.* **2007**, *129*, 3510.

(317) Kong, A.; Han, X.; Lu, X. *Org. Lett.* **2006**, *8*, 1339.

(318) In a synthesis of austamides. Baran, P. J.; Corey, E. J. *J. Am. Chem. Soc.* **2002**, *124*, 7904.

(319) Huang, Q.; Fazio, A.; Dai, G.; Campo, M. A.; Larock, R. C. *J. Am. Chem. Soc.* **2004**, *126*, 7460.

(320) For a discussion of the mechanism, see Kesharwani, T.; Larock, R. C. *Tetrahedron* **2008**, *64*, 6090.

(321) Kawasaki, S.; Satoh, T.; Miura, M.; Nomura, M. *J. Org. Chem.* **2003**, *68*, 6836.

(322) Bertrand, M. B.; Wolfe, J. P. *Org. Lett.* **2007**, *9*, 3073.

(323) Lafrance, M.; Gorelsky, S. I.; Fagnou, K. *J. Am. Chem. Soc.* **2007**, *129*, 14570.

(324) Catellani, M.; Frignani, F.; Rangoni, A. *Angew. Chem., Int. Ed. Engl.* **1997**, *36*, 119.

(325) For a mechanistic study, see Rudolph, A.; Rackelmann, N.; Lautens, M. *Angew. Chem. Int. Ed.* **2007**, *46*, 1485.

(326) a) Mariampillai, B.; Aliot, J.; Li, M.; Lautens, M. *J. Am. Chem. Soc.* **2007**, *129*, 15372. b) Mariampillai, B.; Alberico, D.; Bidau, V.; Lautens, M. *J. Am. Chem. Soc.* **2006**, *128*, 14436.

(327) For examples, see a) Ferraccioli, R.; Giannini, C.; Molteni, G. *Tetrahedron: Asymmetry* **2007**, *18*, 1475. b) Alberico, D.; Rudolph, A.; Lautens, M. *J. Org. Chem.* **2006**, *72*, 775. c) Martins, A.; Alberico, D.; Lautens, M. *Org. Lett.* **2006**, *8*, 4827 and references therein.

(328) Motti, E.; Rossetti, M.; Bocelli, G.; Catellani, M. *J. Organomet. Chem.* **2004**, *689*, 3741.

(329) Catellani, M.; Motti, E.; Minardi, M. *Chem. Commun.* **2000**, 157.

(330) Wilhelm, T.; Lautens, M. *Org. Lett.* **2005**, *7*, 4053.

(331) For a review, see Ritleng, V.; Sirlin, C.; Pfeffer, M. *Chem. Rev.* **2002**, *102*, 1731.

(332) a) Asaumi, T.; Matsuo, T.; Fukuyama, T.; Ie, Y.; Kakiuchi, F.; Chatani, N. *J. Org. Chem.* **2004**, *69*, 4433. b) Ie, Y.; Chatani, N.; Ogo, T.; Marshall, D. R.; Fukuyama, T.; Kakiuchi, F.; Murai, S. *J. Org. Chem.* **2000**, *65*, 1475.

(333) Thalji, R. K.; Ellman, J. A.; Bergman, R. G. *J. Am. Chem. Soc.* **2004**, *126*, 7192.

(334) Kakiuchi, F.; Kan, S.; Igi, K.; Chatani, N.; Murai, S. *J. Am. Chem. Soc.* **2003**, *125*, 1698.

(335) For reviews, see a) Broene, R. D. Reductive Dimerization of Alkenes and Alkynes. In *Comprehensive Organometallic Chemistry II*; Abel, E. W., Stone, F. G. A., Wilkinson, G., Eds.; Pergamon: Oxford, U.K., 1995; Vol. 12, pp 326–347. b) Hanzawa, Y.; Ito, H.; Taguchi, T. *Synlett* **1995**, 299. c) Negishi, E.-i.; Takahashi, T. *Acc. Chem. Res.* **1994**, *27*, 124.

(336) Swanson, D. R.; Rousset, C. J.; Negishi, E.-i.; Takahashi, T.; Seki, T.; Saburi, M.; Uchida, Y. *J. Org. Chem.* **1989**, *54*, 3521.

(337) a) Fagan, P. J.; Nugent, W. A. *J. Am. Chem. Soc.* **1988**, *110*, 2310. See also b) Hara, R.; Nishihara, Y.; Landre, P. D.; Takahashi, T. *Tetrahedron Lett.* **1997**, *38*, 447.

(338) a) Nugent, W. A.; Taber, D. F. *J. Am. Chem. Soc.* **1989**, *111*, 6435. b) Negishi, E.-i.; Holmes, S. J.; Tour, J. M.; Miller, J. A.; Cederbaum, F. E.; Swanson, D. R.; Takahashi, T. *J. Am. Chem. Soc.* **1989**, *111*, 3336. c) Negishi, E.-i.; Miller, S. R. *J. Org. Chem.* **1989**, *54*, 6014.

(339) In a synthesis of phorbol. Wender, P. A.; Rice, K. D.; Schnute, M. E. *J. Am. Chem. Soc.* **1997**, *119*, 7897.

(340) In a synthesis of epi-β-bulnesene. Negishi, E.-i.; Ma, S.; Sugihara, T.; Noda, Y. *J. Org. Chem.* **1997**, *62*, 1922.

(341) In a synthesis of (–)-dendrobine. Uesaka, N.; Saitoh, F.; Mori, M.; Shibasaki, M.; Okamura, K.; Date, T. *J. Org. Chem.* **1994**, *59*, 5633.

(342) a) Luker, T.; Whitby, R. J. *Tetrahedron Lett.* **1995**, *36*, 4109. b) Probert, G. D.; Whitby, R. J.; Coote, S. J. *Tetrahedron Lett.* **1995**, *36*, 4113. c) Gordon, C. J.; Whitby, R. J. *Synlett* **1995**, 77. d) For the related insertion of propargyl chloride, see Gordon, G. J.; Whitby, R. J. *Chem. Commun.* **1997**, 1045.

(343) Baldwin, I. R.; Whitby, R. J. *Chem. Commun.* **2003**, 2786.

(344) Makabe, M.; Sato, Y.; Mori, M. *J. Org. Chem.* **2004**, *69*, 6238.

(345) Jensen, M.; Livinghouse, T. *J. Am. Chem. Soc.* **1989**, *111*, 4495.

(346) a) Urabe, H.; Suzuki, K.; Sato, F. *J. Am. Chem. Soc.* **1997**, *119*, 10014. See also b) Urabe, H.; Takeda, T.; Hideura, D.; Sato, F. *J. Am. Chem. Soc.* **1997**, *119*, 11295. c) Garcia, A. M.; Maseareñas, J. L.; Castedo, L.; Mouriño, A. *J. Org. Chem.* **1997**, *62*, 6353.

(347) For a review, see Sato, F.; Urabe, H.; Okamoto, S. *Chem. Rev.* **2000**, *100*, 2835.

(348) In a synthesis of α,25-hydroxyvitamin D₃. Hanazawa, T.; Koyama, A.; Wada, T.; Morishige, E.; Okamoto, S.; Sato, F. *Org. Lett.* **2003**, *5*, 523.

(349) Gao, Y.; Harada, K. Sato, F. *Chem. Commun.* **1996**, 533.

(350) Fairfax, D.; Stein, M.; Livinghouse, T.; Jensen, M. *Organometallics* **1997**, *16*, 1523.

(351) In a synthesis of (+)-laudanosine and (–)-xylopine. Muhajedin, D. Doye, S. *Eur. J. Org. Chem.* **2005**, 2689.

(352) a) Buchwald, S. L.; Watson, B. T.; Wannamaker, M. W.; Dewan, J. C. *J. Am. Chem. Soc.* **1989**, *111*, 4486. b) Coles, N.; Whitby, R. J.; Blagg, J. *Syn. Lett.* **1992**, 143.

(353) For a review, see Kulinkovich, O. G.; de Meijere, A. *Chem. Rev.* **2000**, *100*, 2789.

(354) In a synthesis of β-araneosene. Kingsbury, J. S.; Corey, E. J. *J. Am. Chem. Soc.* **2005**, *127*, 13813.

(355) Lee, J.; Kim, H.; Cha, J. K. *J. Am. Chem. Soc.* **1996**, *118*, 4198.

(356) Lee, J.; Cha, J. K. *J. Org. Chem.* **1997**, *62*, 1584.

(357) Knight, K. S.; Waymouth, R. M. *J. Am. Chem. Soc.* **1991**, *113*, 6268.

(358) Takahashi, T.; Seki, T.; Nitto, Y.; Saburi, M.; Rousset, C. J.; Negishi, E.-i. *J. Am. Chem. Soc.* **1991**, *113*, 6266.

(359) a) Hoveyda, A. H.; Xu, Z. *J. Am. Chem. Soc.* **1991**, *113*, 5079. b) Hoveyda, A. H.; Xu, Z.; Morken, J. P.; Houri, A. F. *J. Am. Chem. Soc.* **1991**, *113*, 8950. c) Hoveyda, A. H.; Morken, J. P.; Houri, A. F.; Xu, Z. *J. Am. Chem. Soc.* **1992**, *114*, 6692.

(360) For a review, see Hoveyda, A. H.; Morken, J. P. *Angew. Chem., Int. Ed. Engl.* **1996**, *35*, 1262.

(361) Xu, Z.; Johannes, C. W.; Houri, A. F.; La, D. S.; Cogan, D. A.; Hofichina, G. E.; Hoveyda, A. H. *J. Am. Chem. Soc.* **1997**, *119*, 10302.

(362) Visser, M. S.; Heron, N. M.; Didiuk, M. T.; Segal, J. F.; Hoveyda, A. H. *J. Am. Chem. Soc.* **1996**, *118*, 4291 and references therein.

(363) Visser, M. S.; Heron, N. M.; Didiuk, M. T.; Sagal, J. F.; Hoveyda, A. H. *J. Am. Chem. Soc.* **1996**, *118*, 4291.

(364) For a review, see Tamao, K.; Kobayashi, K.; Ito, Y. *Syn. Lett.* **1992**, 539.

(365) In a synthesis of isodomoic acid G. Ni, Y.; Amarasinghe, K. K. D.; Ksebati, B.; Montgomery, J. *Org. Lett.* **2003**, *5*, 3771.

(366) In a synthesis of prostaglandin PGF₂ₐ. Sato, Y.; Takimoto, M.; Mori, M. *Chem. Pharm. Bull.* **2000**, *48*, 1753.

(367) For a review, see Montgomery, J. *Angew. Chem. Int. Ed.* **2004**, *43*, 3890.

(368) Molinaro, C.; Jamison, T. F. *J. Am. Chem. Soc.* **2003**, *125*, 8076.

(369) In a synthesis of 7-O-methyldehydrpinguisenol. Harada, K.; Tonoi, Y.; Kato, H.; Fukuyama, Y. *Tetrahedron Lett.* **2002**, *43*, 3829.

(370) Cadran, N.; Cariou, K.; Herve, G.; Aubert, C.; Fensterbank, L.; Malacria, M.; Marco-Contelles, J. *J. Am. Chem. Soc.* **2004**, *126*, 3408.

(371) Trost, B. M.; Toste, F. D. *J. Am. Chem. Soc.* **2002**, *124*, 5025.

(372) In a synthesis of amphidinolide P. Trost, B. M.; Papillon, J. N. P.; Nussbaumer, T. *J. Am. Chem. Soc.* **2005**, *127*, 17921.

(373) In a synthesis of amphidinolide A. Trost, B. M.; Harrington, P. E.; Chisholm, J. D.; Wrobelski, S. T. *J. Am. Chem. Soc.* **2005**, *127*, 13589.

(374) Yu, Z.-X. Wender, P. A. Houk, K. N. *J. Am. Chem. Soc.* **2004**, *26*, 9154.

(375) Trost, B. M.; Shen, H. C.; Horne, D. B.; Toste, F. D.; Steimetz, B. G.; Koradin, C. *Chem.—Eur. J.* **2005**, *11*, 2577.

(376) In a synthesis of (+)-dictamnol. Wender, P. A.; Fuji, M.; Husfeld, C. O.; Love, J. A. *Org. Lett.* **1999**, *1*, 137.

5

Synthetic Applications of Transition Metal Carbonyl Complexes[1]

5.1 Introduction

Carbon monoxide coordinates to virtually all transition metals, and an enormous number of complexes with a bewildering array of structures are known. However, most synthetically useful reactions of metal carbonyl complexes, those in which the carbon monoxide ligand is more than a spectator ligand, involve the first row, homoleptic ("all CO") carbonyl complexes shown in **Figure 5.1**. Carbon monoxide is among the best π-acceptor ligands, stabilizing low oxidation states and high electron density on the metal. Many neutral M(0) carbonyl complexes can be reduced to very low *formal* oxidation states (–I through –III), and the resulting anions are powerful nucleophiles. The ease of migratory insertion of carbon monoxide into metal–carbon σ-bonds makes metal carbonyls attractive reagents and catalysts for the introduction of

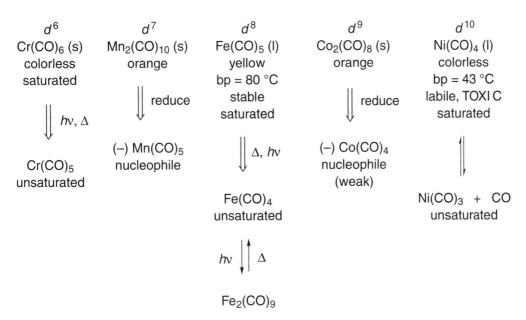

Figure 5.1. Homoleptic Metal Carbonyls

carbonyl groups into organic substrates, and the reversibility of this process is central to *decarbonylation* processes. The general reactions that metal carbonyls undergo are shown in **Figure 5.2**. Specific examples are discussed in the following sections.

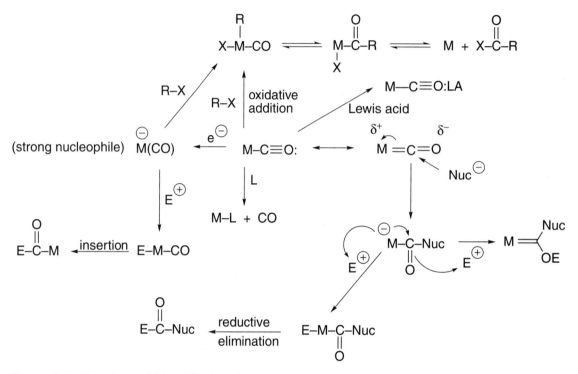

Figure 5.2. Reactions of Metal Carbonyls

5.2 Coupling Reactions of Iron Carbonyls

A number of synthetically useful procedures involving iron pentacarbonyl have been developed. It is not very labile, however, and since it is coordinatively saturated, one CO must be lost before reactions can take place. As a consequence, reactions involving $Fe(CO)_5$ frequently require heating, sonication, or photolysis to generate the coordinatively unsaturated $Fe(CO)_4$ intermediate. The species $Fe(CO)_4$ can be generated under much milder conditions from the gold/orange dimer, $Fe_2(CO)_9$, prepared by irradiation of $Fe(CO)_5$ in acetic acid. Very gentle heating of solutions (suspensions) of $Fe_2(CO)_9$ releases $Fe(CO)_4$ for reactions in which substrates or products are thermally or photochemically unstable.

Reaction of α,α'-disubstituted-α,α'-dibromoketones[2] or bis(sulfonyl)ketones[3] with $Fe(CO)_4$ generates an iron(II) oxallyl cation (**Eq. 5.1**), which undergoes a variety of synthetically interesting [3+4] cycloadditions to give seven-membered rings (**Eq. 5.2**). Generation of the iron–oxallyl cation is usually done in situ, and is likely to involve oxidative addition (nucleophilic attack) of the bromoketone to the electron-rich iron(0) species, to give an iron(II) C-enolate. Rearrangement to the O-enolate, followed by loss of bromide, gives the reactive oxallyl cation. A variety of dienes undergo cycloaddition, including five-membered aromatic heterocycles, to give bridged bicyclic systems (**Eq. 5.3**).

Eq. 5.1

Eq. 5.2

Eq. 5.3

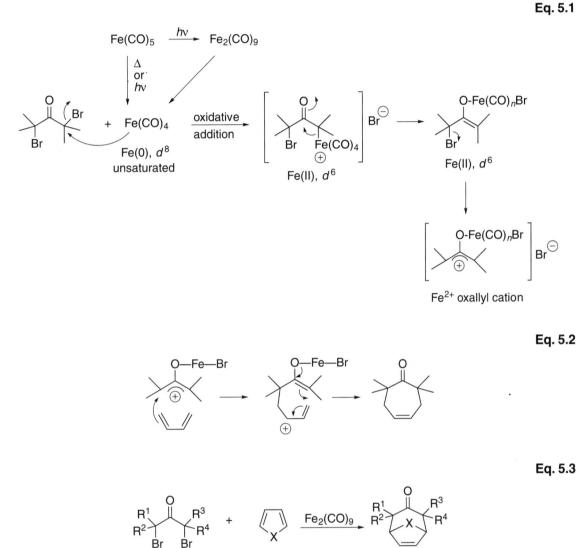

X = CH_2, O, NAc

α-Substitution on the bromoketone is required to stabilize the positive charge in the complex, so dibromoacetone itself does not react. However, tetrabromoacetone is reactive, permitting the introduction of the unsubstituted acetone fragment after a subsequent reductive debromination (**Eq. 5.4**).[4] This strategy has been used to synthesize tropane alkaloids.[5] Cycloheptatrienone undergoes cycloaddition to the carbonyl group instead of the alkene, probably because of the highly stabilized cation resulting from initial attack at oxygen (**Eq. 5.5**).[6]

Eq. 5.4

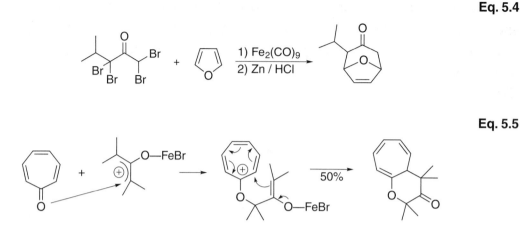

Eq. 5.5

Iron oxallyl cation complexes also undergo cycloaddition reactions with electron-rich alkenes, such as styrene or enamines, to produce cyclopentanones (**Eq. 5.6**).[7] Even relatively nonnucleophilic alkenes such as N-tosyl enamines react cleanly (**Eq. 5.7**).[8]

Eq. 5.6

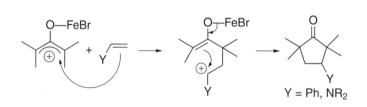

$Y = Ph, NR_2$

Eq. 5.7

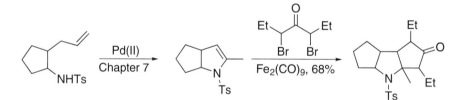

5.3 Carbonylation Reactions[9]

By far the most common synthetic reactions of transition metal carbonyls are those that introduce a carbonyl group into the product. A large number of industrial processes, including hydroformylation (oxo), hydrocarboxylation, and the Monsanto acetic acid process, are predicated on the facility with which CO inserts into metal–carbon bonds. Carbonylation reactions are also useful in the synthesis of fine chemicals.

Although nickel carbonyl couples allylic halides in polar solvents, in less polar solvents in the presence of methanol they are converted to β,γ-unsaturated esters (**Eq. 5.8**). Under these conditions, CO insertion/methanol cleavage is faster than reaction with allyl halides (coupling), which requires very polar solvents.[10]

Eq. 5.8

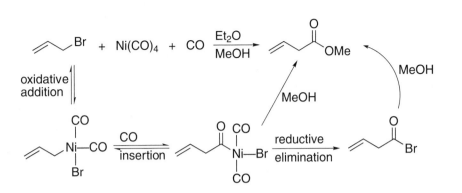

Alkenyl and aryl iodides are converted to esters by treatment with nickel carbonyl and a base in methanol (**Eq. 5.9**).[11] The mechanism is unknown but can be rationalized as an oxidative addition–insertion–nucleophilic cleavage process. Intramolecular versions have also been developed (**Eq. 5.10**),[12] including multiple CO insertions (**Eq. 5.11**).[13]

Eq. 5.9

Eq. 5.10

Eq. 5.11

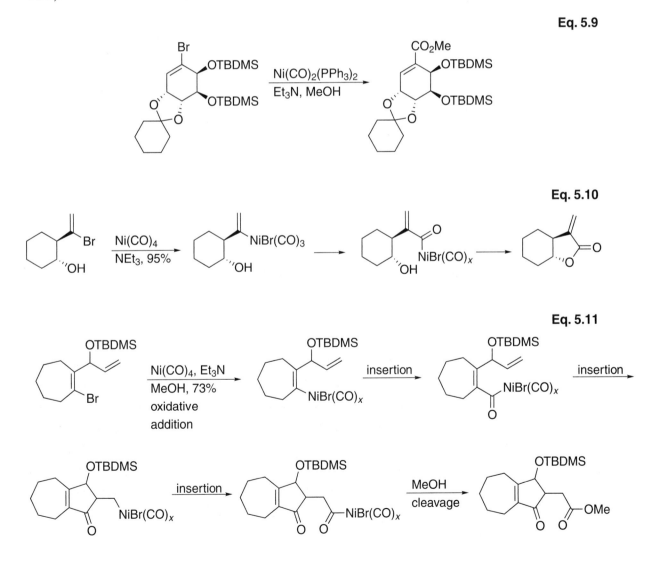

Iron carbonyl promotes a very unusual cyclocarbonylation of allyl epoxides to produce β- or γ-lactones, depending on conditions.[14] The process is somewhat complex, and involves the treatment of alkenyl epoxides with $Fe(CO)_4$, generated from $Fe(CO)_5$ by heat, sonication, or irradiation, or from $Fe_2(CO)_9$ under milder conditions (**Eq. 5.12**). Iron tetracarbonyl is a good nucleophile, and can attack (oxidative addition) the vinyl epoxide in an S_N2' manner to give a cationic σ-alkyl iron(II) carbonyl complex [path (a) in **Eq. 5.12**]. Cationic metal carbonyl complexes are prone to nucleophilic attack at a carbonyl group, in this case producing a "ferrilactone" complex [path (b)], best represented as the η^3-allyl complex, which is stable and isolable. Oxidation of this complex with cerium(IV) leads to the β-lactone [path (c)], presumably by an "oxidatively driven reductive elimination" from the five-membered η^1-allyl complex. In contrast, decomposition of the η^3-allyl complex by treatment with high pressures of carbon monoxide at elevated temperatures produces the γ-lactone, formally by reductive elimination from the *seven*-membered η^1-allyl complex.

Eq. 5.12

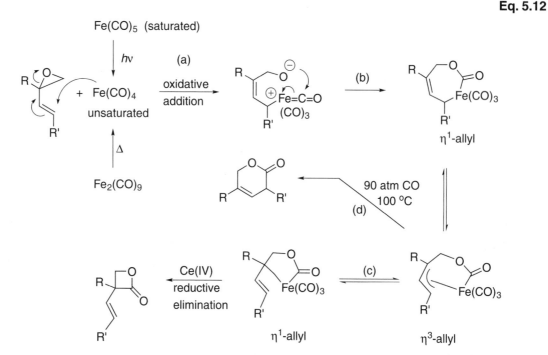

The η^3-allyliron tricarbonyl lactone moiety is a large and relatively stable group that can influence the stereochemical outcome of reactions performed even some distance away from the metal. In this manner, highly stereoselective reactions, such as aldol condensations and hydride reductions, have been achieved. This chemistry has been used in the total synthesis of a variety of natural products[15] (**Eqs. 5.13**[16] and **5.14**[17]). The isomers formed in **Eq. 5.13** are readily separated and only the major isomer was used in the decomplexation step. Reactions under high carbon monoxide pressure to give γ-lactones have also been used in the synthesis of natural products such as malyngolide (**Eq. 5.15**).[18]

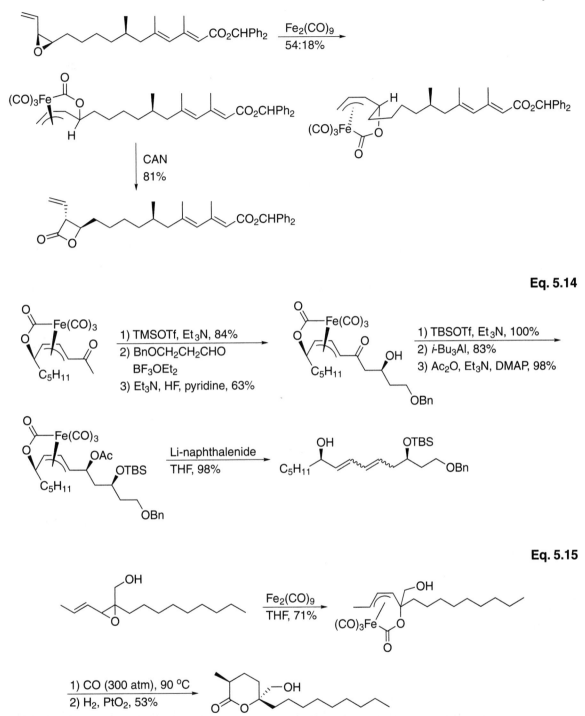

Eq. 5.13

Eq. 5.14

Eq. 5.15

These η^3-allyliron lactone complexes can also be used to synthesize β-lactams. In the presence of Lewis acids, amines attack the allyl complex in an S_N2' manner to produce a rearranged η^3-allyliron lactam complex. Oxidation of this complex produces β-lactams (**Eq. 5.16**).[19] The same η^3-allyliron lactam complexes are available directly from the amino alcohol, and it is likely that the conversion of the lactone complex to the lactam complex proceeds through the amino alcohol.

Eq. 5.16

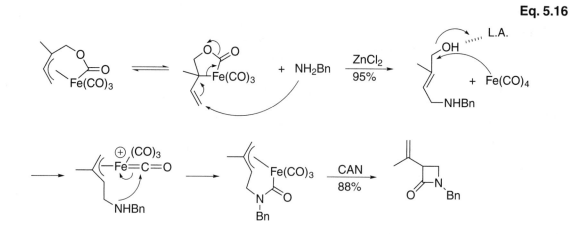

Related reactions using alkenyl-substituted aziridines (**Eq. 5.17**)[20] and cyclopropanes (**Eq. 5.18**)[21] have also been developed. In the latter case, the η^3-allyliron tricarbonyl spontaneously undergoes insertion of CO and reductive elimination to form a six-membered ring.

Eq. 5.17

Eq. 5.18

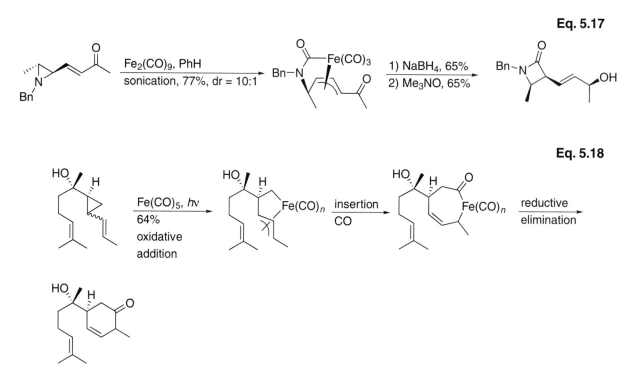

In the preceding carbonylations, the first step is coordination of $Fe(CO)_4$ to the alkene, then nucleophilic attack of the zero-valent metal on the substrate (i.e., S_N2-like oxidative addition; see Chapter 2). Anionic metal carbonyl complexes are even stronger nucleophiles, and effect a number of synthetically useful carbonylation reactions.

Probably the most intensively studied and highly developed (for organic synthesis) anionic metal carbonyl species is disodium tetracarbonylferrate, $Na_2Fe(CO)_4$, known as Collman's reagent.[22] This d^{10}, Fe(–II) complex is easily prepared by the reduction of the commercially available iron pentacarbonyl by sodium benzophenone ketyl in dioxane at reflux. (This complex is also available commercially.)

$Na_2Fe(CO)_4$ is useful in synthesis specifically because of its high nucleophilicity and the ease of the CO insertion reaction in this system. **Figure 5.3** summarizes this chemistry. Organic halides and tosylates react with $Na_2Fe(CO)_4$ with typical S_N2 kinetics (i.e., second order), stereochemistry (i.e., inversion), and order of reactivity (i.e., $CH_3 > RCH_2 > R'RCH$; RI > RBr > ROTs > RCl; vinyl, aryl, inert) to produce coordinatively saturated anionic d^8 alkyliron(0) complexes [path (a) in Figure 5.3]. These undergo protolytic cleavage to produce the corresponding hydrocarbon, an overall reduction of the halide [path (b)]. In the presence of excess carbon monoxide or added triphenylphosphine, migratory insertion takes place to form the acyl complex [path (c)], accessible directly by the reaction of $Na_2Fe(CO)_4$ with acid halides [path (d)]. This

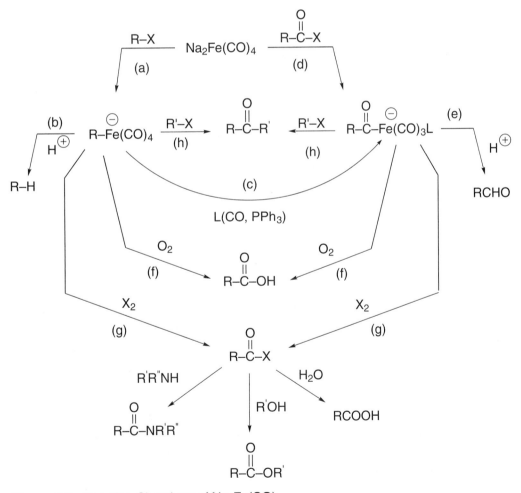

Figure 5.3. Reaction Chemistry of $Na_2Fe(CO)_4$

acyl complex reacts with acetic acid to produce aldehydes, providing a high-yield conversion of alkyl and acyl halides to aldehydes [path (e)]. Oxidative cleavage of either the acyl- or alkyliron(0) complex by oxygen [path (f)] or halogen [path (g)] produces carboxylic acid derivatives. Finally, the acyliron(0) complex itself is still sufficiently nucleophilic to react with reactive alkyl iodides to give unsymmetric ketones in excellent yield [path (h)]. Interestingly, the alkyliron(0) complex reacts in a similar fashion with alkyl iodides to produce unsymmetric ketones, inserting CO along the way [path (h)]. Thus, this chemistry provides routes from alkyl and acid halides to alkanes, aldehydes, ketones, and carboxylic acid derivatives.

Collman's reagent is quite specific for halides. Ester, ketone, nitrile, and alkene functionality is tolerated elsewhere in the molecule. In mixed halides (i.e., chloro–bromo compounds), the more reactive halide is the exclusive site of reaction. The major limitation is the high basicity (pK_b about that of OH$^-$) of Na$_2$Fe(CO)$_4$, which results in competing elimination reactions with tertiary and secondary substrates. Additionally, allylic halides having alkyl groups δ to the halide fail as substrates, since stable 1,3-diene–iron complexes form preferentially. Finally, since the migratory insertion fails when the R group contains adjacent electronegative groups, such as halogen or alkoxy groups, the syntheses involving insertion are limited to simple primary or secondary substrates.

The chemistry described above has been subjected to very close experimental scrutiny, and as a consequence, the mechanistic features are understood in detail. Both the alkyliron(0) complex and the acyliron(0) complex have been isolated as air-stable, crystalline [(Ph$_3$P)$_2$N]$^+$ salts, fully characterized by elemental analysis and IR and NMR spectra. These salts undergo the individual reactions detailed in Figure 5.3. Careful kinetic studies indicate that ion-pairing effects dominate the reactions of Na$_2$Fe(CO)$_4$, and account for the observed 2×10^4 rate increase (tight ion pair [NaFe(CO)$_4$]$^-$ versus solvent-separated ion pair [Na$^+$:S:Fe(CO)$_4$]$^-$ as the kinetically active species) in going from THF to N-methylpyrrolidone. The rate law, substrate order of reactivity, stereochemistry at carbon, and activation parameters (particularly the large negative entropy of activation) are consistent with an S$_N$2-type oxidative addition, with no competing one-electron mechanism. The migratory insertion step is also subject to ion-pairing effects, and is accelerated by tight ion pairing [Li$^+$ > Na$^+$ > Na–crown$^+$ > (Ph$_3$P)$_2$N$^+$]. The insertion reaction is overall second order, first order in both NaRFe(CO)$_4$ and added ligand, with about a 20-fold difference in rate over the range of ligands from Ph$_3$P (slowest) to Me$_3$P (fastest).[23]

An interesting variation of this chemistry[24] results in the production of ethyl ketones from organic halides, Na$_2$Fe(CO)$_4$, and ethene (**Eq. 5.19**). The reaction is thought to proceed via the acyliron(0) complex, which inserts ethene into the acyl–metal bond and rapidly rearranges to the α-metalloketone, which then cannot further insert ethene since the R group now contains an adjacent electronegative acyl group. Intramolecular versions of this can be used to synthesize cyclic ketones (**Eq. 5.20**),[25] but in some cases the process is not regioselective (**Eq. 5.21**).[26]

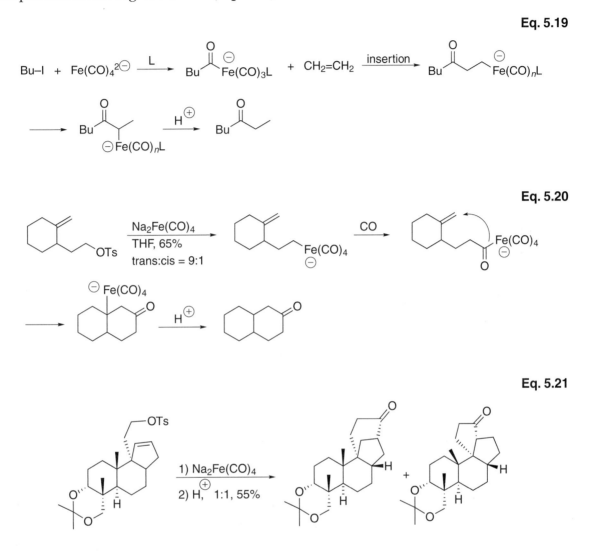

Eq. 5.19

Eq. 5.20

Eq. 5.21

Anionic nickel acyl complexes, generated in situ by the reaction of organolithium reagents with the toxic, volatile nickel carbonyl, effect a number of synthetically useful transformations, including the acylation of allylic halides, 1,4-acylation of conjugated enones, and 1,2-addition to quinones (**Eq. 5.22**).[27] Probably due to the toxicity of the reagent, only a few groups have used it, although one nice synthetic application has been reported (**Eq. 5.23**).[28]

The same general reaction chemistry can be achieved by replacing nickel carbonyl with the orange, air-stable, nonvolatile crystalline solid Co(PPh$_3$)(CO)$_2$(NO)[29] complex, which is much more easily handled. [This complex is isoelectronic with Ni(CO)$_4$, a nickel(0), d^{10} complex, since the nitrosyl ligand is formally a cation, making this a cobalt(-I), d^{10} complex.] Acyl complexes formed from organolithium reagents and Cr(CO)$_6$ also participate in 1,4-acylations of electron-deficient alkenes.[30]

Eq. 5.22

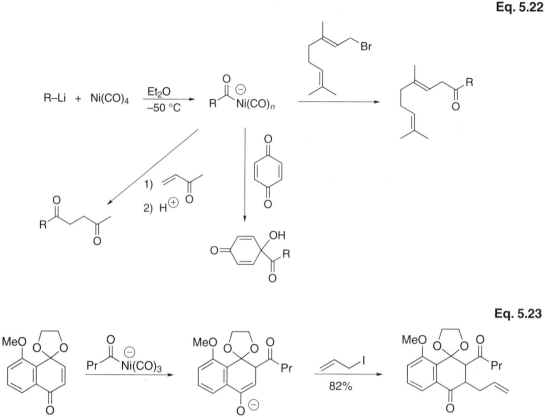

Eq. 5.23

5.4 Decarbonylation,[31] Hydroformylation, and Hydroacylation Reactions

Much of this chapter has been devoted to the consideration of methods for the introduction of carbonyl groups into organic substrates via a "migratory insertion" reaction, in which a metal alkyl or aryl group migrates to an adjacent, coordinated CO to produce a σ-acyl complex, which is then cleaved to produce the organic carbonyl compound. The reverse process, in which organic carbonyl compounds (specifically aldehydes and acid chlorides) are decarbonylated, is also possible, and often quite useful in organic synthesis. Although a number of transition metal complexes can function as decarbonylation agents, by far the most efficient is $RhCl(PPh_3)_3$, known as Wilkinson's catalyst. This complex reacts with alkyl, aryl, and alkenyl aldehydes under mild conditions to produce the corresponding hydrocarbon and the very stable $RhCl(PPh_3)_2(CO)$, as in **Eq. 5.24**. The course of this stoichiometric decarbonylation can easily be monitored by the change in color of the solution, from the deep red of $RhCl(PPh_3)_3$ to the canary yellow of $RhCl(PPh_3)_2(CO)$. *trans*-Alkylcinnamaldehydes are decarbonylated with retention of alkene geometry, producing cis-substituted styrenes.[32] More impressively, decarbonylation of aliphatic aldehydes is highly stereoselective with overall retention of stereochemistry[33] (**Eq. 5.25**).[34]

Eq. 5.24

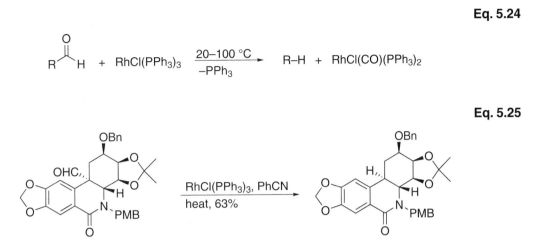

Eq. 5.25

The mechanism proposed for this decarbonylation is depicted in **Eq. 5.26**. It involves, as a first step, oxidative addition of the aldehyde to the rhodium complex. The next step is the reverse of the carbonyl insertion reaction (i.e., deinsertion)—that is, migration of the alkyl group from the carbonyl to the metal. Migratory insertion is a reversible process, and can be driven in either direction by the appropriate choice of reaction conditions. In addition, it has been shown in other cases that this reversible transformation proceeds with retention of the stereochemistry of the migrating alkyl group; thus, there is retention of stereochemistry in the decarbonylation of chiral aldehydes. The last step is the reductive elimination of the alkane (R–H) and the production of RhCl(CO)L$_2$. This last step is irreversible [alkanes do not oxidatively add to RhCl(CO)L$_2$], so it drives the entire process to completion. Since RhCl(CO)L$_2$ is much less reactive in oxidative addition reactions than RhClL$_3$ (CO is a π-acceptor and withdraws electron density from the metal), it does not react with aldehydes under these mild conditions, and the decarbonylation described above is stoichiometric.

Eq. 5.26

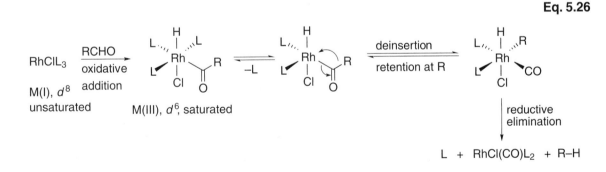

When the reaction is carried out at elevated temperatures (typically < 100°C), both RhClL$_3$ and RhCl(CO)L$_2$ function as decarbonylation catalysts, presumably because RhCl(CO)L$_2$ will oxidatively add aldehydes under these more severe conditions. If β-hydrogens are present in the substrate, β-elimination will predominate at these temperatures, thus forming an alkene. Quite complex aldehydes can be decarbonylated without problems[35] (**Eqs. 5.27,**[36] **5.28,**[37] and **5.29**[38]). Catalytic decarbonylations at lower temperatures have been accomplished using Wilkinson's catalyst in the presence of diphenylphosphoryl azide,[39] $Rh(dppp)_2BF_4$,[40] or $Rh(CO)(triphos)SbF_6$.[41]

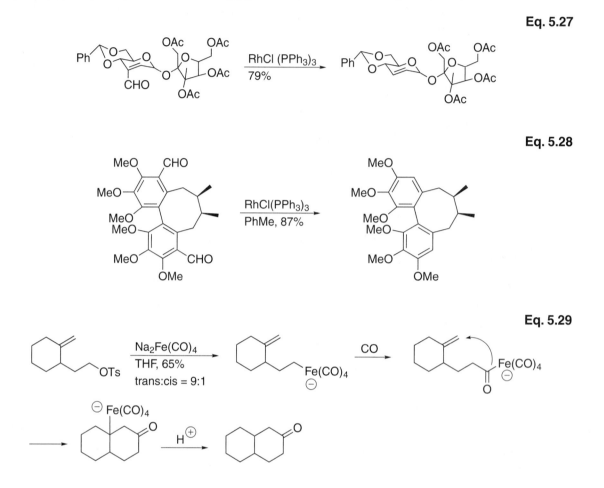

Eq. 5.27

Eq. 5.28

Eq. 5.29

Acid chlorides also undergo decarbonylation upon treatment with $RhCl(PPh_3)_3$, but the reaction suffers several complications not encountered with aldehydes. The reaction of acid chlorides is most straightforward with substrates that have no β-hydrogens. In these cases, decarbonylation occurs smoothly to give the alkyl chloride. In contrast to aldehydes, the initial oxidative adduct is quite stable, and has been isolated and fully characterized in several instances. Heating this adduct leads to decarbonylation via a mechanism thought to be strictly analogous to that involved in aldehyde decarbonylation. However, the decarbonylation of acid halides differs from that of aldehydes in several respects. Decarbonylation of optically active acid chlorides with $RhCl(PPh_3)_3$ produces racemic alkyl chlorides. This is in direct contrast to aldehydes, which undergo decarbonylation with a high degree of retention.

Acid halides with β-hydrogens undergo decarbonylation by $RhCl(PPh_3)_3$ to produce primarily alkenes resulting from β-hydride elimination from the σ-alkyl intermediate, rather than reductive elimination (**Eq. 5.30**). Branched acid chlorides that can undergo β-elimination in several directions give mixtures of products with the most substituted alkenes predominating.

Eq. 5.30

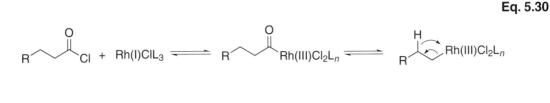

Contrary to reports in the older literature, aroyl chlorides and α,β-unsaturated acid chlorides do *not* undergo decarbonylation to give the corresponding unsaturated chloride, but rather produce stable arylmetal complexes or quaternary phosphonium salts.[42]

In certain cases, the acylmetal complex can undergo alkene- and alkyne-insertion reactions faster than decarbonylation (**Eqs. 5.31**[43] and **5.32**[44]). Note the asymmetric induction observed using a chiral catalyst. This acylation reaction is mostly limited to intramolecular cases due to competing decarbonylation. Interesting and potentially useful variations are the intramolecular acylations of cyclopropyl–enals and –dienals, affording seven- and eight-membered rings (**Eqs. 5.33**[45] and **5.34**[46]), respectively. The role of ethene in **Eq. 5.34** is unclear, although a higher yield of product was obtained in the presence of ethene.

Eq. 5.31

Eq. 5.32

Eq. 5.33

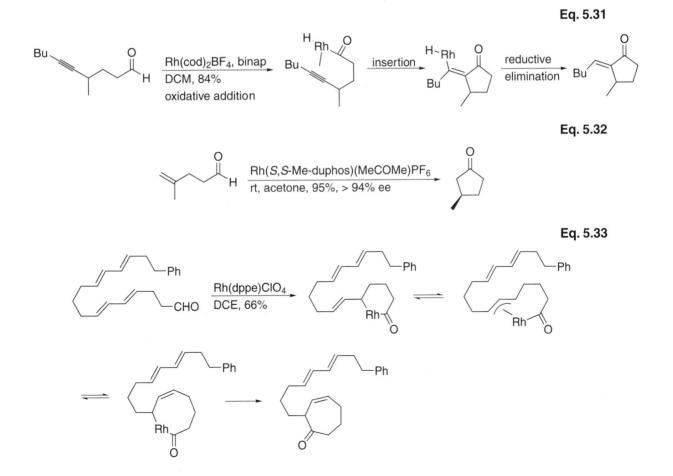

Eq. 5.34

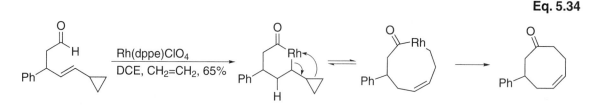

In the presence of an excess of carbon monoxide (i.e., high pressure), alkenes can be transformed into aldehydes (hydroformylation) via an acyl complex.[47] A variety of transition metals catalyze hydroformylations, including rhodium, cobalt, and platinum. The mechanism has been studied in detail and involves steps very similar (but reverse) to the decarbonylation reactions discussed previously (**Eq. 5.35**).[48] Wilkinson's catalyst has been used extensively and, in the presence of a mixture of hydrogen gas and carbon monoxide, the saturated complex $RhH(CO)_2L_2$ is formed. Loss of a CO ligand and coordination of an alkene affords a new saturated complex. A migratory insertion, CO coordination, migratory insertion, and reductive cleavage of the acyl complex furnishes the product and regenerates the catalyst.

Eq. 5.35

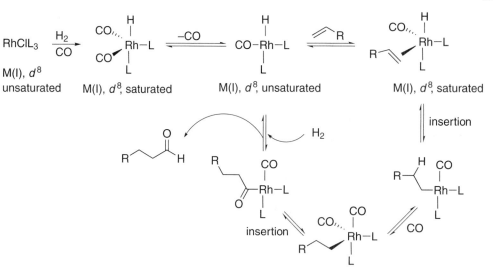

Considering the truly staggering number of publications dealing with every aspect of this reaction and its importance as an industrial process for the preparation of simple aliphatic aldehydes, it has found surprisingly little use in the total synthesis of more complex molecules. A few examples are shown in **Eqs. 5.36**,[49] **5.37**,[50] and **5.38**.[51] Asymmetric hydroformylations have been achieved in the presence of chiral ligands.

Eq. 5.36

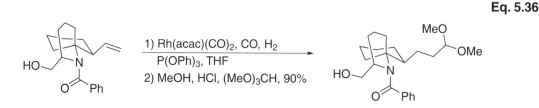

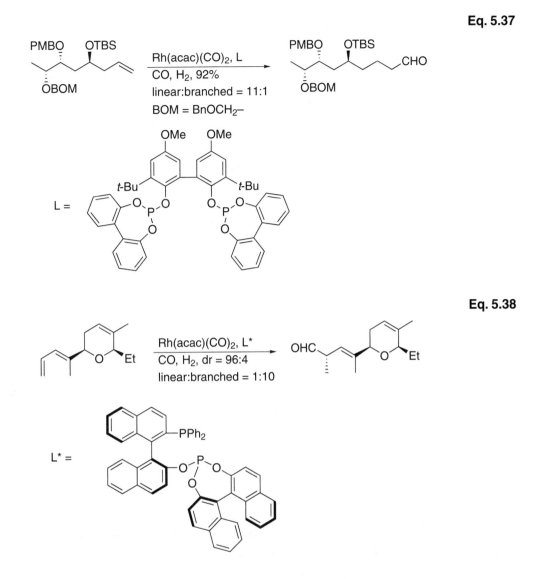

Eq. 5.37

Eq. 5.38

5.5 Metal Acyl Enolates[52]

Many of the metal–acyl species discussed above are quite stable, and can be isolated and handled easily. One such complex, FeCp(COCH₃)(CO)(PPh₃), has been developed into a useful synthetic reagent. This complex has the following interesting characteristics: (1) it is easy to prepare and handle, (2) it is chiral at iron and can be resolved, and (3) the protons α to the acyl group are acidic and the corresponding metal–acyl enolate undergoes reaction with a variety of electrophiles (**Eq. 5.39**). Because the complex is chiral at iron and one face is hindered by the triphenylphosphine ligand, reactions of these acyliron enolates occur with very high stereoselectivity (**Eq. 5.40**). The observed absolute stereochemistry results from alkylation of the *E*-enolate with the carbonyl group and O⁻ anti from the less-hindered face. Oxidative cleavage of the resulting complex produces the carboxylic acid derivative in good yield with high ee.[53]

Eq. 5.39

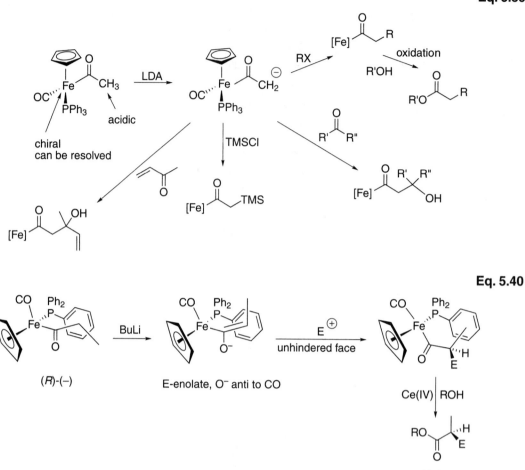

Eq. 5.40

This chemistry has been used extensively in the synthesis of optically active compounds. Two such cases are shown in **Eqs. 5.41**[54] and **5.42**.[55] As might be expected, α,β-unsaturated iron–acyl complexes also undergo reactions in a highly diastereoselective manner. Both conjugate addition–enolate trapping (**Eq. 5.43**)[56,57] and γ-deprotonation–α-alkylation (**Eq. 5.44**)[58] enjoy a high degree of stereocontrol. Vinylogous iron–acyl complexes also undergo a variety of alkylation reactions (**Eq. 5.45**),[59] although these have not yet been used in total synthesis.

Eq. 5.41

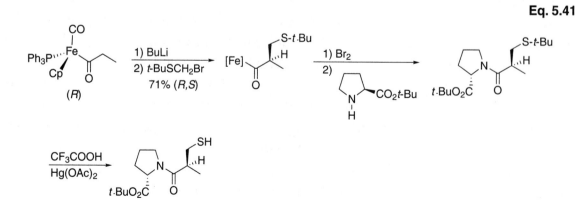

Eq. 5.42

Eq. 5.43

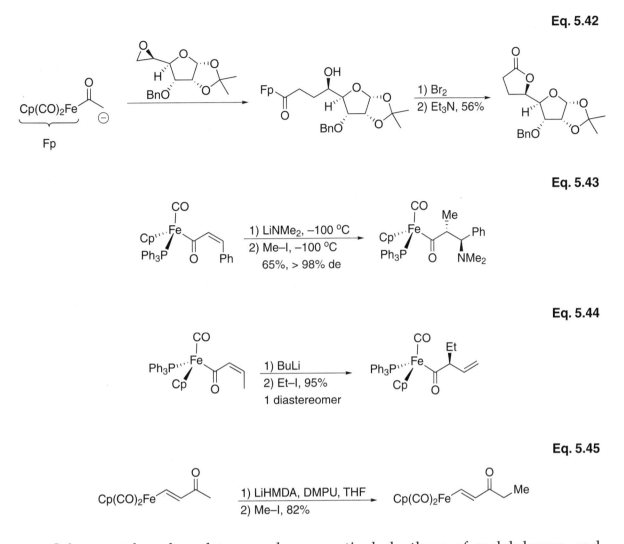

Eq. 5.44

Eq. 5.45

Other metal–acyl enolate complexes, particularly those of molybdenum and tungsten, have been prepared and their reaction chemistry studied (e.g., **Eq. 5.46**[60]). However, the focus of these studies has been more organometallic/mechanistic than synthetic, so the potential of these systems in the synthesis of complex molecules is yet to be realized.

Eq. 5.46

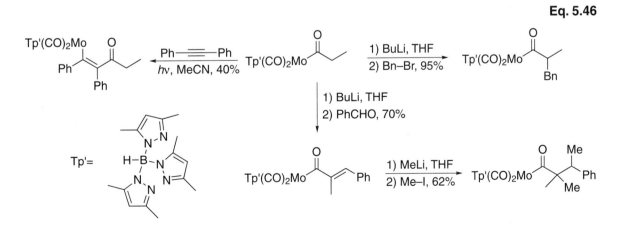

5.6 Bridging Acyl Complexes

The reaction of triiron dodecacarbonyl with thiols, followed by conjugated acid halides, produces diiron bridged acyl complexes (**Eq. 5.47**).[61] These complexes undergo highly exo-selective Diels–Alder reactions with dienes under very mild conditions. Decomplexation by oxidative cleavage of the bridging acyl species produces the corresponding thioester.[62] Complexation to iron activates the enone towards Diels–Alder addition, since methyl crotonate undergoes reaction with the same dienes only at much higher temperatures and with lower selectivity.

Eq. 5.47

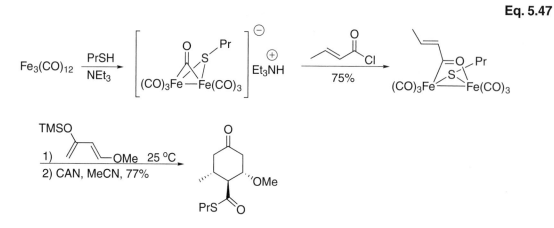

In contrast, nitrones undergo endo-selective 1,3-dipolar cycloaddition to these bridging acyl species.[63] By using a chiral thiol to prepare the bridging acyl species, reasonable asymmetric induction has been achieved (**Eq. 5.48**).[64]

Eq. 5.48

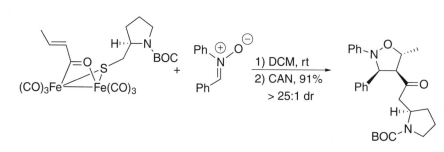

References

(1) For a review, see Bates, R. W. Transition Metal Carbonyl Compounds. In *Comprehensive Organometallic Chemistry II*; Abel, E. W., Stone, F. G. A., Wilkinson, G., Eds.; Plenum: Oxford, U.K., 1995; Vol. 12, pp 349–386.

(2) a) For a review, see Noyori, R. *Acc. Chem. Res.* **1979**, *12*, 61.

(3) Hardinger, S. A.; Bayne, C.; Kantorowski, E.; McClellan, R.; Larres, L.; Nuesse, M.-A. *J. Org. Chem.* **1995**, *60*, 1104.

(4) Noyori, R.; Sato, T.; Hayakawa, Y. *J. Am. Chem. Soc.* **1978**, *100*, 2561.

(5) Hayakawa, Y.; Baba, Y.; Makino, S.; Noyori, R. *J. Am. Chem. Soc.* **1978**, *100*, 1786.

(6) Ishizu, T.; Mori, M.; Kanematsu, K. *J. Org. Chem.* **1981**, *46*, 526.

(7) Hayakawa, Y.; Yokoyama, K.; Noyori, R. *J. Am. Chem. Soc.* **1978**, *100*, 1791; 1799.

(8) Hegedus, L. S.; Holden, M. *J. Org. Chem.* **1985**, *50*, 3920.

(9) Colquhoun, H. M.; Thompson, D. J.; Twigg, M. V. *Carbonylation*; Plenum: New York, 1991.

(10) For a theoretical study of the mechanism, see Bottoni, A.; Miscione, G. P.; Novoa, J. J.; Prat-Resina, X. *J. Am. Chem. Soc.* **2003**, *125*, 10412.

(11) In a synthesis of 6β-hydroxyshikimic acid. Blacker, A. J.; Booth, R. J.; Davies, G. M.; Sutherland, J. K. *J. Chem. Soc., Perkin Trans 1* **1995**, 2861.

(12) a) Semmelhack, M. F.; Brickner, S. J. *J. Org. Chem.* **1981**, *46*, 1723. b) Llebaria, A.; Delgado, A.; Camps, F.; Moreto, J. M. *Organometallics* **1993**, *12*, 2825.

(13) Llebaria, A.; Camps, F.; Moreto, J. N. *Tetrahedron* **1993**, *49*, 1283.

(14) For a review, see Ley, S. V.; Cox, L. R.; Meek, G. *Chem. Rev.* **1996**, *96*, 423.

(15) Bates, R. W.; Fernandez-Moro, R.; Ley, S. V. *Tetrahedron Lett.* **1991**, *32*, 2651.

(16) In a synthesis of the cholesterol inhibitor 1233A. Bates, R. W.; Fernandez-Megia, E.; Ley, S. V.; Rück-Braun, K. Tilbrook, D. M. G. *J. Chem. Soc., Perkin Trans. 1* **1999**, 1917.

(17) In a synthesis of (–)-gloesporone. Ley, S. V.; Cleator, E.; Harter, J.; Hollowood, C. *J. Org. Biomolec. Chem.* **2003**, *1*, 3263.

(18) Horton, A. M.; Ley, S. V. *J. Organomet. Chem.* **1985**, *285*, C17.

(19) a) Annis, G. D.; Hebblethwaite, E. M.; Hodgson, S. T.; Hollingshead, D. M.; Ley, S. V. *J. Chem. Soc., Perkin Trans. 1* **1983**, 2851. b) Horton, A. M.; Hollinshead, D. M.; Ley, S. V. *Tetrahedron* **1984**, *40*, 1737.

(20) Ley, S. V.; Middleton, B. *Chem. Commun.* **1998**, 1995.

(21) In a synthesis of (–)-delobanone. Taber, D. F.; Bui, G.; Chen, B. *J. Org. Chem.* **2001**, *66*, 3423.

(22) Collman, J. P. *Acc. Chem. Res.* **1975**, *8*, 342.

(23) Collman, J. P.; Finke, R. G.; Cawse, J. N.; Brauman, J. I. *J. Am. Chem. Soc.* **1977**, *99*, 2515; *J. Am. Chem. Soc.* **1978**, *100*, 4766.

(24) Cooke, M. P., Jr.; Parlman, R. M. *J. Am. Chem. Soc.* **1975**, *97*, 6863.

(25) Merour, J. Y.; Roustan, J. L.; Charrier, C.; Collin, J.; Benaim, J. *J. Organomet. Chem.* **1973**, *51*, C24.

(26) In a synthesis of aphidicoline. a) McMurry, J. E.; Andrus, A.; Ksander, G. M.; Musser, J. H.; Johnson, M. A. *J. Am. Chem. Soc.* **1979**, *101*, 1330. b) *Tetrahedron* **1981**, *27 Supplement*, 319.

(27) a) Corey, E. J.; Hegedus, L. S. *J. Am. Chem. Soc.* **1969**, *91*, 4926. b) See also Hermanson, J. R.; Gunther, M. L.; Belletire, J. L.; Pinhas, A. R. *J. Org. Chem.* **1995**, *60*, 1900.

(28) In a synthesis of nanaomycin and deoxyfrenolicin. Semmelhack, M. F.; Keller, L.; Sato, T.; Spiess, E. J.; Wulff, W. *J. Org. Chem.* **1985**, *50*, 5566.

(29) Hegedus, L. S.; Perry, R. J. *J. Org. Chem.* **1985**, *50*, 4955.

(30) Söderberg, B. C.; York, D. C.; Harriston, E. A.; Caprara, H. J.; Flurry, A. H. *Organometallics* **1995**, *14*, 3712.

(31) For a review, see Tsuji, J. DeCarbonylation Reactions Using Transition Metal Compounds. In *Organic Syntheses via Metal Carbonyls*; Wender, I., Pino, P., Eds.; Wiley: New York, 1977; Vol. 2, pp 595–654.

(32) Tsuji, J.; Ohno, K. *Tetrahedron Lett.* **1967**, 2173.

(33) a) Walborsky, H. M.; Allen, L. E. *Tetrahedron Lett.* **1970**, 823. b) Walborsky, H. M.; Allen, L. E. *J. Am. Chem. Soc.* **1971**, *93*, 5465.

(34) In a synthesis of 7-deoxypancrastatin. Zhang, H.; Padwa, A. *Tetrahedron Lett.* **2006**, *47*, 3905.

(35) Jung, M. E.; Rayle, H. L. *J. Org. Chem.* **1997**, *62*, 4601.

(36) Iley, D. E.; Fraser-Reid, B. *J. Am. Chem. Soc.* **1975**, *97*, 2563.

(37) In a synthesis of gomisin J. Tanaka, M.; Ohshima, T.; Mitsuhashi, H.; Maruno, M.; Wakamatsu, T. *Tetrahedron* **1995**, *51*, 11693.

(38) In a synthesis of 3-demethoxyerythratidinone. Allin, S. M.; Streetley, G. B.; Slater, M.; James, S. L. *Tetrahedron Lett.* **2004**, *45*, 5493.

(39) O'Connor, J. M.; Ma, J. *J. Org. Chem.* **1992**, *57*, 5075.

(40) Banwell, M. G.; Coster, M. J.; Karunarate, O. P.; Smith, J. A. *J. Chem. Soc., Perkin Trans. 1* **2002**, 1622.

(41) Beck, C. M.; Rathmill, S. E.; Park, Y. J.; Chen, J.; Crabtree, R. H. *Organometallics* **1999**, *18*, 5311.

(42) a) Kampmeier, J. A.; Rodehorst, R. M.; Philip, J. B., Jr. *J. Am. Chem. Soc.* **1981**, *103*, 1847. b) Kampmeier, J. A.; Mahalingam, S. *Organometallics* **1984**, *3*, 489. c) Kampmeier, J. A.; Harris, S. H.; Rodehorst, R. M. *J. Am. Chem. Soc.* **1981**, *103*, 1478.

(43) Takeishi, K.; Sugishima, K.; Sasaki, K.; Tanaka, K. *Chem. Eur. J.* **2004**, *10*, 5681.

(44) a) Barnhart, R. W.; Wang, X.; Noheda, P.; Bergens, S. H.; Whelan, J.; Bosnich, B. *J. Am. Chem. Soc.* **1994**, *116*, 1821. b) Barnhart, R. W.; McMorran, D. A.; Bosnich, B. *Chem. Commun.* **1997**, 589.

(45) Oonishi, Y.; Taniuchi, A.; Mori, M.; Sato, Y. *Tetrahedron Lett.* **2006**, *47*, 5617.

(46) Aloise, A. D.; Layton, M. E.; Shair, M. D. *J. Am. Chem. Soc.* **2000**, *122*, 12610.

(47) For a review, see Ojima, I.; Tsai, C.-Y.; Tzamarioudaki, M.; Bonafoux, D. *Org. React.* **2000**, *56*, 1.

(48) For a recent review, see Kamer, P. C. J.; van Rooy, A.; Schoemaker, G. C.; van Leeuwen, P. W. N. M. *Coord. Chem. Rev.* **2004**, *248*, 2409.

(49) In a synthesis of lepadiformine. Sun, P.; Weinreb, S. M. *Org. Lett.* **2001**, *3*, 3507.

(50) In a synthesis toward bryostatins. Keck, G. E.; Truong, A. P. *Org. Lett.* **2005**, *7*, 2149.

(51) In a synthesis of (+)-ambruticin. Liu, P.; Jacobsen, E. N. *J. Am. Chem. Soc.* **2001**, *123*, 10772.

(52) For reviews, see a) Davies, S. G. *Pure Appl. Chem.* **1988**, *60*, 13. b) Blacksburn, B. K.; Davies, S. G.; Sutton, K. H.; Whittaker, M. *Chem. Soc. Rev.* **1988**, *17*, 147.

(53) For a review, see Davies, S. G.; Bashiardes, G.; Beckett, R. P.; Coote, S. J.; Dordor-Hedgecock, I. M.; Goodfellow, C. L.; Gravatt, G. L.; McNally, J. P.; Whittaker, M. *Philos. Trans. R. Soc. London, Ser. A* **1988**, *326*, 619.

(54) In a synthesis of (*S*,*S*)-captopril. Bashiardes, G.; Davies, S. G. *Tetrahedron Lett.* **1987**, *28*, 5563.

(55) Davies, S. G.; Kellie, H. M.; Polywka, R. *Tetrahedron: Asymmetry* **1994**, *5*, 2563.

(56) Davies, S. G.; Dordor-Hedgecock, I. M.; Easton, R. J. C.; Preston, S. C.; Sutton, K. H.; Walker, J. C. *Bull Soc. Chim. Fr.* **1987**, 608.

(57) Davies, S. G.; Dupont, J.; Easton, R. J. C.; Ichihara, O.; McKenna, J. M.; Smith, A. D.; de Sousa, J. A. A. *J. Organomet. Chem.* **2004**, *689*, 4184.

(58) Davies, S. G.; Easton, R. J. C.; Gonzalez, A.; Preston, S. C.; Sutton, K. H.; Walker, J. C. *Tetrahedron* **1986**, *42*, 3987.

(59) Mattson, M. N.; Helquist, P. *Organometallics* **1992**, *11*, 4.

(60) Rusik, C. A.; Collins, M. A.; Gamble, A. S.; Tonker, T. L.; Templeton, J. L. *J. Am. Chem. Soc.* **1989**, *111*, 2550.

(61) Seyferth, D.; Archer, C. M.; Ruschke, D. P.; Cowle, M.; Hilts, R. W. *Organometallics* **1991**, *10*, 3363 and references therein.

(62) Gilbertson, S. R.; Zhao, X.; Dawson, D. P.; Marshall, K. L. *J. Am. Chem. Soc.* **1993**, *115*, 8517.

(63) Gilbertson, S. R.; Dawson, D. P.; Lopez, O. D.; Marshall, K. L. *J. Am. Chem. Soc.* **1995**, *117*, 4431.

(64) Gilbertson, S. R.; Lopez, O. D. *J. Am. Chem. Soc.* **1997**, *119*, 3399.

Synthetic Applications of Transition Metal Carbene Complexes

6.1 Introduction

Carbene complexes—complexes having formal metal-to-carbon double bonds—are known for metals across the entire transition series, although relatively few have been developed as useful reagents for organic synthesis. The two bonding extremes for carbene complexes are represented by the *electrophilic*, heteroatom-stabilized "Fischer" carbenes and the *nucleophilic*, methylene or alkylidene, "Schrock" carbenes. Between these two extremes lie the "Grubbs-" (and also different "Schrock-") type carbenes, which are particularly useful for alkene and alkyne metathesis reactions. Although they share common structural and reactivity features, their chemistry is sufficiently different to be discussed separately.

6.2 Electrophilic Heteroatom Stabilized "Fischer" Carbene Complexes[1]

The most intensively developed carbene complexes for use in organic synthesis are the electrophilic Fischer carbene complexes of the group 6 metals Cr, Mo, and W. These are readily synthesized by the reaction of the air-stable, crystalline metal hexacarbonyl with a range of organolithium reagents (**Eq. 6.1**). One of the six equivalent CO groups undergoes attack to produce the stable, anionic lithium acyl "ate" complex, in which the negative charge is stabilized and extensively delocalized into the remaining five, π-accepting, electron-withdrawing CO groups. These "ate" complexes are normally isolated as the stable tetramethylammonium salt, which can be prepared on a large scale and stored for months without substantial decomposition. Treatment with hard alkylating agents such as methyl triflate or trimethyloxonium salts (methyl Meerwein's reagent), with alkyl halides under phase-transfer conditions,[2] with sulfur ylides,[3] or with dimethyl sulfate[4] results in alkylation at oxygen, producing the alkoxycarbene complex in excellent yield.

Eq. 6.1

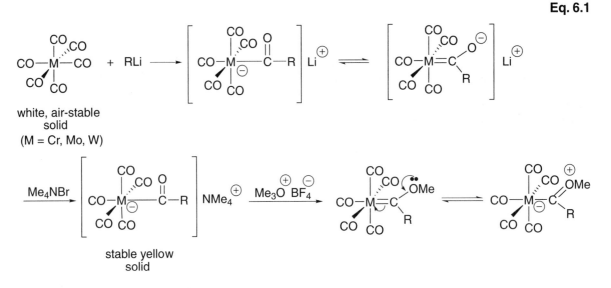

Although synthesis via organolithium reagents is the most common approach to these carbene complexes, they can be generated by any reaction that produces anionic acyl complexes. Chromium hexacarbonyl is easily reduced to the dianion by sodium naphthalenide or potassium/graphite intercolate. Treatment with acid chlorides generates the anionic acyl complex, which again undergoes O-alkylation to produce carbene complexes.[5] Cyclic alkoxycarbene complexes result from intramolecular O-alkylation,[6] while aminocarbene complexes are produced from amides,[7] utilizing TMSCl to assist in the elimination of oxygen (**Figure 6.1**).

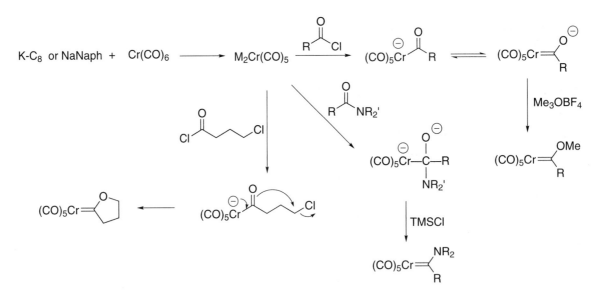

Figure 6.1. Synthesis of Chromium Carbene Complexes from $M_2Cr(CO)_5$

Other, less general routes to carbene complexes involve the reaction of functionalized organozinc reagents with active $Cr(CO)_5$ species (**Eq. 6.2**)[8] and the reactions of 4- and 5-hydroxy-1-alkynes[9] with active $Cr(CO)_5$ species (**Eqs. 6.3**[10] and **6.4**[11]).

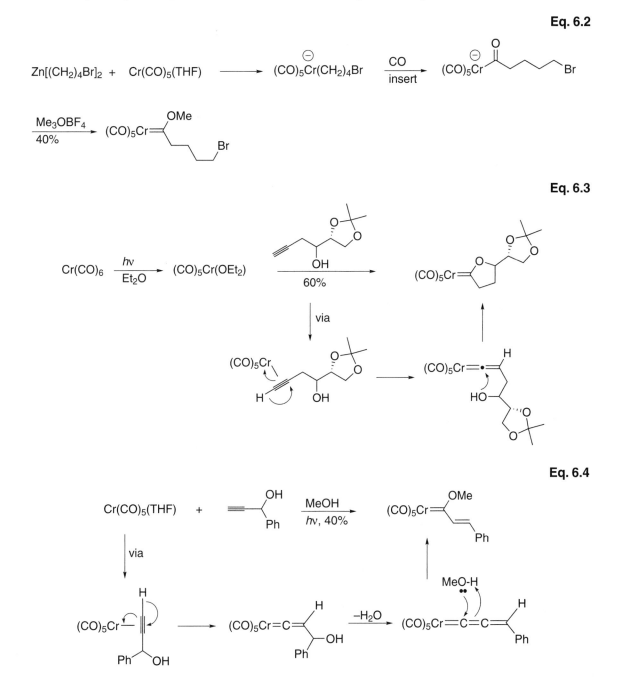

Eq. 6.2

Eq. 6.3

Eq. 6.4

These group 6 carbene complexes are yellow to red crystalline solids, and are easily purified by crystallization or chromatography on silica gel. As solids they are quite air stable and easy to handle. In solution, they are slightly air sensitive, particularly in the presence of light, and reactions are best carried out under an inert atmosphere. The heteroatom is required for stability (the diphenyl carbene complex can be made, but decomposes above –20°C), and this stability results from extensive delocalization of

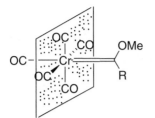

Figure 6.2. The "CO Wall" in Group 6 Carbene Complexes

the lone pair on the heteroatom into the strongly electron-withdrawing metal pentac-arbonyl fragment. This delocalization is evidenced by restricted rotation (14–25 kcal/mol) about the heteroatom–carbene carbon bond, and can be correlated to ^{53}Cr NMR chemical shifts.[12] For delocalization, the orbital of the lone pair must be colinear with the p orbital of the carbene carbon; because the chromium pentacarbonyl fragment presents the carbene carbon with a "wall" of CO's, α-branching of the substituents on the carbene carbon results in steric congestion that may prevent efficient overlap and compromise the stability of the carbene complex (**Figure 6.2**). A practical consequence of this is the difficulty experienced in preparing branched carbene complexes for use in the synthesis of complex molecules.

Electrophilic carbene complexes have a very rich chemistry and undergo reaction at several sites. Fischer carbene complexes are coordinatively saturated, metal(0), d^6 complexes that undergo ligand exchange (CO loss) by a dissociative process (**Eq. 6.5**). Since loss of CO is a prerequisite for substrate coordination and subsequent reaction, this exchange process is central to most synthetically useful reactions. It can be driven thermally ($t_{1/2}$ for CO exchange = 5 minutes at 140°C) or photochemically, and most organic reactions of carbene complexes involve one of these modes of activation.

Eq. 6.5

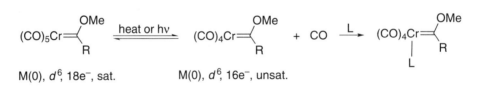

M(0), d^6, 18e⁻, sat.   M(0), d^6, 16e⁻, unsat.

Because CO groups are strongly electron withdrawing, the metal–carbene carbon bond is polarized such that the carbene carbon is electrophilic and generally subject to attack by a wide range of sterically unencumbered nucleophiles (**Eq. 6.6**). In most cases the resulting tetrahedral intermediate is unstable, and the alkoxy group is ejected, producing a new carbene complex. This is one of the best ways to prepare Fischer carbene complexes containing heteroatoms other than oxygen, and the process is quite analogous to the "transesterification" of organic esters. In fact, in many of their reactions, the analogy between alkoxycarbene complexes and organic esters is remarkable. This reaction is inefficient for alkoxides. A better solution is to prepare the reactive (and unstable) acyloxy complexes by the reaction of tetramethylammonium ate complexes with an acid chloride. These complexes are analogous to anhydrides and react with a variety of nucleophiles to form new carbene complexes (**Eq. 6.7**).[13]

Eq. 6.6

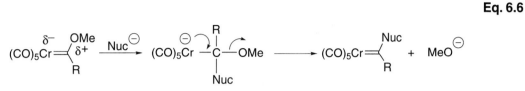

Nuc = R′O, NH$_3$, RNH$_2$
small R$_2$NH, RSH

Eq. 6.7

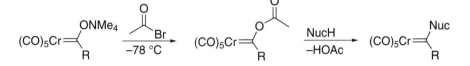

Nucleophilic attack of ketone enolates or hydride[14] on the carbene carbon of Fischer alkoxycarbene complexes furnishes new and unstable complexes that undergo further transformations to give complex organic molecules. Three examples are given in **Eqs. 6.8**,[15] **6.9**,[16] and **6.10**.[17] The last example is particularly interesting in that three carbon monoxide molecules from Cr(CO)$_6$ have been incorporated in the product.

Eq. 6.8

Eq. 6.9

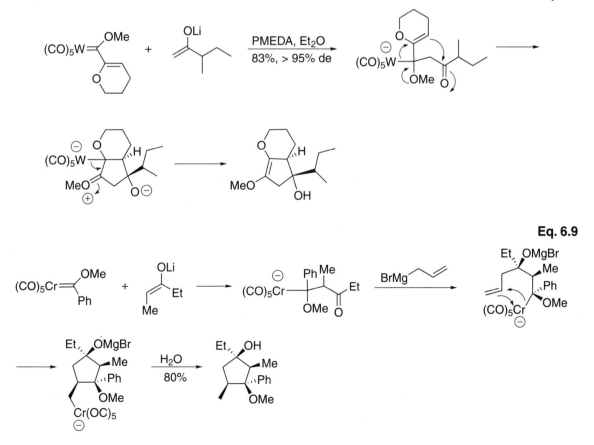

Eq. 6.10

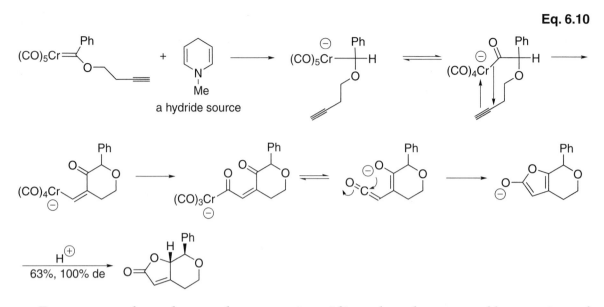

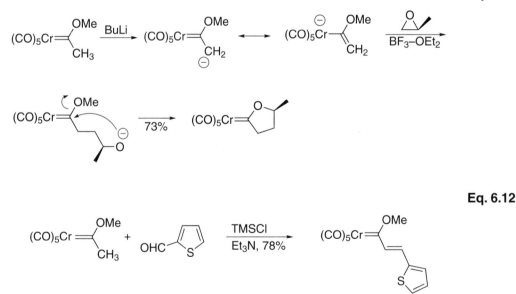

Protons α- to the carbene carbon are quite acidic and can be removed by a variety of bases to give an "enolate" anion stabilized by delocalization into the metal carbonyl fragment (**Eq. 6.11**).[18] The pK_a value for the methyl protons of alkoxy-substituted carbenes is approximately 12 in THF.[19] The pK_a values for group 6 carbenes increase in the order MeS < MeO < Me$_2$N, a reflection of the increasing strength of π-donation with the same order, but they are relatively unaffected by a change in the transition metal. These anions are only weakly nucleophilic, but in the presence of Lewis acid catalysts they react with epoxides[20] and aldehydes (**Eq. 6.12**)[21] to produce homologated carbene complexes.

Eq. 6.11

Eq. 6.12

These reactions are important, since they make possible the elaboration of carbene fragments for ultimate incorporation into organic substrates. Alkylation of these anions with active halides often results in messy polyalkylation. The use of triflates mitigates this problem, as does replacing one of the acceptor CO ligands with a donor phosphine ligand.[22] This makes the chromium fragment a poorer acceptor, decreases the acidity of the α-protons and increases the reactivity of the anion. Similarly, aminocarbene complexes produce more reactive α-anions than do alkoxycarbenes (compare the α-proton acidity of amides vs. esters, N being a stronger donor than O), and are more easily α-alkylated (below). α-Anions of aminocarbene complexes also alkylate aldehydes (**Eq. 6.13**)[23] and add 1,4 to conjugated enones (**Eq. 6.14**).[24] With optically active amino groups, high diastereoselectivity can be achieved.[25]

Eq. 6.13

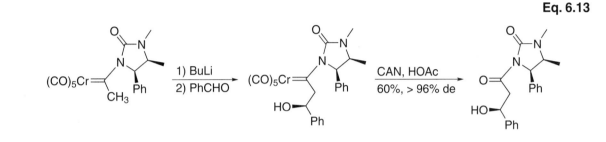

Eq. 6.14

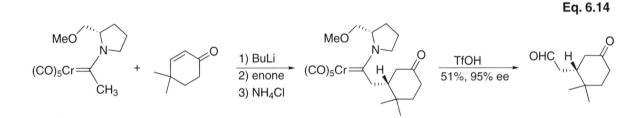

Strong bases that irreversibly deprotonate carbene complexes are often required. Weaker bases, such as pyridine, catalyze the decomposition of alkoxycarbene complexes into enol ethers, by reversible α-deprotonation, followed by reprotonation on the metal and reductive elimination (**Eq. 6.15**).

Eq. 6.15

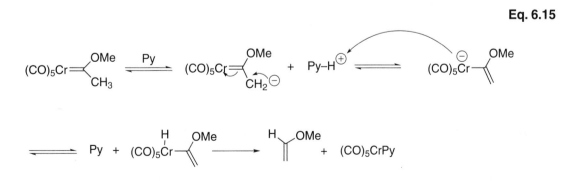

Although originally viewed as a nuisance, this process is quite useful for the synthesis of complex enol ethers from alkoxycarbene complexes.[26] When combined with the formation of cyclic alkoxycarbene complexes from alkynols (**Eq. 6.3**),[8b] quite an efficient synthesis of cyclic enol ethers results (**Eqs. 6.16**[27] and **6.17**).[28] The reaction even works with alkynyl amines (**Eq. 6.18**)[29] and has been made modestly catalytic[30] (**Eq. 6.19**).[31]

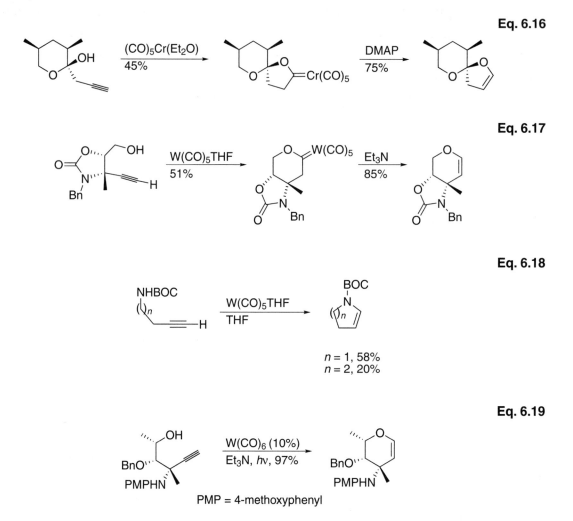

Eq. 6.16

Eq. 6.17

Eq. 6.18

$n = 1, 58\%$
$n = 2, 20\%$

Eq. 6.19

PMP = 4-methoxyphenyl

σ-Alkyne–tungsten complexes undergo formation of related cationic carbene complexes upon treatment with aldehydes in the presence of a Lewis acid. The reaction probably involves the formation of a cationic alkenylidene complex, followed by intramolecular addition of the alcohol, and a proton-assisted cleavage of the carbon–oxygen bond (**Eq. 6.20**).[32] In addition to the facile decomplexation by air oxidation, thus affording unsaturated lactones, these carbenes undergo 1,1-diaddition of Grignard and hydride reagents, 1,4-diaddition of cuprates, and cyclopropanation of the exocyclic alkene with diazomethane. The cyclic carbene complexes can be further elaborated by treatment with a base to give 1-σ-tungsten-1,3-diene complexes.[33] Very facile room temperature [4+2] cycloadditions with nitriles produce fused pyridines after oxidative decomplexation (**Eq. 6.21**).[34]

Eq. 6.20

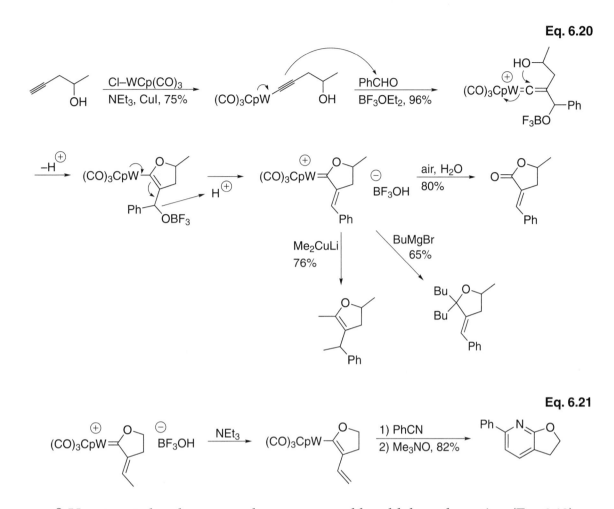

Eq. 6.21

α,β-Unsaturated carbene complexes, prepared by aldol condensation (**Eq. 6.12**) or from alkenyllithium reagents, undergo many reactions common to unsaturated esters, but are considerably more reactive. For instance, Diels–Alder reactions are 2×10^4 faster with carbene complexes than with esters, permitting very facile elaboration of carbene complexes (**Eq. 6.22**).[35]

Eq. 6.22

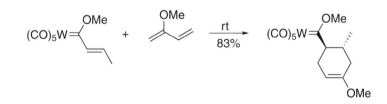

This, again, is a consequence of the electron-withdrawing ability of the M(CO)$_5$ fragment.[36] Intramolecular versions are facile (**Eq. 6.23**),[37] even using the carbene moiety as the diene component (**Eq. 6.24**),[38] and stereoselectivity is often high (**Eq. 6.25**).[39] Alkynyl carbene complexes are also highly reactive towards Diels–Alder cycloaddition, and highly functionalized complexes result when heteroatom dienes are used

(**Eq. 6.26**).[40] Alkynyl carbene complexes having a chiral amino group participate in diastereoselective reactions (**Eq. 6.27**).[41]

Eq. 6.23

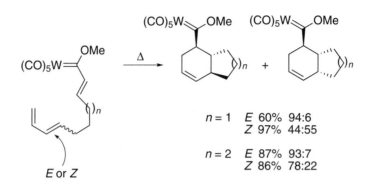

$$n = 1 \quad E \quad 60\% \quad 94{:}6$$
$$Z \quad 97\% \quad 44{:}55$$

$$n = 2 \quad E \quad 87\% \quad 93{:}7$$
$$Z \quad 86\% \quad 78{:}22$$

Eq. 6.24

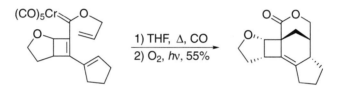

Eq. 6.25

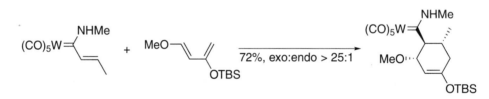

Eq. 6.26

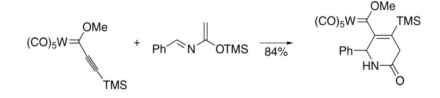

Eq. 6.27

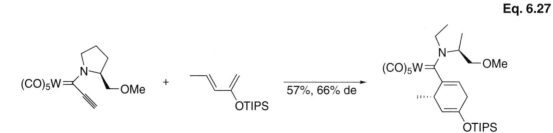

Unsaturated carbene complexes also participate in 1,3-dipolar cycloadditions with, for example, nitrones[42] and diazoalkanes.[43] Diastereoselective reactions can be achieved using chiral carbene complexes (**Eq. 6.28**).[44] Electron-rich alkenes undergo [2+2] cycloadditions (**Eq. 6.29**)[45] and this can be coupled with a benzannulation to give benzocyclobutenes via carbon monoxide insertion (**Eq. 6.30**).[38]

Eq. 6.28

Eq. 6.29

Eq. 6.30

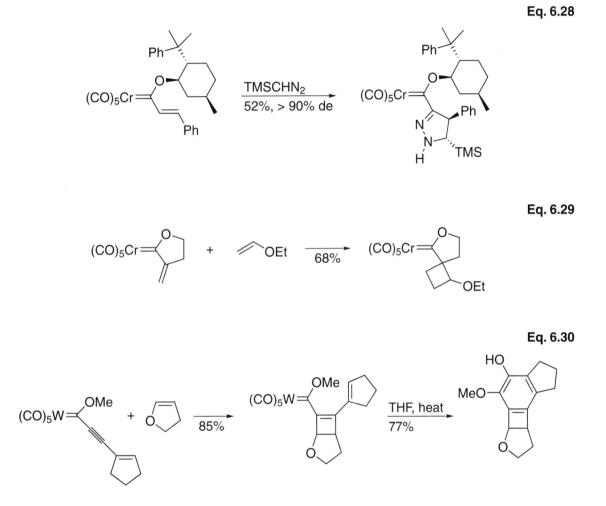

Michael addition reactions to conjugated carbene complexes are also efficient.[46] In fact, in the exchange of the alkoxy group by amines with α,β-unsaturated carbene complexes, Michael addition of the amine is the major competing pathway.

Michael addition provides a very useful route for elaborating carbene complexes for ultimate use in synthesis. Stabilized enolates add readily, and the resulting carbene anion is inert to further attack by nucleophiles, permitting further functionalization of the newly introduced carbonyl group (**Eq. 6.31**).[47] Chiral carbene complexes can be used in asymmetric Michael additions (**Eq. 6.32**).[48] The electron-withdrawing capability of the $Cr(CO)_5$ moiety is such that even furan rings can be used as the Michael acceptor (**Eq. 6.33**).[49]

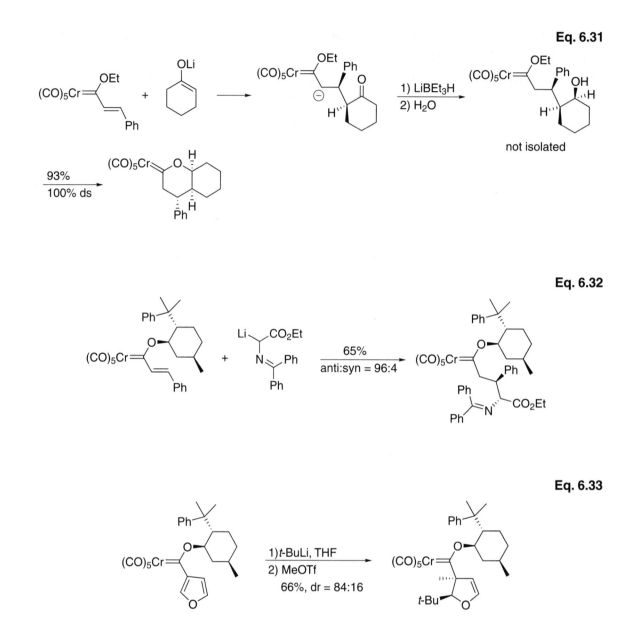

Eq. 6.31

Eq. 6.32

Eq. 6.33

The reaction of alkynyl carbene complexes with β-dicarbonyl enolates results in Michael addition, followed by displacement of the alkoxy group by the enolate oxygen. Reaction of these complexes with enamines results in a remarkable cyclization (**Eq. 6.34**).[50] In addition to carbon nucleophiles, sulfur, oxygen, and nitrogen nucleophiles add to alkynyl carbene complexes to give β-substituted alkenyl complexes.

Eq. 6.34

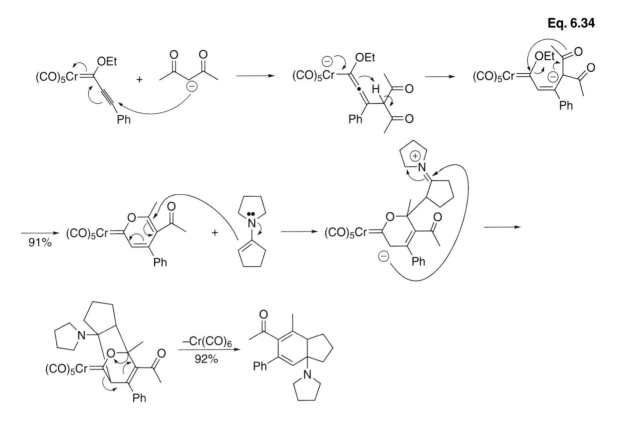

Once elaborated, the carbene fragment can be converted to organic products in a number of ways. Oxidation with CAN, DMSO, tertiary amine oxides, or dimethyldioxirane (**Eq. 6.35**)[51,52] generates the carboxylic acid derivative, as does oxidation of the tetramethylammonium salt in the presence of a nucleophile.[53] However, the metal–carbon double bond can be, and has been, put to better use. A plethora of interesting annulation reactions has been developed wherein the carbon–metal double bond reacts with alkenes or alkynes to give multifunctionalized products.[54]

Eq. 6.35

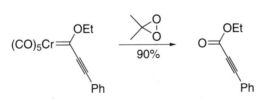

One of the first organic reactions of group 6 carbene complexes studied was the cyclopropanation of electrophilic alkenes (**Eq. 6.36**).[55] The reaction is carried out by simply heating mixtures of the carbene complex and alkene, although often at elevated temperatures. Moderate to good yields of the cyclopropane are obtained. With 1,2-disubstituted alkenes, the stereochemistry of the alkene is maintained in the product, but mixtures of stereoisomers at the (former) carbene carbon are obtained. Dienes, if they are not too highly substituted, are monocyclopropanated (**Eq. 6.37**).[56] 1,3-Dienoic esters cyclopropanate exclusively at the double bond remote (γ,δ) from the ester.[57] In many cases, in addition to the cyclopropane product, varying amounts of a byproduct corresponding to vinyl C–H insertion are obtained (**Eq. 6.38**).[58]

Eq. 6.36

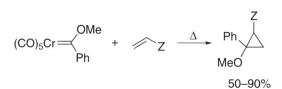

$$Z = CO_2Me, CONMe_2, CN, P(O)(OMe)_2, SO_2Ph$$

Eq. 6.37

Eq. 6.38

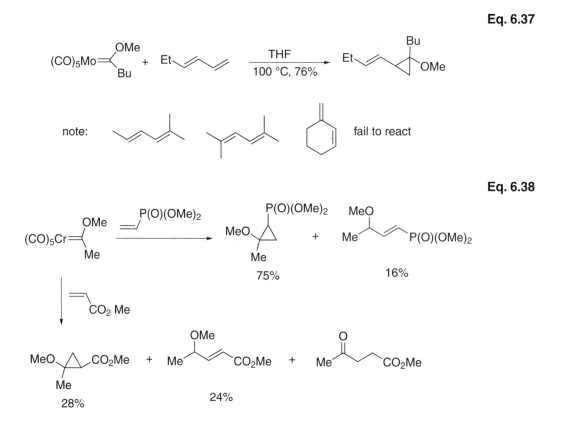

Although cyclopropanes are the expected product of the reaction of free organic carbenes and alkenes, there is ample evidence that free organic carbenes are not involved in the cyclopropanation reactions of transition metal carbene complexes. Rather, the route in **Figure 6.3** is followed. Cyclopropanation is suppressed by CO pressure, indicating that loss of CO to generate a vacant coordination site is required. Coordination of the alkene [(a) in **Figure 6.3**], followed by formal [2+2] cycloaddition to the metal–carbon double bond (b), would generate a metallocyclobutane (c). Depending on the regioselectivity of this cycloaddition step, the alkene substituent may end up either α or β to the metal. Reductive elimination (d) gives the cyclopropane. The observed "carbon–hydrogen" insertion byproduct could be formed by β-hydride elimination (e), then reductive elimination (f) from the metallacyclobutane.

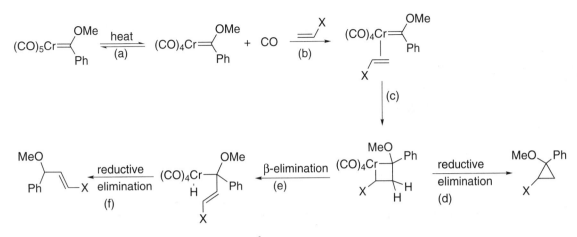

Figure 6.3. Mechanism of Cyclopropanation

Evidence for metallacyclobutane intermediates comes from the attempted cyclopropanation of electron-rich alkenes. Under normal reaction conditions, cyclopropanation does not result; rather, a new alkene resulting from metathesis (alkene group interchange) is obtained (**Eq. 6.39**). Alkene metathesis is a well-known process for carbene complexes, and has found many uses in synthesis (see Section 6.5). It is thought to occur by the cycloaddition/cycloreversion process shown.

Eq. 6.39

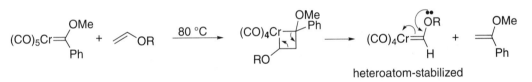

Because the metathesis of electron-rich alkenes produces heteroatom-stabilized carbene complexes, whereas electron-poor alkenes would produce unstabilized carbene complexes, metathesis is the favored pathway with electron-rich substrates. Electron-rich alkenes can be cyclopropanated by unstable acyloxycarbene complexes, generated in situ by O-acylation of tetramethyl ammonium (but not lithium, because of tight ion pairing) "ate" complexes (**Eq. 6.40**),[59] or under high (> 100 atm) pressures of CO, which seems to suppress metathesis.

Eq. 6.40

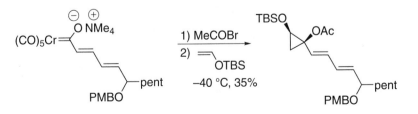

In contrast to electron-poor and electron-rich alkenes, simple aliphatic alkenes (with few exceptions[60]) do not undergo intermolecular cyclopropanation by chromium alkoxycarbenes under any conditions. In general, however, intramolecular reactions are considerably more favorable, and aryl*alken*oxycarbene complexes undergo extremely facile intramolecular cyclopropanation of the unactivated alkene (see below).

Carbene complexes with structurally complex alkoxy groups are not directly available by the standard synthesis (**Figure 6.1**) because the corresponding alkyl triflates or oxonium salts are difficult to synthesize. Furthermore, they cannot be made by alkoxy–alkoxy interchange, since these reactions (in contrast to alkoxy–amine exchange) are slow, inefficient, and beset with side reactions. However, acyloxycarbene complexes, generated in situ by the reaction of acid halides with ammonium acylate complexes, react cleanly, even with structurally complex alcohols, to produce alkoxycarbene complexes in excellent yield (**Eq. 6.41**).[61]

Eq. 6.41

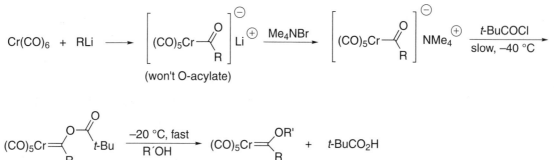

When homoallylic alcohols were exchanged into arylacyloxycarbenes, the carbene complex could not be isolated. Instead, spontaneous, intramolecular cyclopropanation occurred, even under modest pressures of carbon monoxide (**Eq. 6.42**).[62] This is a general phenomenon, and a variety of cyclopropanated tetrahydrofurans, pyrans, and pyrrolidines are available by this route. An intramolecular cyclopropanation–Cope rearrangement is shown in **Eq. 6.43**.[63]

Eq. 6.42

Eq. 6.43

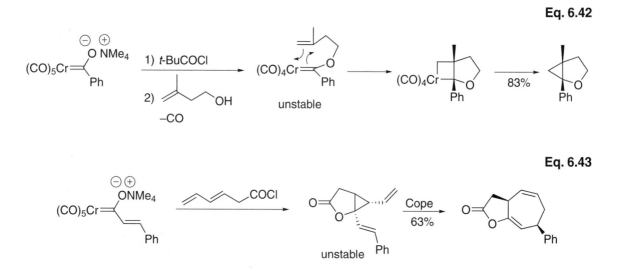

This increase in reactivity is entropic; when the alkene is more than three carbons removed from the oxygen, cyclopropanation does not occur under conditions sufficiently severe to decompose the complex. However, more remote alkenes *will* cyclopropanate if they are conformationally held in proximity to the carbene system[64] (**Eq. 6.44**).[65]

Eq. 6.44

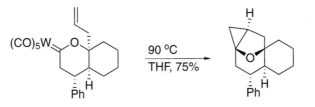

Alkynes cyclo-add to carbene complexes more readily than do alkenes. Treatment of enynes with chromium carbene complexes produces bicyclic compounds in excellent yield, through a process of alkyne cycloaddition/electrocyclic ring opening/alkene cycloaddition/reductive elimination (**Eq. 6.45**).[66] Many variations of this theme are possible and have been tried, including tethering the alkene to the carbene[67] (**Eq. 6.46**),[68] the alkyne to the carbene (**Eq. 6.47**),[69] both the alkyne and alkene to the carbene,[70] and the alkyne to a 1,3 diene (**Eq. 6.48**).[71] All lead to a remarkable increase in complexity in a single reaction. The alkyne cycloaddition–metathesis intermediate can be intercepted by a carbonyl group, as shown in **Eq. 6.49**.[72]

Eq. 6.45

Eq. 6.46

Eq. 6.47

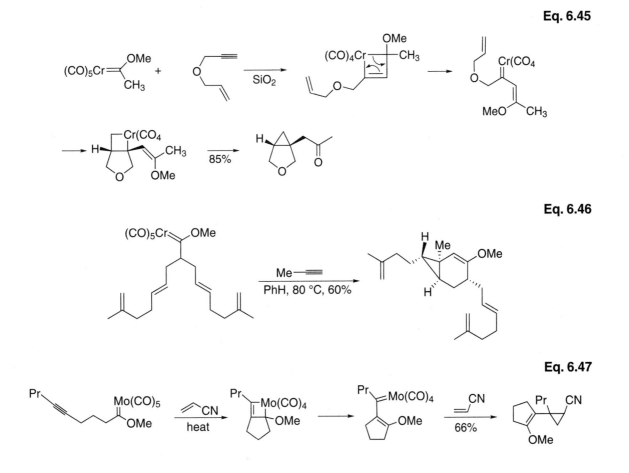

Eq. 6.48

Eq. 6.49

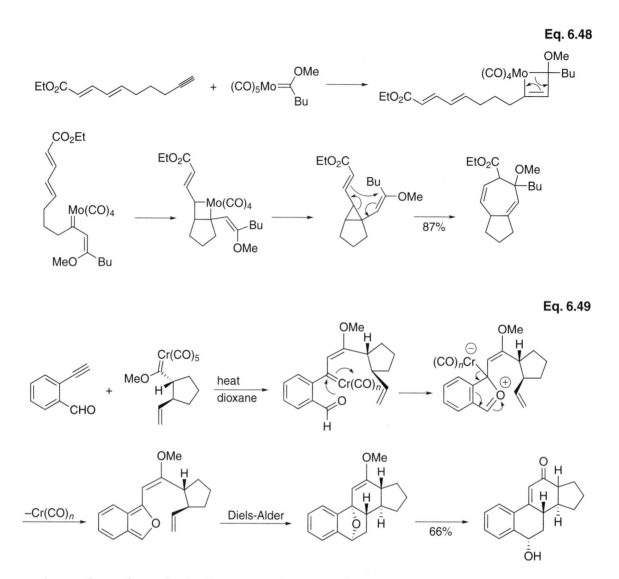

A number of synthetically interesting and diverse reactions of Fischer carbene complexes wherein carbon monoxide has been inserted into the product have been developed. A synthetically very attractive application is the Dötz benzannulation reaction, a thermal reaction of alkenyl and aryl alkoxycarbene complexes with alkynes, to produce hydroquinone derivatives. The general transformation is shown in **Eq. 6.50**.[73] Although the overall sequence is complex, the mechanism is reasonably well understood, and involves many of the standard processes of organometallic chemistry. Again, the reaction begins with a thermal loss of one CO to generate a vacant coordination site. Coordination of the alkyne, followed by cycloaddition, generates the metallacyclobutene, which inserts CO to give a metallacyclopentenone. Fragmentation of this metallacycle generates a metal-bound η^6-vinyl ketene, for which there is excellent experimental evidence (an X-ray study in one case and isolation of the vinylketene with sterically hindered alkynes[74]). Cyclization of this vinyl ketene, followed by enolization, produces the hydroquinone derivative as its $Cr(CO)_3$ complex (not shown; see Chapter 10). The free ligand is obtained by oxidative removal of chromium by exposure to air and light. Depending on the metal, the solvent, and particularly the heteroatom, the CO insertion step may be supplanted by an electrocyclic ring-opening step, producing indenes rather than hydroquinones. This is particularly common for aminocarbene complexes.[75]

Eq. 6.50

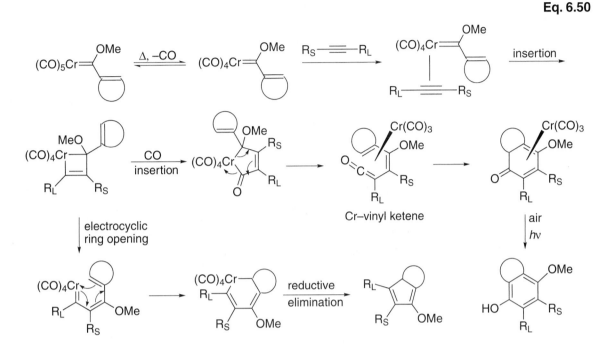

R_L = large substituent; R_S = small substituent

Because of the stability and the ease of handling of group 6 carbene complexes, and the many ways they can be elaborated, highly functionalized, structurally complex carbene complexes are relatively easy to prepare, making their use in organic synthesis appealing. The Dötz reaction has been extensively utilized to make quinone derivatives. Selected examples are shown in **Eqs. 6.51**,[76] **6.52**,[77] **6.53**,[78] and **6.54**.[79] The hydroquinone monoalkyl ether is readily oxidized to a quinone using CAN.

Eq. 6.51

Eq. 6.52

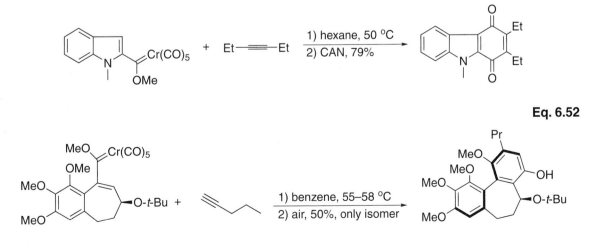

Eq. 6.53

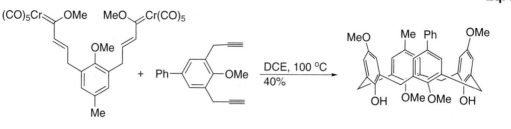

Eq. 6.54

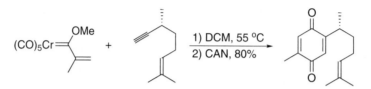

As is the case with cyclopropanation, a cascade of different reactions can be arranged so as to make many bonds in a single reaction sequence. Such a case is shown in **Eq. 6.55**,[80] in which a Diels–Alder cycloaddition is followed by an alkyne metathesis, and then a Dötz reaction, all in a one-pot reaction.

Eq. 6.55

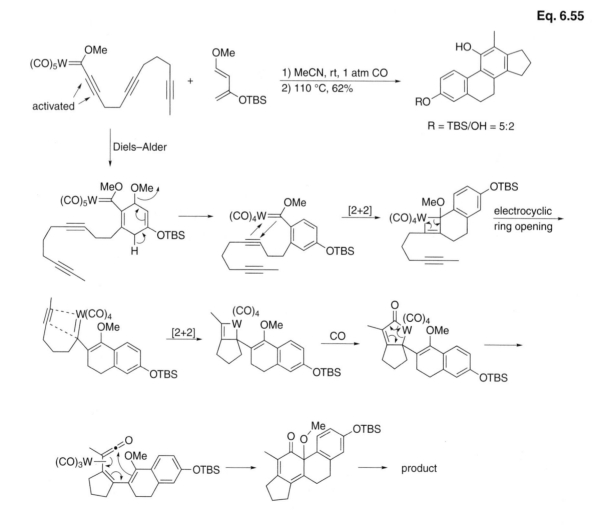

The observation that the photolysis of chromium carbene complexes with visible light resulted in insertion of CO into the metal–carbon double bond to generate species with ketene-like reactivity (see below)[81] led to the development of a very clever alternative to the Dötz benzannulation process (**Eq. 6.56**).[82] By synthesizing cis-dienyl carbene complexes, the key vinylketene intermediate in the Dötz reactions could be generated by a photochemically-driven CO insertion, resulting in an efficient benzannulation process. Thermal or photochemical insertions of isonitriles into the metal–carbon double bond similarly produces alkenylketeneimine intermediates, which also cyclize readily (**Eq. 6.57**).[83] Very nice applications of both of these processes are shown in **Eqs. 6.58**[84] and **6.59**.[85]

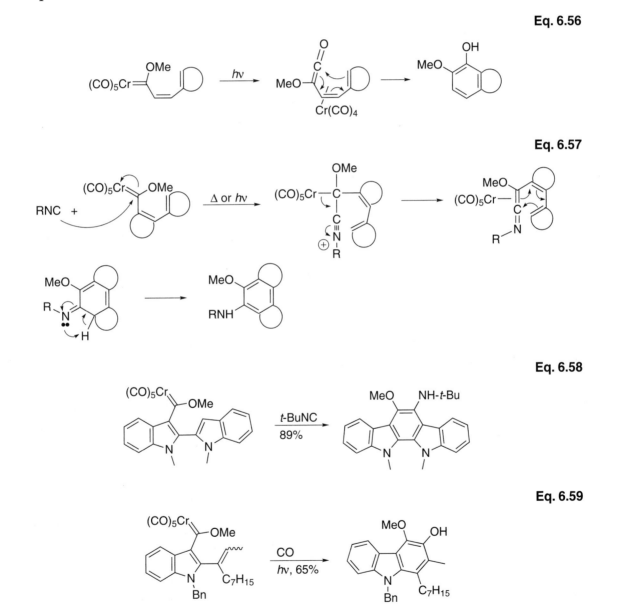

Eq. 6.56

Eq. 6.57

Eq. 6.58

Eq. 6.59

The Dötz reaction is also quite sensitive to the nature of the heteroatom on the carbene complex. Although alkoxycarbene complexes usually give hydroquinone derivatives, aminocarbene complexes usually fail to insert CO,[86] and give indane derivatives instead (see **Eq. 6.50**).[87] When carried out intramolecularly, the reaction is remarkably diastereoselective (**Eq. 6.60**).[88] When combined with Diels–Alder cycloaddition, complex systems can be produced efficiently (**Eq. 6.61**).[89]

Eq. 6.60

Eq. 6.61

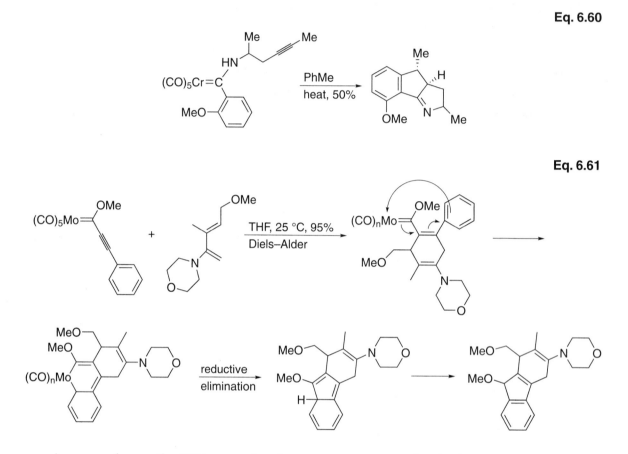

As seen above, the Dötz reaction has many steps, each of which can be diverted along other pathways, leading to other products. The main problem in the development of this chemistry was to develop conditions that led to a single major product, since the reaction is quite sensitive to both the reaction conditions and the structural features of the reactants. For example, the ketene intermediate can be intercepted by pendant nucleophiles, such as alcohols (**Eq. 6.62**),[90] amines, and carbonyls (**Eq. 6.63**).[91]

Eq. 6.62

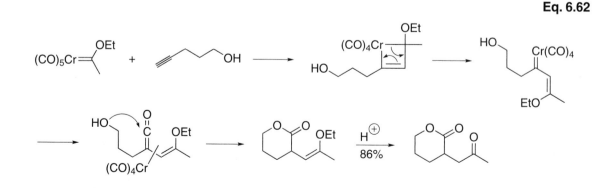

Eq. 6.63

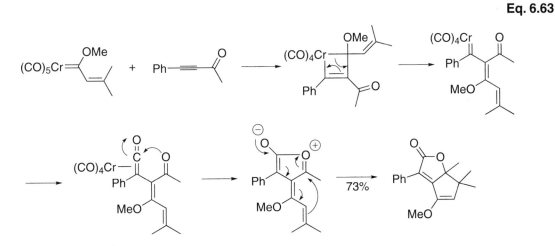

When cyclopropylcarbene complexes are subjected to the typical Dötz reaction, totally unexpected products result (**Eq. 6.64**).[92] Although the mechanism is not well understood, it is likely that things proceed as usual to the metallacyclopentenone stage [(b) in **Eq. 6.64**], at which point, the sp^2-like cyclopropyl group inserts as if it were an alkene (cyclopropyl groups are readily cleaved by a variety of transition metals) to give a metallacyclooctadienone (c). This then re-inserts to give a bicyclic system [(d); shown as occurring in two steps, for clarity], which fragments to give the cyclopentadienone and an alkene (e). The dienone is reduced by the low-valent chromium fragment in the presence of water. Intramolecular versions also work.[93] When the metal is changed from chromium to the larger molybdenum and tungsten, seven-membered rings form, apparently from direct reductive elimination from the metallacyclooctadienone (**Eqs. 6.65** and **6.66**).[94] (Note that isomerization of the double bonds via 1,5-hydride shifts to give the most stable product occurs under the relatively severe reaction conditions.)

Eq. 6.64

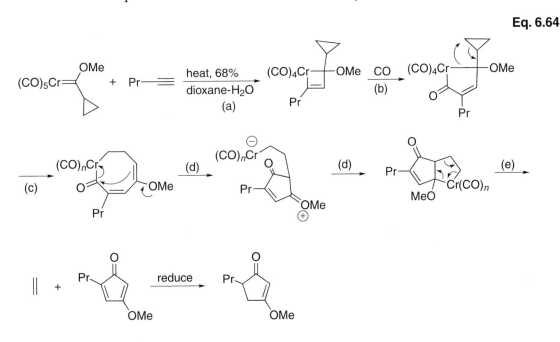

Eq. 6.65

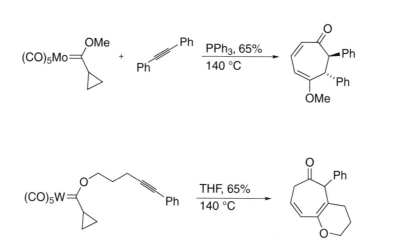

Eq. 6.66

Iron carbene complexes analogous to the group 6 Fischer carbenes are easily prepared, either by the Fischer route (**Eq. 6.1**) or from Collman's reagent [$Na_2Fe(CO)_4$] (**Eq. 6.67**). In the thermal reaction with alkenes, no cyclopropanes are formed; rather, coupling products predominate (**Eq. 6.68**).[95] With alkynes, iron alkoxycarbenes react by a completely different route from group 6 analogs, inserting *two* CO's and giving a mixture of equilibrating pyrones (**Eq. 6.69**)![96]

Eq. 6.67

Eq. 6.68

Eq. 6.69

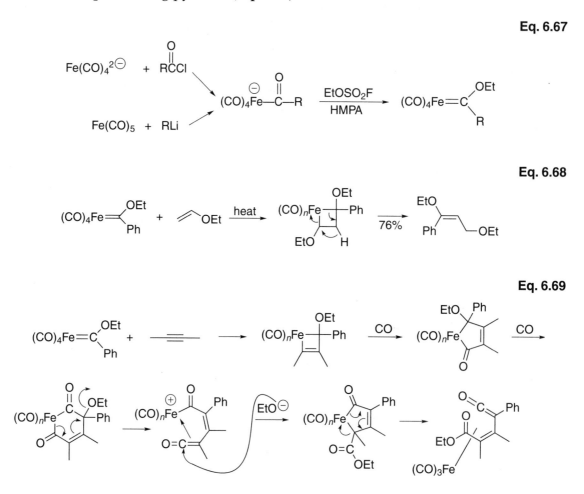

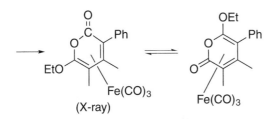

In addition to thermal reactions, chromium heteroatom carbene complexes have a very rich photochemistry that results in a number of synthetically useful processes. All of these complexes are colored and have absorptions in the 350–450 nm range. This visible absorption has been assigned as an allowed metal-to-ligand charge transfer (MLCT) band.[97] From the molecular orbital diagram of the complex, the HOMO is metal–d-orbital centered, and the LUMO is carbene-carbon–p-orbital centered. Thus, photolysis into the MLCT results in a *formal*, reversible, one-electron oxidation of the metal, since, in the absorption process, an electron is removed from the d-centered HOMO to the p-centered LUMO.[98] One of the best ways to drive insertion of carbon monoxide into metal–carbon bonds is to oxidize the metal (Chapters 2 and 4) and, indeed, photolysis of chromium carbene complexes with visible light appears to drive a reversible insertion of CO into the metal–carbon double bond, producing a metal-lacyclopropanone, best represented as a metal-bound ketene (**Eq. 6.70**).[99] Since excited states of transition metal complexes are usually too short-lived to undergo intermolecular reactions, this is almost surely a ground state complex which, if not trapped, de-inserts CO, regenerating the carbene complex. Thus, many carbene complexes can be recovered unchanged after extended periods of photolysis in the absence of reagents that react with ketenes, but are completely consumed in a few hours when photolyzed in the presence of such reagents. Despite several photolysis studies, no *spectroscopic* evidence for the formation of ketenes has been found. The photodriven reactions of Fischer carbene complexes display ketene-like reactivity.

Eq. 6.70

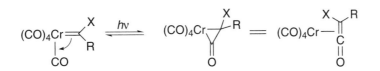

From a synthetic point of view,[100] this is very useful, since it permits the generation of unusual, electron-rich alkoxy-, amino-, and thioketenes not easily available by conventional routes, under very mild conditions (i.e., visible light, any solvent, Pyrex vessels, and no additional reagents). The ketene is metal-bound, and generated in low concentrations, thus normal ketene side reactions, such as dimerization or the incorporation of more than one ketene unit into the product, are not observed. However, the species thus generated undergo most of the normal reactions of ketenes, and herein lies their utility.

The reaction of ketenes with imines to produce β-lactams (the Staudinger reaction) has been extensively developed for the synthesis of this important class of antibiotics. Most biologically active β-lactams are optically active, and have an amino group α- to the lactam carbonyl. To prepare these from chromium carbene complex chemistry, an optically active complex having a hydrogen on the carbene carbon is necessary. These are not available from the conventional Fischer synthesis, since anionic metal formyl complexes are strong reducing agents and transfer hydrides to electrophiles. The requisite optically active aminocarbene complex can be made efficiently by an alternative route involving the reaction of $K_2Cr(CO)_5$ with the appropriate amide, followed by addition of TMSCl to promote the displacement of the oxygen (**Figure 6.1**).[101]

This optically active aminocarbene complex undergoes efficient reaction with a very wide range of imines to give optically active β-lactams in excellent chemical yield, and with very high diastereoselectivity (**Eq. 6.71**).[102] The absolute stereochemistry of the α-position is determined by the chiral auxiliary on nitrogen, which transfers its absolute configuration to this position (i.e., $R \to R$ and $S \to S$), while the relative (i.e., cis–trans) stereochemistry between the two new chiral centers is determined by the imine.[103] When the reaction is carried out under a modest (3–6 atm) pressure of CO, chromium hexacarbonyl can be recovered and reused. The chiral auxiliary is easily removed by hydrolysis of the acetonide, followed by either reductive (H_2, Pd/C) or oxidative (IO_4^-) cleavage of the amino alcohol. Alternatively, the oxazolidine can be transformed into an oxazolidinone and this compound can be alkylated with retention of stereochemistry using potassium hexamethyldisilazide and an electrophile.[104] This chemistry has been used in a formal synthesis of (+)-1-carbacephalothin (**Eq. 6.72**).[105]

Eq. 6.71

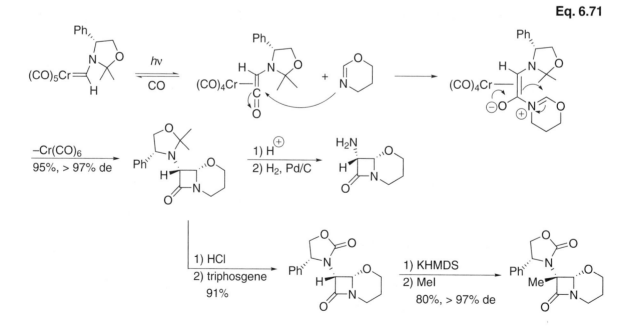

Eq. 6.72

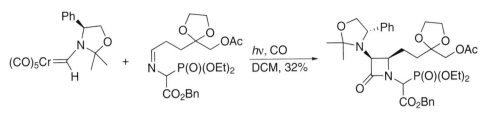

Imidazolines are particularly interesting substrates for this β-lactam forming process, since they fail to form β-lactams when allowed to react with ketenes or with ester enolates. However, photochemical reaction with chromium carbene complexes produce azapenams, a new class of β-lactam, in good yield (**Eq. 6.73**).[106,107] This is one of several cases for which chromium carbene photochemistry and classical ketene chemistry differ in detail.

Eq. 6.73

$$(CO)_5Cr \underset{Me}{\overset{OMe}{=\!\!<}} \ + \ \text{(imidazoline, Cbz, N, Me, CO}_2\text{Me)} \quad \xrightarrow[\substack{2)\ H_2\ Pd/C \\ 70\%,\ dr > 99\%}]{1)\ h\nu} \quad \text{(azapenam, MeO, Me, H, N, O, Me, CO}_2\text{Me)}$$

Ketenes also undergo facile [2+2] cycloaddition reactions with alkenes, producing cyclobutanones, with a high degree of stereoselectivity, favoring the more hindered cyclobutanone ("masochistic stereoinduction"[108]). Photolysis of chromium alkoxycarbene complexes in the presence of a wide variety of alkenes produces cyclobutanones in high yields, with roughly the same stereoselectivity observed using conventionally generated ketenes (**Eq. 6.74**).[109] As expected, the reaction is restricted to relatively electron-rich alkenes, and electron-deficient alkenes fail altogether. It is also restricted to alkoxycarbene complexes or aminocarbene complexes having aryl groups on nitrogen to decrease the nucleophilicity of the ketene.[110] Intramolecular versions also work well, giving access to a variety of bicyclic systems. Even unconventional cycloadditions proceed in remarkably good yield[111] (**Eq. 6.75**).[112] With optically active alkenes, high diastereoselectivity is observed (**Eq. 6.76**).[113] The resulting optically active cyclobutanones undergo Baeyer–Villager oxidation/elimination to produce butenolides, useful synthetic intermediates.

Eq. 6.74

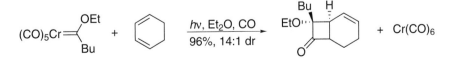

Eq. 6.75

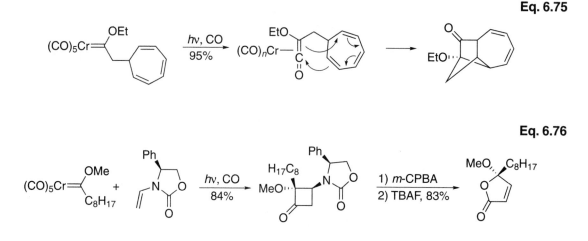

Eq. 6.76

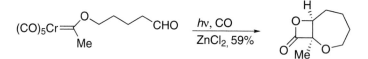

Photolysis of alkoxy chromium carbenes in the presence of an aldehyde and a Lewis acid[114] or base[115] affords β-lactones. Intramolecular reactions are usually more efficient (**Eq. 6.77**). Ketenes in general are insufficiently reactive and usually require a catalyst.

Eq. 6.77

Ketenes react with alcohols or amines to give carboxylic acid derivatives. Similarly, the photolysis of chromium carbene complexes in the presence of alcohols produces esters in excellent yield. Use of aminocarbene complexes, prepared from amides (**Figure 6.1**), provides a very direct synthesis of α-amino acid derivatives from amides, a transformation not easily achieved using conventional methodology.[116] Lactams, including β-lactams, undergo this transformation as well, producing cyclic amino acids.

Optically active α-amino acids are an exceptionally important class of compounds, not only because of their central biological role as the fundamental units of peptides and proteins, but also because of their role in the development of new pharmaceutical agents. By synthesizing optically active aminocarbene complexes and taking advantage of their α-carbon reactivity, a variety of natural (*S*) and unnatural (*R*) amino acids can be synthesized (**Eq. 6.78**).[117] With aldehydes as electrophiles, homoserines could be efficiently synthesized.[118]

Eq. 6.78

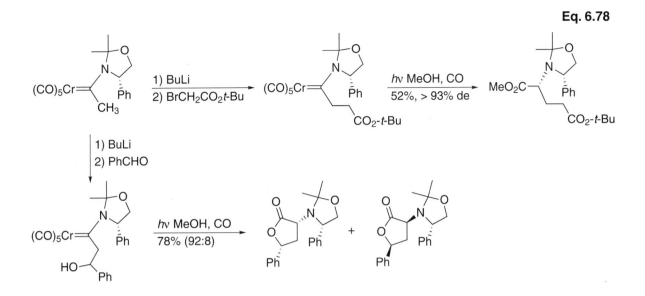

By replacing the alcohol with an optically active α-amino acid ester, dipeptides can be synthesized in a process that forms the peptide bond and the new stereogenic center on the carbene-complex-derived amino acid fragment in a single step (**Eq. 6.79**).[119] This permits the direct introduction of natural (or unnatural) amino acid fragments into peptides utilizing visible light as the coupling agent. In this case, "double diastereoselection" is observed with the (R,S,S) dipeptide being the matched and the (S,R,S) dipeptide being the mismatched pair. Even sterically hindered α,α-dialkyl- and N-alkyl-α,α-dialkylamino acid esters couple in good yield,[120] as do polystyrene-bound,[121] and soluble, PEG-bound[122] amino acid residues.

Eq. 6.79

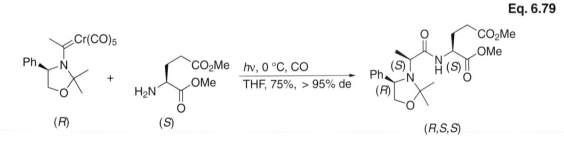

Finally, transmetallation of group 6 Fischer carbene complexes to other metals can be accomplished either photochemically or thermally (**Eq. 6.80**[123]).[124] More interestingly, palladium[125] and copper[126] complexes catalyze the facile dimerization of alkoxychromium carbenes (**Eq. 6.81**). A plausible mechanism involving an initial transmetallation from chromium to palladium (or copper) has been proposed. Reactions of α,β-unsaturated alkoxychromium carbenes with alkynes[127] or allenes[128] produce five-membered rings in the presence of rhodium or nickel catalysts (**Eq. 6.82**). The regioselectivity of the addition of the allene was modulated by the catalyst used. All of these reactions probably proceed via transmetallation of the carbene ligand.

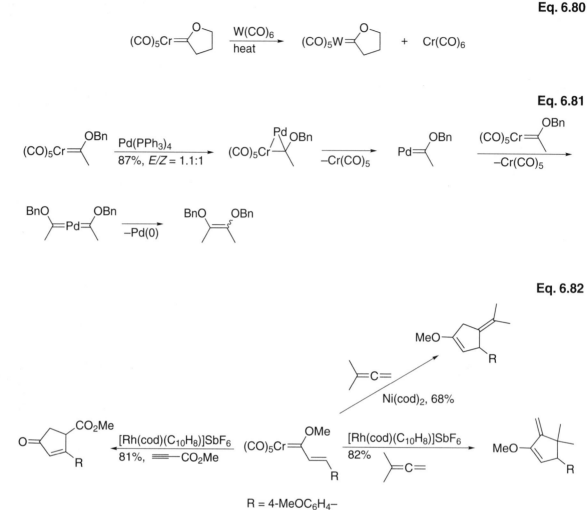

Eq. 6.80

Eq. 6.81

Eq. 6.82

R = 4-MeOC$_6$H$_4$–
C$_{10}$H$_8$ = naphthalene

6.3 Electrophilic, Nonstabilized Carbene Complexes[129]

Group 6 carbene complexes lacking a heteroatom on the carbene carbon are unstable and difficult to prepare. As a consequence, they have not been extensively utilized in synthesis, although they do cyclopropanate alkenes.[130] Of particular interest as cyclopropanation reagents are hydrido carbenes formed by the reaction of alkoxycarbenes with dihydropyridines. The hydrido carbenes are unstable and form pyridine ylides. Thermal reactions of the ylides with alkenes produce cyclopropanated products probably via the hydrido carbene (**Eq. 6.83**).[131] An example of an intramolecular cyclopropanation is shown in **Eq. 6.84**.

Eq. 6.83

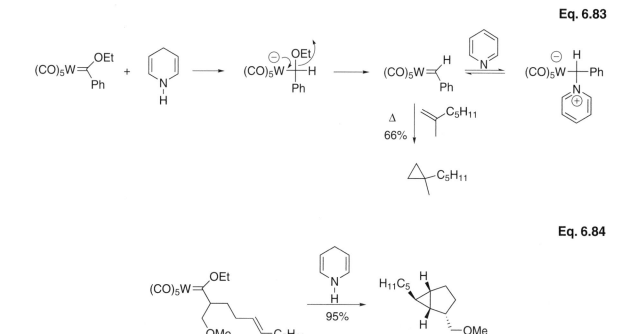

Eq. 6.84

In contrast to group 6 carbenes, a number of stable, easily prepared iron carbonyl complexes are precursors to unstabilized electrophilic carbene complexes. When these precursor complexes are treated with appropriate reagents in the presence of alkenes, cyclopropanation results. Among the most convenient of these carbene precursors are the methylthiomethyl complexes, easily prepared on a large scale by the reaction of $CpFe(CO)_2Na$ with CH_3SCH_2Cl.[132] S-Methylation, followed by warming in the presence of an excess of an electron-rich alkene, produces the unstable cationic carbene complex which then cyclopropanates the alkene. The corresponding methoxymethyl complex behaves similarly (**Eq. 6.85**). Perhaps the most versatile approach relies upon the reaction of the strongly nucleophilic $CpFe(CO)_2Na$ with aldehydes[133] or ketals[134] to produce the requisite α-oxoiron complexes. These undergo α-elimination to produce cationic carbene complexes when treated with alkylating or silylating agents (**Eq. 6.86**).

Eq. 6.85

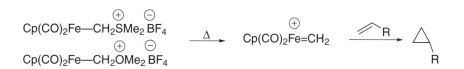

This cyclopropanation is somewhat limited. Only relatively unfunctionalized, electron-rich alkenes undergo the reaction, and relatively few carbene groups—namely, "CH_2", "$CHCH_3$", "$C(CH_3)_2$", "$CHPh$", and "CHC_3H_5"—can be transferred. With substituted carbene complexes, mixtures of cis and trans disubstituted cyclopropanes are obtained.

Eq. 6.86

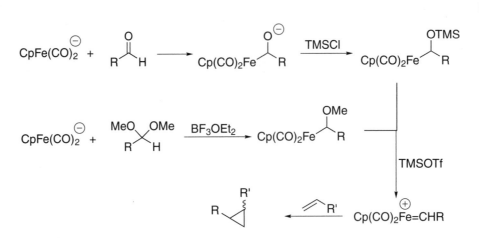

By replacing one of the carbon monoxide groups with a phosphine, the iron–carbene complex precursors become chiral at iron, and can be resolved. Additional asymmetry can be introduced by using an optically active phosphine, giving complexes chiral both at iron and phosphorus. These optically active carbene complexes cyclopropanate alkenes with quite good enantioselectivity, but only modest cis–trans selectivity (**Eq. 6.87**). The mechanism of this asymmetric cyclopropanation has been extensively studied.[135] Because chiral metal complexes induce asymmetry in the products, "free" carbenes cannot be involved. Unexpectedly, perhaps, metallacyclobutanes also are not involved. From careful stereochemical and labeling studies, the mechanism shown in **Eq. 6.88** has been established. It involves attack of the alkene on the electrophilic carbene complex to generate an electrophilic center at the γ-carbon. If this carbon bears an electron-donating group, this intermediate is sufficiently long-lived to permit rotation about the β–γ-bond, leading to loss of the original alkene stereochemistry. This γ-cationic center is then attacked by the Fe–C bond from the back side, resulting in inversion at this center. Consistent with this mechanism, iron–carbene intermediates can effect the cationic cyclization of polyenes (**Eq. 6.89**),[136] as well as carbene-like C–H insertion reactions (**Eq. 6.90**).[137]

Eq. 6.87

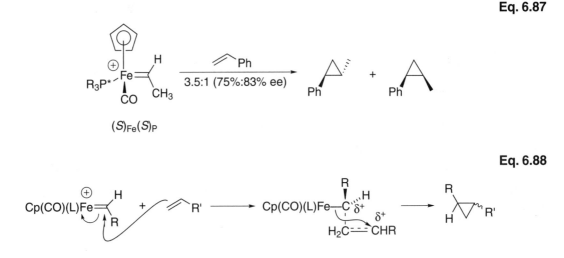

Eq. 6.89

Eq. 6.90

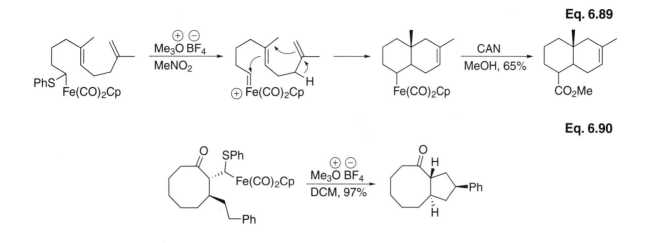

6.4 Metal-Catalyzed Decomposition of Diazo Compounds Proceeding Through Unstabilized Electrophilic Carbene Intermediates

Transition metals catalyze a number of important reactions of organic diazo compounds, among which the cyclopropanation of alkenes is one of the most synthetically useful (**Eq. 6.91**).[138] A wide variety of transition metal complexes are efficient catalysts, among which copper(I) triflate, palladium(II) salts, rhodium(II) acetate, and $Rh_6(CO)_{16}$ are most commonly used. The reaction is most effective when relatively stable α-diazocarbonyl compounds are used, since the major competing reaction is decomposition of the diazo compound itself. Alkenes ranging from electron-rich enol ethers to electron-poor α,β-unsaturated esters undergo catalytic cyclopropanation with diazo compounds, although alkenes with electron-withdrawing substituents are likely to react by a mechanism different from that of other alkenes, and β-substitution drastically suppresses the reaction. The order of reactivity for alkenes is electron rich > "neutral" >> electron poor and cis > trans. Alkene geometry is maintained, but there is little stereoselectivity for the diazo-derived group. Dienes, alkynes, and aromatic rings are also cyclopropanated. Detailed mechanistic studies have implicated the intermediacy of metal carbene complexes formed by the electrophilic addition of the metal catalyst to the diazo compound in this cyclopropanation.[139]

Eq. 6.91

$$\text{CH}_2=\text{CHX} + \text{N}_2\text{CHCO}_2\text{Et} \xrightarrow{\text{ML}_n} \text{(cyclopropane: EtO}_2\text{C, X)}$$

$ML_n = Rh_2(OAc)_4, CuCl–P(O-i-Pr)_3, Rh_6(CO)_{16}, PdCl_2(PhCN)_2$

$X = BrCH_2, ClCH_2, PhO, Bu, OAc, OEt, OBu, i-Pr, t-Bu, CH_2=CPh$

$CH_2=CMe, CH_2=C-t-Bu, CH=CHOMe, CH=CHCl, CH=CHPh, CH=CHMe$

From a synthetic point of view, rhodium(II) acetate is by far the most-used catalyst, because it is commercially available, easy to handle, and reacts quickly and efficiently. Many functional groups are tolerated, making elaborate cyclopropanes available (e.g., **Eq. 6.92**).[140] Intramolecular versions are quite efficient, producing polycyclic ring systems (**Eqs. 6.93**[141] and **6.94**[142]). By using rhodium(II) or copper (I) or (II) complexes in the presence of an optically active ligand, high asymmetric induction can be achieved (**Eq. 6.95**).[143,144,145] With polyenes, the choice of catalyst effects regioselectivity (**Eq. 6.96**).[146]

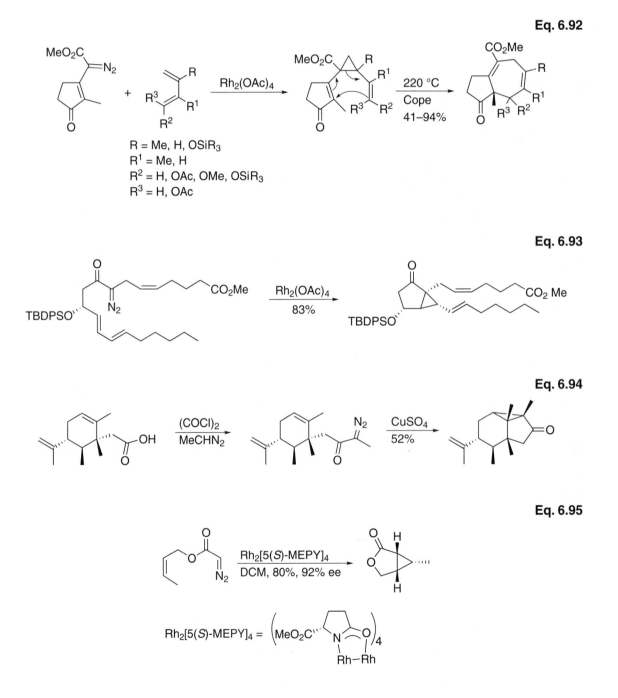

Eq. 6.92

R = Me, H, OSiR₃
R¹ = Me, H
R² = H, OAc, OMe, OSiR₃
R³ = H, OAc

Eq. 6.93

Eq. 6.94

Eq. 6.95

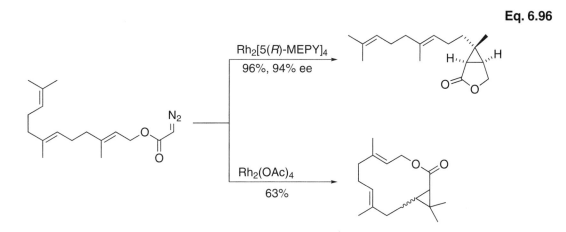

Eq. 6.96

Aromatic rings also participate in cyclopropanation reactions with diazocompounds, and complex organic molecules have been prepared (**Eqs. 6.97**[147] and **6.98**[148]).

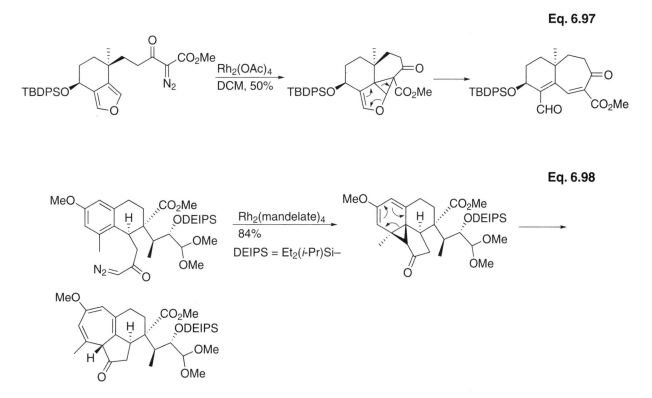

Eq. 6.97

Eq. 6.98

When substrates have additional unsaturation appropriately situated, metathesis "cascade" cyclizations ensue (**Eq. 6.99**),[149] much like the corresponding reactions of chromium carbene complexes (**Eqs. 6.45–6.49**). This provides further evidence for the intermediacy of metal carbene complexes in these cyclopropanations, since such metathesis-derived products are unlikely for other types of carbenoid species.

Eq. 6.99

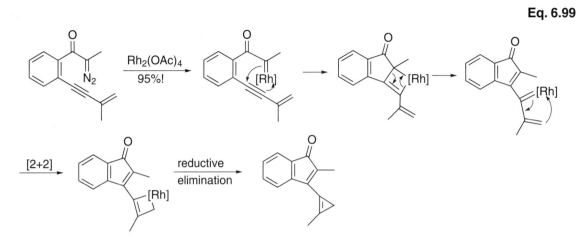

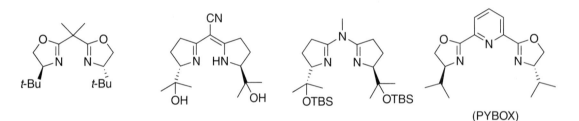

As stated above, a very large number of transition metals catalyze the cyclopropanation of alkenes by diazocompounds and a large number of chiral ligands have been developed.[150,151] Some, normally used in conjunction with Cu^{2+} or Ru^{2+},[152] are shown in **Figure 6.4**.

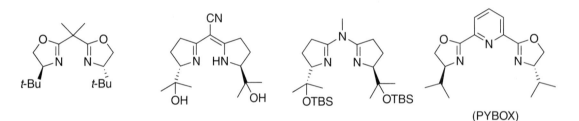

Figure 6.4. Chiral Ligands for Asymmetric Cyclopropanation

The cyclopropanation of alkenes by the metal-catalyzed decomposition of diazoalkanes has many parallels to related reactions of stable carbene complexes, and the two systems are clearly mechanistically related. However, two reactions uncommon for metal carbene complexes—namely, X–H insertion and ylide formation—are observed with the metal-catalyzed decomposition of diazoalkanes, and both are synthetically useful.

Rhodium(II) catalyzes the insertion of diazoalkanes into C–H, O–H, N–H, S–H, and Si–H bonds in a process that is very unlikely to involve free carbenes.[153] Detailed mechanisms involving metal carbene complexes have not yet been elucidated, but are likely to involve electrophilic attack of the metal-bound carbene carbon on the X–H bond. The general order of reactivity for C–H insertion is $2° > 1° \approx 3°$, but is catalyst dependent.[154] As expected for electrophilic attack, electron-withdrawing groups deactivate C–H bonds and donating groups activate them. Intramolecular insertions favor the formation of five-membered rings, and fluorinated carboxylate ligands on rhodium favor the insertion with aromatic C–H bonds.[155] Very high asymmetric induction can be observed when chiral metal complex catalysts are used (**Eqs. 6.100**[156] and **6.101**[157]), strongly implicating an intimate association of the metal with the carbene fragment during bond formation.

Eq. 6.100

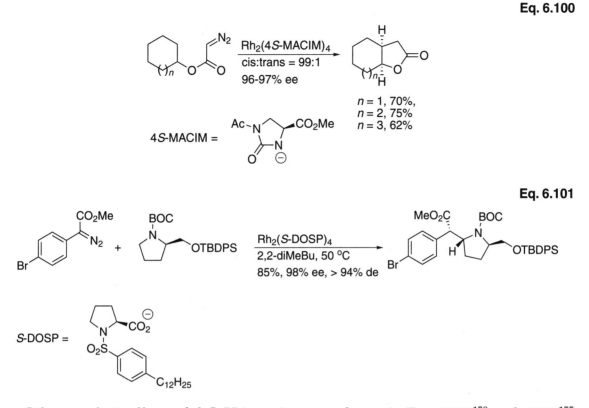

Other synthetically useful C–H insertions are shown in **Eqs. 6.102**[158] and **6.103**.[155] Some examples in total synthesis are shown in **Eqs. 6.104**[159] and **6.105**.[160] Rhodium(II) and copper salts also catalyze N–H (**Eq. 6.106**),[161] O–H (**Eq. 6.107**),[162] and Si–H[163] insertions that are synthetically useful.

Eq. 6.101

Eq. 6.102

Eq. 6.103

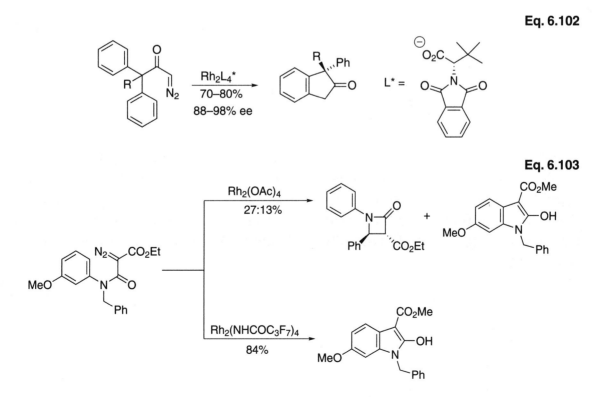

Eq. 6.104

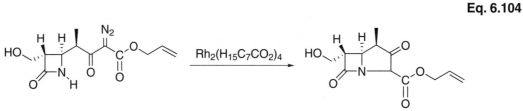

Eq. 6.105

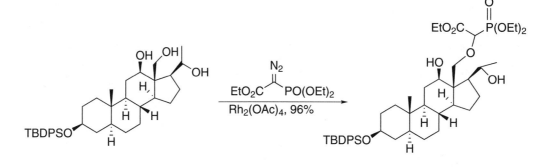

Eq. 6.106

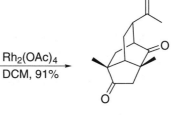

Eq. 6.107

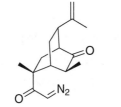

Finally, transition metals catalyze the reaction between diazo compounds and heteroatom lone pairs to produce ylides, which have a very rich chemistry in their own right (**Eq. 6.108**).[164] The specific reactivity depends on the nature of the ylide formed. For example, sulfur ylides undergo facile [2,3]-rearrangements (**Eq. 6.109**)[165] or Stevens rearrangement (**Eq. 6.110**),[166] as do nitrogen ylides[167] (**Eq. 6.111**)[168] and oxygen ylides[169] (**Eq. 6.112**).[170]

Eq. 6.108

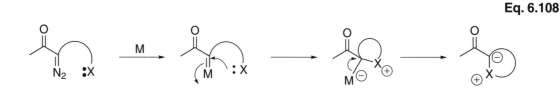

Eq. 6.109

Eq. 6.110

Eq. 6.111

Eq. 6.112

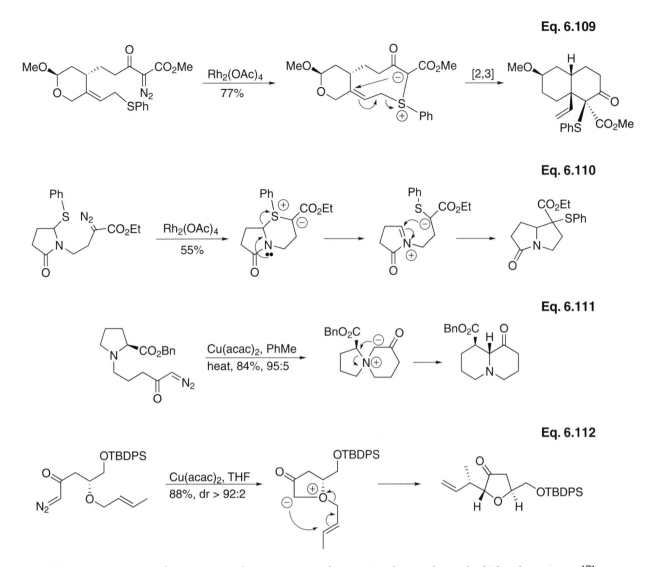

Most impressive from a synthetic point of view is the carbonyl ylide chemistry,[171] which is accessible by metal-catalyzed diazo decomposition, primarily because carbonyl ylides have a rich dipolar cycloaddition chemistry. This allows the construction of very complex ring systems directly and efficiently, as illustrated by the synthesis of the aspidosperma alkaloid ring system in a single step (**Eq. 6.113**).[172] When coupled with metathesis chemistry (**Eq. 6.74**),[171] remarkable transformations occur in a single pot (**Eq. 6.114**).[173] Two examples of applications in organic total synthesis are shown in **Eqs. 6.115**[174] and **6.116**.[175]

Eq. 6.113

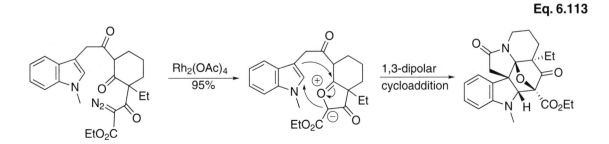

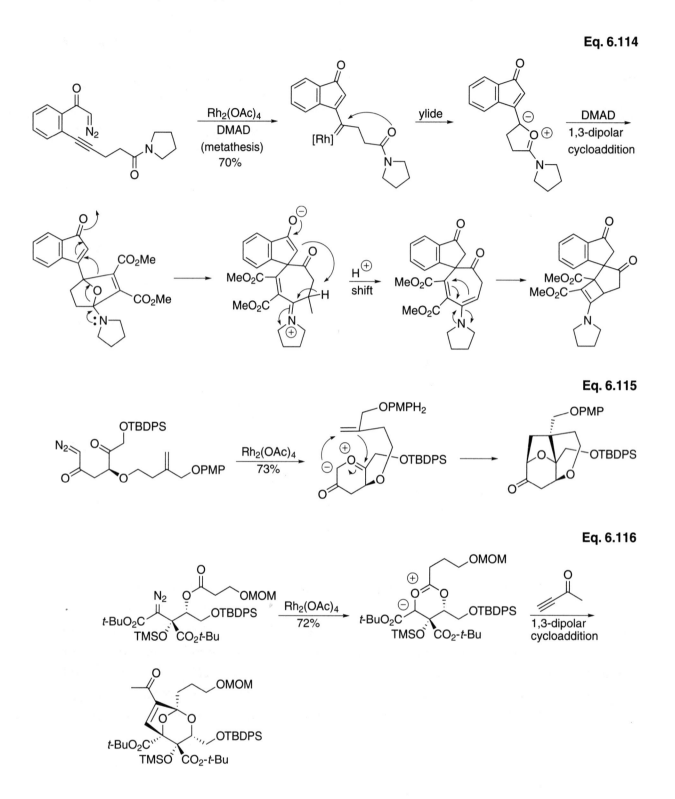

Eq. 6.114

Eq. 6.115

Eq. 6.116

6.5 Metathesis Processes of Electrophilic Carbene Complexes[176]

Alkene metathesis is a process by which metal carbene complexes catalyze a methylene group interchange via a cycloaddition–cycloreversion process (**Eq. 6.117**).

Eq. 6.117

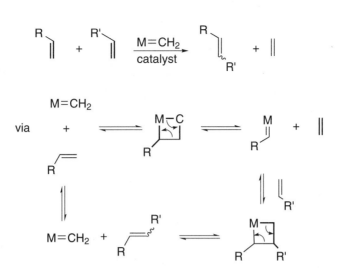

Metathesis has long been of importance industrially, but the high reactivity of the catalysts, coupled with their relative intolerance to functional groups, drastically limited their use in organic synthesis. However, the development of new catalysts that are relatively easy to handle and that perform metathesis efficiently in the presence of a wide array of functional groups has increased the use of this process in complex syntheses.

Although a substantial number of catalysts have been developed, the catalysts of most use in organic synthesis are shown in **Figure 6.5**. They are the Schrock molybde-

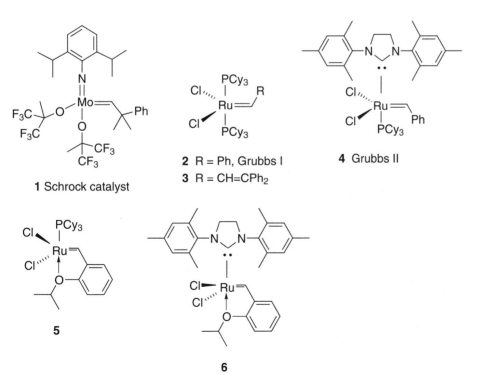

Figure 6.5. Commonly Used Metathesis Catalysts

num metathesis catalyst **1**,[177] the Grubbs ruthenium catalysts **2**[178] and **3**,[179] the Grubbs second-generation catalyst **4**,[180] and the recyclable Hoveyda–Grubbs catalysts **5**[181] and **6**.[182] They share many common features, and their reaction chemistry overlaps considerably. In some cases, however, there is a clear preference for one over the other. The Grubbs and Hoveyda catalysts are, in general, easier to prepare and handle, tolerate a wider array of polar functional groups, and are more stable towards air and moisture. The Schrock catalyst has significantly higher metathesis activity, and often is more efficient with sterically hindered alkenes.[183] Together they have had a *major* impact on synthetic methodology by permitting the use of the alkene functional group as a carbon–carbon bond-forming moiety.

One of the earliest applications of alkene metathesis was to polymer synthesis, and initially these new metathesis catalysts were also applied to polymers in ring-opening metathesis polymerizations (ROMP) (**Eq. 6.118**).[184] Their tolerance for functionality permitted the synthesis of highly functionalized polymers, even under aqueous conditions (**Eq. 6.119**).[185] Metathesis is also finding increased application to materials synthesis (**Eq. 6.120**).[186]

Eq. 6.118

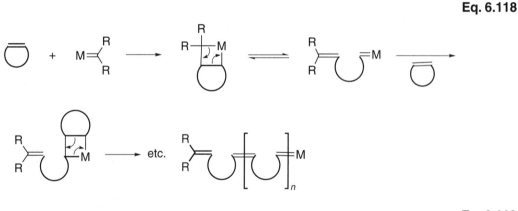

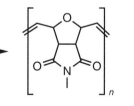

Eq. 6.119

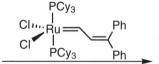

Eq. 6.120

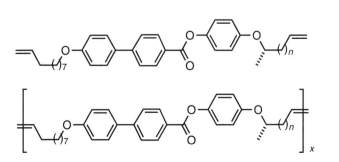

However, it is the application of *ring-closing* metathesis (RCM) that has had the greatest impact on organic synthesis.[187] Although the synthetic community was slow to appreciate the power of ring-closing metathesis, and much of the early work was done by the developers of the catalysts themselves, once it caught on, the use of ring-closing metathesis exploded. Currently, a seemingly limitless variety of carbocyclic and heterocyclic systems and ring sizes, including macrocycles, can be prepared using RCM. Equally impressive is the functional group compatibility with the catalyst systems employed, and a truly staggering array of examples has been reported in the literature. A few of the more spectacular examples of RCM applied in the synthesis of complex organic molecules are shown in **Eqs. 6.121,**[188] **6.122,**[189] **6.123,**[190] **6.124,**[191] and **6.125.**[192]

Eq. 6.121

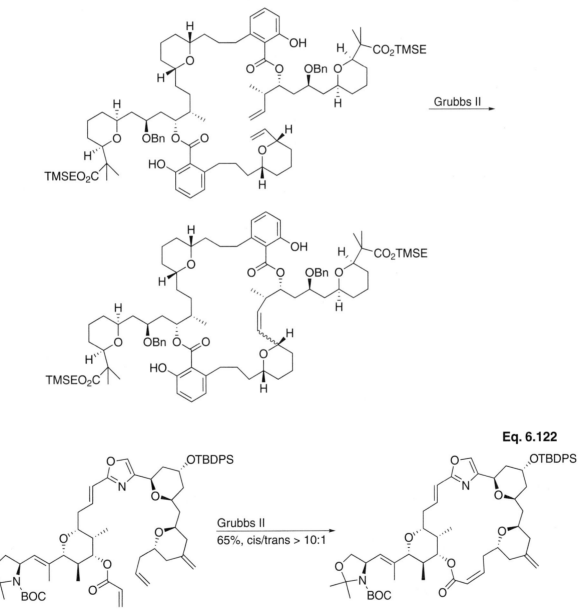

Eq. 6.122

Eq. 6.123

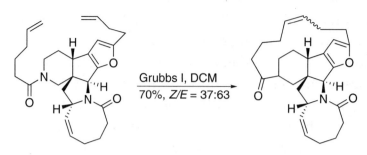

Eq. 6.124

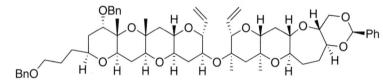

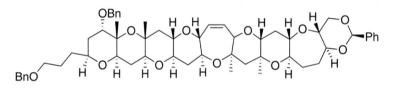

Eq. 6.125

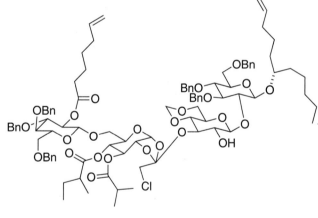

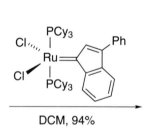

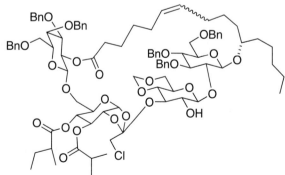

Alkene–alkene cross metathesis was until recently considered useful only for the homodimerization of selected alkenes or was seen as a nuisance in RCM reactions. However, new catalysts and better models for predicting the chemo- and stereoselectivity[193] have been developed and the cross metathesis of two different alkenes is now a viable methodology in organic synthesis of complex molecules (**Eqs. 6.126**[194] and **6.127**[195]).

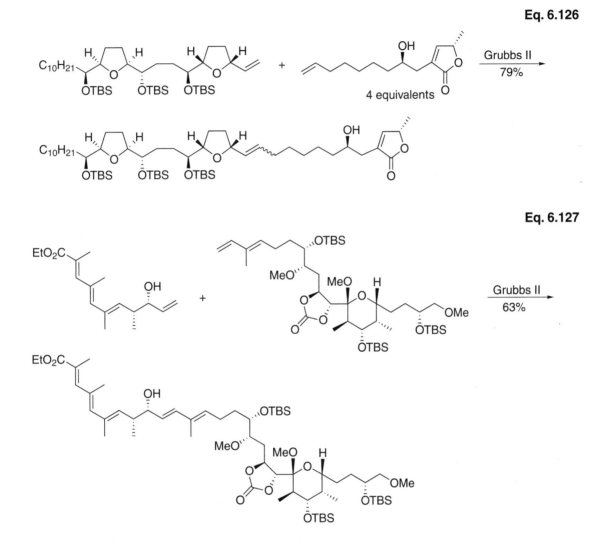

Eq. 6.126

Eq. 6.127

The closely related alkene–alkyne metathesis is a useful reaction for the preparation of 1,3-diene systems.[196] The mechanism of the reaction depends on the substrate and a mechanistic representation of the intramolecular reaction forming small- to medium-sized rings is shown in **Eq. 6.128**. The mechanism is based on a ^{13}C- and ^{2}D-labelling study, indicating an ene-then-yne metathesis sequence.[197] Recent calculations in support of this mechanism also indicated that ruthenacyclobutenes were not part of the catalytic cycle.[198] Further complicating the mechanism is the observation that added ethene accelerates the reaction and increases the yield of product (**Eq. 6.129**).

This has been attributed to a shift to a second catalytic cycle wherein the release of the product is aided by the cycloaddition of ethene followed by cyclorevision. In some cases, both exo- and endocyclization modes are possible, producing two different ring sizes (**Eq. 6.130**).[199] Further complicating the mechanistic picture is the observation of products supporting an yne-then-ene mechanism for some substrates.[200]

Eq. 6.128

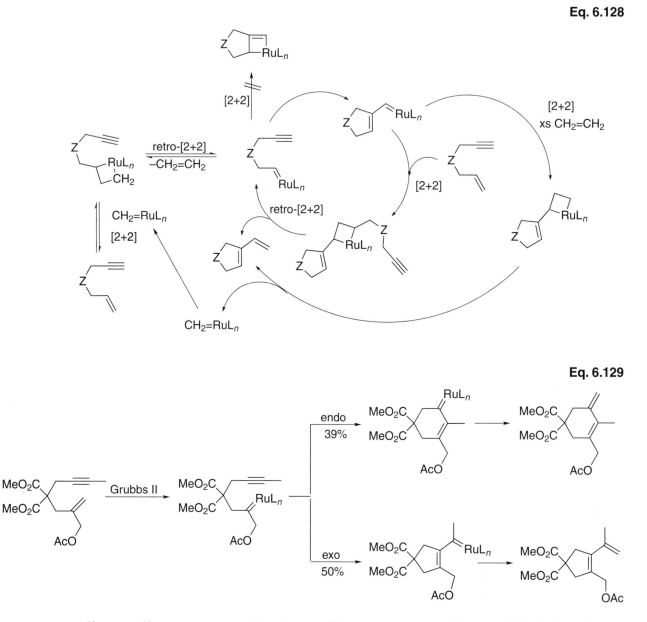

Eq. 6.129

Alkyne–alkene macrocyclizations offer a more complex mechanistic picture wherein the endo–exo selectivity for the alkenylalkylidene–ruthenium complex formation depends on the substrate and the length of the tether between the reacting ends. For large rings, endocyclization is observed (**Eq. 6.130**).[201] Macrocyclizations in the presence of ethene further complicate the picture. In this case, the intermolecular alkyne–ethene metathesis is faster compared to the intramolecular reaction, affording a triene that can be isolated and, if submitted to metathesis conditions, afford a cyclic product via RCM (**Eq. 6.130**).

Eq. 6.130

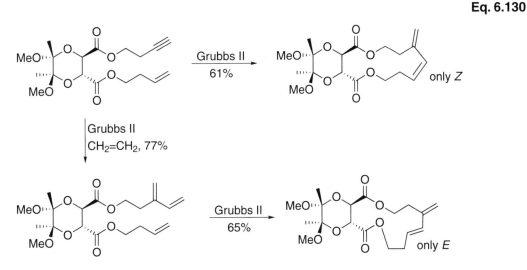

Metathesis reactions between two alkynes are mechanistically similar to alkene–alkene metathesis, but the standard catalysts used in RCM do not catalyze the related alkyne–alkyne transformation.[202] Cross metathesis and the ring-closing metathesis of alkynes can be achieved using molybdenum and tungsten catalysts, such as $Mo(CO)_6$ activated with a phenol (e.g., 4-chlorophenol or 2-fluorophenol), the carbyne complex $(t\text{-BuO})_3WCCMe_3$, and a number of $Mo[(t\text{-Bu})N(Ar)]_3$ complexes activated by the addition of dichloromethane (**Figure 6.6**). The active catalyst in the latter case is the chloro complex. Alkyne–alkyne metathesis reactions proceed via a metal carbyne complex[203] and sequential [2+2] cycloaddition–[2+2] cyclorevisions (**Eq. 6.131**).

Eq. 6.131

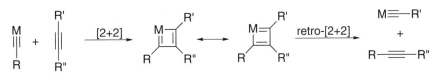

Terminal alkynes are poor substrates that are often incompatible with the catalyst. This problem is elegantly overcome by the use of methyl-substituted alkynes,

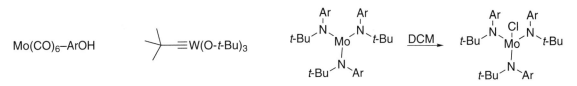

Figure 6.6. Alkyne–Alkyne Metathesis Catalysts

thus affording the readily removed 2-butyne as byproduct. An intermolecular and an intramolecular application of this methodology in total synthesis are shown in **Eqs 6.132**[204] and **6.133**.[205]

Eq. 6.132

Eq. 6.133

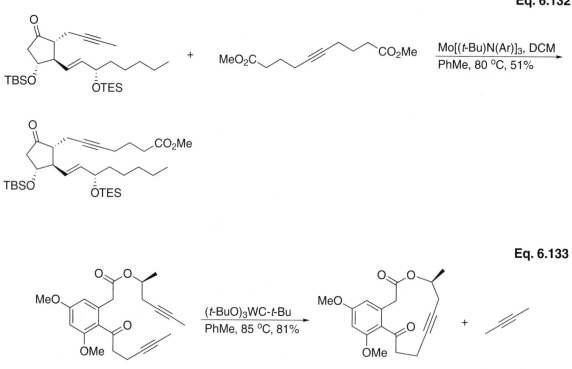

Perhaps the ultimate in metathesis reactions are relayed metatheses wherein a number of carbon–carbon bonds are formed in a precise and usually predictable manner. Examples of elegant relay metathesis reactions include a ring-opening cross metathesis (ROCM) (**Eq. 6.134**),[206] a ring-opening ring-closing metathesis (RORCM) (**Eq. 6.135**),[207] and a related ene-yne-ene RORCM (**Eq. 6.136**).[208]

Eq. 6.134

Eq. 6.135

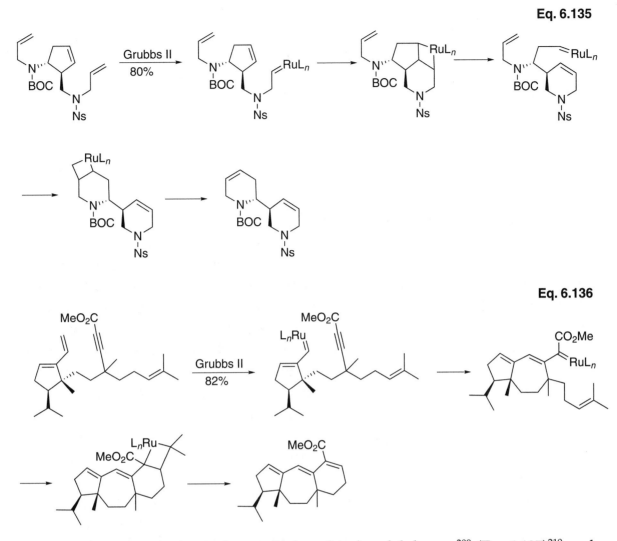

Eq. 6.136

Recently, asymmetric versions utilizing chiral molybdenum[209] (**Eq. 6.137**)[210] and ruthenium (**Eq. 6.138**)[211] catalysts have been introduced.

Eq. 6.137

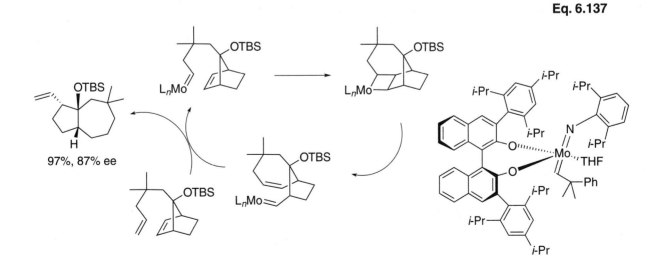

Eq. 6.138

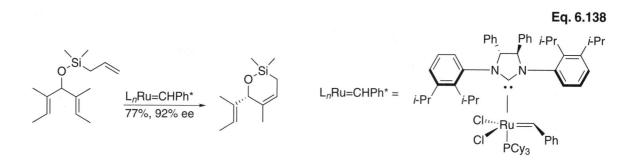

6.6 Nucleophilic "Schrock" Carbene Complexes[212]

Nucleophilic carbene complexes represent the opposite extreme of reactivity and bonding from the electrophilic Fischer carbene complexes. Nucleophilic carbene complexes are usually formed by metals at the far left of the transition series, in high oxidation states (usually d^0), with strong donor ligands, such as alkyl or cyclopentadienyl, and no acceptor ligands. The carbene ligand is usually simply the $=CH_2$ group. These are so different from electrophilic carbene complexes that the carbene ligand in these complexes is formally considered a four-electron, dinegative ligand, $[H_2C::]^{2-}$. These complexes often have Wittig-reagent-like reactivity and are often represented as $M^+-CH_2^- \leftrightarrow M=CH_2$, much like ylides. However, nucleophilic carbene complexes also form metallacycles, and both of these aspects of reactivity are useful in synthesis.

The most extensively studied nucleophilic carbene complex is "Tebbe's" reagent, prepared by the reaction of trimethylaluminum with titanocene dichloride (**Eq. 6.139**).[213] In the presence of pyridine, this complex is synthetically equivalent to "$Cp_2Ti=CH_2$," and it is very efficient in converting carbonyl groups to methylenes (**Eq. 6.140**).[214] The reaction is thought to proceed via an oxometallacycle that can fragment to give the alkene and a very stable titanium(IV) oxo species (**Eq. 6.141**).

Eq. 6.139

$$Cp_2TiCl_2 + AlMe_3 \longrightarrow Cp_2Ti\begin{matrix} CH_2 \\ Cl \end{matrix}Al\begin{matrix} Me \\ Me \end{matrix} \xrightarrow{\text{pyridine}} \text{"}Cp_2Ti=CH_2\text{"}$$

Tebbe's reagent

Eq. 6.140

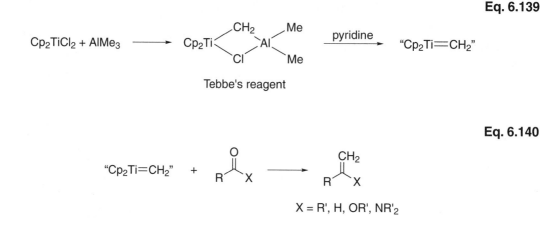

X = R', H, OR', NR'$_2$

Eq. 6.141

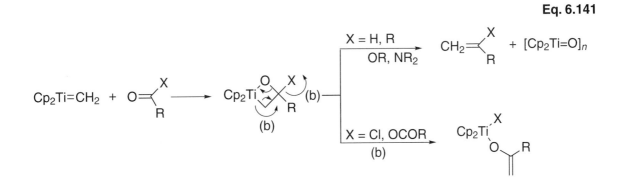

The reaction not only works well with aldehydes and ketones, but it also converts esters to enol ethers and amides to enamines! With acid halides and anhydrides, titanium enolates are produced instead. Tebbe's reagent does not enolize ketones and is particularly efficient for methylenating sterically hindered, enolizable ketones which are not methylenated efficiently by Wittig reagents.[215] In addition, these reagents do not racemize chiral centers α to carbonyl carbons. The reagent tolerates a very high degree of functionality (**Eqs. 6.142**[216] and **6.143**[217]).

Eq. 6.142

Eq. 6.143

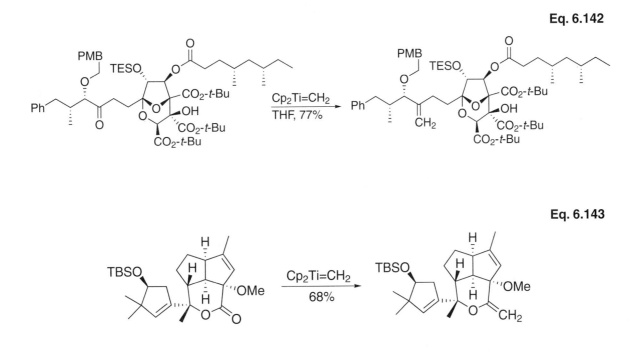

In addition to methylenating carbonyl compounds, Tebbe's reagent reacts with alkenes to form titanacycles. These can be cleaved by a variety of reagents (**Eq. 6.144**), or can be used to synthesize allenes (**Eq. 6.145**) by a cycloaddition–cyclorevision sequence.[218] Methylenation, combined with metathesis, can be a powerful synthetic tool (**Eq. 6.146**).[219]

Eq. 6.144

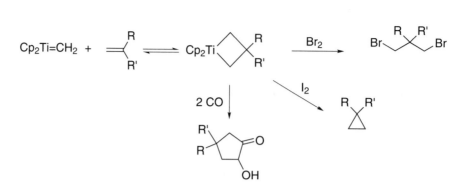

Eq. 6.145

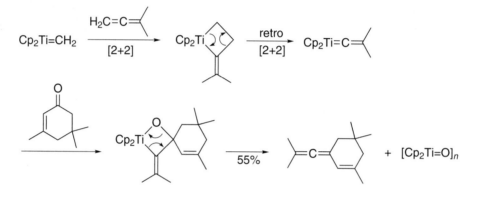

Eq. 6.146

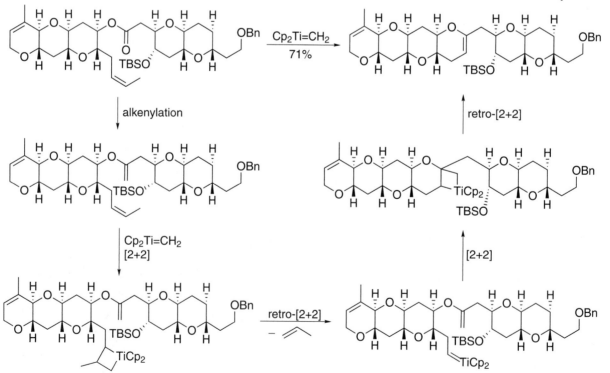

Alkylidene and alkenylidene titanium complexes can also be prepared by the reaction of $Cp_2Ti[P(OEt)_3]_2$ with dithioacetals or 1,1-dichloroalkenes, respectively. Alkylidene titanium complexes undergo intermolecular cyclopropanations with alkenes[220] and intramolecular metathesis reactions with alkenes (**Eq. 6.147**).[221] They also alkenylate aldehydes, ketones, esters, and carbamates. Alkenylidene complexes react with aldehydes and ketones to form 1,2-dienes and with alkynes to form 1,3-dienes (**Eq. 6.148**).[222]

Eq. 6.147

Eq. 6.148

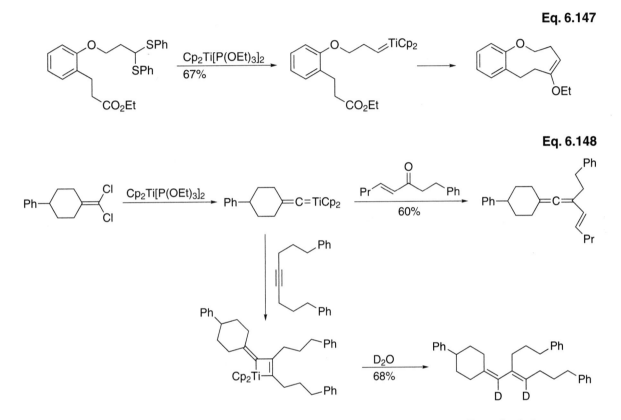

A more easily prepared and handled methylenating reagent is dimethyl titanocene, Cp_2TiMe_2 ("Petasis reagent"), generated by the simple treatment of titanocene dichloride with methyllithium or MeMgX. Gentle heating of this complex (~60°C) in the presence of ketones, aldehydes, or esters results in methylation with yields and selectivities similar to those observed with Tebbe's reagent.[223,224] The dialkyltitanium reagents are more versatile compared to Tebbe's reagent, in that benzylidene (PhCH=),[225] trimethylsilylmethylene (TMSCH=),[226] allenelylidene (CR$_2$=C=, **Eq. 6.149**),[227,228] and methylenecyclopropane[229154] can be transferred, while Tebbe's reagent is restricted to the simple methylene group. Examples of syntheses employing Cp_2TiMe_2 are shown in **Eqs. 6.150**[230] and **6.151**.[231]

Eq. 6.149

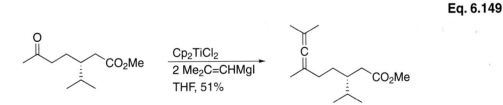

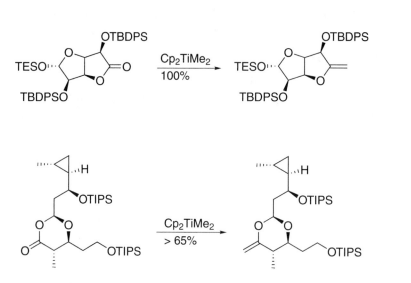

Eq. 6.150

Eq. 6.151

Methylenation can also be achieved using titanium tetrachloride, zinc dust, and dibromomethane or diiodomethane (**Eq. 6.152**).[232] The addition of a catalytic amount of lead accelerates the reaction.[233] Related zirconium reagents, made in situ from, for example, dichlorozirconocene or zirconium tetrachloride, zinc dust, and diiodomethane, also alkenylate aldehydes and ketones (**Eq. 6.153**).[234] The true nature of these methylenation reagents is unknown.

Eq. 6.152

Eq. 6.153

The reaction of aldehydes with iodoform in the presence of a stoichiometric amount of chromium dichloride ("Takai–Utimoto alkenylation") affords *E*-iodoalkenes in high yields and high *E/Z*-ratios at ambient temperature. Other 1,1-diiodoalkanes can be used, but often they afford low yields of products. Of synthetic important are reactions of $Bu_3SnCHBr_2$ and $Me_3SiCHBr_2$, which introduce $Bu_3SnCH=$ and $Me_3SiCH=$ groups. The mechanism and true nature of the intermediate chromium complex of this reaction is unclear. The reaction is catalytic in chromium if a stoichiometric amount of samarium or samarium diiodide is present.[235] A very important variation of the reactions of 1,1-dihalides are reactions of trihalomethanes with chromium dichloride, yielding E-haloalkanes.[236] Some examples of this reaction in total synthesis are shown in **Eqs. 6.154**[237] and **6.155**.[238]

Eq. 6.154

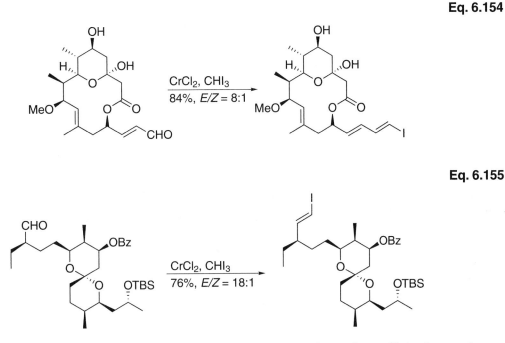

Finally, related chemistry wherein a metal-bound (at least formally) nitrene intermediate is formed has started to emerge.[239] The nitrene complexes are formed in situ by reaction with aryl- or alkylsulfonyl imino phenyliodinane (RSO$_2$N=IPh). These reagents can, in turn, be more conveniently prepared by the oxidation of aryl- or alkylsulfonamides with diacetoxyiodobenzene.

The nitrene intermediates undergo aziridination of alkenes (**Eq. 6.156**).[240] Perhaps more synthetically interesting are the more recently developed carbon–hydrogen bond insertion reactions. Insertions into allylic and benzylic carbon–hydrogen bonds are particularly facile. Examples include an asymmetric intermolecular insertion (**Eq. 6.157**)[241] and an application of nitrene insertion in synthesis (**Eq.6.158**).[242]

Eq. 6.156

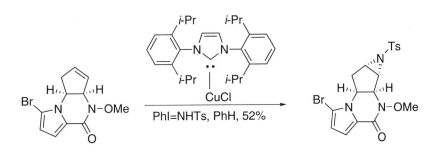

Eq. 6.157

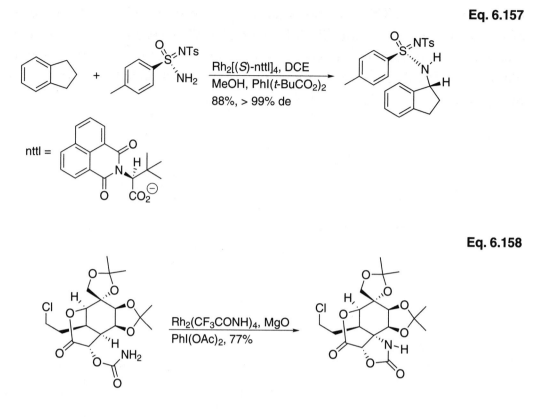

Eq. 6.158

References

(1) For reviews, see a) Wu, Y.-T.; Kurahashi, T.; de Meijere, A. *J. Organomet. Chem.* **2005**, *690*, 5900. b) Barluenga, J.; Fernandez-Rodrigues, M. A.; Aguilar, E. *J. Organomet. Chem.* **2005**, *690*, 539. c) Barluenga, J.; Tomas, M. *Chem. Rev.* **2004**, *104*, 2259. d) *Transition Metal Carbene Complexes*; Seyferth, D., Ed.; Verlag Chemie: Weinheim, 1983.

(2) Hoye, T. R.; Chen, K.; Vyvyan, J. R. *Organometallics* **1993**, *12*, 2806.

(3) Matsuyama, H.; Nakamura, T.; Iyoda, M. *J. Org. Chem.* **2000**, *65*, 4796.

(4) Bao, J.; Wulff, W. D.; Dominy, J. B.; Fumo, M. J.; Grant, E. B.; Rob, A. C.; Whitcomb, M. C.; Yeung, S.-M.; Ostrander, R. L.; Rheingold, A. L. *J. Am. Chem. Soc.* **1996**, *118*, 3392.

(5) Semmelhack, M. F.; Lee, G. R. *Organometallics* **1987**, *6*, 1839.

(6) Casey, C. P.; Brunswald, W. P. *J. Organomet. Chem.* **1976**, *118*, 309.

(7) a) Hegedus, L. S.; Schwindt, M. A.; DeLombaert, S.; Imwinkelried, R. *J. Am. Chem. Soc.* **1990**, *112*, 2264. b) Schwindt, M. A.; Lejon, T.; Hegedus, L. S. *Organometallics* **1990**, *9*, 2814.

(8) Stadtmüller, H.; Knochel, P. *Organometallics* **1995**, *14*, 3863.

(9) a) Dötz, K.H.; Sturm, W.; Alt, H. G. *Organometallics* **1987**, *6*, 1424. b) Schmidt, B.; Kocienski, P.; Reid, G. *Tetrahedron* **1996**, *52*, 1617.

(10) Weyerhausen, B.; Nieger, M.; Dötz, K. H. *Organometallics* **1998**, *17*, 1602.

(11) Cosset, C.; Del Rio, I.; Le Bozec, H. *Organometallics* **1995**, *14*, 1938.

(12) Hafner, A.; Hegedus, L. S.; deWeck, G.; Hawkins, B.; Dötz, K. H. *J. Am. Chem. Soc.* **1988**, *110*, 8413.

(13) a) Conner, J. A.; Jones, E. M. *J. Chem. Soc., Chem. Commun.* **1971**, 570. b) Semmelhack, M. F.; Bozell, J. J. *Tetrahedron Lett.* **1982**, *23*, 2931. c) Hafner, A.; Hegedus, L. S.; deWeck, G.; Hawkins, B.: Dötz, K. H. *J. Am. Chem. Soc.* **1988**, 110, 8413.

(14) Rudler, H.; Parlier, A.; Durand-Reville, T.; Martin-Vaca, B.; Audouin, M.; Garrier, E.; Certal, V, Vaissermann, J. *Tetrahedron* **2000**, *56*, 5001.

(15) Barluenga, J.; Alonso, J.; Fananas, F. J. *J. Am. Chem. Soc.* **2003**, *125*, 2610.

(16) Barluenga, J.; Perez-Sanches, I.; Rubio, E.; Florez, J. *Angew. Chem. Int. Ed.* **2003**, *42*, 5860.

(17) Rudler, H.; Parlier, A.; Certal, V.; Lastennet, G.; Audouin, M.; Vaissermann, J. *Eur. J. Org. Chem.* **2004**, 2471.

(18) a) Casey, C. P. Metal-Carbene Complexes in Organic Synthesis. In *Transition Metal Organometallics in Organic Synthesis*; Alper, H., Ed. Academic Press: New York, 1976; Vol. 1, pp 190–223. b) Bernasconi, C. F.; Sun, W. *J. Am. Chem. Soc.* **1993**, *115*, 12526. (c) For a review on the physical organic chemistry of Fischer carbene complexes, see Bernasconi, C. F. *Chem. Soc. Rev.* **1997**, *26*, 299.

(19) Bernasconi, C. F.; Ruddat, V. *J. Am. Chem. Soc.* **2002**, *124*, 14968.

(20) Lattuada, L.; Licandro, E.; Maiorana, S.; Molinari, H.; Papagni, A. *Organometallics* **1991**, *10*, 807.

(21) Aumann, R.; Heinen, H. *Chem. Ber.* **1987**, *120*, 537.

(22) a) Xu, Y-C.; Wulff, W. D. *J. Org. Chem.* **1987**, *52*, 3263. b) Armin, S. R.; Sarkar, A. *Organometallics* **1995**, *14*, 547.

(23) Powers, T. S.; Shi, Y.; Wilson, K. J.; Wulff, W. D. *J. Org. Chem.* **1994**, *59*, 6882.

(24) a) Anderson, B. A.; Wulff, W. D.; Rahm, A. *J. Am. Chem. Soc.* **1993**, *115*, 4602. b) Baldoli, C.; Del Butero, P.; Licandro, E.; Maiorana, S.; Papagni, A.; Zannoti-Gerosa, A. *J. Organomet. Chem.* **1995**, *486*, 1995. c) Shi, Y.; Wulff, W. D.; Yap, G. P. A.; Rheingold, A. *Chem. Commun.* **1996**, 2600.

(25) For a review on asymmetric synthesis with Fischer complexes, see Wulff, W. D. *Organometallics* **1998**, *17*, 3116.

(26) a) McDonald, F. E.; Connolly, C. B.; Gleason, M. M.; Towne, T. B.; Treiber, K. D. *J. Org. Chem.* **1993**, *58*, 6952. b) McDonald, F. E.; Schultz, C. C. *J. Am. Chem. Soc.* **1994**, *116*, 9363.

(27) In a synthesis of salinomycin. Kocienski, P. J.; Brown, R. C. D.; Pommier, A.; Proctor, M. Schmidt, B. *J. Chem. Soc., Perkin Trans. 1* **1998**, 9.

(28) McDonald, F. E.; Zhu, H. H.-Y. *Tetrahedron* **1997**, *53*, 11061.

(29) McDonald, F. E.; Chatterjee, A .K. *Tetrahedron Lett.* **1997**, *38*, 7687.

(30) McDonald, F. E.; Gleason, M. M. *J. Am. Chem. Soc.* **1996**, *118*, 6648.

(31) In a synthesis of vancosamine. Cutchins, W. W.; McDonald, F. E. *Org. Lett.* **2002**, *4*, 749.

(32) Liang, K.-W.; Li, W.-T.; Peng, S.-M.; Wang, S.-L.; Liu, R.-S. *J. Am. Chem. Soc.* **1997**, *119*, 4404.

(33) Huang, H.-L.; Liu, R.-S. *J. Org. Chem.* **2003**, *68*, 805.

(34) Li, W.-T.; Lai, F.-C.; Lee, G.-H.; Peng, S.-M.; Liu, R.-S. *J. Am. Chem. Soc.* **1998**, *120*, 4520.

(35) a) Wang, S. L. B.; Wulff, W. D. *J. Am. Chem. Soc.* **1990**, *112*, 4550. b) Wulff, W. D.; Bauta, W. E.; Kaesler, R. W.; Lankford, P. J.; Miller, R. A.; Murray, C. K.; Yang, D. C. *J. Am. Chem. Soc.* **1990**, *112*, 3642.

(36) Alkenyl carbene complexes were estimated to have reactivity comparable to maleic anhydride. Adam, H.; Albrecht, T.; Sauer, J. *Tetrahedron Lett.* **1994**, *35*, 557.

(37) Wulff, W. D.; Power, T. S. *J. Org. Chem.* **1993**, *58*, 2381.

(38) Barluenga, J.; Aznar, F.; Palomero, M. A. *J. Org. Chem.* **2003**, *68*, 537.

(39) Anderson, B.; Wulff, W. D.; Power, T. S. *J. Am. Chem. Soc.* **1992**, *114*, 10784.

(40) Barluenga, J.; Tomas, M.; Ballesteros, A.; Santamaria, J.; Suarez-Sobrino, A. *J. Org. Chem.* **1997**, *62*, 9229.

(41) Rahm, A.; Rheingold, A. L.; Wulff, W. D. *Tetrahedron* **2000**, *56*, 4951.

(42) Loft, M. S.; Mowlem, T. J.; Widdowson, D. A. *J. Chem. Soc., Perkin Trans. 1* **1995**, 97.

(43) Barluenga, J.; Fernandez-Mari, F.; Viado, A.L.; Aguilar, E.; Olano, B. *J. Chem. Soc., Perkin Trans. 1* **1997**, 2267.

(44) Barluenga, J.; Fernandez-Mari, F.; Viado, A. L.; Aguilar, E.; Olano, B. *J. Chem. Soc., Perkin Trans. 1* **1997**, 2267.

(45) Dötz, K. H.; Koch, A. W.; Weyershausen, B.; Hupfer, H.; Nieger, M. *Tetrahedron* **2000**, *56*, 4925.

(46) For a review of the addition of nucleophiles to α,β-unsaturated Fischer carbene complexes, see Barluenga, J.; Florez, J. Fananas, F. J. *J. Organomet. Chem.* **2001**, *624*, 5.

(47) a) Aoki, S.; Fujimura, T.; Nakamura, E. *J. Am. Chem. Soc.* **1992**, *114*, 2985. b) Nakamura, E.; Tanaka, K.; Fujimura, T.; Aoki, S.; Williard, P. G. *J. Am. Chem. Soc.* **1993**, *115*, 9015.

(48) Ezquerra, J.; Pedregal, C.; Merino, I.; Florez, J.; Barluenga, J. *J. Org. Chem.* **1999**, *64*, 6554.

(49) Barluenga, J.; Nandy, S. K.; Laxmi, Y. R. S.; Suarez, J. R.; Merino, I.; Florez, J.; Garcia-Granda, S.; Montejo-Bernardo, J. *Chem.—Eur. J.* **2003**, *9*, 5725.

(50) Aumann, R.; Meyer, A. G.; Fröhlich, R. *J. Am. Chem. Soc.* **1996**, *118*, 10853.

(51) Gibert, M.; Ferrer, M.; Lluch, M.; Sanchez-Baeze, F.; Messeguer, A. *J. Org. Chem.* **1999**, *64*, 1591.

(52) Lluch, A.M.; Jordi, L.; Sanchez-Baeza, F.; Ricart, S.; Camps, F.; Messeguer, A.; Moreto, J. M. *Tetrahedron Lett.* **1992**, *33*, 3021.

(53) Söderberg, B. C.; Bowden, B. A. *Organometallics* **1992**, *11*, 2220.

(54) For reviews, see a) Herndon, J. W. *Tetrahedron* **2000**, *56*, 1257. b) Barluenga, J.; Fernandez-Rodrigues, M. A.; Aguilar, E. *J. Organomet. Chem.* **2005**, *690*, 539.

(55) For reviews, see a) Dötz, K. H. *Angew. Chem., Int. Ed. Engl.* **1984**, *23*, 587. (b) Reissig, H.-U. *Organomet. in Synth.* **1989**, *2*, 311. For other studies, see c) Reissig, H.-U. *Organometallics* **1990**, *9*, 3133. d) Doyle, M. P. Transition Metal Carbene Complexes: Cyclopropanation. In *Comprehensive Organometallic Chemistry II*; Abel, E. W., Stone, F. G. A., Wilkinson, G., Eds.; Pergamon: Oxford, U.K., 1995; Vol. 12, pp 387–395.

(56) Harvey, D. F.; Lund, K. P. *J. Am. Chem. Soc.* **1991**, *113*, 8916.

(57) Buchert, M.; Hoffmann, M. Reissig, H.-U. *Chem. Ber.* **1995**, *128*, 605.

(58) Wienand, A.; Reissig, H.-U. *Angew. Chem., Int. Ed. Engl.* **1990**, *29*, 1129.

(59) Murray, C. K.; Yang, D. C.; Wulff, W. D. *J. Am. Chem. Soc.* **1990**, *112*, 5660.

(60) a) Barluenga, J.; Fernandez-Acebes, A.; Trabanco, A. A.; Florez, J. *J. Am. Chem. Soc.* **1997**, *119*, 7591. b) Barluenga, J.; Lopez, S.; Trabanco, A. A.; Florez, J. *Chem. Eur. J.* **2001**, *7*, 4723.

(61) Semmelhack, M. F.; Bozell, J. J. *Tetrahedron Lett.* **1982**, *23*, 2931.

(62) Söderberg, B. C.; Hegedus, L. S. *Organometallics* **1990**, *9*, 3113.

(63) Barluenga, J.; Aznar, F.; Guiterrez, I.; Martin, J. A. *Org. Lett.* **2002**, *4*, 2719.

(64) Barluenga, J.; Montserrat, J. M.; Florez, J. *J. Chem. Soc., Chem. Commun.* **1993**, 1068.

(65) Barluenga, J.; Dieguez, A.; Rodriguez, F.; Florez, J.; Fananas, J. *J. Am. Chem. Soc.* **2002**, *124*, 9056.

(66) a) For a review, see Harvey, D. F.; Sigano, D. M. *Chem. Rev.* **1996**, *96*, 271. b) Hoye, T. R.; Suriano, J.A. *Organometallics* **1992**, *11*, 2044; **1989**, *8*, 2670. c) Mori, M.; Watanuki, S. *J. Chem. Soc., Chem. Commun.* **1992**, 1082. d) Katz, T. J.; Yang, G. X. Q. *Tetrahedron Lett.* **1991**, *32*, 5895.

(67) Alvarez, C.; Parlier, A.; Rudler, H.; Yefsah, R.; Daran, J. C.; Knobler, C. *Organometallics* **1989**, *8*, 2253.

(68) In a synthesis of carabrone. Hoye, T. R.; Vyuyan, J. R. *J. Org. Chem.* **1995**, *60*, 4184.

(69) a) Harvey, D. F.; Brown, M. F. *J. Am. Chem. Soc.* **1990**, *112*, 7806. b) *Tetrahedron Lett.* **1991**, *32*, 2871. c) *Tetrahedron Lett.* **1991**, 5223, 6311.

(70) a) Harvey, D. F.; Brown, M. F. *J. Org. Chem.* **1992**, *57*, 5559. b) Harvey, D. F.; Lund, K. P.; Neil, D. A. *J. Am. Chem. Soc.* **1992**, *114*, 8424.

(71) a) Harvey, D. F.; Lund, K. P. *J. Am. Chem. Soc.* **1991**, *113*, 5066. b) Harvey, D. F.; Grenzer, E. M. *J. Org. Chem.* **1996**, *61*, 159.

(72) Ghorai, J. W. Herndon, Y.-F. Lam, *Org. Lett.* **2001**, *3*, 3535.

(73) a) Dötz, K. H. *Angew. Chem., Int. Ed. Engl.* **1984**, *23*, 587. b) For a review, see Wulff, W. D. Transition Metal Carbene Complexes: Alkyne and Vinyl Ketene Chemistry. In *Comprehensive Organometallic Chemistry II*; Abel, E. W., Stone, F. G. A., Wilkinson, G., Eds.; Pergamon: Oxford, U.K., 1995; Vol. 12, pp 469–548.

(74) In a synthesis of vitamin K_1. Dötz, K. H.; Mühlemeier, J. *Angew. Chem., Int. Ed. Engl.* **1982**, *21*, 929.

(75) a) Wulff, W. D.; Bax, B. M.; Branwold, T. A.; Chan, K. S.; Gilbert, A. M.; Hsung, R. P. *Organometallics* **1994**, *13*, 102. b) Wulff, W. D.; Gilbert, A. M.; Hsung, R. P.; Rahm, A. *J. Org. Chem.* **1995**, *60*, 4566.

(76) Bauta, W. E.; Wulff, W. D.; Pavkovic, S. F.; Zaluzec, E. J. *J. Org. Chem.* **1989**, *54*, 3249.

(77) Vorogushin, A. V.; Wulff, W. D.; Hansen, H.-J. *J. Am. Chem. Soc.* **2002**, *124*, 6512.

(78) Gopalsamuthiram, V.; Wulff, W. D. *J. Am. Chem. Soc.* **2004**, *126*, 13936.

(79) In a synthesis of (–)-curcuquinone. Minatti, A.; Dötz, K. H. *J. Org. Chem.* **2005**, *70*, 3745.

(80) Bao, J.; Wulff, W. D.; Dragisch, V.; Wenglowsky, S.; Ball, R. G. *J. Am. Chem. Soc.* **1994**, *116*, 7616.

(81) Hegedus, L. S.; deWeck, G.; D'Andrea, S. *J. Am. Chem. Soc.* **1988**, *110*, 2122.

(82) a) Merlic, C. A.; Xu, D. *J. Am. Chem. Soc.* **1991**, *113*, 7418. b) Merlic, C. A.; Xu, D.; Gladstone, B. G. *J. Org. Chem.* **1993**, *58*, 538.

(83) Merlic, C. A.; Burns, E. E.; Xu, D.; Chen, S. Y. *J. Am. Chem. Soc.* **1992**, *114*, 8722.

(84) Merlic, C. A.; McInnes, D. M.; You, Y. *Tetrahedron Lett.* **1997**, *38*, 6787.

(85) In a synthesis of carbazoquinocin C. Rawat, M.; Wulff, W. D. *Org. Lett.* **2004**, *6*, 329.

(86) Hydroquinone derivatives have been observed in some cases as the major product using aminocarbenes. Wulff. W. D.; Gilbert, A. M.; Hsung, R. P.; Rahm, A. *J. Org. Chem.* **1995**, *60*, 4566.

(87) a) Yamashita, A. *Tetrahedron Lett.* **1986**, *27*, 5915. For reviews on aminocarbene complex chemistry, see b) Grotjahn, D. B.; Dötz, K. H. *Synlett* **1991**, 381. c) Schwindt, M. P.; Miller, J. R.; Hegedus, L. S. *J. Organomet. Chem.* **1991**, *413*, 143.

(88) a) Dötz, K. H.; Schäfer, T. O.; Harms, K. *Synthesis* **1992**, 146. b) *Angew. Chem., Int. Ed. Engl.* **1990**, *29*, 176.

(89) Barluenga, J.; Aznar, F.; Barluenga, S. *J. Chem. Soc., Chem. Commun.* **1995**, 1973.

(90) Ishibashi, T.; Ochifuji, N.; Mori, M. *Tetrahedron Lett.* **1996**, *37*, 6165.

(91) Brandvold, T. A.; Wulff, W. D. *J. Am. Chem. Soc.* **1999**, *112*, 1645.

(92) a) Turner, S. U.; Senz, U.; Herndon, J. W.; McMullen, L. A. *J. Am. Chem. Soc.* **1992**, *114*, 8394. b) Hill, D. K.; Herndon, J. W. *Tetrahedron Lett.* **1996**, *37*, 1359. c) Herndon, J. W.; Patel, P. P. *Tetrahedron Lett.* **1997**, *38*, 59.

(93) Herndon, J. W.; Matasi, J. J. *J. Org. Chem.* **1990**, *55*, 786.

(94) Herndon, J. W.; Zora, M.; Patel, P. P. *Tetrahedron* **1993**, *49*, 5507.

(95) Semmelhack, M. F.; Tamura, R. *J. Am. Chem. Soc.* **1983**, *105*, 4099; 6750.

(96) Semmelhack, M. F.; Tamura, R.; Schnatter, W.; Springer, J. *J. Am. Chem. Soc.* **1984**, *106*, 5363.

(97) Foley, H. C.; Strubinger, L .M.; Targos, T. S.; Geoffroy, G. L. *J. Am. Chem. Soc.* **1983**, *105*, 3064.

(98) Geoffroy, G. L. *Adv. Organomet. Chem.* **1985**, *24*, 249.

(99) a) Hegedus, L. S.; DeWeck, G.; D'Andrea, S. *J. Am. Chem. Soc.* **1988**, *110*, 2122. b) For a review on metal–ketene complexes, see Geoffroy, G. L.; Bassner, S. L. *Adv. Organomet. Chem.* **1988**, *28*, 1.

(100) For a review, see Hegedus, L. S. *Tetrahedron* **1997**, *53*, 4105.

(101) a) Imwinkelried, R.; Hegedus, L. S. *Organometallics* **1988**, *7*, 702. b) Schwindt, M. A.; Lejon, T.; Hegedus, L. S. *Organometallics* **1990**, *9*, 2814.

(102) Hegedus, L. S.; Imwinkelried, R.; Alarid-Sargent, M.; Dvorak, D.; Satoh, Y. *J. Am. Chem. Soc.* **1990**, *112*, 1109.

(103) For a discussion of stereoselectivity in these reactions, see Hegedus, L. S.; Montgomery, J.; Narukawa, Y.; Snustad, D. C. *J. Am. Chem. Soc.* **1991**, *113*, 5784.

(104) Colson, J.-P.; Hegedus, L. S. *J. Org. Chem.* **1993**, *58*, 5918.

(105) Narukawa, Y.; Juneau, K. N.; Snustad, D. C.; Miller, D. B.; Hegedus, L. S. *J. Org. Chem.* **1992**, *57*, 5453.

(106) a) Betschart, C.; Hegedus, L. S. *J. Am. Chem. Soc.* **1992**, *114*, 5010. b) Es-Sayed, M.; Heiner, T.; de Meijere, A. *Synlett* **1993**, 57.

(107) Hsiao, Y.; Hegedus, L. S. *J. Org. Chem.* **1997**, *62*, 3586.

(108) Valenti, E.; Pericas, M. A.; Moyano, A. *J. Org. Chem.* **1990**, *55*, 3582.

(109) Söderberg, B. C.; Hegedus, L. S.; Sierra, M. A. *J. Am. Chem. Soc.* **1990**, *112*, 4364.

(110) Söderberg, B. C.; Hegedus, L. S. *J. Org. Chem.* **1991**, *56*, 2209.

(111) Aumann, R.; Krüger, C.; Goddard, R. *Chem. Ber.* **1992**, *125*, 1627.

(112) In a synthesis of tetrahydrocerulenin. Miller, M. A.; Hegedus, L. S. *J. Org. Chem.*

(113) a) Hegedus. L. S.; Bates, R. W.; Söderberg, B. C. *J. Am. Chem. Soc.* **1991**, *113*, 923. b) Reed, A. D.; Hegedus, L. S. *J. Org. Chem.* **1995**, *60*, 3717. c) Reed, A. D.; Hegedus, L. S. *Organometallics* **1997**, *16*, 2313.

(114) Colson, J.-P.; Hegedus, L. S. *J. Org. Chem.* **1994**, *59*, 4972.

(115) Merlic, C. A.; Doroh, B. C. *J. Org. Chem.* **2003**, *68*, 6056.

(116) For a review, see Hegedus, L. S. *Acc. Chem. Res.* **1995**, *28*, 299.

(117) Hegedus, L. S.; Schwindt, M. A.; DeLombaert, S.; Imwinkelried, R. *J. Am. Chem. Soc.* **1990**, *112*, 2264.

(118) Schmeck, C.; Hegedus, L. S. *J. Am. Chem. Soc.* **1994**, *116*, 9927.

(119) Miller, J. R.; Pulley, S. R.; Hegedus, L. S.; DeLombaert, S. *J. Am. Chem. Soc.* **1992**, *114*, 5602.

(120) Debuisson, C.; Fukumoto, Y.; Hegedus, L. S. *J. Am. Chem. Soc.* **1995**, *117*, 3697.

(121) Pulley, S. R.; Hegedus, L. S. *J. Am. Chem. Soc.* **1993**, *115*, 9037.

(122) a) Zhu, J.; Hegedus, L. S. *J. Org. Chem.* **1995**, *60*, 5831. b) Zhu, J.; Deur, C.; Hegedus, L. S. *J. Org. Chem.* **1997**, *62*, 7704.

(123) Casey, C. P.; Anderson, R. L. *J. Chem. Soc., Chem. Commun.* **1975**, 895.

(124) For a review, see Gomez-Gallego, M.; Mancheno, M. J.; Sierra, M. A. *Acc. Chem. Res.* **2005**, *38*, 44.

(125) Sierra, M. A.; del Amo, J. C.; Mancheno, M. J.; Gomez-Gallego, M. *J. Am. Chem. Soc.* **2001**, *123*, 851.

(126) Barluenga, J.; Barrio, P.; Vicente, R.; Lopez, L. A.; Tomas, M. *J. Organomet. Chem.* **2004**, *689*, 3793.

(127) Barluenga, J.; Vicente, R.; Lopez, L. A.; Rubio, E.; Tomas, M.; Alvarez-Rua, C. *J. Am. Chem. Soc.* **2006**, *126*, 470.

(128) Barluenga, J.; Vicente, R.; Barrio, P.; Lopez, L. A.; Tomas, M. *J. Am. Chem. Soc.* **2004**, *126*, 5974.

(129) For a review, see Brookhart, M.; Studabaker, W. B. *Chem. Rev.* **1987**, *87*, 411.

(130) Fischer, H.; Hoffmann, J. *Chem. Ber.* **1991**, *124*, 981.

(131) Rudler, H.; Audouin, M.; Parlier, A.; Martin-Vaca, B.; Goumont, R.; Durand-Reville, T.; Vaissermann, J. *Am. Chem. Soc.* **1996**, *118*, 12045.

(132) Mattson, M. N.; Bays, J. P.; Zakutansky, J.; Stolarski, V.; Helquist, P. *J. Org. Chem.* **1989**, *54*, 2467.

(133) Vargas, R. M.; Theys, R. D.; Hossain, M. M. *J. Am. Chem. Soc.* **1992**, *114*, 777.

(134) Theys, R. D.; Hossain, M. M. *Tetrahedron Lett.* **1992**, *33*, 3447.

(135) a) Brookhart, M.; Liu, Y.; Goldman, E. W.; Timmers, D. A.; Williams, G. D. *J. Am. Chem. Soc.* **1991**, *113*, 927. b) Brookhart, M.; Liu, Y. *J. Am. Chem. Soc.* **1991**, *113*, 939.

(136) Baker, C. T.; Mattson, M. N.; Helquist, P. *Tetrahedron Lett.* **1995**, *36*, 7015.

(137) Ishii, S.; Helquist, P. *Synlett* **1997**, 508.

(138) For reviews, see a) Maas, G. *Chem. Soc. Rev.* **2004**, *33*, 183. b) Davies, H. M. L.; Antoulinakis, E. G. *Org. React.* **2001**, *57*, 1. c) Adams, J.; Spero, D. M. *Tetrahedron* **1991**, *47*, 1765. d) Doyle, M. P. Transition Metal Carbene Complexes: Cyclopropanation. In *Comprehensive Organometallic Chemistry II*; Abel, E. W., Stone, F. G. A., Wilkinson, G., Eds.; Pergamon: Oxford, U.K., 1995; Vol. 12, pp 387–395. e) Singh, V. K.; Dattagupta, A.; Sebar, G. *Synthesis* **1997**, 137. f) Reissig, H.-U. *Angew. Chem. Int., Ed. Engl.* **1996**, *35*, 971.

(139) a) Doyle, M. P.; Dorow, R. L.; Buhro, W. E.; Griffin, J. H.; Tamblyn, W. H.; Trudell, M. L. *Organometallics* **1984**, *3*, 44. b) Doyle, M. P.; Griffin, J. H.; Bagheri, V.; Dorow, R. L. *Organometallics* **1984**, *3*, 53.

(140) a) Davies, H. M. L.; Clark, T. J.; Smith, H. D. *J. Org. Chem.* **1991**, *56*, 3817. b) Cantrell, W. R., Jr.; Davies, H. M. L. *J. Org. Chem.* **1991**, *56*, 723. c) Davies, H. M. L.; Clark, T. J.; Kinomer, G. F. *J. Org. Chem.* **1991**, *56*,

6440. d) Davies, H. M. L.; Hu, B. *Heterocycles* **1993**, *35*, 385.

(141) In a synthesis of 8-epi-PGF$_{2\alpha}$. Taber, D. F.; Herr, R. J.; Gleave, D. M. *J. Org. Chem.* **1997**, *62*, 194.

(142) In a synthesis of pinguisenol. Srikrishna, A.; Vijaykumar, D. *J. Chem. Soc., Perkin Trans. 1* **2000**, 2583.

(143) In a synthesis of (+)-ambruticin S. Kirkland, T. A.; Colucci, J.; Geraci, L. S.; Marx, M. A.; Schneider, M.; Kaelin, D. E., Jr.; Martin, S. F. *J. Am. Chem. Soc.* **2001**, *123*, 12432.

(144) For reviews, see a) Davies, H. M. L.; Beckwith, R. E. J. *Chem. Rev.* **2003**, *103*, 2861. b) Lebel, H.; Maroux, J.-F.; Molinaro, C.; Charette, A. B. *Chem. Rev.* **2003**, *103*, 977. c) Doyle, M. P.; McKervey, M. A. *Chem. Commun.* **1997**, 983. d) Doyle, M. P.; Forbes, D. C. *Chem. Rev.* **1998**, *98*, 911. e) Doyle, M. P.; Protopopova, M. N. *Tetrahedron* **1998**, *54*, 7979.

(145) a) Doyle, M. P.; Austin, R. E.; Bailey, A. S.; Dwyer, M. P.; Dyatkin, A. B.; Kwan, M. M. Y.; Liras, S.; Oalmann, C. J. Pieters, R. J; Protopopova, M. N.; Raab, C. E.; Roos, G. H. P.; Zhou, Q.-L.; Zhou, Q.-L.; Martin, S. F. *J. Am. Chem. Soc.* **1995**, *117*, 5763. b) Doyle, M. P.; Dyatkin, A. B.; Kalinin, A. V.; Ruppar, D. A. *J. Am. Chem. Soc.* **1995**, *117*, 11021.

(146) a) For a review, see Padwa, A.; Austin, D.J. *Angew. Chem., Int. Ed. Engl.* **1994**, *33*, 1797. b) Rogers, D. H.; Yi, E. C.; Poulter, C. D. *J. Org. Chem.* **1995**, *60*, 941.

(147) In a synthesis toward guanacastepenes. Hughes, C. C.; Kennedy-Smith, J. J.; Trauner, D. *Org. Lett.* **2003**, *5*, 4113.

(148) In a synthesis of harringtonolide. Zhang, H.; Appels, D. C.; Hockless, D. C. R.; Mander, L. N. *Tetrahedron Lett.* **1998**, *39*, 6577.

(149) a) Padwa, A.; Austin, D. J.; Xu, S. L. *Tetrahedron Lett.* **1991**, *32*, 4103. b) Hoye, T.; Dinsmore, C. J. *Tetrahedron Lett.* **1991**, *32*, 3755. c) Hoye, T. R.; Dinsmore, C. J. *J. Am. Chem. Soc.* **1991**, *113*, 4343. d) Padwa, A.; Austin, D. J.; Xu, S. L. *J. Org. Chem.* **1992**, *57*, 1330.

(150) Evans, D. H.; Woerpel, K. A.; Hinman, M. M.; Faul, M. M. *J. Am. Chem. Soc.* **1991**, *113*, 726. b) Evans, D. A.; Woerpel, K. A.; Scott. M. J. *Angew. Chem., Int. Ed. Engl.* **1992**, *31*, 430.

(151) For a review, see Pfaltz, A. *Acc. Chem. Res.* **1993**, *20*, 339.

(152) Nishiyama, H.; Itoh, Y.; Matsumoto, H.; Park, S.-B.; Itoh, K. *J. Am. Chem. Soc.* **1994**, *116*, 2223.

(153) For reviews, see a) Davies, H. M. L.; Nikolai, J. *Org. Biomolec. Chem.* **2005**, *3*, 4176. b) Davies, H. M. L.; Oystein, L. *Synthesis* **2004**, 2595. c) Doyle, M. P. Transition Metal Carbene Complexes: Diazodecomposition Ylide and Insertion Chemistry. In *Comprehensive Organometallic Chemistry II*; Abel, E. W., Stone, F. G. A., Wilkinson, G., Eds.; Pergamon: Oxford, U.K., 1995; Vol. 12, pp 421–468.

(154) For a review of ligand effects on C–H insertions, see Padwa, A. *Angew. Chem., Int. Ed. Engl.* **1994**, *33*, 1797.

(155) Miah, S.; Slawin, A. M. Z.; Moody, C. J.; Sheehan, S. M.; Marino, J. P., Jr.; Semones, M. A.; Padwa, A. *Tetrahedron* **1996**, *52*, 2489.

(156) Doyle, M. P.; Dyatkin, A. B.; Roos, G. H. P.; Cañas, F.; Pierson, D. A.; van Barten, A.; Müller, P.; Polleux, P. *J. Am. Chem. Soc.* **1994**, *116*, 4507.

(157) Davies, H. M. L.; Venkataramani, C.; Hansen, T.; Hopper, D. W. *J. Am. Chem. Soc.* **2003**, *125*, 6462.

(158) Watanabe, N.; Ogawa, T.; Ohlake, Y.; Ikegami, S.; Hashimoto, S.-i. *Synlett* **1996**, 85.

(159) In a synthesis of 9-isocyanoneopupkeanane. Srikrishna, A.; Satyanarayana, G. *Tetrahedron* **2005**, *61*, 8855.

(160) In a synthesis of the core of zaragozic acids. Wardrop, D. J.; Velter, A. I.; Forslund, R. E. *Org. Lett.* **2001**, *3*, 2261.

(161) Ruediger, E. H.; Solomon, C. *J. Org. Chem.* **1991**, *56*, 3183.

(162) Bhandaru, S.; Fuchs, P. L. *Tetrahedron Lett.* **1995**, *36*, 8347.

(163) Buck, R. T.; Coe, D. M.; Drysdale, M. J.; Ferris, L.; Haigh, D.; Moody, C. J.; Pearson, N. D.; Sanghera, J. B. *Tetrahedron: Asymmetry* **2003**, *14*, 791,

(164) For reviews, see a) Hodgson, D. M.; Pierard, F. Y. T. M.; Stupple, P. A. *Chem. Soc. Rev.* **2001**, *30*, 50. b) Padwa, A.; Krumpe, K. E. *Tetrahedron* **1992**, *48*, 5385. c) Padwa, A.; Hornbuckle, S. F. *Chem. Rev.* **1991**, *91*, 263. d) Padwa, A. *Acc. Chem. Res.* **1991**, *24*, 22.

(165) Kido, F.; Abiko, T.; Kato, M. *J. Chem. Soc., Perkin Trans. 1* **1995**, 2989.

(166) Kametani, T.; Yakawa, H.; Honda, T. *J. Chem. Soc., Chem. Commun.* **1986**, 651.

(167) West, F. G.; Naidu, B. N.; Tester, R. W. *J. Org. Chem.* **1994**, *59*, 6892.

(168) In a synthesis of (–)-epilupinine. Naidu, B. N.; West, F. G. *Tetrahedron* **1997**, *53*, 16565.

(169) West, F. G.; Naidu, B. N. *J. Org. Chem.* **1994**, *59*, 6051.

(170) In a synthesis of the A-ring of gambieric acid. Clark, J. S.; Fessard, T. C.; Wilson, C. *Org. Lett.* **2004**, *6*, 1773.

(171) For reviews, see a) Padwa, A. *J. Organomet. Chem.* **2005**, *690*, 5533. b) Padwa, A.; Weingarten, M. D. *Chem. Rev.* **1996**, *96*, 223.

(172) Padwa, A.; Price, A. T. *J. Org. Chem.* **1998**, *63*, 556.

(173) Padwa, A.; Kassir, J. M.; Semones, M. A.; Weingarten, M. D. *Tetrahedron Lett.* **1993**, *34*, 7853.

(174) In a synthesis of polygaloides A and B. Nakamura, S.; Sugano, Y.; Kikuchi, F.; Hashimoto, S. *Angew. Chem. Int. Ed.* **2006**, *45*, 6532.

(175) In a synthesis of zaragozic acid C. Nakamura, S.; Hirata, Y.; Kurosaki, T.; Anada, M.; Kataoka, O.; Kitagaki, S.; Hsahimoto, S. *Angew. Chem. Int. Ed.* **2003**, *42*, 5351.

(176) For reviews, see a) Deshmukh, P. H.; Blechert, S. *J. Chem. Soc., Dalton Trans.* **2007**, 2479. b) Chattopadhyay, S. K.; Karmakar, S.; Biswas, T.; Majumdar, K. C.; Rahman, H.; Roy, B. *Tetrahedron* **2007**, *63*, 3919. c) Aitken, S. G.; Abell, A. D. *Aust. J. Chem.* **2005**, *58*, 3. d) Nicolaou, K. C.; Bulger, P. G.; Sarlah, D. *Angew. Chem. Int. Ed.* **2005**, *44*, 4490. e) Schuster, M.; Blechert, S. *Angew. Chem., Int. Ed. Engl.* **1997**, *36*, 2036.

(177) For reviews, see a) Schrock, R. R.; Czekelius, C. *Adv. Synth. Catal.* **2007**, *349*, 55. b) Schrock, R. R. *Acc. Chem. Res.* **1990**, *23*, 158.

(178) Belderrain, T. R.; Grubbs, R. H. *Organometallics* **1997**, *16*, 400.

(179) Wilhelm, T. E.; Belderrain, T. R.; Brown, S. N.; Grubbs, R. H. *Organometallics* **1997**, *16*, 3867.

(180) Scholl, M.; Ding, S.; Lee, C. W.; Grubbs, R. H. *Org. Lett.* **1999**, *1*, 953.

(181) Kingsbury, J. S.; Harrity, J. P. A.; Bonitatebus, P. J., Jr.; Hoveyda, A. H. *J. Am. Chem. Soc.* **1999**, *121*, 791.

(182) Garber, S. B.; Kingsbury, J. S.; Gray, B. L.; Hoveyda, A. H. *J. Am. Chem. Soc.* **2000**, *122*, 8168.

(183) Kirkland, T. A.; Grubbs, R. H. *J. Org. Chem.* **1997**, *62*, 7311.

(184) For a review, see Moore, J. S. Transition Metals in Polymer Synthesis. Ring Opening Metathesis Polymerization and Other Transition Metal Polymerization Techniques. In *Comprehensive Organometallic Chemistry II*; Abel, E. W., Stone, F. G. A., Wilkinson, G., Eds.; Pergamon: Oxford, U.K., 1995; Vol. 12, pp 1209–1233.

(185) Lynn, D. M.; Kanaoka, S.; Grubbs, R. H. *J. Am. Chem. Soc.* **1996**, *118*, 784.

(186) Walba, D. M.; Keller, P.; Shao, R.; Clark, N. A.; Hillmeyer, M.; Grubbs, R. H. *J. Am. Chem. Soc.* **1996**, *118*, 2740.

(187) For a review, see Dieters, A.; Martin, S. F. *Chem. Rev.* **2004**, *104*, 2199.

(188) In a synthesis of (+)-SCH 351448. Kang, E. J.; Cho, E. J.; Lee, Y. E.; Ji, M. K.; Shin, D. M.; Chung, Y. K.; Lee, E. *J. Am. Chem. Soc.* **2004**, *126*, 2680.

(189) In a synthesis of the phorboxazole A macrolide. Wang, B.; Forsyth, C. *J. Org. Lett.* **2006**, *8*, 5223.

(190) In a synthesis of nakadomarin A. Nagata, T.; Nakagawa, M.; Nishida, A. *J. Am. Chem. Soc.* **2003**, *125*, 7484.

(191) In a synthesis of gambierol. Kadota, I.; Takamura, H.; Sato, K.; Ohno, A.; Matsuda, K.; Yamamoto, Y. *J. Am. Chem. Soc.* **2003**, *125*, 46.

(192) In a synthesis of woodrosin I. Fürstner, A.; Jeanjean, F.; Razon, P. *Angew. Chem. Int. Ed.* **2002**, *41*, 2097.

(193) For a general model for selectivity, see Chatterjee, A. K.; Choi, T.-L.; Sanders, D. P.; Grubbs, R. H. *J. Am. Chem. Soc.* **2003**, *125*, 11360.

(194) In a synthesis of (+)-*cis*-sylvaticin. Donohoe, T. J.; Harris, R. M.; Burrows, J.; Parker, J. *J. Am. Chem. Soc.* **2006**, *128*, 13704.

(195) In a synthesis of apoptolidinone. Crimmins, M. T.; Christie, H. S.; Chaudhary, K.; Long, A. *J. Am. Chem. Soc.* **2005**, *127*, 13810.

(196) For reviews, see a) Diver, S. T. *Coord. Chem. Rev.* **2007**, *251*, 671. b) Mori, M. *Adv. Synth. Catal.* **2007**, *349*, 121. c) Villar, H.; Fring, M.; Bolm, C. *Chem. Soc. Rev.* **2007**, *36*, 55. d) Mori, M.; Hansen, E. L.; Lee, D. *Acc. Chem Res.* **2006**, *39*, 509. e) Diver, S. T.; Giessert, A. J. *Chem. Rev.* **2004**, *104*, 1317.

(197) Lloyd-Jones, G. C.; Margue, R. G.; de Vries, J. G. *Angew. Chem. Int. Ed.* **2005**, *44*, 7442.

(198) Lippstreu, J. J.; Straub, B. F. *J. Am. Chem. Soc.* **2005**, *127*, 7444.

(199) Kitamura, T.; Sato, Y.; Mori, M. *Chem. Commun.* **2001**, 1258.

(200) Dieltiens, Moonen, K.; Stevens, C. V. *Chem.—Eur. J.* **2007**, *13*, 203.

(201) Hansen, E. C.; Lee, D. *J. Am. Chem. Soc.* **2004**, *126*, 15074.

(202) For reviews, see a) Zhang, W.; Moore, J. S. *Adv. Synth. Catal.* **2007**, *349*, 93. b) Fürstner, A.; Davies, P. A. *Chem. Commun.* **2005**, 2307.

(203) Wengrovius, J. H.; Sancho, J.; Schrock, R. R. *J. Am. Chem. Soc.* **1981**, *103*, 3932.

(204) In a synthesis of PGE$_2$ methyl ester. Fürstner, A.; Mathes, C.; Grela, K. *Chem. Commun.* **2001**, 1057.

(205) In a synthesis of (+)-citreofuran. Fürstner, A.; Castanet, A.-S.; Radkowski, K.; Lehmann, C. W. *J. Org. Chem.* **2003**, *68*, 1521.

(206) In a synthesis of (+)-cylindramide A. Hart, A. C.; Phillips, A. J. *J. Am. Chem. Soc.* **2006**, *128*, 1094.

(207) In a synthesis of (+)-astrophylline. Schaudt, M.; Blechert, S. *J. Org. Chem.* **2003**, *68*, 2913.

(208) In a synthesis of guanacastepene. Boyer, F.-D.; Hanna, I.; Ricard, L. *Org. Lett.* **2004**, *6*, 1817.

(209) a) Fujimura, O.; Grubbs, R. H. *J. Org. Chem.* **1998**, *63*, 825. b) Alexander, J. B.; La, D. S.; Cefalo, D. R.; Graf, D. D.; Hoveyda, A. H.; Schrock, R. R. *J. Am. Chem. Soc.* **1998**, *120*, 9720.

(210) In a synthesis of (+)-africanol. Weatherhead, G. S.; Cortez, G. A.; Schrock, R. R.; Hoveyda, A. H. *Proc. Natl. Acad. Sci. U.S.A.* **2004**, *101*, 5805.

(211) Funk, T. W.; Berlin, J. M.; Grubbs, R. H. *J. Am. Chem. Soc.* **2006**, *128*, 1840.

(212) For reviews, see a) Hartley, R. C.; McKiernan, G. J. *J. Chem. Soc., Perkin Trans. 1* **2002**, 2763. b) Stille, J. R. Transition Metal Carbene Complexes. Tebbe's Reagent and Related Nucleophilic Alkylidenes. In *Comprehensive Organometallic Chemistry II*; Abel, E. W., Stone, F. G. A., Wilkinson, G., Eds.; Pergamon: Oxford, U.K., 1995; Vol. 12, pp 577–600.

(213) Cannizzo, L. F.; Grubbs, R. H. *J. Org. Chem.* **1985**, *50*, 2386.

(214) Cannizzo, L. F.; Grubbs, R. H. *J. Org. Chem.* **1985**, *50*, 2316.

(215) For a comparison of Tebbes reagent with Ph$_3$PCH$_2$, see Pine, S. H.; Shen, G. S.; Hoang, H. *Synthesis* **1991**, 165.

(216) In a synthesis of zaragozic acid A. Stoermer, D.; Caron, S.; Heathcock, C. H. *J. Org. Chem.* **1996**, *61*, 9115.

(217) Borrelly, S.; Paquette, L. A. *J. Am. Chem. Soc.* **1996**, *118*, 727.

(218) Buchwald, S. L.; Grubbs, R. H. *J. Am. Chem. Soc.* **1983**, *105*, 5490.

(219) Nicolaou, K .C.; Postema, M. H. D.; Claiborne, C. F. *J. Am. Chem. Soc.* **1996**, *118*, 1565.

(220) Takeda, T.; Arai, K.; Shimokawa, H.; Tsubouchi, A. *Tetrahedron Lett.* **2005**, *46*, 775.

(221) Rahim M. A.; Sasaki, H.; Saito, J.; Fujiwara, T.; Takeda, T. *Chem. Commun.* **2001**, 625.

(222) a) Shono, T.; Ito, K.; Tsubouchi, A.; Takeda, T. *Org. Biomolec. Chem.* **2005**, *3*, 2914. b) Shono, T.; Hayata, Y.; Tsubouchi, A. Takeda, T. *Tetrahedron Lett.* **2006**, *47*, 1257.

(223) Petasis, N. A.; Bzowej, E. I. *J. Am. Chem. Soc.* **1990**, *112*, 6392.

(224) DeShong, P.; Rybczynski, P. J. *J. Org. Chem.* **1991**, *56*, 3207.

(225) Petasis, N. A.; Bzowej, E. I. *J. Org. Chem.* **1992**, *57*, 1327.

(226) Petasis, N. A.; Akritopoulou, I. *Synlett* **1992**, 665.

(227) In a synthesis of (+)-aphanamol 1. Wender, P. A. Zhang, L. *Org. Lett.* **2000**, *2*, 2323.

(228) Petasis, N. A.; Hu, Y.-H. *J. Org. Chem.* **1997**, *62*, 782.

(229) Petasis, N. A.; Bzowej, E. I. *Tetrahedron Lett.* **1993**, *34*, 943.

(230) In a synthesis toward halichondrin B. Lambert, W. T.; Hanson, G. H.; Benayoud, F.; Burke, S. D. *J. Org. Chem.* **2005**, *70*, 9382.

(231) In a synthesis of (–)-clavosolide A. Smith, A. B., III; Simov, V. *Org. Lett.* **2006**, *8*, 3315.

(232) In a synthesis of 6-epi-sarsolilide A. Zhang, J.; Xu, X. *Tetrahedron Lett.* **2000**, *41*, 941.

(233) Takai, K.; Kakiuchi, T.; Kataoka, Y.; Utimoto, K. *J. Org. Chem.* **1994**, *59*, 2668.

(234) In a synthesis of amphidinolide T1. Colby, E. A.; O'Brien, K. C.; Jamison, T. F. *J. Am. Chem. Soc.* **2004**, *126*, 998.

(235) Matsubara, S.; Horiuchi, M.; Takai, K.; Utimoto, K. *Chem. Lett.* **1995**, 259.

(236) Takai, K.; Nitta, K.; Utimoto, K. *J. Am. Chem. Soc.* **1986**, *108*, 7408.

(237) In a synthesis of 20,21-diepicallipeltoside A aglycone. Trost, B. M.; Gunzer, J. L.; Dirat, O.; Rhee, Y. H. *J. Am. Chem. Soc.* **2002**, *124*, 10396.

(238) In a synthesis of rutamycin B. White, J. D.; Hanselmann, R.; Jackson, R. W.; Porter, W. J.; Ohba, Y.; Tiller, T.; Wang, S. *J. Org. Chem.* **2001**, *66*, 5217.

(239) For a review, see Müller, P.; Fruit, C. *Chem. Rev.* **2003**, *103*, 2905.

(240) In a synthesis of agelastatin A. Trost, B. M.; Dong, G. *J. Am. Chem. Soc.* **2006**, *128*, 6054.

(241) Liang, C.; Robert-Peillard, F.; Fruit, C.; Müller, P.; Dodd, R. H.; Dauban, P. *Angew. Chem. Int. Ed.* **2006**, *45*, 4641.

(242) In a synthesis of (–)-tetrodotoxin. Hinman, A.; Du Bois, J. *J. Am. Chem. Soc.* **2003**, *125*, 11510.

7

Synthetic Applications of Transition Metal Alkene, Diene, and Dienyl Complexes

7.1 Introduction

Nucleophilic attacks on transition-metal-complexed alkene, diene, and dienyl systems are among the most useful of organometallic processes for the synthesis of complex organic molecules.[1] Not only does complexation reverse the normal reactivity of these functional groups, changing them from nucleophiles to electrophiles, but the process also results in the formation of new bonds between the nucleophile and the alkene carbon and new metal–carbon bonds that can be further elaborated (**Eq. 7.1**). The fundamental features of this process were presented in Chapter 2. Here its uses in organic synthesis are considered.

Eq. 7.1

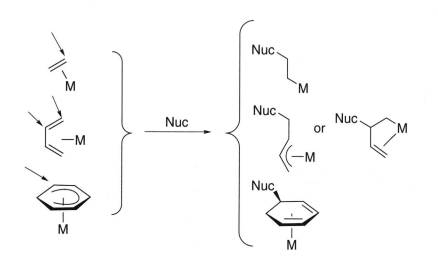

7.2 Metal–Alkene Complexes[1]

a. Palladium(II) Complexes

Alkene complexes of palladium(II) were among the first used to catalyze useful organic transformations and are perhaps the most extensively developed, at least in part because they are both easily generated and highly reactive. Palladium chloride is the most common catalyst precursor. It is a commercially available red-brown, chloro-bridged oligomer that is insoluble in most organic solvents (**Figure 7.1**). However, the oligomer is easily disrupted by treatment with alkali metal chlorides, such as LiCl or NaCl, providing the considerably more soluble, hygroscopic monomeric palladate, M_2PdCl_4. Nitriles, particularly acetonitrile and benzonitrile, also give air-stable, easily handled, soluble monomeric solids, which are the most convenient sources of palladium(II). The treatment of soluble palladium(II) salts with alkenes results in rapid, reversible coordination of the alkene to the metal [path (a) in **Figure 7.2**].

Ethylene and terminal monoalkenes coordinate most strongly, followed by cis and trans internal alkenes. Although these alkene complexes can be isolated if desired, they are normally generated in situ and used without isolation (**Figure 7.2**). Once complexed, the alkene becomes *generally* reactive towards nucleophiles ranging from Cl⁻ through Ph⁻, a range of about 10^{36} in basicity. Attack occurs from the face opposite the metal (i.e., trans attack) and at the more-substituted carbon, resulting in the formation of a carbon–nucleophile bond and a metal–carbon bond [path (b) in **Figure 7.2**]. This is the opposite regiochemistry and stereochemistry from that obtained by the insertion of alkenes into metal–carbon σ-bonds. The regiochemistry is that corresponding to attack at the alkene position which can *best* stabilize positive charge and, in a sense, the highly electrophilic palladium(II) is behaving like a very expensive and selective proton towards the alkene.

The newly formed σ-alkylpalladium(II) complex has a very rich chemistry. Warming above –20°C results in β-hydride elimination [path (c) in **Figure 7.2**], generating an

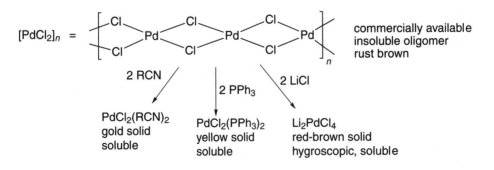

also Pd(OAc)₂ soluble

Figure 7.1. Palladium(II) Sources

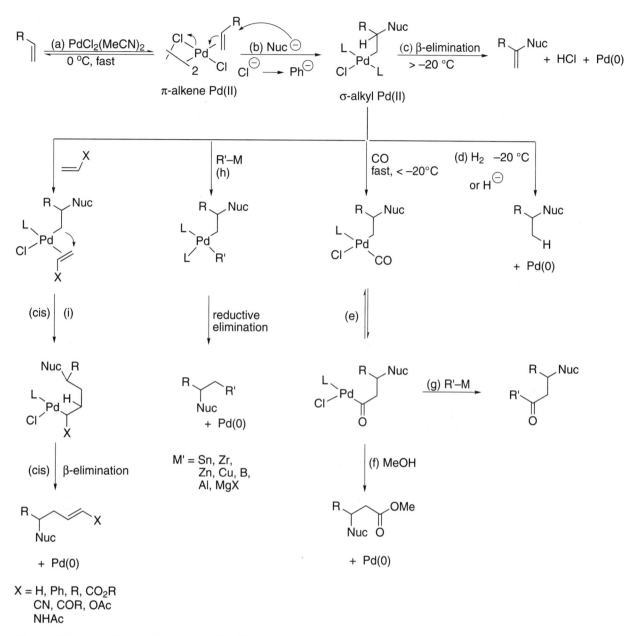

Figure 7.2. Reactions of Palladium(II) Alkene Complexes

alkene which is the product of a formal nucleophilic substitution of a hydrogen. This σ-alkylpalladium(II) complex is easily reduced, either by hydrides or hydrogen gas at one atmosphere, resulting in a formal nucleophilic addition to the alkene [path (d)]. Carbon monoxide insertion is fast, even at low temperatures, and successfully competes with β-elimination. Thus, σ-acyl species are easily produced simply by exposing the reaction solution to an atmosphere of carbon monoxide [path (e)]. Cleavage of this acyl species with methanol produces esters [path (f)], while treatment with main group organometallics leads to ketones via a *carbonylative coupling* [path (g)]. The initially formed σ-alkylpalladium(II) complex itself undergoes transmetallation from many main group organometallics, resulting in overall *nucleophilic addition–/alkylation* [path (h)].

In principle, alkenes can insert into metal–carbon σ-bonds, but in practice this usually cannot be achieved when β-hydrogens are present, since β-elimination is faster. In exceptional cases, however, alkenes can insert into σ-alkylpalladium(II) complexes prepared from nucleophilic attack on alkenes, resulting in an overall *nucleophilic addition–alkenylation* [path (i) in **Figure 7.2**]. In all of the preceding processes, palladium(II) is required to activate the alkene, but palladium(0)—from reductive elimination—is the product of the reaction. Thus, for catalysis, a procedure to oxidize Pd(0) back to Pd(II) in the presence of reactants and products is needed. Many systems have been developed ranging from the classic O_2/$CuCl_2$ redox system to more recently developed benzoquinone, or quinone/reoxidant. The choice of reoxidant is usually dictated by the stability of the products toward the oxidant.

Palladium(II) alkene complexes are particularly well behaved in their reactions with oxygen nucleophiles, such as water, alcohols,[2] and carboxylates, and a large number of useful synthetic transformations are based on this process. One of the earliest is the Wacker process,[3] the palladium(II)-catalyzed "oxidation" of ethene to acetaldehyde for which the key step is the nucleophilic attack of water on metal-coordinated ethene. Although this particular (industrial) process is of little interest for the synthesis of fine chemicals, when applied to terminal alkenes it provides a very efficient synthesis of methyl ketones.[4] The mechanism of this reaction has classically been depicted as an alkene coordination–anti-hydroxypalladation–β-hydride elimination sequence. However, it is more complex and depends on the substrate and the additives (**Eq. 7.2**). For the nucleophilic addition step, a mechanism involving a syn addition of water is considered operating at low chloride concentrations and an anti addition at high chloride concentrations.[5] Product formation by β-hydride elimination is usually depicted in catalytic cycles, but a 1,2-hydride shift has also been suggested. The latter was shown to be the predominant termination step in a labelling study.[6] The reaction is specific for terminal alkenes, since they coordinate much more strongly than internal alkenes, and internal alkenes are tolerated elsewhere in the substrate (**Eq. 7.3**).[7]

A variety of remote functional groups are tolerated (**Eqs. 7.4**[8] and **7.5**[9]), provided they are not strong ligands for palladium(II) (e.g., amines), in which case the catalyst is poisoned. Copper(II) chloride/oxygen is usually used to reoxidize the palladium, since the product alcohols are stable to these reagents. A significant improvement with high substrate compatibility was recently reported wherein oxygen was used as the sole oxidant and DMA was used as the solvent.[10]

Eq. 7.2

Eq. 7.3

Eq. 7.4

Eq. 7.5

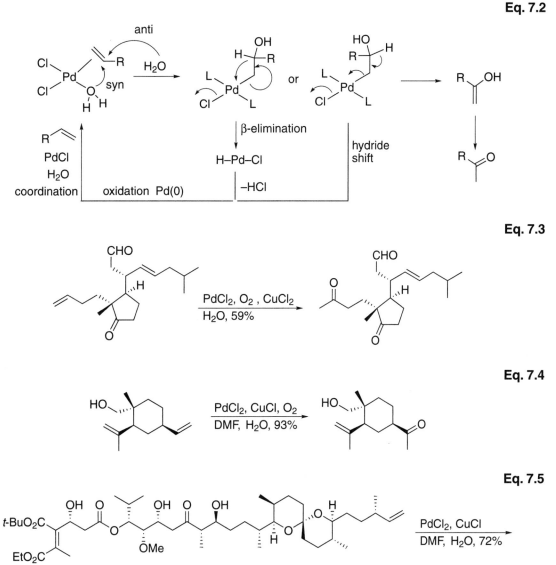

In all of these cases, nucleophilic attack occurred at the secondary position, as expected. However, if the substrate has adjacent functional groups that can coordinate to palladium to form a five- or six-membered palladacyclic intermediate, the attack of the nucleophile may be directed to the terminal position (**Eqs. 7.6**[11] and **7.7**[12]). Adjacent directing groups can even activate internal alkenes to undergo Wacker oxidation (**Eqs. 7.8**[13] and **7.9**[14]), although this is a rare occurrence.

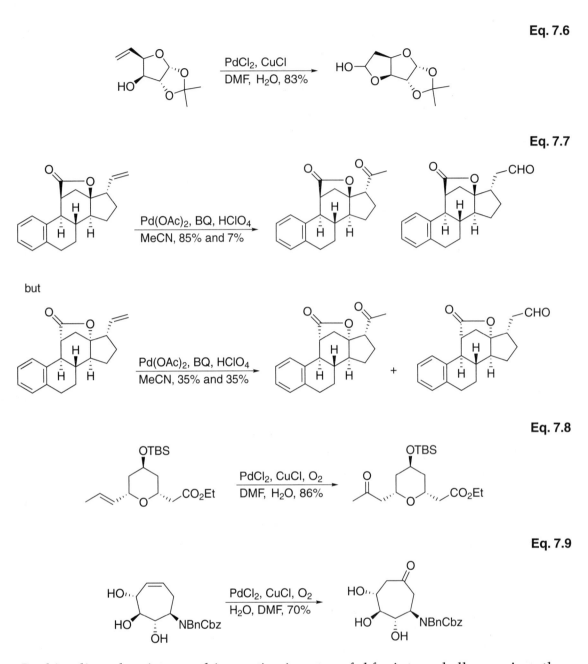

but

Lacking ligand assistance, this reaction is not useful for internal alkenes since they react only slowly, if at all. However, α,β-unsaturated ketones and esters are cleanly converted to β-keto compounds under somewhat specialized conditions (**Eq. 7.10**).[15]

Eq. 7.10

Alcohols also add cleanly to alkenes in the presence of palladium(II), but the intermolecular version has found little use in synthesis. A recent example of an intermolecular *di*alkoxylation is shown in **Eq. 7.11**.[16] The aromatic alcohol is crucial for the reaction, because substituting the OH for H or OMe results in the formation of Wacker-type products.

Eq. 7.11

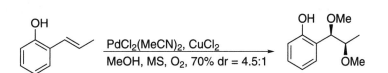

In contrast, the intramolecular version provides a very attractive route to oxygen heterocycles.[17] For primary alcohols, the mechanism has been shown to proceed via a syn-alkoxypalladation[18] and not the commonly accepted anti addition mechanism. Phenols undergo intramolecular syn-alkoxylations in the absence of chloride ions and mainly anti-alkoxypalladation in the presence of chloride.[19] In contrast, carboxylates only undergo anti-carboxypalladations. In intramolecular reactions, internal and even trisubstituted alkenes undergo attack and the regioselectivity is often determined more by ring size than by substitution (**Eqs. 7.12**[20] and **7.13**[21]). In the presence of chiral ligands, asymmetry can be induced (**Eq. 7.14**[22]).

Eq. 7.12

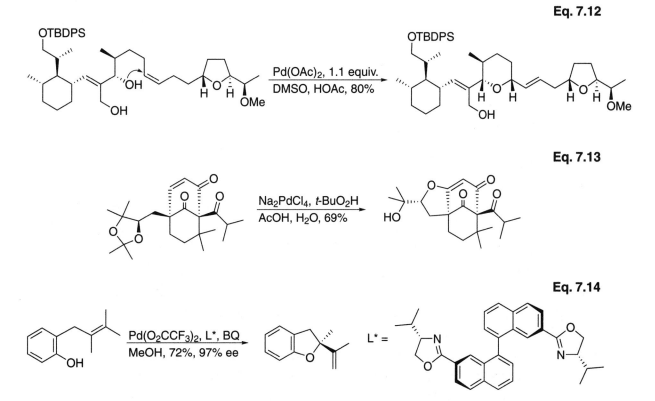

Eq. 7.13

Eq. 7.14

Reoxidation is necessary in these palladium(II)-catalyzed processes because "HPdX" is produced in the β-elimination step (**Figure 7.2**) and rapidly loses HX (via reductive elimination) to produce palladium(0). If some group other than hydride, such as Cl⁻ or

OH⁻, can be β-eliminated, subsequent reductive elimination to form palladium(0) (and Cl₂ or HOCl) does not occur, and the resulting PdCl₂ or Pd(OH)(Cl) can re-enter the catalytic cycle directly, obviating the need for oxidants (**Eq. 7.15**[23]).[24] Because CO insertion into Pd–C bonds is so facile, and competes effectively with β-hydride elimination, the efficient alkoxycarbonylation of alkenes has been achieved (**Eq. 7.16**[25]).

Eq. 7.15

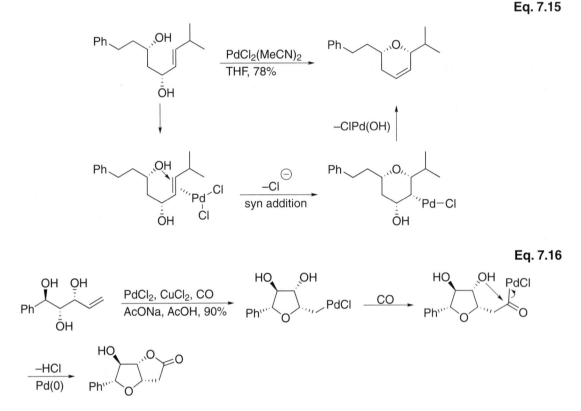

Eq. 7.16

If β-elimination is blocked, even alkenes can insert (**Eq. 7.17**),[26] although this is rare.[27] Under exceptional circumstances, alkenes can be made to insert into σ-alkylpalladium(II) complexes having β-hydrogens. Such a case is shown in **Eq. 7.18**.[28] The initial step is palladium(II)-assisted alcohol attack on the complexed enol ether, generating a σ-alkylpalladium(II) complex, which has a β-hydrogen potentially accessible for β-elimination. However, β-elimination requires a syn-coplanar relationship and, if the adjacent alkene coordinates to palladium, rotation to achieve this geometry is suppressed. At the same time, coordination of the alkene permits it to insert into the σ-metal–carbon bond, which is indeed what occurs, to form the bicyclic system. The insertion is a syn process, and the σ-alkylpalladium(II) complex resulting from this insertion cannot achieve a syn-coplanar geometry with any of the β-hydrogens, since the ring system is rigid. In this special circumstance, *inter*molecular alkene insertion occurs, giving the bicyclic product in excellent yield. This process forms one carbon–oxygen and two carbon–carbon bonds in a one-pot procedure. Complex compounds can be prepared via related multiple insertion reactions (**Eq. 7.19**).[29]

Eq. 7.17

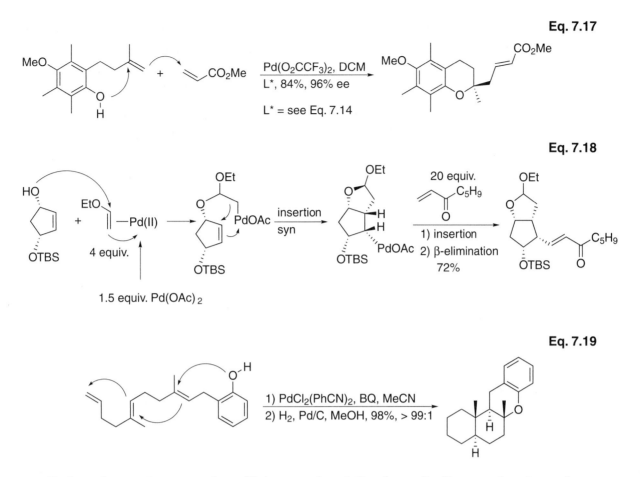

Eq. 7.18

Eq. 7.19

Carboxylate anions are also efficient nucleophiles for palladium-assisted attack on alkenes.[30] Intramolecular versions are efficient (**Eq. 7.20**)[31] and, as is often the case for intramolecular reactions, even trisubstituted alkenes are attacked.[32]

Eq. 7.20

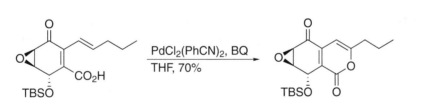

Amines also attack palladium-bound alkenes, but catalytic intermolecular amination of simple unactivated alkenes has not been achieved, since both the amine and the product enamine are potent ligands for palladium and act as catalyst poisons. However, this is a very active area of research. The stoichiometric amination of simple alkenes can be achieved in good yield, but three equivalents of amine are required, and the resulting σ-alkylpalladium(II) complex must be reduced to give the free amine.[33] A careful study of intramolecular amination elucidated some of the difficulties (**Eq. 7.21**).[34] Treatment of the cyclopentyl aminoalkene with palladium(II) chloride resulted in the formation of a stable, chelated aminoalkene complex, confirming that one problem is indeed the strong coordination of amines to Pd(II). N-Acylation of the amine dramatically decreased its basicity and coordinating ability, and this acetamide did not irreversibly bind to palladium(II). Instead, intramolecular amination of the alkene ensued. However, the resulting σ-alkylpalladium(II) complex was stabilized by chelation to the amide oxygen, preventing β-hydride elimination, and thus catalysis. Finally, by placing the strongly electron-withdrawing sulfonyl group on nitrogen, neither the substrate nor the product coordinated strongly enough to palladium(II) to prevent catalysis, and efficient cyclization was achieved (**Eq. 7.22**).

Eq. 7.21

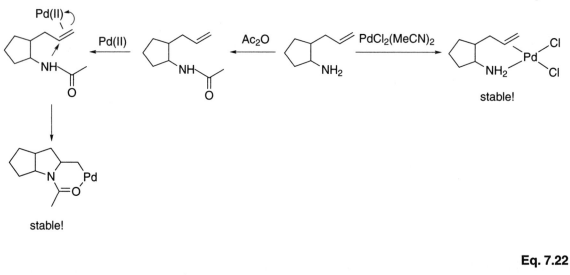

Eq. 7.22

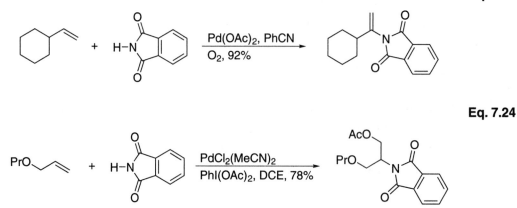

A more recent development is the catalytic *intermolecular aminations* of alkenes using carboxamides, carbamates, or sulfonamides as nucleophiles.[35] An example is shown in **Eq. 7.23**.[36] The initially formed σ-complex can be intercepted by acetate from PhI(OAc)$_2$ (**Eq. 7.24**)[37] or insert into a pendant alkene (**Eq. 7.25**).[38]

Eq. 7.23

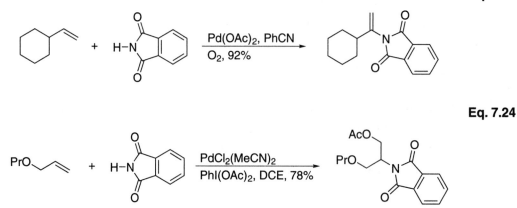

Eq. 7.24

Eq. 7.25

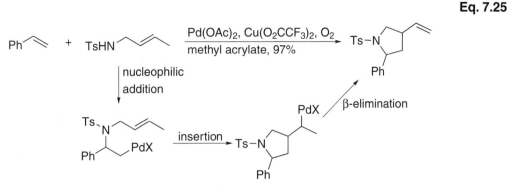

Aromatic amines are about 10^6 less basic than aliphatic amines, and this factor permits the efficient intramolecular amination of alkenes by these amines without the need for N-acylation or sulfonation, although these N-functionalized substrates also cyclize efficiently (**Eqs. 7.26**[39] and **7.27**[40]). A variety of functionalized indoles can be made by this procedure. Even *intermolecular* palladium-catalyzed aminations of alkenylarenes have been developed using alkyl-[41] and arylamines (see Chapter 4).[42]

Eq. 7.26

Eq. 7.27

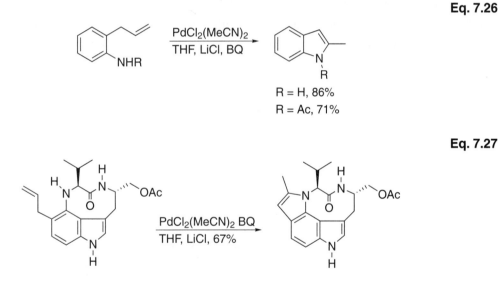

A very mild and general catalyst system for the cyclization of both aliphatic and aromaticic alkenyl tosylamides has recently been developed.[43] It consists of palladium acetate in DMSO and relies upon oxygen as a reoxidant. In most cases the regioselectivity parallels that of the previous systems (**Eq. 7.28**), but ortho-allyl-N-tosylanilines cyclize to give dihydroquinolines (i.e., attack at the less-substituted alkene terminus) rather than indoles. Air can also be used as the reoxidant (**Eq. 7.29**).[44] As with oxygen nucleophiles, if the alkene to be aminated has an allylic leaving group, reoxidation is unnecessary and catalytic cyclizations proceed smoothly (**Eq. 7.30**[45]).[46]

Eq. 7.28

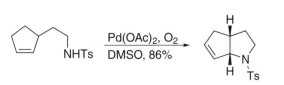

Eq. 7.29

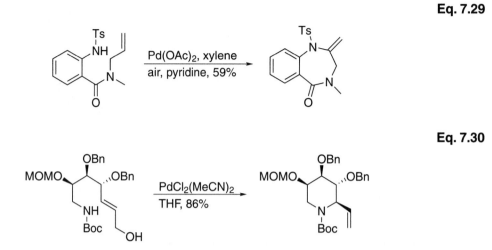

Eq. 7.30

The σ-alkylpalladium(II) complex from the amination of alkenes can easily be trapped by CO[47] (**Eqs. 7.31**[48] and **7.32**[49]). With the careful choice of conditions, even alkenes will insert reasonably efficiently, although β-hydride elimination competes even under the best of circumstances (**Eq. 7.33**).[50] Although there are nine sites at which the catalyst can (and almost certainly does) coordinate, only one leads to cyclization. Provided that the catalyst is not irreversibly bound to one of the other sites, catalysis ensues.

Eq. 7.31

Eq. 7.32

Eq. 7.33

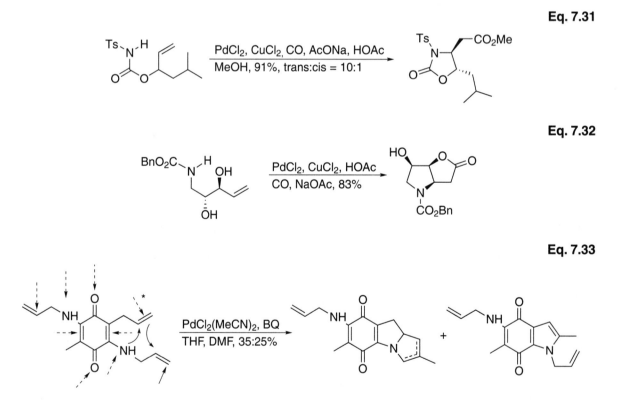

The use of carbon nucleophiles results in yet another set of problems. Carbanions, even stabilized ones, are easily oxidized, and palladium(II) salts, as well as many other transition metal salts, tend to oxidatively dimerize carbanions, preventing their attack on alkenes. Even if this can be prevented, catalysis is (as yet) impossible to achieve since it is necessary to find an oxidizing agent capable of reoxidizing the catalytic amount of Pd(0) present to Pd(II), but at the same time unable to oxidize the large amount of the carbanion substrate present. Stoichiometric alkylation of alkenes by stabilized carbanions can be achieved by very careful choice of conditions (**Eq. 7.34**).[51] Preforming the alkene complex at 0°C, followed by cooling to –78°C and addition of two equivalents of triethylamine, generates an alkene–amine complex. The coordinated amine directs the carbanion away from the metal and to the alkene, suppressing metal-centered, electron-transfer redox chemistry, and promoting alkylation of the alkene. As usual, attack occurs at the more-substituted carbon of the alkene, producing a σ-alkylpalladium(II) complex that can undergo reduction, β-elimination, or CO insertion. The Pd(0) produced can be recovered and reoxidized separately, but not concurrently with alkylation, so the process is stoichiometric in palladium, and unlikely to find extensive use in synthesis.

Eq. 7.34

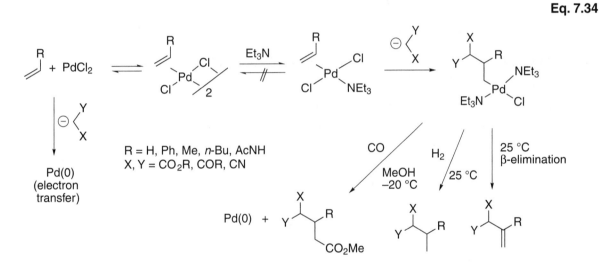

The exceptions to this will be cases in which this chemistry permits the efficient synthesis of products that would be considerably more difficult to prepare by classical means. Such a situation is seen in the synthesis toward (+)-thienamycin (**Eq. 7.35**).[52] In this case, one-pot alkylation–acylation of an optically active ene carbamate produced the keto diester in 70% chemical yield, and with complete control of the stereochemistry at the chiral center α to nitrogen. In this case, palladium was used to form two C–C and one O–C bond, and to generate a new chiral center. When palladium(II)-assisted alkylation of this alkene was combined with carbonylative coupling to tin reagents, even more complex structures were efficiently produced (**Eq. 7.36**).[53] In this case, palladium(II) promotes stereospecific alkylation of the alkene, then insertion of CO into the metal–carbon σ-bond, followed by transmetallation from tin to palladium, and finally reductive elimination. In this case three C–C bonds and a chiral center are efficiently generated in a one-pot reaction.

Eq. 7.35

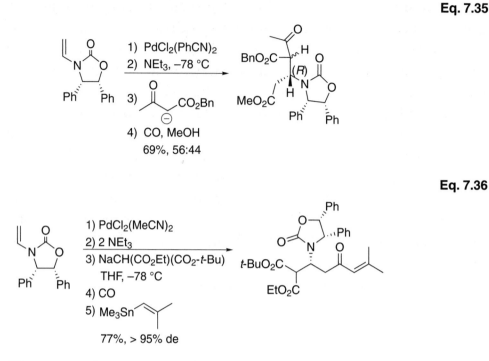

Eq. 7.36

Palladium(II) acetate also promotes the intramolecular alkylation of alkenes by silylenol ethers (**Eq. 7.37**[54]).[55] It is possible to make this process proceed catalytically, since silylenol ethers themselves are not oxidatively dimerized by Pd(II). This has been realized using oxygen as the oxidant in dimethylsulfoxide[56] (**Eq. 7.38**)[57].

Eq. 7.37

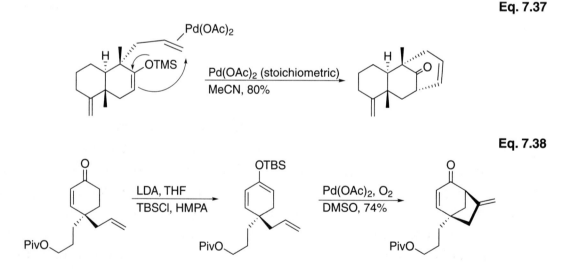

Eq. 7.38

The reactions of silylenol ethers usually take a different course when the enol ether is derived from a ketone (or aldehyde) other than methyl ketones. In this case, palladium(II) catalyzes an oxidation to give α,β-unsaturated ketones (the "Saegusa oxidation").[58] The mechanism has not been studied in detail, but probably involves the formation of an oxo–η^3-allyl complex, followed by a β-hydride elimination (**Eq. 7.39**). Benzoquinone was initially used as a reoxidant to complete the catalytic cycle, but only a few turnovers of palladium(II) were observed. Despite this limitation, the reaction is a valuable tool in organic synthesis since it tolerates a variety of functional groups (**Eq. 7.40**[59]). A significant improvement is the use of oxygen as the oxidant in dimethylsulfoxide[60] (**Eq. 7.41**[61]).

Eq. 7.39

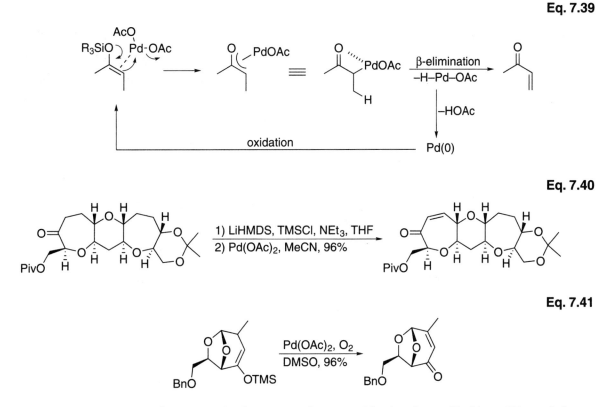

Eq. 7.40

Eq. 7.41

In contrast to enolates, enols do not undergo oxidation by palladium(II) and these nucleophiles can be employed in inter-[62] and intramolecular addition reactions with alkenes.[63] Depending on the tether length between the enol and the alkene, either a cycloalkane or a cycloalkene is formed (**Eq. 7.42**).[64] A reoxidant is not needed in the former case since the catalyst is regenerated in the reaction. It is unclear why β-elimination products are not observed in these reactions. For the longer tether, copper dichloride is used to oxidize palladium(0) to palladium(II) (**Eq. 7.43**).

Eq. 7.42

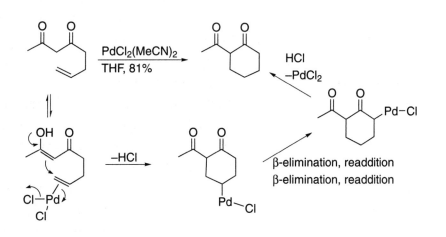

Eq. 7.43

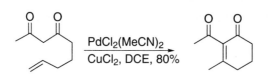

Aromatic rings can also be activated by palladium and added to alkenes. A mechanism involving an initial insertion of palladium into a carbon–hydrogen bond, followed by alkene insertion and reductive elimination, has been suggested (**Eq. 7.44**).[65] Ethyl nicotinate is a ligand in this reaction. The intramolecular reaction is a potentially useful method for the formation of fused heterocyclic compounds, as exemplified in a synthesis of (+)-dragmacidin F (**Eq. 7.45**)[66].

Eq. 7.44

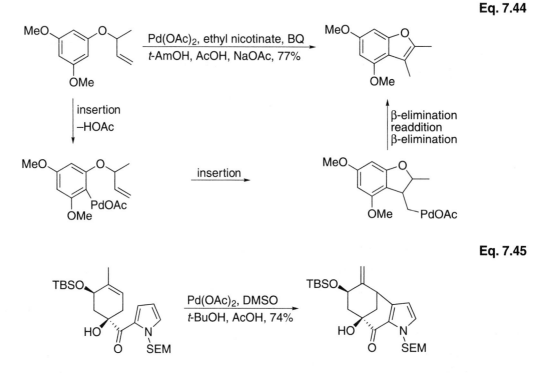

Eq. 7.45

The hydroalkylation reaction described in **Eq. 7.42** is unusual for palladium since β-elimination is the most common termination step. In contrast, a variety of transition metals in +1 to +3 oxidation states catalyze hydrofunctionalizations of alkenes. These include gold,[67] platinum,[68,69] ruthenium, rhodium, and iridium catalyzing hydroaminations, hydroalkoxylations, hydroalkylations, hydroarylations,[70] and hydroalkenylations. Plausible mechanisms for these reactions can be divided into two groups—namely, coordination of the alkene followed by nucleophilic addition, and carbon–hydrogen bond insertion followed by alkene insertion (c.f. **Eq. 7.44**). The mechanisms are different compared to the palladium-catalyzed reactions in that no redox chemistry is observed. The σ-metal intermediate formed undergoes protolytic cleavage and not β-hydride elimination (**Eq. 7.46**).[71] The true mechanisms of the hydrofunctionalizations are not yet known. A recent study indicated that nonmetallic Lewis acids such as TfOH generated from the catalysts may actually be the true catalytic species in some cases.[72]

Eq. 7.46

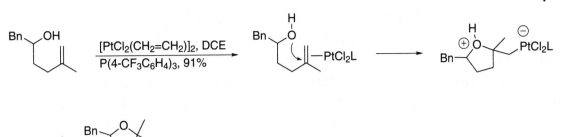

Late transition-metal-catalyzed intramolecular alkene hydroarylations are currently the most developed reactions. Particularly useful are reactions of heteroaromatic compounds (**Eq. 7.47**).[73] Asymmetric reactions are possible using a chiral ligand (**Eq. 7.48**).[74]

Eq. 7.47

Eq. 7.48

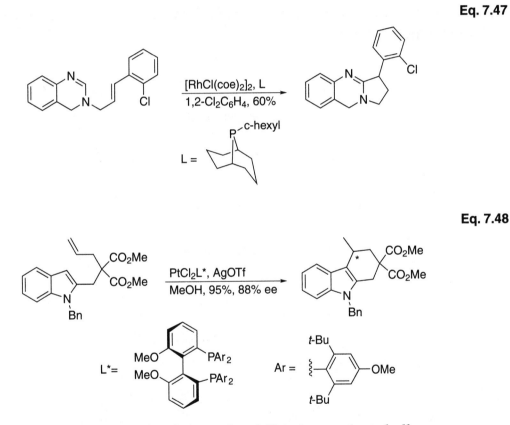

Intermolecular additions of carbon nucleophiles to unactivated alkenes are more challenging. The use of gold(I and III),[75] silver(I) (**Eq. 7.49**),[76] and platinum(II)[77] catalysts have more recently been described and high yields of saturated products can be obtained.

Eq. 7.49

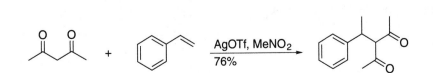

Allenes can be hydrofunctionalized using the same catalysts. The mechanism likely involves allene coordination followed by nucleophilic addition to give a σ-allyl complex followed by protonolysis.[78] Addition to the allene termini is usually observed in intramolecular reactions (**Eqs. 7.50**[79] and **7.51**[80]). In the presence of a palladium catalyst, the nucleophile adds to the central carbon of the allene to generate an η³-allyl complex that undergoes further transformations (see Chapter 9). A very interesting change in chemoselectivity is observed by simple tuning the oxidation state of the gold complex.[81] For gold(III) complexes, a [1,2]-bromide shift was observed, forming 3-bromofuranes. A possible mechanism involves coordination of the metal to the carbonyl carbon, formation of a bromonium ion, and nucleophilic ring opening (**Eq. 7.52**). In contrast, using a gold(I) complex resulted in formation of isomeric 2-bromofuranes (**Eq. 7.53**). This result can be rationalized as a nucleophilic addition to a gold-coordinated allene, forming a σ-gold complex, isomerization to a gold carbene, and finally a [1,2]-hydride shift. A related activation of alkynes by gold and platinum complexes is discussed in Chapter 8.

Eq. 7.50

Eq. 7.51

Eq. 7.52

Eq. 7.53

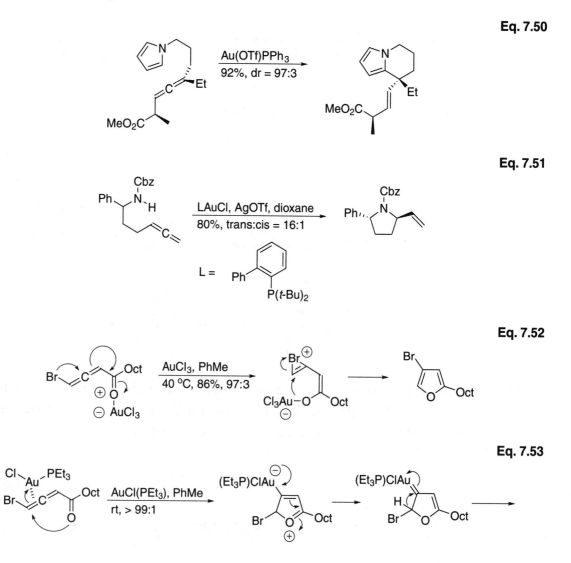

The departure from a β-elimination termination (see Chapter 4) can also be seen in cycloisomerizations of 1,6- and 1,7-dienes in the presence of platinum catalysts. When platinum complexes are used, the σ-platinum intermediates decompose to form bicyclo-[3.1.0] and bicyclo-[4.1.0] products (**Eq. 7.54**).[82]

Eq. 7.54

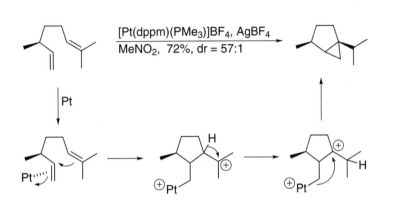

Palladium(II) complexes catalyze the rearrangement of a very broad array of allylic systems, in a process which involves intramolecular nucleophilic attack on a Pd(II)-complexed alkene as the key step (**Eq. 7.55**).[83] Allyl acetates rearrange under mild conditions without complication from skeletal rearrangements or the formation of other side products. With optically active allyl acetates, rearrangement occurs with complete transfer of chirality (**Eqs. 7.56**[84] and **7.57**[85]). The absolute stereochemistry observed at the newly formed chiral center must result from the coordination of palladium to the face of the alkene opposite the acetate, and attack of the alkene from the face opposite the metal, as expected for the mechanism shown in **Eq. 7.55**.

Eq. 7.55

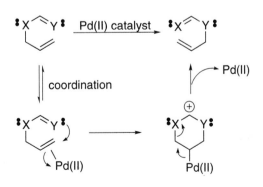

Eq. 7.56

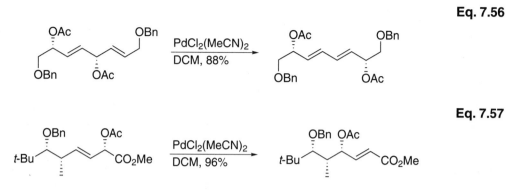

Eq. 7.57

Palladium(II) complexes catalyze the Cope rearrangement of 1,5-dienes under very mild conditions, leading to a roughly 10^{10} rate enhancement over the thermal process (**Eq. 7.58**).[86] There are some limitations to the catalyzed process, in that C-2 must bear a substituent to help stabilize the developing positive charge, and C-5 must bear a hydrogen so that efficient complexation to Pd(II) results. Recall that trisubstituted alkenes complex poorly to Pd(II), and that gem-disubstituted alkenes undergo attack at the substituted position, making Cope rearrangements disfavored for these systems under Pd(II) catalysis. Again, with chiral substrates, complete chirality transfer is observed.[87] Palladium(II) complexes also catalyze the oxy-Cope rearrangement (**Eq. 7.59**).[88]

Eq. 7.58

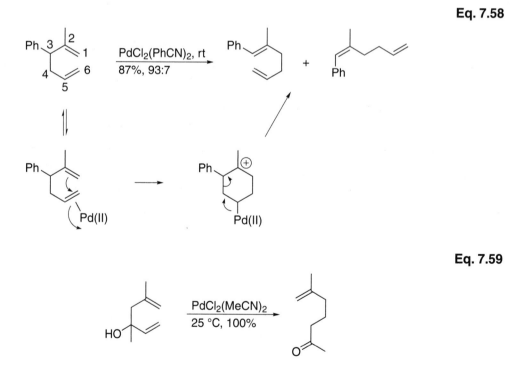

Eq. 7.59

Many other allylic systems undergo this palladium(II)-catalyzed allylic transposition. Trichloromethylimidates rearrange cleanly, and with complete transfer of chirality (the "Overman rearrangement") (**Eq. 7.60**[89]).[90,91] O-Allylimidates[92] and (allyloxy) iminodiazaphospholidines (**Eq. 7.61**)[93] rearrange in a similar fashion and, in the presence of chiral ligands, modest (50–60%) asymmetric induction is observed.[94] O-Allyl oximes rearrange to nitrones, which can be efficiently utilized in 1,3-dipolar cycloaddition reactions (**Eq. 7.62**).[95]

Eq. 7.60

Eq. 7.61

Eq. 7.62

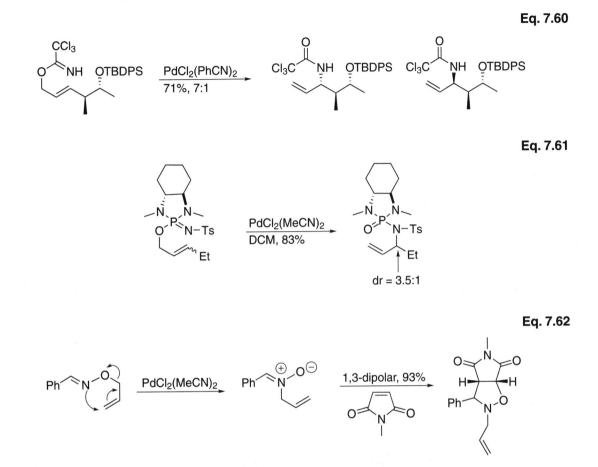

Palladium catalyzes ring expansions of 1-alkenyl-1-cyclobutanoles to give five-membered rings.[96] The reaction is triggered by the coordination of palladium to the alkene followed by a ring expansion that relieves ring strain (**Eq. 7.63**).[97]

Eq. 7.63

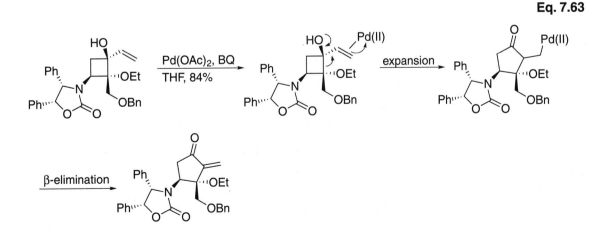

b. Iron(II) Complexes

Cationic alkene complexes of iron(II) are easily made by alkene exchange with the CpFe(CO)$_2$(isobutene)$^+$ complex, which is a stable gold-yellow solid, made by the reaction of Fp$^-$ [where Fp = CpFe(CO)$_2$] with methallyl bromide, followed by γ-protonation with HBF$_4$ (see Chapter 2) (**Eq. 7.64**).[98]

Eq. 7.64

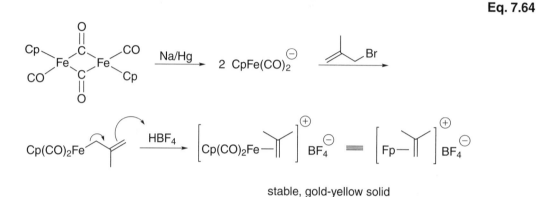

stable, gold-yellow solid

Once complexed, the alkene becomes reactive towards nucleophilic attack, sharing most of the reactivity, regiochemical, and stereochemical features of the neutral alkene–palladium(II) complexes discussed in Section 7.2a. However, in contrast to the palladium analogs, the resulting σ-alkyliron(II) complexes are very stable, and iron must be removed in a separate chemical step (usually oxidation). As an added complication, oxidative cleavage of Fp–alkyl complexes usually results in insertion of CO into the metal–carbon σ-bond, producing carbonylated products (**Eq. 7.65**). For these reasons, iron–alkene complexes have been used in synthesis considerably less often than those of palladium. However, the reaction chemistry of electron-rich alkenes complexed to iron is quite unusual, and synthetically useful.

Eq. 7.65

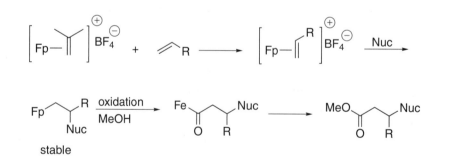

stable

When complexed to the Fp cation, dimethoxyethene is quite reactive towards nucleophiles, and both methoxy groups can be replaced, making this iron–alkene complex essentially a vinylidene dication (CH$^+$=CH$^+$) equivalent.[99] Treatment with optically active 2,3-butandiol generates the Fp$^+$–dioxene complex in good yield (**Eq. 7.66**). The optically active dioxene is easily freed by treatment with iodide, providing a simple route to optically active compounds not easily prepared otherwise. More interestingly, ketone enolates attack the dioxene complex from the face opposite the metal to give the stable Fp–alkyl complex with high diastereoselectivity. Stereoselective reduction of the carbonyl group with L-Selectride, followed by oxidative removal of the metal, produces the tricyclic furan in excellent yield and with high diastereoselectivity. In this case, oxidation of the iron does not result in CO insertion, but rather it simply makes the iron a good leaving group (**Eq. 7.66**).[100]

Eq. 7.66

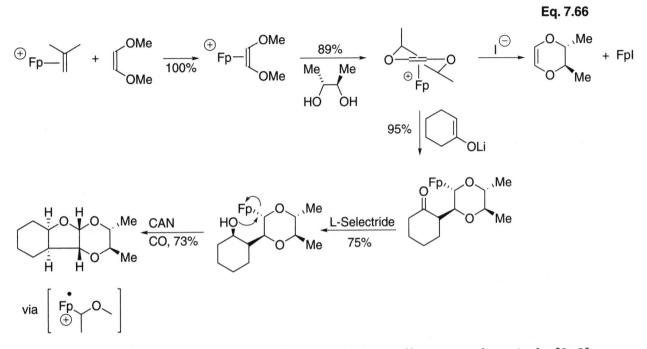

An unusual class of reactions involving cationic iron–alkene complexes is the [3+2] cycloaddition reaction that occurs between electron-poor alkenes and η^1-allyl Fp complexes (**Eq. 7.67**).[101] In this case, the η^1-allyl–Fp complex attacks the electrophilic alkene, producing a stabilized enolate and a cationic Fp–alkene complex. The enolate then attacks the cationic alkene complex, closing the ring, and generating a Fp–alkyl complex. Oxidative removal of the iron promotes CO insertion, and generates the highly functionalized cyclopentane ring system.

Eq. 7.67

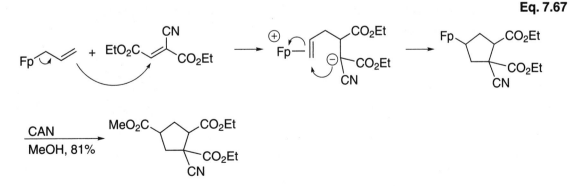

The same chemical features that make cationic iron–alkene complexes reactive towards nucleophiles makes them unreactive towards electrophiles, and complexation of alkenes to Fp⁺ can protect them against an array of electrophilic reactions. Complexation occurs preferentially at the less-substituted alkene, and at alkenes rather than alkynes, so selective protection is possible (**Eqs. 7.68, 7.69,** and **7.70**).[102] The alkene is freed of iron after reaction by treatment with iodide.

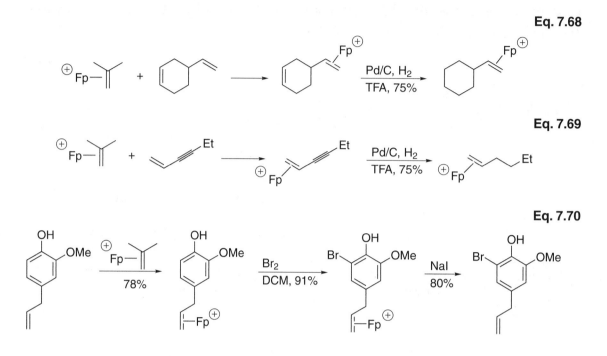

Eq. 7.68

Eq. 7.69

Eq. 7.70

7.3 Metal–Diene Complexes

a. Fe(CO)$_3$ as a 1,3-Diene Protecting Group[103]

Heating Fe$_2$(CO)$_9$ with dienes produces iron carbonyl complexes of conjugated dienes (**Eq. 7.71**). Nonconjugated dienes often rearrange to give conjugated diene complexes in the presence of Fe$_2$(CO)$_9$ or Fe(CO)$_5$. These iron tricarbonyl complexes of dienes are very stable: they fail to undergo Diels–Alder reactions, and undergo facile Friedel–Crafts acylations without decomplexation from the metal (**Eq. 7.72**).[104] Although they can be made to react with strong nucleophiles (see Section 7.3b), complexation of the diene segment is most often used to protect it.

Eq. 7.71

Eq. 7.72

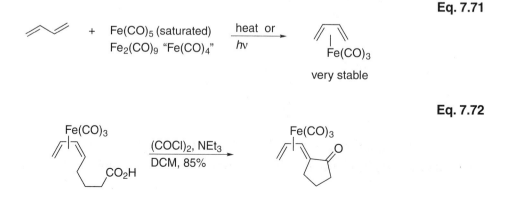

For example, dienals are highly reactive and polymerize readily, making reactions at the aldehyde difficult and inefficient. However, treatment with $Fe_2(CO)_9$ forms the stable diene complex, in which the aldehyde reacts normally. When the reactions are complete, the iron is oxidatively removed, giving the metal-free organic compound. Particularly notable is the range of reactions the iron tricarbonyl group will withstand. These include Wittig, aldol, Pd/C–H_2 reduction, addition of Grignard and organozinc reagents, Swern oxidation, and even the cis hydroxylation of an alkene with osmium tetroxide (**Eq. 7.73**).[105] Note the exclusive dihydroxylation from the face opposite the iron tricarbonyl group. The complexation also protects the diene from cyclopropanations using diazocompounds.[106]

Eq. 7.73

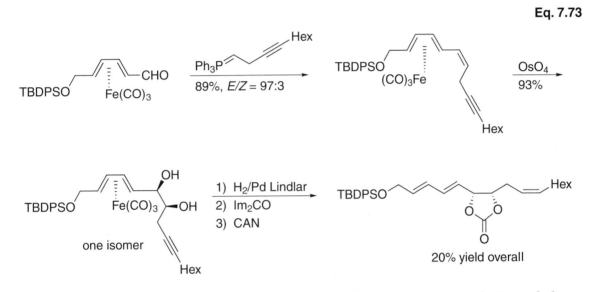

Because iron occupies a single face of these diene complexes, substituted diene complexes are intrinsically chiral, can be resolved, and the presence of the iron has a substantial stereochemical influence (**Eqs. 7.74,**[107] **7.75,**[108] **and 7.76**[109]).

Eq. 7.74

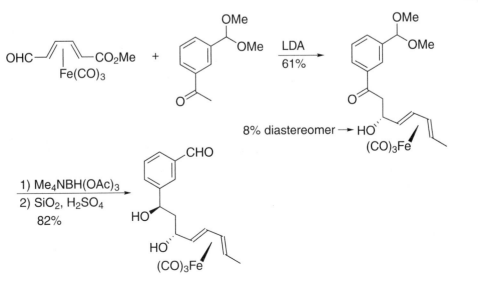

Eq. 7.75

Eq. 7.76

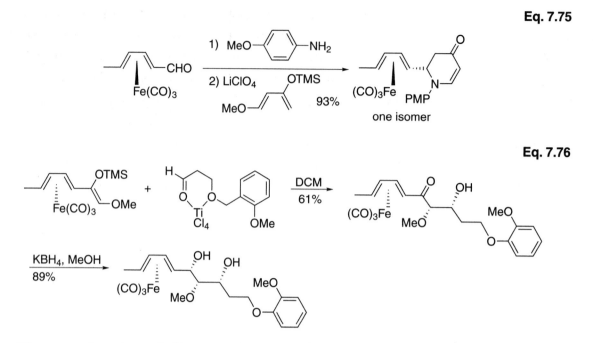

The complexation of dienes to iron tricarbonyl fragments can also be used to both activate and stabilize allylic positions, as well as to exert stereochemical control (**Eqs. 7.77**[110] and **7.78**[111]). The stereochemical control can be used in an iterative fashion via Lewis acid assisted 1,2-migration of iron tricarbonyl (**Eq. 7.79**).[112] Upon treatment with a base, diene iron tricarbonyl complexes of trienes can be isomerized via 1,3-migration of the metal to the thermodynamically more stable diene complex (**Eqs. 7.80 and 7.81**).[113] The migration occurs on the same "face" of the polyene system. The mechanism of this migration is unclear.

Adjacent cations are stabilized by the iron tricarbonyl group. This can be utilized in the synthesis of complex molecules via cationic cyclization (**Eq. 7.82**)[114].

Eq. 7.77

Eq. 7.78

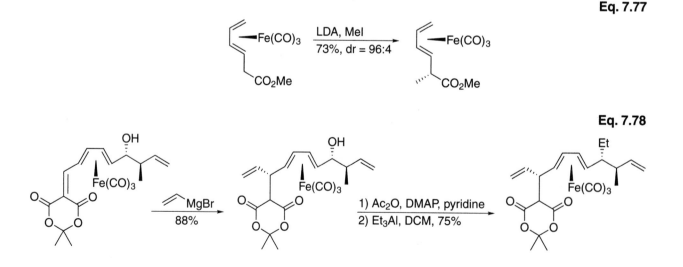

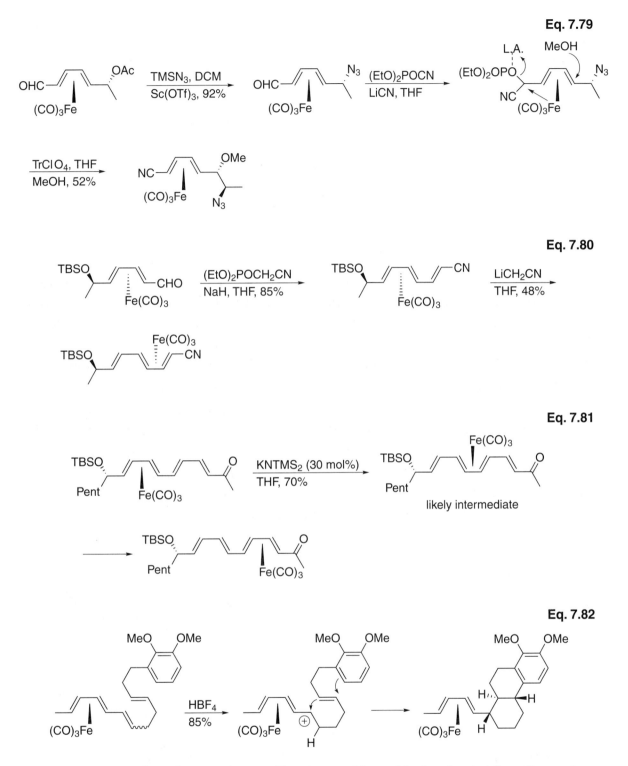

The iron tricarbonyl group is usually removed by oxidative decomplexation using, for example, CAN, FeCl$_3$, CuCl$_2$, amine oxides, hydrogen peroxide, or Pb(OAc)$_4$, leaving the diene intact. In contrast, Raney nickel not only removes the iron tricarbonyl group but also reduces the diene to an alkene (**Eq. 7.83**).[115] This reductive decomplexation does not occur for iron tricarbonyl cycloheptatriene complexes, where only reduction of the noncomplexed alkene is observed.[116]

Eq. 7.83

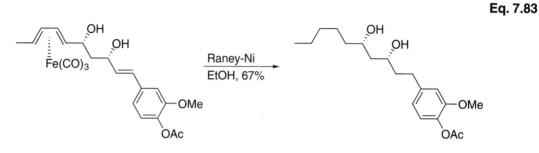

Complexation of the cyclohepta-3,5-dienone permitted α-dialkylation with clean cis stereochemistry (**Eq. 7.84**).[117] In this case, the iron fragment prevents isomerization of the enolate and directs the alkyl groups to the face opposite the metal. Reduction of the resulting α,α'-dimethylketone also occurs stereoselectively, but because of the flanking methyl groups, attack occurs from the same face as the metal. Similarly, two of the three conjugated double bonds of cyclohepta-2,4,6-trien-1-ol could be complexed to iron, permitting oxidation or hydroboration of the uncomplexed alkene with very high stereoselectivity (**Eq. 7.85**).[118] Nucleophilic additions to tropanone iron tricarbonyl complexes are also stereospecific, with the nucleophile adding to the side opposite the metal (**Eq. 7.86**).[119] Cyclohexadiene complexes having pendant unsaturation undergo a thermally or photochemically induced spiroannulation with migration of the iron–diene group. An example of this reaction and a plausible mechanistic explanation are shown in **Eq. 7.87**.[120] A second cyclization is possible if a diene is attached to the cyclohexadiene complex (**Eq. 7.88**).[121]

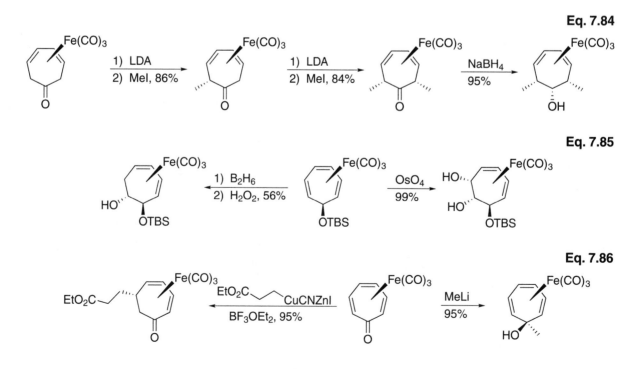

Eq. 7.87

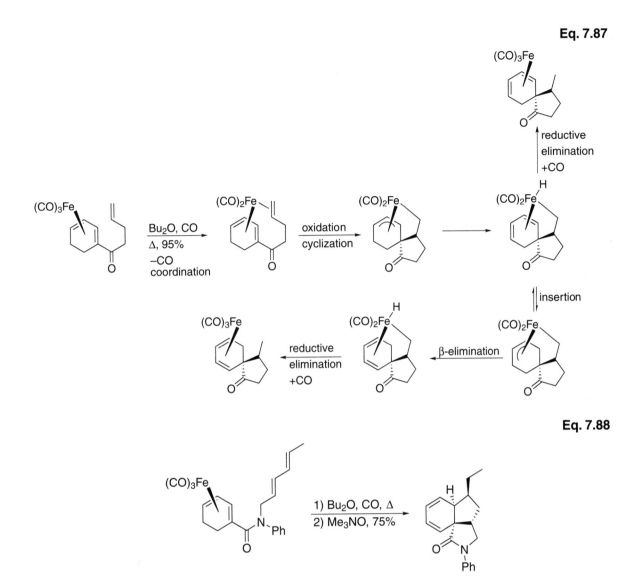

Eq. 7.88

Similar activation and directing effects have been achieved using complexation of dienes to molybdenum(II) (**Eq. 7.89**).[122] In this case, the cationic diene complex was converted to the neutral η^3-allylmolybdenum complex by allylic proton abstraction. This protected one of the double bonds of the diene, permitting functionalization of the other. Cycloheptadiene underwent a similar series of reactions.

Eq. 7.89

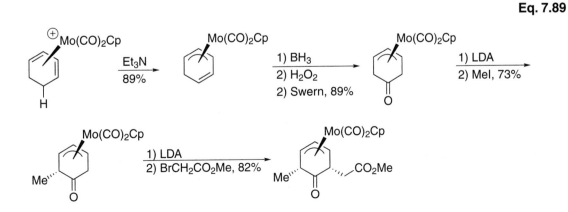

Finally, complexation to iron can stabilize normally inaccessible tautomers of aromatic compounds, permitting quite unusual synthetic transformations (**Eqs. 7.90**[123,124,125] and **7.91**[126]).

Eq. 7.90

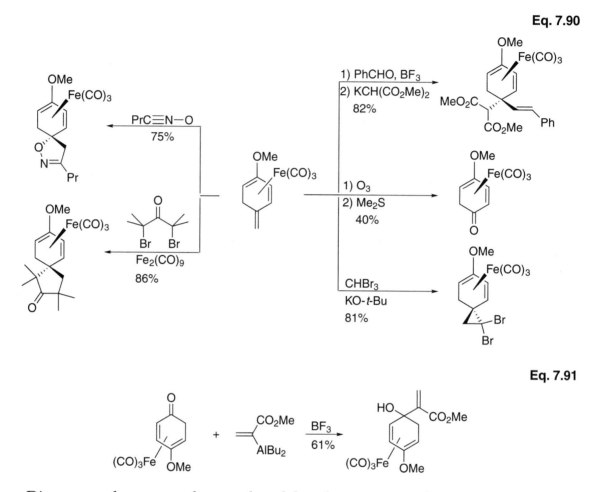

Eq. 7.91

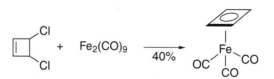

Diene complexes are also produced by the reaction of 1,4-dihalo-2-butenes with $Fe_2(CO)_9$. An especially useful example of this reaction is the preparation of (η^4-cyclobutadiene)iron tricarbonyl from dichlorocyclobutene (**Eq. 7.92**).[127] This reaction illustrates the stabilization of extremely unstable organic molecules by complexation to appropriate transition metals. Free cyclobutadiene itself has only fleeting existence, since it rapidly dimerizes. However, (η^4-cyclobutadiene)iron tricarbonyl is a very stable complex.

Eq. 7.92

The complexed cyclobutadiene undergoes a variety of electrophilic substitution reactions, including Friedel–Crafts acylation, formylation, chloromethylation, and aminomethylation. The keto group of the acylated cyclobutadieneiron tricarbonyl complex can be reduced by hydride reagents without decomposition.[128] On the other hand, oxidation of the complex with cerium(IV), triethylamine oxide, or pyridine *N*-oxide frees the cyclobutadiene ligand for use in synthesis. In the presence of a tethered alkene, the free cyclobutadiene undergoes a [2+2] reaction, forming bicyclo[2.2.0] ring systems of interest in organic synthesis. When combined with metathesis (Chapter 6), this chemistry provides a rapid entry into cyclooctadiene ring systems (**Eq. 7.93**).[129] Alkyne-substituted complexes participate in a [2+2+1] cycloaddition, reminiscent of a Pauson–Khand reaction (see Chapter 8), upon decomplexation with CAN in the presence of carbon monoxide (**Eq. 7.94**).[130]

Eq. 7.93

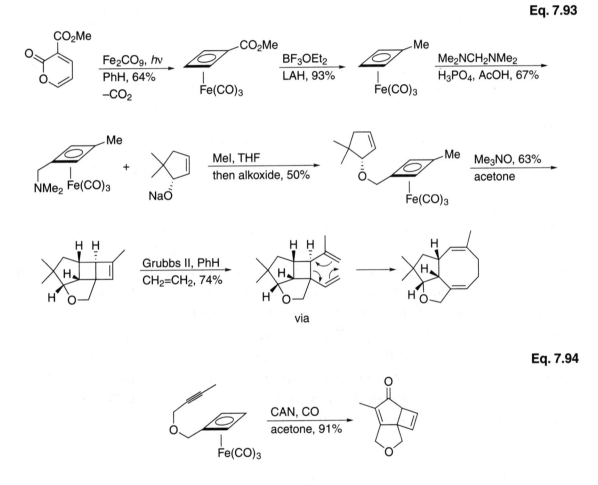

Eq. 7.94

b. Nucleophilic Attack on Metal–Diene Complexes[131]

Although the complexation of dienes to electron-deficient metal fragments stabilizes these ligands towards electrophilic attack, it activates them towards nucleophilic attack, a reaction mode unavailable to the uncomplexed ligand. Neutral η^4-1,3-diene iron tricarbonyl complexes are reactive only towards very strong nucleophiles, and attack can occur at either the terminal or internal position of the diene. Kinetically, attack at the terminal position to form an η^3-allyliron complex is favored. However, this is a reversible process, and thermodynamically, attack at an internal position, generating a η^1-alkyl-η^2- alkene iron complex, is favored (**Eq. 7.95**).[132] As usual, attack occurs from the face opposite the metal. Protolytic cleavage of the "thermodynamic" iron complex gives the alkylation product, while treatment with carbon monoxide results in insertion into the iron–carbon σ-bond. Protolytic cleavage of this σ-acyl complex produces the trans aldehyde, while treatment with an electrophile in the presence of a ligand leads to acylation.[133]

Eq. 7.95

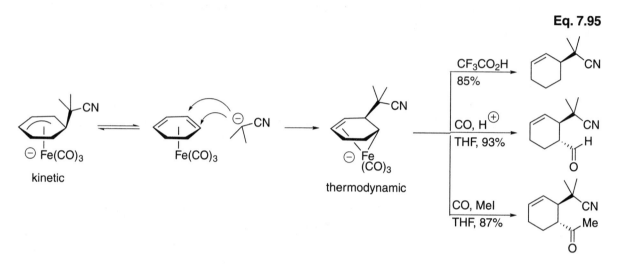

The situation is a little more complicated with acyclic dienes. Complexed butadiene itself is converted cleanly to the corresponding cyclopentanone by a process involving nucleophilic attack [path (a) in **Eq. 7.96**], followed by CO insertion (b), followed by alkene insertion (c).[134] However, substitution on the diene suppresses the alkene insertion step, and the amount of the "normal" cleavage product (e.g., aldehyde) increases as substitution increases [(c) in **Eq. 7.97**]. This alkene insertion, followed by the migration of iron to the position α to the carbonyl group, is precedented in the insertion of ethene into iron complexes generated from $Fe(CO)_4^{2-}$. In the absence of CO, the σ-alkyliron complex undergoes reaction with a range of electrophiles, resulting in overall 1,2-difunctionalization of the diene (**Eq. 7.98**).[135]

Eq. 7.96

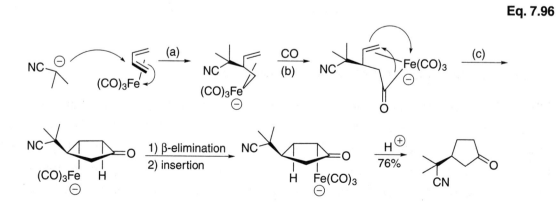

Eq. 7.97

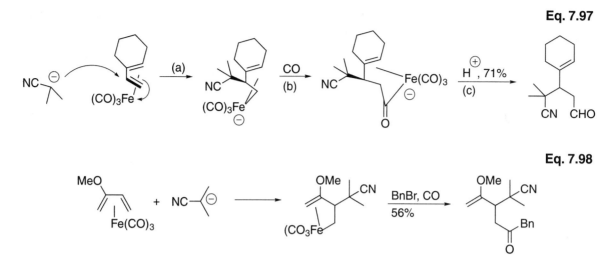

Eq. 7.98

By making the metal fragment more electrophilic, the reactivity of metal–diene complexes toward nucleophiles can be increased. This is often achieved by replacing a neutral metal fragment by a cationic one, such as the $CpMo(CO)_2^+$ fragment. These complexes are relatively easy to synthesize, although several steps are required (**Eq. 7.99**).[136]

Eq. 7.99

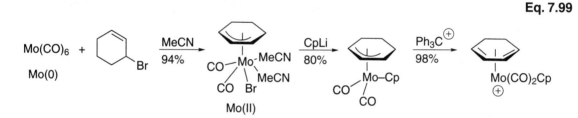

These complexes undergo reaction with a much wider range of nucleophiles, exclusively at the terminal position, to generate a neutral η^3-allylmolybdenum complex. Hydride abstraction from this complex regenerates a cationic diene complex, which again undergoes nucleophilic attack from the face opposite the metal, resulting in overall cis-1,4 difunctionalization of the diene (**Eq. 7.100**).[137] Cyclohepta-1,3-diene complexes undergoes a similar series of transformations.

Eq. 7.100

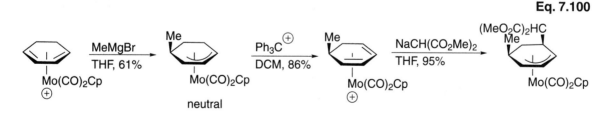

The major limitation, to date, is finding an efficient way to remove the molybdenum from the final η^3-allyl complex. Modest success has been achieved by conversion of the neutral η^3-allylmolybdenum carbonyl complex to the cationic nitrosyl (i.e., exchange of neutral CO for NO$^+$), thus permitting nucleophilic attack on the η^3-allyl complex (**Eq. 7.101**),[138] by oxidative coupling and aromatization using very active MnO_2 (**Eq. 7.102**),[139] or by oxidizing the η^3-allylmolybdenum complex with iodonium trifluoroacetate (**Eq. 7.103**).[140] Of particular synthetic interest is the application of this methodology to heterocyclic dienes[141] (**Eqs. 7.104**[142] and **7.105**[143])

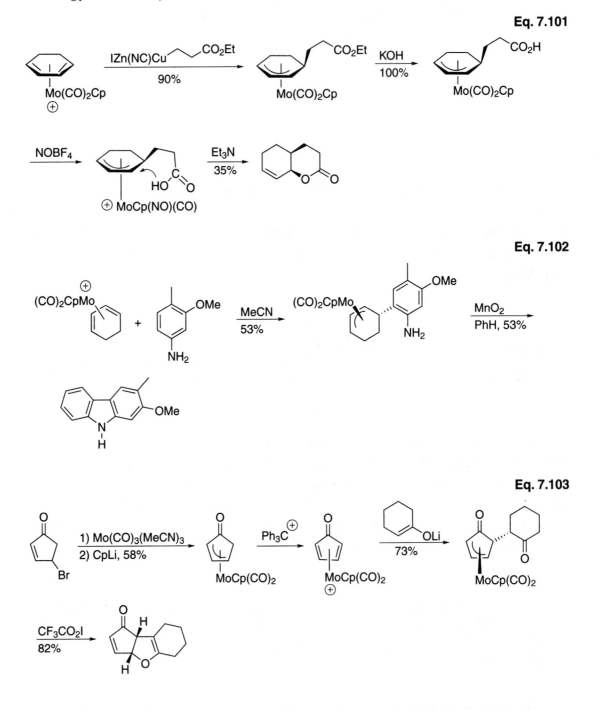

Eq. 7.101

Eq. 7.102

Eq. 7.103

Eq. 7.104

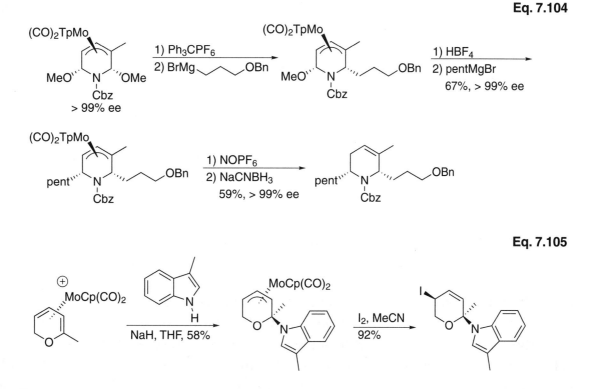

Eq. 7.105

c. Nucleophilic Attack on Cationic Metal–Dienyl Complexes[144]

Cationic *dienyl* complexes of iron are among the most extensively used complexes in organic synthesis. They are generally reactive towards a wide range of nucleophiles, and the presence of iron permits a high degree of both stereo- and regiocontrol. Cyclohexadienyl complexes are the most studied, at least in part because a wide array of substituted cyclohexadienes are readily available from the Birch reduction of aromatic compounds. The cyclohexadienyliron complexes are prepared by simply treating either the conjugated or nonconjugated diene with $Fe_2(CO)_9$ (**Eq. 7.106**). Treatment of the resulting neutral diene complex with trityl cation (hydride abstraction) produces the relatively air- and moisture-stable cationic cyclohexadienyliron complex. A very wide range of nucleophiles attack this complex at the terminus of the dienyl system from the face opposite the metal, regenerating the neutral diene complex. Treatment with trityl cation removes hydride from the unsubstituted position of the diene complex, and the resulting dienyl complex again reacts with nucleophiles at this position from the face opposite the metal, giving a cis-1,2-disubstituted cyclohexadiene. The iron is easily removed by treatment with amine oxides, giving the free organic compound (**Eq. 7.106**).

Eq. 7.106

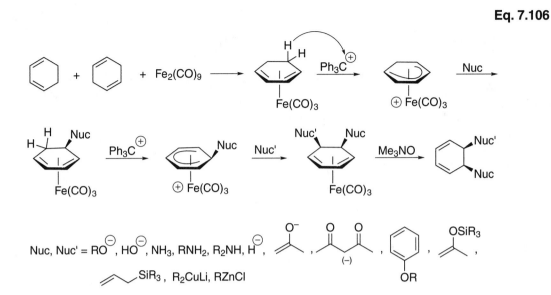

$$\text{Nuc, Nuc'} = RO^{\ominus},\ HO^{\ominus},\ NH_3,\ RNH_2,\ R_2NH,\ H^{\ominus},$$

$$\text{SiR}_3,\ R_2CuLi,\ RZnCl$$

Substituted complexes can be prepared and undergo reaction with high regioselectivity. The diene complex derived from 4-methylanisole undergoes hydride abstraction α to the methoxy group, giving a single regioisomer of the dienyl complex. Nucleophilic attack occurs exclusively at the *methyl* terminus of the dienyl system, apparently because the other end is electronically deactivated by the strongly electron-donating methoxy group. Oxidative removal of the iron, followed by hydrolysis, gives the 4,4-disubstituted cyclohexenone (**Eqs. 7.107**[145] and **7.108**[146]).[147]

Cationic dienyliron complexes are potent electrophiles, and even undergo electrophilic aromatic substitution reactions with electron-rich arenes.[148,149,150] This has been extensively used to prepare carbazole alkaloids (**Eqs. 7.109**[151] and **7.110**[152]).[153]

Eq. 7.107

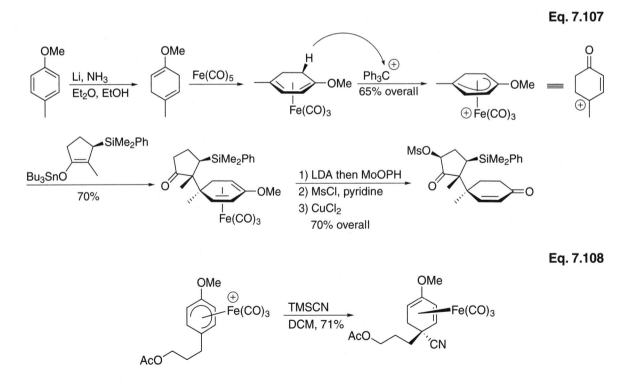

Eq. 7.108

Eq. 7.109

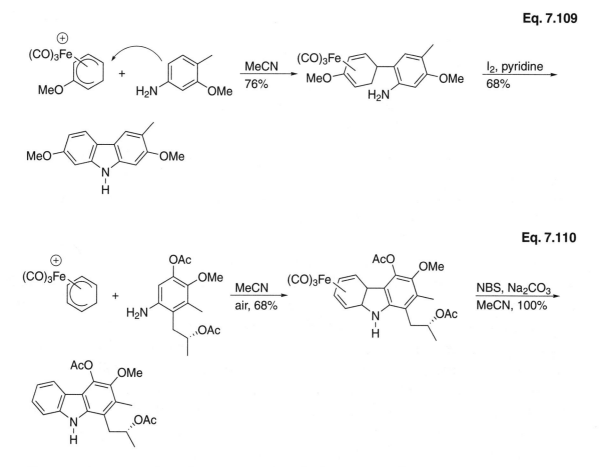

Eq. 7.110

Because iron tricarbonyl occupies a single face of the η^5-dienyl complex, unsymmetrically substituted complexes are intrinsically chiral and can be resolved (**Eq. 7.111**).[154] Since addition of nucleophiles is stereospecific, these complexes are of use in asymmetric synthesis. Functional groups can direct complexation, as seen with the optically active diol from microbial oxidation in **Eq. 7.112**.[155] Efficient asymmetric complexation of dienes with $Fe(CO)_5$ using a chiral diamine is also possible (**Eq. 7.113**).[156]

Eq. 7.111

Eq. 7.112

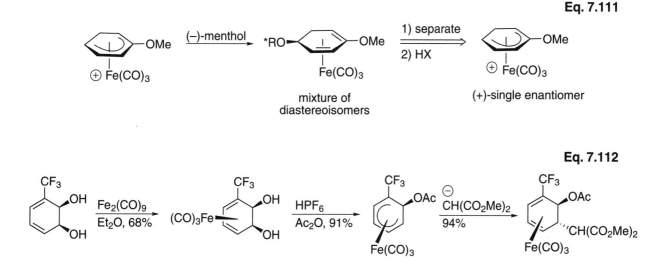

Eq. 7.113

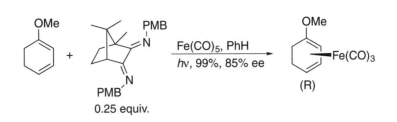

The preceding chemistry is not restricted to cyclohexadienes. The same degree of activation, regioselectivity, and stereoselectivity is achieved with cycloheptadiene[157] (**Eq. 7.114**)[158]. Cyclooctatetraene iron tricarbonyl or dicarbonyltriphenylphosphine complexes form interesting cyclopropanated cationic complexes upon protonation (**Eq. 7.115**).[159] The regioselectivity of nucleophilic addition is different depending on the spectator ligands on iron. In contrast, acyclic η^5-dienyl complexes have been much less studied, and have only rarely been used in synthesis. They are more difficult to prepare, and often react with lower regioselectivity than do their cyclic analogs. However, once formed, they are generally reactive towards nucleophiles (**Eqs. 7.116**[160] and **7.117**[161]). [162] The regioselectivity depends on the nucleophile and the groups on the dienyl chain. Ester-substituted pentadienyl complexes react with stabilized carbon nucleophiles, Grignard reagents, and organolithium reagents, adding to the β-position relative to the ester.[163] In contrast, cuprates add to the terminal position (**Eq. 7.118**).[164] The products from internal addition afford cyclopropanated products upon oxidative decomplexation.

Eq. 7.114

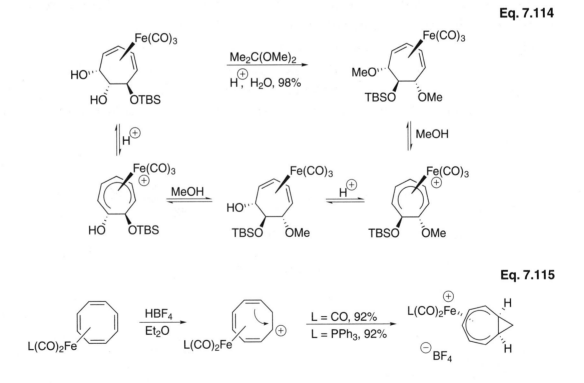

Eq. 7.115

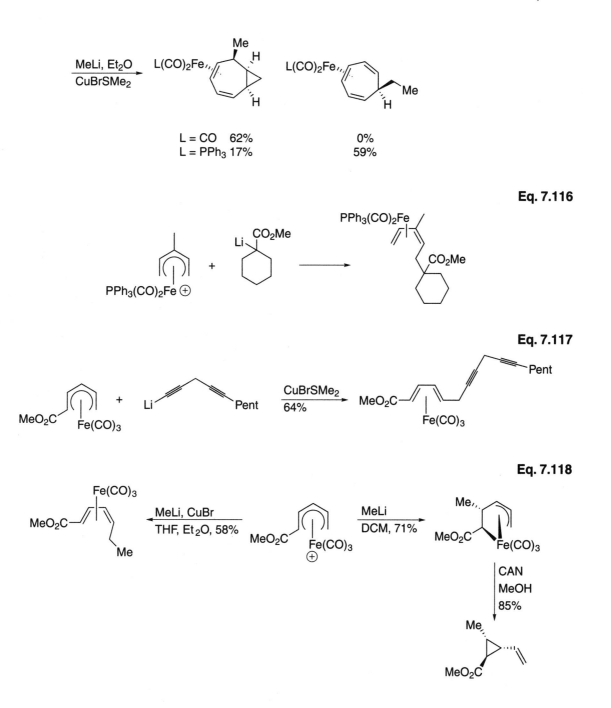

Eq. 7.116

Eq. 7.117

Eq. 7.118

d. Metal-Catalyzed Cycloaddition Reactions[165]

Many types of cycloaddition reactions [*n*+*m*+...], which do not proceed under normal thermal or photochemical reaction conditions, proceed readily in the presence of appropriate transition metal catalysts. These include [2+2], [3+2], homo-Diels–Alder, [4+2], [5+2], [4+4], [6+2], [6+4], and many others. Almost certainly none of these metal-catalyzed cycloadditions are concerted, but rather they proceed via metal–carbon σ-bonded species, and most can be viewed as "reductive coupling" reactions wherein the metal is oxidized in the initial steps of the reaction, much like the low-valent zirconium couplings discussed in Chapter 4. Few have been studied mechanistically, and the mechanisms presented below are for the most part hypothetical. Only the synthetically interesting variants are discussed here. Alkyne–alkene–carbon monoxide [2+2+1] cycloadditions to form cyclopentenones (i.e., Pauson–Khand reactions) and some related carbonylative cycloadditions are discussed in Chapter 8.

Although nickel- and ruthenium-catalyzed [2+2] cycloadditions of alkenes and alkynes to strained alkenes such as norbornene have long been known, only relatively simple compounds have been synthesized using this chemistry. However, much more complex systems are produced by palladium-catalyzed intramolecular [2+2] cycloadditions between alkenes and alkynes (**Eq. 7.119**).[166] The reaction almost surely proceeds through a metallacyclopentene, with palladium being in the unusual (but not unattainable) +4 oxidation state. The range of substrates is quite limited for this unusual process. Allene–ynes undergo related [2+2] cycloadditions catalyzed by a variety of transition metals, including palladium (**Eq. 7.120**).[167]

Eq. 7.119

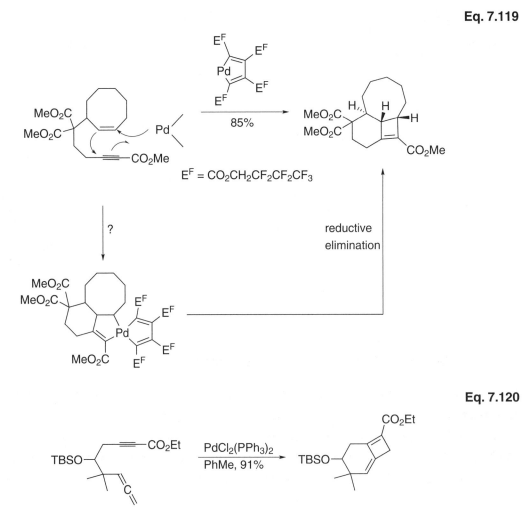

$E^F = CO_2CH_2CF_2CF_2CF_3$

Eq. 7.120

The [3+2] cycloadditions proceeding through oxallyliron species are discussed in Chapter 5, and those proceeding through trimethylene methane complexes are discussed in Chapter 9. An interesting example involving neither of these is shown in **Eq. 7.121**, which is thought to be initiated by palladacyclopentene formation as in **Eq. 7.119**, followed by reversion to form a vinylcarbenepalladium complex, which then undergoes a [3+2] cycloaddition (actually, [4+2] if you count palladium), followed by reductive elimination.[168]

Eq. 7.121

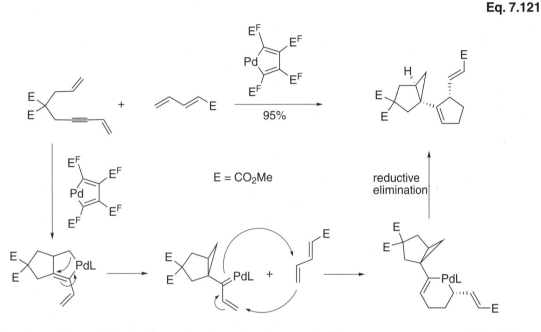

Metal-catalyzed [2+2+2] cyclotrimerizations of alkynes are presented in Chapter 8. Here, the homo-Diels–Alder [2+2+2] cycloaddition is considered (**Eq. 7.122**). Nickel(0)[169] and cobalt(0)[170] complexes catalyze this process, and high enantioselectivity has been observed by using chiral diphosphine ligands with the cobalt catalyst systems.[171]

Eq. 7.122

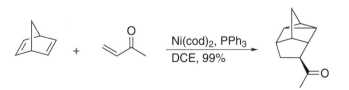

The standard Diels-Alder [4+2] cycloaddition has long been studied and is a staple in the arsenal of synthetic chemists. However, this process normally is most efficient for reactions between electronically *dissimilar* dienes and dienophiles. Normally difficult [4+2] cycloadditions between similar dienes and dienophiles can be efficiently catalyzed by a variety of low-valent metal complexes. Again, these are likely to proceed via metallacycles (**Eq. 7.123**).

Eq. 7.123

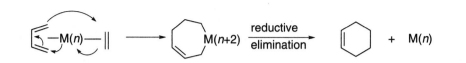

Nickel(0) (**Eq. 7.124**)[172] and rhodium(I)[173] (**Eq. 7.125**)[174] complexes are most efficient for this process and, in the presence of chiral ligands, reasonable asymmetric induction can be achieved (**Eq. 7.126**).[175] Allenes can also be the dienophile in [4+2] cycloadditions with dienes. The double bond selectivity on the allene can be modulated by the choice of catalyst (**Eq. 7.127**).[176] This would be a very hard feat to accomplish using uncatalyzed conditions. Thermal [4+2] cycloaddition of the same substrate resulted in a complex mixture of products. The powerful ability to change the chemoselectivity of a given reaction by simply modifying the catalyst is shown in **Eq. 7.128**, where either a [4+2] or a [2+2+1] product can be obtained.[177]

Eq. 7.124

Eq. 7.125

Eq. 7.126

Eq. 7.127

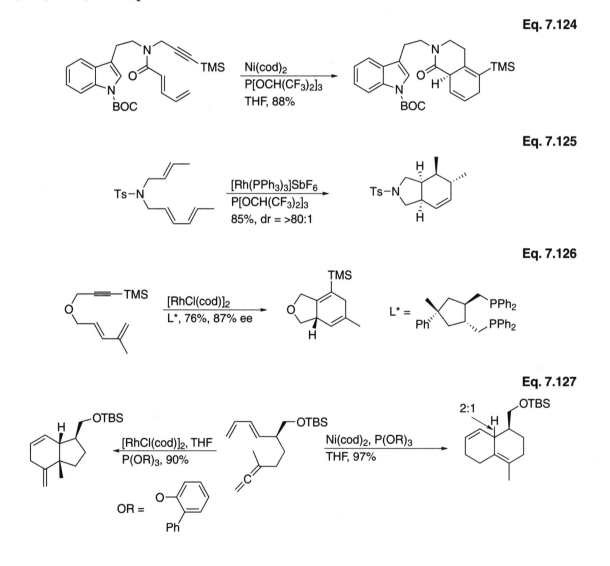

Eq. 7.128

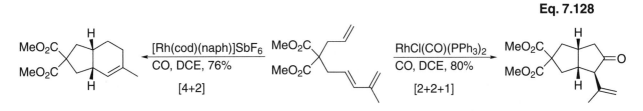

Nickel(0)-catalyzed [4+4] cycloadditions evolved from early studies of metal-catalyzed cyclooligomerizations of butadiene. Only recently, however, has the process been applied in complex organic synthesis.[178] The mechanism is likely to involve a "reductive" dimerization of two complexed dienes, followed by reductive elimination (**Eq. 7.129**).[179] It has proven particularly useful in its intramolecular variant (**Eq. 7.130**[180]).[181,182] Cyclooctane derivatives can be obtained by a [4+2+2] cycloaddition (**Eqs. 7.131,**[183] **7.132,**[184] and **7.133**[185]).

Eq. 7.129

Eq. 7.130

Eq. 7.131

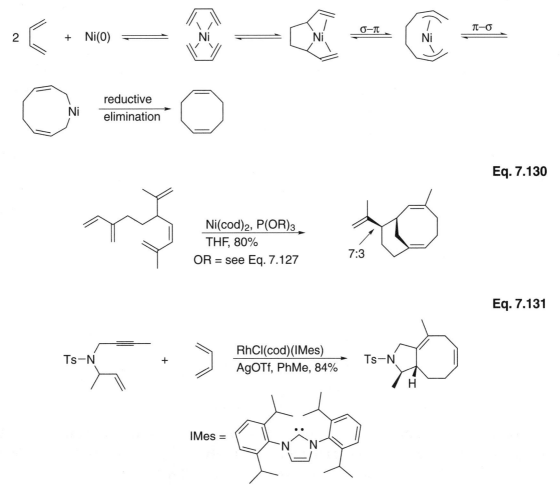

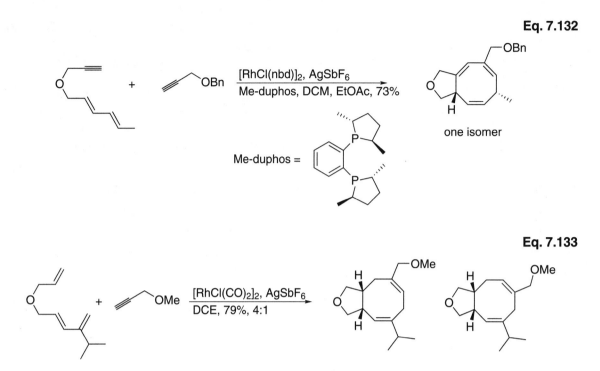

Eq. 7.132

Eq. 7.133

Rhodium and ruthenium[186] complexes catalyze cycloadditions of alkenylcyclopropanes with alkynes or allenes to give complex organic molecules. Two mechanisms have been proposed, differing in the timing of the cleavage of the cyclopropane ring, forming either a σ,η^3-complex [path (a) in **Eq. 7.134**] or a metallacyclopentene [path (b)]. A theroetical study of the mechanism indicated that the formation of a σ,η^3-complex was energetically favored.[187]

Eq. 7.134

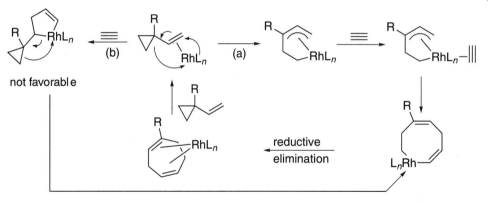

Very facile intramolecular reactions that form fused cyclic compound have been reported. An interesting reversal of regioselectivity of the cyclopropane ring cleavage can be realized by choosing the appropriate catalyst (**Eq. 7.135**)[188]

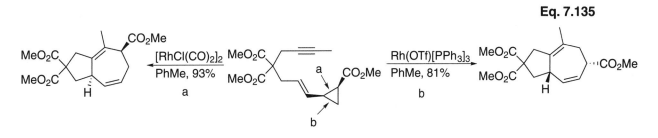

Eq. 7.135

Excellent asymmetric induction (up to 99% ee) has been reported using chiral ligands.[189] Examples of an alkyne–alkenylcyclopropane cyclization and an allene–alkenylcyclopropane cyclization are shown in **Eqs. 7.136**[190] and **7.137**,[191] respectively. Related intermolecular reactions of cyclopropyl imines with alkynes yield dihydroazepines (**Eq. 7.138**).[192]

Eq. 7.136

Eq. 7.137

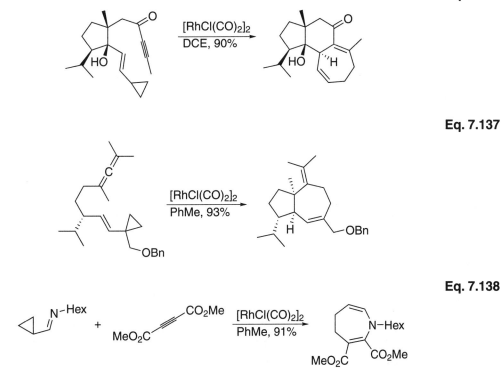

Eq. 7.138

The insertion of carbon monoxide into the metallacyclooctane intermediate, followed by reductive elimination, affords cyclooctanone derivatives (**Eq. 7.139**).[193] The product undergoes intramolecular aldol condensation upon hydrolytic workup. The use of strained rings in cycloadditions is not limited to cyclopropanes; [6+2] cycloadditions of alkenylcyclobutanones and allenylcyclobutanes to afford eight-membered rings have also been developed (**Eqs. 7.140**[194] and **7.141**[195]).

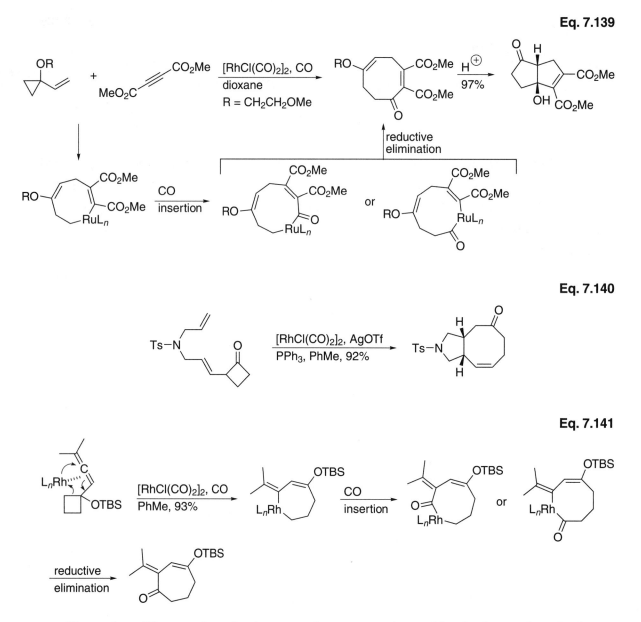

Eq. 7.139

Eq. 7.140

Eq. 7.141

Chromium(0)-complexed trienes undergo a variety of both thermal and photochemical higher-order cycloaddition reactions.[196] Both inter- and intramolecular [6+2] cycloadditions of alkenes and alkynes to chromium cycloheptatriene complexes are efficient (**Eqs. 7.142,**[197] **7.143,**[198] **7.144,**[199] and **7.145**[200]), as are intermolecular[201] (**Eq. 7.146**)[202] and intramolecular (**Eq. 7.147**)[203] [6+4] cycloadditions.

Eq. 7.142

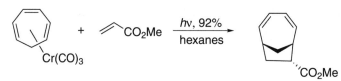

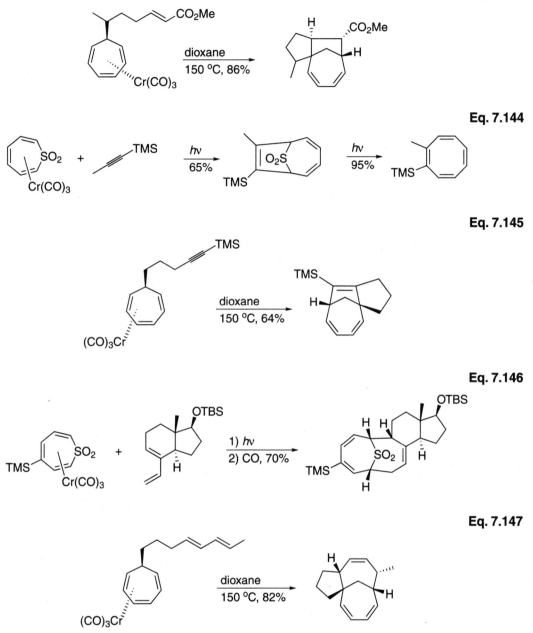

Eq. 7.143

Eq. 7.144

Eq. 7.145

Eq. 7.146

Eq. 7.147

Under appropriate conditions, the [6+2] and [6+4] cycloadditions can be catalyzed by chromium complexes (**Eq. 7.148**).[204,205] Again, these are unlikely to be concerted processes, but rather proceed via metallacyclo intermediates (**Figure 7.3**). Cobalt also catalyzes [6+2] cycloadditions of cycloheptatriene and alkynes.[206] Remarkable multicomponent reactions of chromium tricarbonyl heptatrienes are possible, affording quite complex polycyclic compounds (**Eqs. 7.149**[207] and **7.150**[208]). Finally, chromium tricarbonyl cycloheptatriene complexes participate in [6+3] cycloadditions with azirines (**Eq. 7.151**).[209]

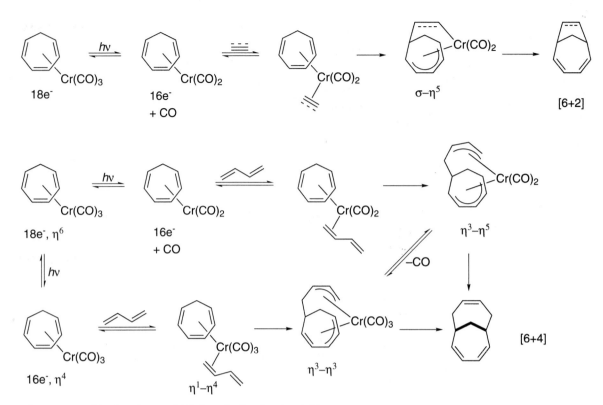

Figure 7.3. Mechanisms of Higher-Order Cycloadditions

Eq. 7.148

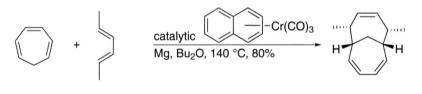

Eq. 7.149

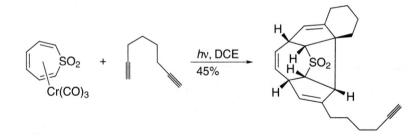

Eq. 7.150

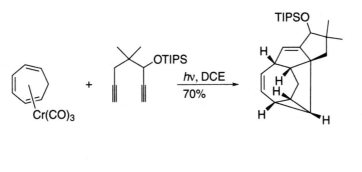

Eq. 7.151

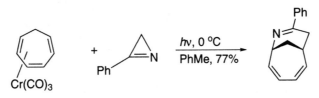

References

(1) For reviews, see a) McDaniel, K. F. Transition Metal Alkene, Diene, and Dienyl Complexes: Nucleophilic Attack on Alkene Complexes. In *Comprehensive Organometallic Chemistry II*; Abel, E. W., Stone, F. G. A., Wilkinson, G., Eds.; Pergamon: Oxford, U.K., 1995; Vol. 12, pp 601–622. b) Hegedus, L. S. Palladium in Organic Synthesis. In *Organometallics in Synthesis; a Manual*, 2nd ed.; Schlosser, M., Ed.; Wiley: Chichester, U.K., 1998.

(2) For a review, see Muzart, J. *Tetrahedron* **2005**, *61*, 5955.

(3) For a review, see Takacs, J. M.; Jiang, X.-t. *Curr. Org. Chem.* **2003**, *7*, 36.

(4) Tsuji, J. *Synthesis* **1984**, 369.

(5) Hayashi, T.; Yamasaki, K.; Mimura, M.; Uozumi, Y. *J. Am. Chem. Soc.* **2004**, *126*, 3036.

(6) Cornell, C. N.; Sigman, M. S. *J. Am. Chem. Soc.* **2005**, *127*, 2796.

(7) In a synthesis toward de-AB-cholesta-8(14),22-dien-9-one. Takahashi, T.; Ueno, H.; Miyazawa, M. Tsuji, J. *Tetrahedron Lett.* **1985**, *26*, 4463.

(8) In a synthesis of b-elemene. Kim, D.; Lee, J.; Chang, J.; Kim, S. *Tetrahedron* **2001**, *57*, 1247.

(9) In a synthesis of tautomycin. Oikawa, M.; Ueno, T.; Oikawa, H.; Ichihara, A. *J. Org. Chem.* **1995**, *60*, 5048.

(10) Mitsudome, T.; Umetani, T.; Nosaka, N.; Mori, K.; Mizugaki, T.; Ebitani, K.; Kaneda, K. *Angew. Chem. Int. Ed.* **2006**, *45*, 481.

(11) In a synthesis of *trans*-(+)-lauridiol. Gadikota, R. R.; Keller, A. I.; Callam, C. S.; Lowary, T. L. *Tetrahedron: Asymmetry* **2003**, *14*, 737.

(12) Pellisier, H.; Michellys, P.-Y.; Santelli, M. *Tetrahedron* **1997**, *53*, 7577.

(13) In a synthesis of leuscandrolide A. Fettes, A.; Carreira, E. M. *Angew. Chem. Int. Ed.* **2002**, *41*, 4098.

(14) In a synthesis of calystegine alkaloids. Skaanderup, P. R.; Madsen, R. *J. Org. Chem.* **2003**, *68*. 2115.

(15) Tsuji, J. Nagashima, H.; Hori, K. *Chem. Lett.* **1980**, 257.

(16) Schultz, M. J.; Sigman, M. S. *J. Am. Chem. Soc.* **2006**, *128*, 1460.

(17) For reviews, see a) Muzart, J. *Tetrahedron* **2005**, *61*, 5955. b) Hosokawa, T.; Murahashi, S-i. *Acc. Chem. Res.* **1990**, *23*, 49.

(18) Trend, R. M.; Ramtohul, Y. K.; Stolz, B. M. *J. Am. Chem. Soc.* **2005**, *127*, 17778.

(19) Hayashi, T.; Yamasaki, K.; Mimura M.; Uozumi, Y. *J. Am. Chem. Soc.* **2004**, *126*, 3036.

(20) In a synthesis toward tetronomycin. Semmelhack, M. F.; Epa, W. R.; Cheung, A. W.-H.; Gu, Y.; Kim, C.; Zhang, C.; Lew, W. *J. Am. Chem. Soc.* **1994**, *116*, 7455.

(21) In a synthesis toward garsubellin A. Usuda, H.; Kanai, M.; Shibasaki, M. *Org. Lett.* **2002**, *4*, 859.

(22) Uozumi, Y.; Kato, K.; Hayashi, T. *J. Am. Chem. Soc.* **1997**, *119*, 5063.

(23) Uenichi, J.; Ohmi, M. Ueda, A. *Tetrahedron: Asymmetry* **2005**, *16*, 1299.

(24) a) Tenagha, A.; Kammerer, F. *Synlett* **1996**, 576. b) See also Saito, S.; Hara, T.; Takahashi, N.; Hirai, M.; Moriwake, T. *Synlett* **1992**, 237.

(25) In a synthesis of (+)-goniothalesdiol. Babjak, M.; Kapitan, P.; Gracza, T. *Tetrahedron* **2005**, *61*, 2471.

(26) In a synthesis of vitamin E. Tietze, L. F.; Sommer, K. M.; Zinngrebe, J.; Stecker, F. *Angew. Chem. Int. Ed.* **2005**, *44*, 257.

(27) Semmelhack, M. F.; Epa, W. R. *Tetrahedron Lett.* **1993**, *34*, 7205.

(28) Larock, R. C.; Lee, N. H. *J. Am. Chem. Soc.* **1991**, *113*, 7815.

(29) Koh, J. H.; Mascarenhas, C.; Gagne, M. R. *Tetrahedron* **2004**, *60*, 7405.

(30) In a synthesis of nitidine. Minami, T.; Nishimoto, A.; Hanaoka, M. *Tetrahedron Lett.* **1995**, *36*, 9505.

(31) In a synthesis of ent-EI-1941-2. Shoji, M.; Uno, T.; Hayashi, Y. *Org. Lett.* **2004**, *6*, 4535.

(32) Korte, D. E.; Hegedus, L. S.; Wirth, R. K. *J. Org. Chem.* **1977**, *42*, 1329.

(33) a) Åkermark, B.; Bäckvall, J.-E.; Hegedus, L. S.; Zetterberg, K.; Siirala-Hansen, K.; Sjöberg, K. *J. Organomet. Chem.* **1974**, *72*, 127. b) Åkermark, B.; Åkermark, G.; Hegedus, L. S.; Zetterberg, K. *J. Am. Chem. Soc.* **1981**, *103*, 3037.

(34) Hegedus, L. S.; McKearin, J. M. *J. Am. Chem. Soc.* **1982**, *104*, 2444.

(35) For a review, see Kotov, V.; Scarborough, C. C.; Stahl, S. S. *Inorg. Chem.* **2007**, *46*, 1910.

(36) a) Brice, J. L.; Harang, J. E.; Timokhin, V. I; Anastasi, N. R.; Stahl, S. S. *J. Am. Chem. Soc.* **2005**, *127*, 2868. See also b) Rogers, M. M; Kotov, V.; Chatwichien, J.; Stahl, S. S. *Org. Lett.* **2007**, *9*, 4331. c) Timokhin, V. I.; Stahl, S. S. *J. Am. Chem. Soc.* **2005**, *127*, 17888.

(37) a) Liu, G.; Stahl, S. S. *J. Am. Chem. Soc.* **2006**, *128*, 7179. b) For a closely related intramolecular amination–acetoxylation, see Alexanian, E. J.; Lee, C.; Sorensen, E. J. *J. Am. Chem. Soc.* **2005**, *127*, 7690.

(38) Scarborough, C. C.; Stahl, S. S. *Org. Lett.* **2006**, *8*, 3251.

(39) For a review, see Hegedus, L. S. *Angew. Chem., Int. Ed. Engl.* **1988**, *27*, 1113.

(40) In synthesis of (–)-indolactam V analogs. Irie, K.; Isaka, T.; Iwata, Y.; Yanai, Y.; Nakamura, Y.; Korzumi, F.; Ohigashi, H.; Wender, P.; Satomi, Y.; Nishmo, H. *J. Am. Chem. Soc.* **1996**, *118*, 10733.

(41) Utsunomiya, M.; Hartwig, J. F. *J. Am. Chem. Soc.* **2003**, *125*, 14286.

(42) Johns, A. M.; Utsunomiya, M.; Incarvito, C. D.; Hartwig, J. F. *J. Am. Chem. Soc.* **2006**, *128*, 1828.

(43) a) Rönn, M.; Bäckvall, J.-E.; Andersson, P. G. *Tetrahedron Lett.* **1995**, *36*, 7749. b) Larock, R. C.; Hightower, L. A.; Hasvold, L. A.; Peterson, K. P. *J. Org. Chem.* **1996**, *61*, 3584.

(44) Beccalli, E. M.; Broggini, G.; Paladino, G.; Penoni, A.; Zoni, C. *J. Org. Chem.* **2004**, *69*, 5627.

(45) In a synthesis of 1-deoxymannojirimycin. Yokoyama, H.; Otaya, K.; Kobayashi, H.; Miyazawa, M.; Yamaguchi, S.; Hirai, Y. *Org. Lett.* **2000**, *2*, 2427.

(46) Hirai, Y.; Watanabe, J.; Nozaki, T.; Yokoyama, H.; Yamaguchi, S. *J. Org. Chem.* **1997**, *62*, 776.

(47) For a review, see Tamaru, Y.; Kimura, M. *Synlett* **1997**, 749.

(48) Harayama, H.; Abe, A.; Sakado, T.; Kimura, M.; Fugami, K.; Tanaka, S.; Tamaru, Y. *J. Org. Chem.* **1997**, *62*, 2113.

(49) Hummer, W.; Dubois, E.; Graczca, T.; Jager, V. *Synthesis* **1997**, 634.

(50) Weider, P. R.; Hegedus, L. S.; Asada, H.; D'Andrea, S. V. *J. Org. Chem.* **1985**, *50*, 4276.

(51) a) Hegedus, L. S.; Williams, R. E.; McGuire, M. A.; Hayashi, T. *J. Am. Chem. Soc.* **1980**, *102*, 4973. b) Hegedus, L. S.; Darlington, W. H. *J. Am. Chem. Soc.* **1980**, *102*, 4980.

(52) Montgomery, J.; Wieber, G. M.; Hegedus, L. S. *J. Am. Chem. Soc.* **1990**, *112*, 6255.

(53) In a synthesis of (+)-negamycin. a) Masters, J. J.; Hegedus, L. S.; Tamariz, J. *J. Org. Chem.* **1991**, *56*, 5666. b) Masters, J. J.; Hegedus, L. S. *J. Org. Chem.* **1993**, *58*, 4547. c) Laidig, G.; Hegedus, L. S. *Synthesis* **1995**, 527.

(54) a) Kende, A. S.; Roth, B.; Sanfilippo, P. J. Blacklock, T. J. *J. Am. Chem. Soc.* **1982**, *104*, 5808. b) Kende, A. S.; Roth, B.; Sanfilippo, P. J.; *J. Am. Chem. Soc.* **1982**, *104*, 1784.

(55) Toyota, M.; Nishikawa, Y.; Motoki, K.; Yoshida, N.; Fukumoto, K. *Tetrahedron* **1993**, *49*, 11189.

(56) Toyota, M.; Wada, T.; Fukumoto, K.; Ihara, M. *J. Am. Chem. Soc.* **1998**, *120*, 4916.

(57) In a synthesis of serofendic acids A and B. Toyota, M.; Asano, T.; Ihara, M. *Org. Lett.* **2005**, *7*, 3929.

(58) Ito, Y.; Hirao, T.; Saegusa, T. *J. Org. Chem.* **1978**, *43*, 1011.

(59) In a synthesis of gymnocin A. Tsukano, C.; Ebine, M.; Sasaki, M. *J. Am. Chem. Soc.* **2005**, *127*, 4326.

(60) Larock, R. C.; Hightower, T. R. *Tetrahedron Lett.* **1995**, *36*, 2423.

(61) In a synthesis of (+)-lansonolide A. Yoshimura, T.; Yakushiji, F.; Kondo, S.; Wu, X.; Shindo, M.; Shishido, K. *Org. Lett.* **2006**, *8*, 475.

(62) Wang, X.; Widenhoefer, R. A. *Chem. Commun.* **2004**, 660.

(63) Pei, T.; Widenhoefer, R. A. *J. Am. Chem. Soc.* **2001**, *123*, 11290.

(64) Pei, T.; Wang, X.; Widenhoefer, R. A. *J. Am. Chem. Soc.* **2003**, *125*, 648.

(65) Zhang H.; Ferreira, E. M.; Stolz, B. M. *Angew. Chem. Int. Ed.* **2004**, *43*, 6144.

(66) Garg, N. K.; Caspi, D. D.; Stolz, B. M. *J. Am. Chem. Soc.* **2004**, *126*, 9552.

(67) For a review, see Widenhoefer, R. A.; Han, X. *Eur. J. Org. Chem.* **2006**, 4555.

(68) Liu, C.; Widenhoefer, R. A. *Tetrahedron Lett.* **2005**, *46*, 285.

(69) For a review, see Chianese, A. R.; Lee, S. J.; Gagne, M. R. *Angew. Chem. Int. Ed.* **2007**, *46*, 4042.

(70) For a review see Ritleng, V.; Sirlin, C.; Pfeffer, M. *Chem. Rev.* **2002**, *102*, 1731.

(71) Qian, H.; Han, X.; Widenhoefer, R. A. *J. Am. Chem. Soc.* **2004**, *126*, 9536.

(72) a) Li, Z.; Zhang, J.; Brouwer, C.; Yang, C.-G.; Reich, N. W.; He, C. *Org. Lett.* **2006**, *8*, 4175. b) Rosenfeld, D. C.; Shekhar, S.; Takemiya, A.; Utsunomiya, M.; Hartwig, J. F. *Org. Lett.* **2006**, *8*, 4179.

(73) In a synthesis of vasicoline. Wiedmann, S. H.; Ellmann, J. A.; Bergman, R. G. *J. Org. Chem.* **2006**, *71*, 1969.

(74) Han, X.; Widenhoefer, R. A. *Org. Lett.* **2006**, *8*, 3801.

(75) Yao, X.; Li, C.-J. *J. Am. Chem. Soc.* **2004**, *126*, 6884.

(76) Yao, X.; Li, C.-J. *J. Org. Chem.* **2006**, *70*, 5752.

(77) Zhang, Z.; Wang, X.; Widenhoefer, R. A. *Chem. Commun.* **2006**, 3717.

(78) For a mechanistic study of intramolecular platinum-catalyzed reactions, see Soriano, E.; Marco-Contelles, J. *Organometallics* **2006**, *25*, 4542.

(79) Liu, Z.; Wasmuth, A. S.; Nelson, S. G. *J. Am. Chem. Soc.* **2006**, *128*, 10352.

(80) Zhang, Z.; Liu, C.; Kinder, R. E.; Han, X.; Qian, H.; Widenhoefer, R. A. *J. Am. Chem. Soc.* **2006**, *128*, 9066.

(81) Sromek, A. W.; Rubina, M.; Gevorgyan, V. *J. Am. Chem. Soc.* **2005**, *127*, 10500.

(82) In a synthesis of *cis*-thujane. Feducia, J. A.; Campbell, A. N.; Doherty, M. Q.; Gagne, M. R. *J. Am. Chem. Soc.* **2006**, *128*, 13290.

(83) For a review, see Overman, L. E. *Angew. Chem., Int. Ed. Engl.* **1984**, *23*, 579.

(84) a) Saito, S.; Kuroda, A.; Matsunaga, H.; Ikeda, S. *Tetrahedron* **1996**, *52*, 13919. For earlier examples, see b) Grieco, P. A.; Takigawa, T.; Bongers, S. L.; Tanaka, H. *J. Am. Chem. Soc.* **1980**, *102*, 7587. c) Grieco, P. A.; Tuthill, P. A.; Sham, H. L. *J. Org. Chem.* **1981**, *46*, 5005.

(85) Panek, J. S.; Yang, K. M.; Solomon, J. S. *J. Org. Chem.* **1993**, *58*, 1003.

(86) Overman, L. E.; Knoll, F. M. *J. Am. Chem. Soc.* **1980**, *102*, 865.

(87) Overman, L. E.; Jacobsen, E. J. *J. Am. Chem. Soc.* **1982**, *104*, 7225.

(88) Bluthe, N.; Malacria, M.; Gore, J. *Tetrahedron Lett.* **1983**, *24*, 1157.

(89) Jamieson, A. G.; Sutherland, A.; Willis, C. L. *Org. Biomolec. Chem.* **2004**, *2*, 808.

(90) Metz, P.; Mues, C.; Schoop, A. *Tetrahedron* **1992**, *48*, 1071.

(91) For a review, see Overman, L. E.; Carpenter, N. E. *Org. React.* **2005**, *66*, 1.

(92) Overman, L. E.; Zipp, G. G. *J. Org. Chem.* **1997**, *62*, 2288.

(93) Lee, E. E.; Batey, R. A. *J. Am. Chem. Soc.* **2005**, *127*, 14887.

(94) a) Calter, M.; Hollis, T. K.; Overman, L. E.; Ziller, J.; Zipp, G. G. *J. Org. Chem.* **1997**, *62*, 1449. b) Hollis, T. K.; Overman, L. E. *Tetrahedron Lett.* **1997**, *38*, 8837.

(95) Grigg, R.; Markandu, J. *Tetrahedron Lett.* **1991**, *32*, 279.

(96) a) Nemoto, H.; Nagamochi, M.; Fukumoto, K. *J. Chem. Soc., Perkin Trans. 1* **1993**, 2329. b) In a synthesis of (+)-equilenin. Yoshida, M.; Mohamed, A.-H. I.; Nemoto, H.; Ihara, M. *J. Chem. Soc., Perkin Trans. 1* **2000**, 2629.

(97) Hegedus, L. S.; Ranslow, P. B. *Synthesis* **2000**, 953.

(98) Cutler, A.; Ehntholt, D.; Lennon, P.; Nicholas, K.; Marten, D. F.; Madhavarao, M.; Raghu, S.; Rosan, A.; Rosenblum, M. *J. Am. Chem. Soc.* **1975**, *97*, 3149.

(99) Marsi, M.; Rosenblum, M. *J. Am. Chem. Soc.* **1984**, *106*, 7264.

(100) Rosenblum, M.; Foxman, B. M.; Turnbull, M. M. *Heterocycles* **1987**, *25*, 419.

(101) For a review, see Welker, M. E. *Chem. Rev.* **1992**, *92*, 97.

(102) a) Boyle, P. F.; Nicholas, K. M. *J. Org. Chem.* **1975**, *40*, 2682. b) Nicholas, K. M. *J. Am. Chem. Soc.* **1975**, *97*, 3254.

(103) For a review, see Donaldson, W. A. Transition Metal Alkene Diene and Dienyl Complexes: Complexation of Dienes for Protection. In *Comprehensive Organometallic Chemistry II*; Abel, E. W., Stone, F. G. A., Wilkinson, G., Eds.; Pergamon: Oxford, U.K., 1995; Vol. 12, pp 623–637.

(104) Yeh, M. C.-P.; Chang, S.-C.; Chang, C.-J. *J. Organomet. Chem.* **2000**, *599*, 128.

(105) In a synthesis of (11R,12S)-diHETE. Gigou, A.; Beaucourt, J.-P.; Lellouche, J.-P.; Gree, R. L. *Tetrahedron Lett.* **1991**, *32*, 635.

(106) Schumacher, M.; Miesch, L.; Franck-Neumann, M. *Tetrahedron Lett.* **2003**, *44*, 5393.

(107) In a synthesis toward macrolactin A analogs. Benvengnu, T. J.; Toupet, L. J.; Gree, R. L. *Tetrahedron* **1996**, *52*, 11811.

(108) In a synthesis of the piperidine alkaloid SS20864A. Takemoto, Y.; Ueda, S.; Takeuchi, J.; Baba, Y.; Iwata, C. *Chem. Pharm. Bull.* **1997**, *45*, 1900.

(109) In a synthesis of carbonolide B. Franck-Neumann, M.; Geoffroy, P.; Gumery, F. *Tetrahedron Lett.* **2000**, *41*, 4219.

(110) Wasicak, J. T.; Craig, R. A.; Henry, R.; Dasgupta, B.; Li, H.; Donaldson, W. A. *Tetrahedron* **1997**, *53*, 4185.

(111) In a synthesis of ikarugamycin. Roush, W. R.; Wada, C. K. *J. Am. Chem. Soc.* **1994**, *116*, 2151.

(112) Takemoto, Y.; Yoshikawa, N.; Baba, Y.; Iwata, C.; Tanaka, T.; Ibuka, T.; Ohishi, H. *J. Am. Chem. Soc.* **1999**, *121*, 9143.

(113) Takemoto, Y.; Ishii, K.; Ibuka, T.; Miwa, Y.; Taga, T.; Nakao, S.; Tanaka, T.; Ohishi, H.; Kai, Y.; Kanehisa, N. *J. Org. Chem.* **2001**, *66*, 6116.

(114) a) Pearson, A. J.; Ghidu, V. P. *J. Org. Chem.* **2004**, *69*, 8975. b) For ionization starting from an adjacent alcohol, see Franck-Neumann, M.; Geoffroy, F.; Hanss, *Tetrahedron Lett.* **2002**, *43*, 2277.

(115) In a synthesis of (+)-[6]-gingerdiol. Franck-Neumann, M.; Geoffroy, P.; Bissinger, P.; Adelaide, S. *Tetrahedron Lett.* **2001**, *42*, 6401.

(116) Coquerel, Y.; Depres, J.-P.; Greene, A. E.; Cividino, P.; Court, J. *Synth. Commun.* **2001**, *31*, 1291.

(117) a) Pearson, A. J.; Chang, K. *J. Chem. Soc., Chem. Commun.* **1991**, 394; b) *J. Org. Chem.* **1993**, *58*, 1228.

(118) Pearson, A. J.; Srinivasan, K. *J. Chem. Soc., Chem. Commun.* **1991**, 392.

(119) a) Yeh, M.-C. P.; Hwu, C.-C.; Ueng, C.-H.; Lue, H.-L. *Organometallics* **1994**, *13*, 1788. b) Coquerel, Y.; Depres, J.-P.; Greene, A. E.; Philouze, C. *J. Organomet. Chem.* **2002**, *659*, 176.

(120) a) Pearson, A. J.; Wang, X.; Dorange, I. B. *Org. Lett.* **2004**, *6*, 2535. b) Pearson, A. J.; Dorange, I. B. *J. Org. Chem.* **2001**, *66*, 3140.

(121) Pearson, A. J.; Wang, X. *J. Am. Chem. Soc.* **2003**, *125*, 638.

(122) Pearson, A. J.; Mallik, S.; Mortezaei, R.; Perry, M. W. D.; Shively, R. J., Jr.; Youngs, W. J. *J. Am. Chem. Soc.* **1990**, *112*, 8034.

(123) Ong, C. W.; Chien, T.-L. *Organometallics* **1996**, *15*, 1323.

(124) Ong, C. W.; Wang, J. N.; Chien, T.-L. *Organometallics* **1998**, *17*, 1442.

(125) Han, J. L.; Ong, C. W. *Tetrahedron* **2005**, *61*, 1501.

(126) Hudson, R. D. A.; Osborne, S. A.; Stephenson, G. R. *Tetrahedron* **1997**, *53*, 4095.

(127) Hudson, R. D. A.; Osborne, S. A.; Stephenson, G. R. *Tetrahedron* **1997**, *53*, 4095.

(128) Fitzpatrick, J. D.; Watts, L.; Emerson, G. F.; Pettit, R. *J. Am. Chem. Soc.* **1965**, *87*, 3254.

(129) In a synthesis of (+)-asteriscanolide. Limanto, J.; Snapper, M. L. *J. Am. Chem. Soc.* **2000**, *122*, 8071.

(130) Seigal, B. A.; An, M. H.; Snapper, M. L. *Angew. Chem. Int. Ed.* **2005**, *44*, 4929.

(131) For a review, see Pearson, A. J. Transition Metal Diene and Dienyl Complexes: Nucleophilic Attack on Diene and Dienyl Complexes. In *Comprehensive Organometallic Chemistry II*; Abel, E. W., Stone, F. G. A., Wilkinson, G., Eds.; Pergamon: Oxford, U.K., 1995; Vol. 12, pp 635–685.

(132) a) Semmelhack, M. F.; Herndon, J. W. *Organometallics* **1983**, *2*, 363. b) Semmelhack, M. F.; Herndon, J. W.; Springer, J. P. *J. Am. Chem. Soc.* **1983**, *105*, 2497.

(133) a) Balazs, M.; Stephenson, G. R. *J. Organomet. Chem.* **1995**, *498*, C17. b) For an intramolecular version, see Yeh, M.-C. P.; Chuang, L.-W.; Ueng, C. H. *J. Org. Chem.* **1996**, *61*, 3874.

(134) a) Semmelhack, M. F.; Herndon, J. W.; Liu, J. K. *Organometallics* **1983**, *2*, 1885. b) Semmelhack, M. F.; Le, H. T. M. *J. Am. Chem. Soc.* **1985**, *107*, 1455.

(135) Yeh, M. C. P.; Hwu, C. C. *J. Organomet. Chem.* **1991**, *419*, 341.

(136) Faller, J. W.; Murray, H. H.; White, D. L.; Chao, K. H. *Organometallics* **1983**, *2*, 400.

(137) a) Pearson, A. J.; Khan, M. N. I.; Clardy, J. C.; Ciu-Heng, H. *J. Am. Chem. Soc.* **1985**, *107*, 2748. b) Pearson, A. J.; Khan, M. N. I. *Tetrahedron Lett.* **1984**, *25*, 3507. c) Pearson, A. J.; Khan, M. N. I. *Tetrahedron Lett.* **1985**, *26*, 1407.

(138) a) Yeh, M. C. P.; Tsou, C.-J.; Chuang, C.-N.; Lin, H.-W. *J. Chem. Soc., Chem. Commun.* **1992**, 890. b) Pearson, A. J.; Douglas, A. R. *Organometallics* **1998**, *17*, 1446.

(139) In a synthesis of carbazoles. Knölker, H.-J.; Goesmann, H.; Hofmann, *Synlett* **1996**, 737.

(140) Liebeskind, L. S.; Bombrun, A. *J. Am. Chem. Soc.* **1991**, *113*, 8736.

(141) Hansson, S.; Miller, J. F.; Liebeskind, L. S. *J. Am. Chem. Soc.* **1990**, *112*, 9660.

(142) In a synthesis of (–)-indolizidine 209B. Shu, C.; Alcudia, A.; Yin, J. Liebeskind, L. S. *J. Am. Chem. Soc.* **2001**, *123*, 12477.

(143) Bjurling, E.; Johansson, M. H.; Andersson, C.-M. *Organometallics* **1999**, *18*, 5606.

(144) For a review, see Pearson, A. J. Nucleophiles with Cationic Pentadienyl–Metal Complexes. In *Comprehensive Organic Synthesis*; Trost, B. M., Fleming, I., Semmelhack, M. F., Eds.; Pergamon: Oxford, U.K., 1991; Vol. 4, pp 663–694.

(145) In a synthesis of trichodermol. Pearson, A. J.; O'Brien, M. K. *J. Org. Chem.* 1989, 54, 4663.

(146) In a synthesis toward upenamide. Han, J. L.; Ong, C. W. *Tetrahedron* **2007**, *63*, 609.

(147) a) Pearson, A. J. *Acc. Chem. Res.* **1980**, *13*, 463. b) Pearson, A. J.; Richards, I. C. *Tetrahedron Lett.* **1983**, *24*, 2465.

(148) Potter, G. A.; McCague, R. *J. Chem. Soc., Chem. Commun.* **1992**, 635.

(149) Stephenson, G. R. *J. Organomet. Chem.* **1985**, *286*, C41.

(150) a) Knölker, H.-J.; Bauermeister, M.; Blaser, D.; Boese, R.; Pannek, J.-B. *Angew. Chem., Int. Ed. Engl.* **1989**, *28*, 223. b) Knölker, H.-J.; Bauermeister, M. *J. Chem. Soc., Chem. Commun.* **1990**, 664. c) Knölker, H.-J.; Fröhner, W. *Synlett* **1997**, 1108. d) Knölker, H.-J.; Fröhner, W. *Tetrahedron Lett.* **1998**, *39*, 2537.

(151) In a synthesis of 7-methoxy-*O*-methylmukonal. Kataeva, O.; Krahl, M. P.; Knölker, H.-J. *Org. Biomol. Chem.* **2005**, *3*, 3099.

(152) In a synthesis of neocarazostatin B. Czerwonka, R.; Reddy, K. R.; Baum, E.; Knölker, H.-J. *Chem. Commun.* **2006**, 711.

(153) For a review, see Knölker, H.-J.; Reddy, K. R. *Chem. Rev.* **2002**, *102*, 4303.

(154) a) Bandara, B. M. R.; Birch, A. J.; Kelley, L. F.; Khor, T. C. *Tetrahedron Lett.* **1983**, *24*, 2491. b) Howell, J. A. S.; Thomas, M. J. *J. Chem. Soc., Dalton Trans.* **1983**, 1401. c) Atton, J. G.; Evans, D. J.; Kane-Maguire, L. A. P.; Stephenson, G. R. *J. Chem. Soc., Chem. Commun.* **1984**, 1246.

(155) Pearson, A. J.; Gehrmini, A. M.; Pinkerton, A. A. *Organometallics* **1992**, *11*, 936.

(156) Knölker, H.-J.; Hermann, H.; Herzberg, D. *Chem. Commun.* **1999**, 831.

(157) a) Pearson, A. J.; Kole, S. L.; Yoon, J. *Organometallics* **1986**, *5*, 2075. b) Pearson, A. J.; Ray, T. *Tetrahedron Lett.* **1986**, *27*, 3111. c) Pearson, A. J.; Lai, Y.-S.; Lu, W.; Pinkerton, A. A. *J. Org. Chem.* **1989**, *54*, 3882.

(158) Pearson, A. J.; Katiyar, S. *Tetrahedron* **2000**, *56*, 2297.

(159) Wallock, N. J.; Donaldson, W. A. *J. Org. Chem.* **2004**, *69*, 2997.

(160) Chaudhury, S.; Li, S.; Donaldson, W. A. *Chem. Commun.* **2006**, 2069.

(161) In a synthesis of 5(*R*)-hydroxyeicosatetraenoic acid (HETE) methyl ester. Laabassi, M.; Gree, R. *Bull. Soc. Chim. Fr.* **1992**, *129*, 151.

(162) Donaldson, W. A. *J. Organomet. Chem.* **1990**, *395*, 187.

(163) In a synthesis of macrolactin A. Bärmann, H.; Prahlad, V.; Tao, C.; Yun, Y. K.; Wang, Z.; Donaldson, W. A. *Tetrahedron* **2000**, *56*, 2283.

(164) In a synthesis toward ambruticin. Lukesh, J. M.; Donaldson, W. A. *Chem. Commun.* **2005**, 110.

(165) For reviews, see a) Aubert, C.; Buisine, O.; Malacria, M. *Chem. Rev.* **2002**, *102*, 813. b) Lautens, M.; Klute, W.; Tam, W. *Chem. Rev.* **1996**, *96*, 49. c) Frühauf, H.-W. *Chem. Rev.* **1997**, *97*, 523.

(166) Trost, B. M.; Yanai, M.; Hoogsteen, K. *J. Am. Chem. Soc.* **1993**, *115*, 5294.

(167) Oh, C. H.; Gupta, A. K.; Park, D. I.; Kimj, N. *Chem. Commun.* **2005**, 5670.

(168) a) Trost, B. M.; Hashmi, A. S. K. *Angew. Chem., Int. Ed. Engl.* **1993**, *32*, 1085. b) Trost, B. M.; Hashmi, A. S. K. *J. Am. Chem. Soc.* **1994**, *116*, 2183.

(169) Lautens, M.; Edwards, L. G.; Tam, W.; Lough, A. J. *J. Am. Chem. Soc.* **1995**, *117*, 10276 and references therein.

(170) Lautens, M.; Tam, W.; Lautens, J. C.; Edwards, L. G.; Crudden, C. M.; Smith, A. C. *J. Am. Chem. Soc.* **1995**, *117*, 6863 and references therein.

(171) Lautens, M.; Lautens, J. C.; Smith, A. C. *J. Am. Chem. Soc.* **1990**, *112*, 5627.

(172) a) In a synthesis of yohimbane. Wender, P. A.; Smith, T. E. *J. Org. Chem.* **1996**, *61*, 824. b) Wender, P. A.; Smith, T. E. *J. Org. Chem.* **1995**, *60*, 2962.

(173) a) Gilbertson, S. R.; Hoge, G. S. *Tetrahedron Lett.* **1998**, *39*, 2075. b) Jolly, R. S.; Luedlke, G.; Sheehan, D.; Livinghouse, T. *J. Am. Chem. Soc.* **1990**, *112*, 4965.

(174) O'Mahoney, D. J. R.; Belanger, D. B.; Livinghouse, T. *Org. Biomol. Chem.* **2003**, *1*, 2038.

(175) McKinstry, L.; Livinghouse, T. *Tetrahedron Lett.* **1994**, *50*, 6145.

(176) Wender, P. A.; Jenkins, T. E.; Suzuki, S. *J. Am. Chem. Soc.* **1995**, *117*, 1843.

(177) Wender, P. A.; Croatt, M. P.; Deschamps, N. M. *J. Am. Chem. Soc.* **2004**, *126*, 5948.

(178) For a review, see Siebwith, S. N.; Arnard, N. T. *Tetrahedron* **1996**, *52*, 6251.

(179) Jolly, P. W.; Wilke, G. *The Organic Chemistry of Nickel*; Wiley: New York, 1975; Vol. 2, p. 94.

(180) In a synthesis of salsolene oxide. Wender, P. A.; Croatt, M. P.; Witulski, B. *Tetrahedron* **2006**, *62*, 7505.

(181) Wender, P. A.; Ihle, N. C.; Correa, C. R. D. *J. Am. Chem. Soc.* **1988**, *110*, 5904.

(182) Wender, P. A.; Nuss, J. M.; Smith, D. B.; Swarez-Sobrino, A.; Vågberg, J.; DeCosta, D.; Bordner, J. *J. Org. Chem.* **1997**, *62*, 4908.

(183) a) Evans, P. A.; Baum, E. W.; Fazal, A. N.; Pink, M. *Chem. Commun.* **2005**, 63. b) See also Varela, J. A.; Castedo, L.; Saa, C. *Org. Lett.* **2003**, *5*, 2841.

(184) DeBoef, B.; Counts, W. R.; Gilbertson, S. R. *J. Org. Chem.* **2007**, *72*, 799.

(185) Wender, P. A.; Christy, J. P. *J. Am. Chem. Soc.* **2006**, *128*, 5354.

(186) Trost, B. M.; Shen, H. C.; Horne, D. B.; Toste, F. D.; Steinmetz, B. G.; Koradin, C. *Chem.óEur. J.* **2005**, *11*, 2577.

(187) Yu, Z.-X.; Wender, P. A.; Houk, K. N. *J. Am. Chem. Soc.* **2004**, *126*, 9154.

(188) Wender, P. A.; Dyckman, A. *J. Org. Lett.* **1999**, *1*, 2089.

(189) Wender, P. A.; Haustedt, L. O.; Lim, J.; Love, J. A.; Williams, T. J.; Yoon, J.-Y. *J. Am. Chem. Soc.* **2006**, *128*, 6302.

(190) Wender, P. A.; Bi, F. C.; Brodney, M. A.; Gosselin, F. *Org. Lett.* **2001**, *3*, 2105.

(191) In a synthesis of (+)-aphanamol 1. Wender, P. A.; Zhang, L.; *Org. Lett.* **2000**, *2*, 2323.

(192) Wender, P. A.; Pedersen, T. M.; Scanio, M. J. C. *J. Am. Chem. Soc.* **2002**, *124*, 15154.

(193) Wender, P. A.; Gamber, G. G.; Hubbard, R. D.; Zhang, L. *J. Am. Chem. Soc.* **2002**, *124*, 2876.

(194) Wender, P. A.; Correa, A. G.; Sato, Y.; Sun, R. *J. Am. Chem. Soc.* **2000**, *122*, 7815.

(195) Wender, P. A.; Deschamps, N. M.; Sun, R. *Angew. Chem. Int. Ed.* **2006**, *45*, 3957.

(196) For a review, see Rigby, J. H. *Tetrahedron* **1999**, *55*, 4521.

(197) For a review, see Rigby, J. H. *Acc. Chem. Res.* **1993**, *26*, 579.

(198) Rigby, J. H.; Kirova-Snovei, M. *Tetrahedron Lett.* **1997**, *38*, 8153.

(199) Rigby, J. H.; Warshakoon, N. C. *Tetrahedron Lett.* **1997**, *38*, 2049.

(200) Rigby, J. H.; Kirova, M.; Niyaz, N.; Mohanmadi, F. *Synlett* **1997**, 805.

(201) Rigby, J. H.; Fales, K. R. *Tetrahedron Lett.* **1998**, *39*, 1525.

(202) Rigby, J. H.; Warshakoon, N. C.; Payen, A. J. *J. Am. Chem. Soc.* **1999**, *121*, 8237.

(203) a) Rigby, J. H.; Rege, S. D.; Sandanayaka, V. P.; Kirova, M. *J. Org. Chem.* **1996**, *61*, 843. b) Rigby, J. H.; Hu, J.; Heeg, M. J. *Tetrahedron Lett.* **1998**, *39*, 2265.

(204) Rigby, J. H.; Fiedler, C. *J. Org. Chem.* **1997**, *62*, 6106.

(205) a) Rigby, J. H.; Mann, L. W.; Myers, B. J. *Tetrahedron Lett.* **2001**, *42*, 8773. b) Kündig, E. P.; Robvieux, F.; Kondratenko, M. *Synthesis* **2002**, 2053.

(206) Achard, M.; Tenaglia, A.; Buono, G. *Org. Lett.* **2005**, *7*, 2353.

(207) In a synthesis of 9-epi-pentalenic acid. Rigby, J. H.; Laxmisha, M. S.; Hudson, A. R.; Heap. C. R.; Heeg, M. J. *J. Org. Chem.* **2004**, *69*, 6751.

(208) Rigby, J. H.; Heap, C. R.; Warshakoon, N. C. *Tetrahedron* **2000**, *56*, 3505.

(209) Chaffee, K.; Morcos, H.; Sheridan, J. B. *Tetrahedron Lett.* **1995**, *36*, 1577.

8

Synthetic Applications of Transition Metal Alkyne Complexes

8.1 Introduction

Although virtually all transition metals react with alkynes, relatively few form simple, stable metal–alkyne complexes analogous to the corresponding metal–alkene complexes. This is because many alkyne complexes are quite reactive toward additional alkyne, and react further to produce more elaborate complexes or organic products. Those that are stable tend to be so stable that use in synthesis is precluded by their lack of reactivity. The successful development of transition-metal–alkyne complex chemistry for use in organic synthesis has involved procedures for the management of these extremes in reactivity.

8.2 Nucleophilic Attack on Metal–Alkyne Complexes

In contrast to metal–alkene complexes, nucleophilic attack on metal–alkyne complexes is relatively rare and has found little use in synthesis. Stable $CpFe(CO)_2^+$ alkyne complexes undergo attack by a wide range of nucleophiles, forming very stable σ-alkenyliron complexes, which can be made to insert CO by treatment with Ag(I) (Lewis acid promoted insertion) (**Eq. 8.1**).[1] Although these complexes should have additional reactivity, little has been done with them.

Eq. 8.1

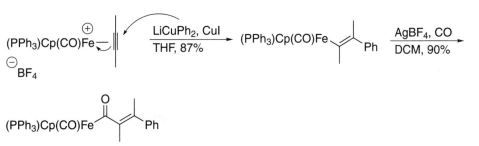

Palladium is by far the most utilized transition metal for additions to alkynes. Intramolecular palladium-catalyzed reactions of alkynes with soft nucleophiles, such as alcohols, amides, amines, and acids, to give, mainly, five- and six-membered rings, can be divided into four broad groups depending on events occurring before or after the cyclization (Figure 8.1).[2] All of the reactions are likely to occur by initial complexation of the metal to the alkyne, followed by cyclization–protonolysis [path (a) in **Figure 8.1**], cyclocarbonylation [path (b)], cyclization–carbonylation [path (c)], or cyclization–coupling [path (d)]. The latter type of reaction includes reactions of aryl and alkenyl halides and triflates.

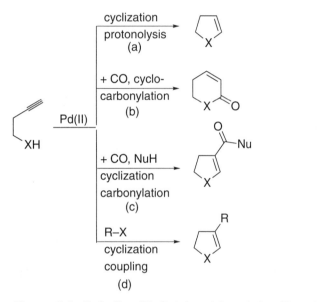

Figure 8.1. Palladium(II)-Catalyzed Annulation Reactions of Alkynes

Palladium(II) complexes catalyze the cyclization of hydroxy-substituted alkynes[3] in a process that probably involves nucleophilic attack on a complexed alkyne[4] (**Eq. 8.2**),[5] similar to that observed with palladium(II)-complexed alkenes (Chapter 7). However, the mechanism of this process has not been studied, and the putative protolytic cleavage of the σ-alkenylpalladium(II) complex by the protonated heterocycle is unusual. Related cyclizations of alkynoic acids afford lactones (**Eq. 8.3**).[6]

Eq. 8.2

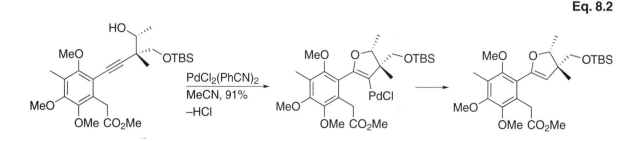

Eq. 8.3

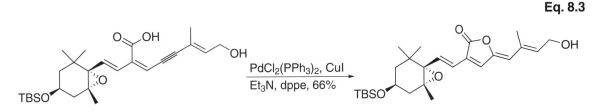

Palladium catalyzes the addition of two alcohols to alkynes to afford spirocyclic compounds. The second addition is probably not palladium-catalyzed since β-elimination side products are not observed (**Eq. 8.4**).[7] Both modes of carbonylation [i.e., paths (b) and (c) in **Figure 8.1**] have been described for hydroxy-alkynes (**Eqs. 8.5**[8] and **8.6**[9]). In the latter case, hydrolysis of the acetate probably occurs prior to cyclization.

Eq. 8.4

Eq. 8.5

Eq. 8.6

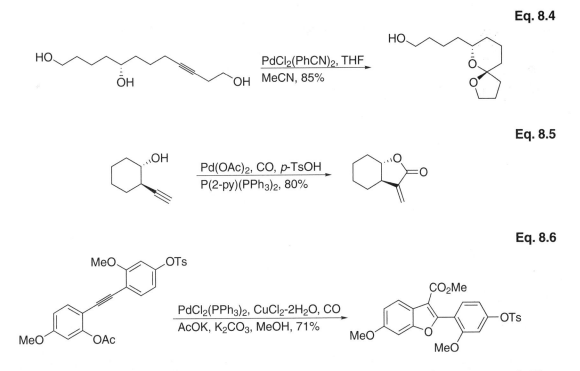

The fourth type of cyclization differs in that palladium(0) catalysts are used. Key steps in the mechanism probably involve an oxidative addition of palladium(0) to the halide/triflate and coordination of the alkyne, followed by nucleophilic addition. The dialkylpalladium complex thus formed finally reductively eliminates, affording the product (**Eq. 8.7**).[10]

Eq. 8.7

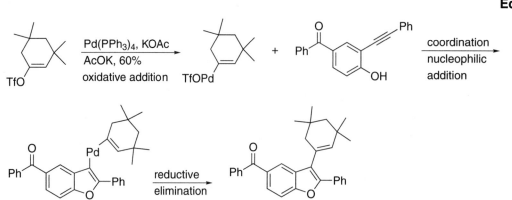

Not only free alcohols (and alkoxides) add to alkynes. Carbonyl groups also add, perhaps via the enol or hemiacetal, undergoing facile cyclization reactions with alkynes. The mechanism has not been studied in detail, but an activation of both the carbonyl and the alkyne by palladium has been suggested.[11] An example of this type of reaction, applied to the synthesis of diacetylenic spiroacetal enol ether natural products, is shown in **Eq. 8.8**.[12]

Eq. 8.8

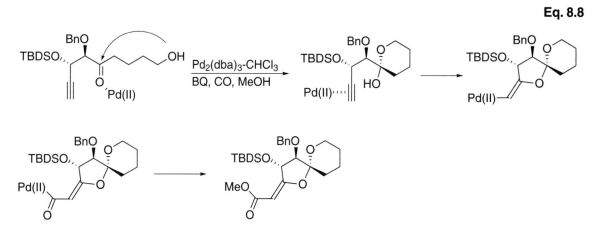

More developed are reactions of amino-alkynes to give nitrogen-containing heterocycles, in particular the reactions affording indoles (**Eq. 8.9**[13]).[14,15,16] Similar to the formation of 2,3-disubstituted benzo[*b*]furans, functionalization of the 3-position of the indole skeleton is possible using organic halides and triflates in the presence of a palladium(0) catalyst (**Eq. 8.10**[17]).[18] Two additional examples of cyclizations to give nitrogen heterocycles are shown in **Eqs. 8.11**[19] and **8.12**.[20]

Eq. 8.9

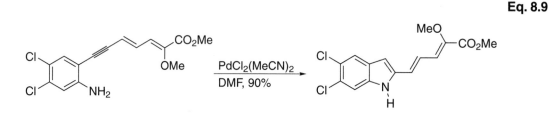

Eq. 8.10

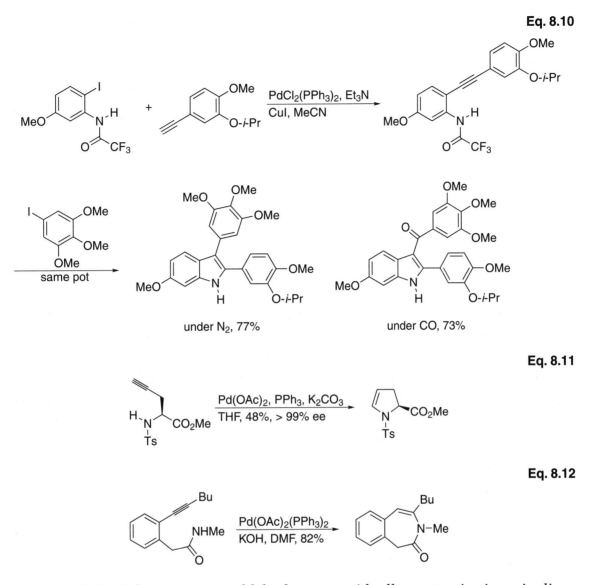

under N$_2$, 77%   under CO, 73%

Eq. 8.11

Eq. 8.12

Imines derived from aromatic aldehydes react with alkynes to give isoquinolines and related heterocycles. Similar to the indole-forming reactions described in **Eq. 8.10**, a flexible functionalization of the 4-position of the isoquinoline ring is possible by the addition of organic halides,[21] by the addition of organic halides in the presence of carbon monoxide,[22] or by Heck-type reactions using alkenes (**Eq. 8.13**).[23] The *N-tert*-butyl group is cleaved in the reaction.

Eq. 8.13

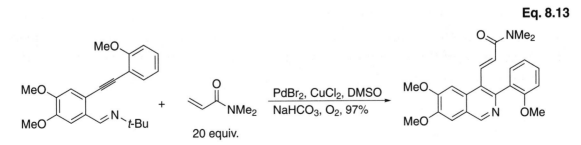

20 equiv.

Intermolecular additions to alkynes are not as developed as the intramolecular reactions. Palladium is usually not the catalyst of choice for intermolecular hydroaminations.[24] However, rhodium, ruthenium and iridium complexes have successfully been employed in intermolecular reactions, nicely complementing palladium (**Eq. 8.14**).[25] Markovnikov orientation of the addition is observed as the sole or major product. The mechanism of the aminations is unclear and likely involves either alkyne coordination–nucleophilic addition, as observed for palladium-catalyzed reactions, or insertion of the metal into the nitrogen–hydrogen bond, followed by migratory alkyne insertion. (See Chapter 4 for early transition-metal-catalyzed hydroaminations.)

Eq. 8.14

Ruthenium catalyzes anti-Markovnikov and Markovnikov additions of carboxylic acids to alkynes.[26] The anti-Markovnikov regioselectivity observed for terminal alkynes can be explained by a mechanism involving the transformation of an η^2-alkyne–metal complex to an η^1-alkenylidene complex (**Eq. 8.15**[27]). Addition of the acid to the electrophilic carbon of the alkenylidene ligand, and protonation of the metal, followed by reductive elimination of the intermediate, produces the observed product. The Z-selectivity of the reaction can be explained by the steric repulsion between the metal and the butyl group. Although not as stereoselective, rhenium complexes also catalyze this reaction, forming anti-Markovnikov addition products.[28]

Eq. 8.15

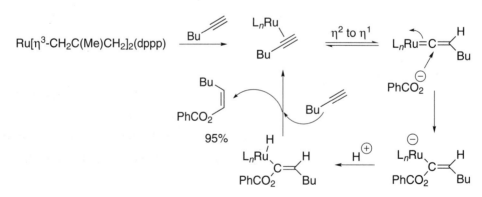

A different mechanism has been proposed for the ruthenium-catalyzed reactions of internal alkynes and terminal alkynes that afford Markovnikov products. A sequence of complexation of the alkyne, addition of the acid to the more electrophilic carbon, protonation, and reductive elimination has been proposed (**Eq. 8.16**[29]).

Eq. 8.16

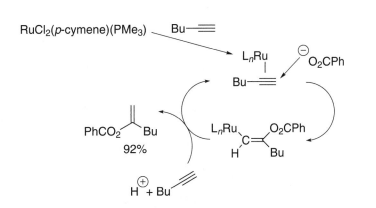

In addition to the ruthenium-catalyzed hydroacyloxylations described above, interesting and synthetically useful reactions of terminal alkynes with alcohols and amines, probably involving alkenylidene–ruthenium intermediates, have been developed.[30] The oxidation state of the product from intramolecular reactions can be modulated by the proper choice of catalyst (**Eq. 8.17**).[31] Proposed mechanisms for the reactions are shown in **Eq. 8.18**.[32] For electron-rich ligands, the oxidation to form lactones can be rationalized as a formation of a σ-bonded ruthenium–dihydropyrane complex, followed by protonation of the alkene to give a cationic ruthenium carbene that is attacked by *N*-hydroxysuccinate [cycle (a) in **Eq. 8.18**]. Protonation of this complex would give the lactone, succinimide, and the active catalyst. In contrast, electron-poor ligands favor ligand exchange with *N*-hydroxysuccinate, protonolysis liberating a dihydropyrane, and ligand exchange to give the active catalyst [cycle (b)].

Eq. 8.17

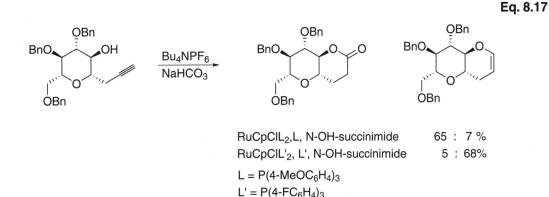

RuCpClL$_2$,L, N-OH-succinimide   65 : 7 %
RuCpClL'$_2$, L', N-OH-succinimide   5 : 68%

L = P(4-MeOC$_6$H$_4$)$_3$
L' = P(4-FC$_6$H$_4$)$_3$

Eq. 8.18

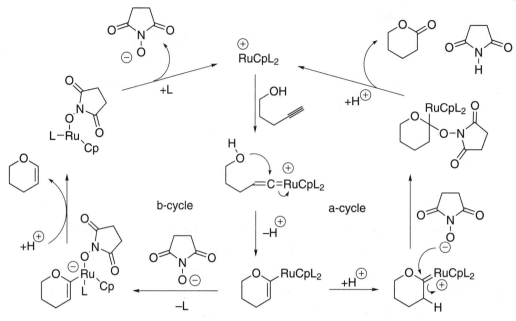

Intermolecular reactions of allylic alcohols with terminal alkynes in the presence of a ruthenium catalyst yield interesting rearranged products. Key steps in a proposed mechanism are the formation of an alkylidene–ruthenium complex, followed by addition of the alcohol to give a ruthenium–carbene (**Eq. 8.19**).[33] The carbene complex undergoes an oxidative [3.3]-sigmatropic rearrangement [at least formally Ru(II)→Ru(IV)] to give an η^1–η^1 or an η^1–η^3 complex. Finally, a reductive elimination yields the product.

Eq. 8.19

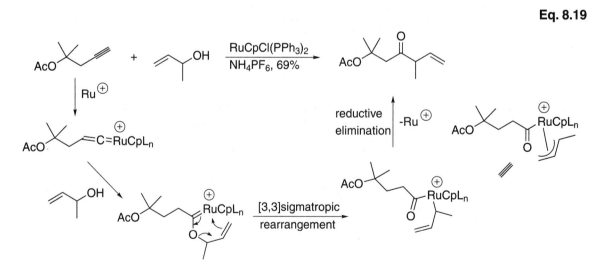

Annulation of haloalkynes via haloalkylidene–gold complexes has been proposed as the mechanism for the synthesis of polycyclic aromatic compounds (**Eq. 8.20**).[34] Related migrations of tin, germanium, and silicon are also possible (**Eq. 8.21**).[35]

Eq. 8.20

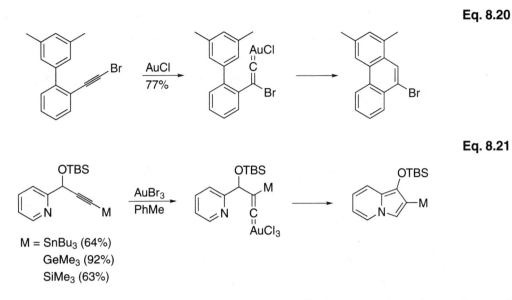

Eq. 8.21

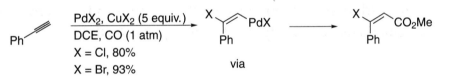

M = SnBu$_3$ (64%)
GeMe$_3$ (92%)
SiMe$_3$ (63%)

Halogens can also be used as nucleophiles in palladium-catalyzed reactions of alkynes (halopalladation). A mechanistic study indicated that the stereochemistry of the addition of chloride to terminal alkynes was cis at low chloride concentrations and trans at high chloride concentrations.[36] This selectivity has been observed in a number of cases. For example, Z-3-chloro/bromo enoates are obtained from terminal alkynes in the presence of carbon monoxide, an alcohol, and an excess of a copper dihalide **(Eq. 8.22)**.[37]

Eq. 8.22

The σ-palladium intermediate can insert into pendant alkenes to give cyclic compounds after elimination and this has been used to synthesize both *E*- and *Z*-α-alkylidene lactones **(Eqs. 8.23**[38] **and 8.24)**.

Eq. 8.23

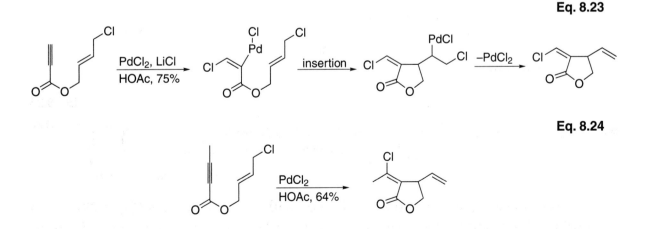

Eq. 8.24

Palladium catalyzes cis-chloro- or -bromopalladations of propargylic alcohols, followed by carbon monoxide insertion and lactonization to give Z-α-haloalkylidene-β-lactones (**Eq. 8.25**).[39] For propargylic amines, E-chloro-β-lactams are obtained (**Eq. 8.26**).[40] The reason for the cis-chloropalladation shown in **Eq. 8.25** at high chloride concentration is not known.

Eq. 8.25

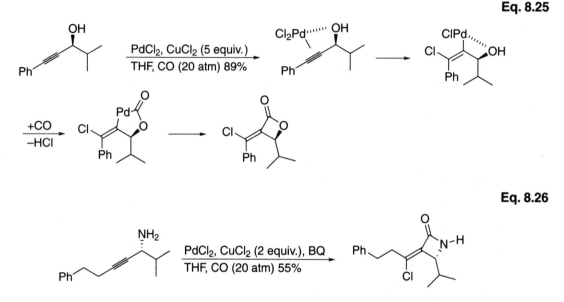

Eq. 8.26

Halo-ruthenations of alkynes form the basis of synthetically interesting reactions of alkynes with ethenyl ketones to give 4-halo-3-alkene-1-ones (**Eq. 8.27**).[41] The ruthenium enolate formed after insertion of the alkene can also be intercepted with an aldehyde in a remarkable four-component coupling (**Eq. 8.28**).[42]

Eq. 8.27

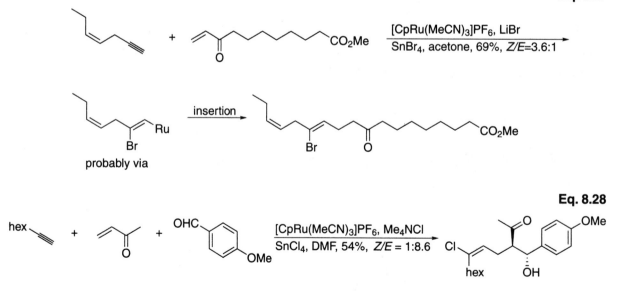

Eq. 8.28

Rhodium-catalyzed reactions of alkynes with allylic chlorides (**Eq. 8.29**)[43], acid chlorides (**Eq. 8.30**),[44] chloroformate esters (**Eq. 8.31**),[45] or ethoxyallyl chloride[46] afford chlorinated products via cis-chlororhodations. Oxidative addition to rhodium(I) is a likely first step in each reaction. No product derived from trans-chlororhodation, even under high chloride concentrations, is observed in **Eq. 8.29**, in contrast to the reactions using palladium (see **Eq. 8.23**).

Eq. 8.29

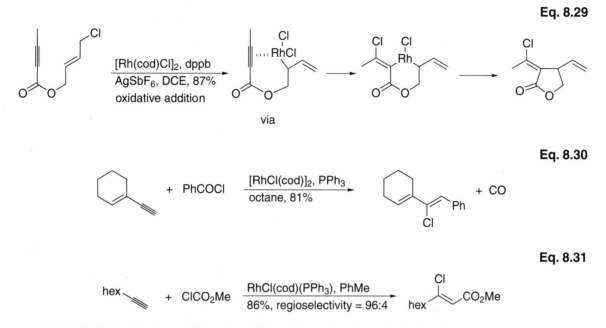

via

Eq. 8.30

Eq. 8.31

Gold(I)/(III), platinum(II), silver(I), and copper (I) complexes also catalyze the Markovinikov-type addition of heteronucleophiles, such as water, alcohols, and amines,[47] to alkynes. The complexes act as Lewis acids and very facile formation of heterocycles is usually observed in intramolecular reactions. A few examples are shown in **Eqs. 8.32**,[48] **8.33**,[49] and **8.34**.[50]

Eq. 8.32

Eq. 8.33

Eq. 8.34

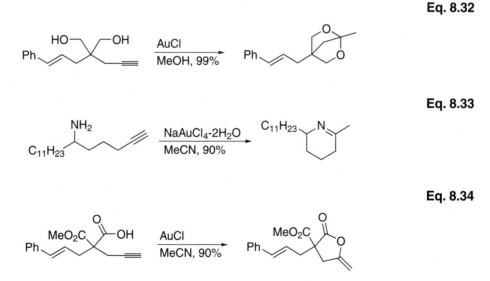

Carbonyl groups can also be used as the nucleophile in addition reactions to alkynes. Further reactions of the metal–oxonium intermediates include [4+2] cycloadditions (**Eq. 8.35**),[51] nucleophilic additions (**Eq. 8.36**),[52] and rearrangements (**Eq. 8.37**).[53] In all cases, proposed key steps are alkyne coordination (activation), nucleophilic addition of the carbonyl oxygen, further reactions as mentioned above, and hydrolytic cleavage of the metal–carbon sigma bond.

Eq. 8.35

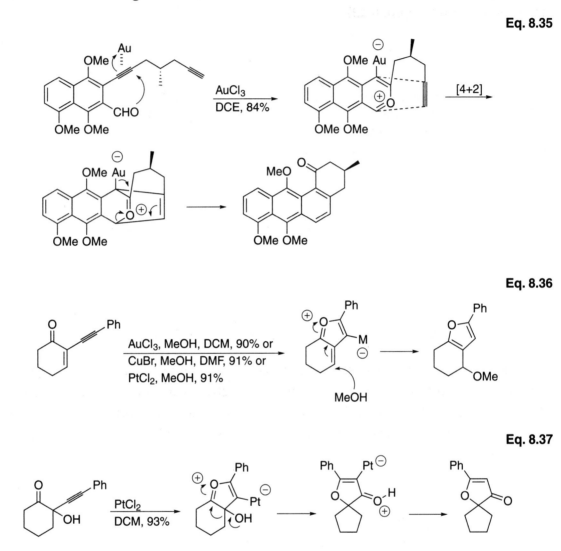

Eq. 8.36

Eq. 8.37

Complexation of transition metal complexes to alkynes also activates the alkyne for the addition of carbon nucleophiles. Palladium, platinum, ruthenium, and gold complexes catalyze hydroarylations[54] of alkynes to give aryl-substituted alkenes. Kinetic isotope effect data[55] and DFT calculations[56] indicate that the mechanism is an electrophilic aromatic substitution and not a carbon–hydrogen bond activation (**Eq. 8.38**).[57] The size of the formed ring can be modulated by proper catalyst choice (**Eq. 8.39**).[58] Gold complexes catalyze double intermolecular additions (**Eq. 8.40**).[59]

Eq. 8.38

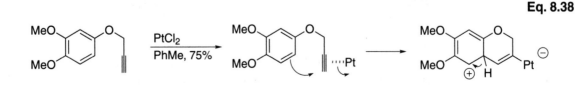

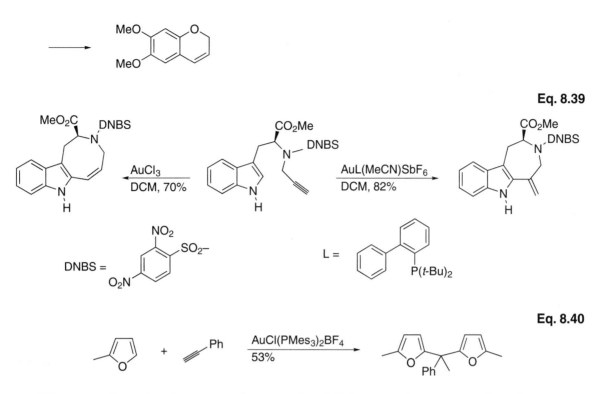

Eq. 8.39

Eq. 8.40

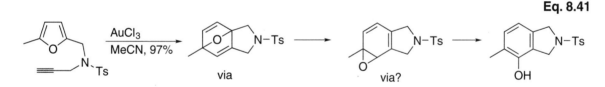

Alkynes tethered to furans undergo cycloaddition reactions to give fused aromatic compounds. Gold complexes are particularly efficient, but rhodium, platinum, palladium, and iridium complexes also catalyze the reaction (**Eq. 8.41**[60]).[61,62] The mechanism of the reaction is unclear but likely involves an initial [4+2] cycloaddition[63] between the furan and the metal-complexed alkyne and the formation of an arene oxide.[64]

Eq. 8.41

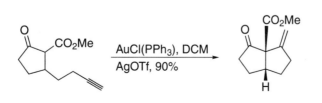

A number of transition metals catalyze very mild and efficient Conia ene reactions of β-keto esters with alkynes (**Eq. 8.42**).[65,66,67] Asymmetric reactions can be realized using chiral ligands (**Eq. 8.43**).[68] Silyl enol ethers and enamine derivatives can be used as nucleophiles (**Eqs. 8.44**[69] and **8.45**[70]). Rhenium complexes catalyze an interesting annulation of silyl dienol ethers to give bicyclo[3.3.0] ring systems (**Eq. 8.46**).[71] A mechanism involving initial nucleophilic addition to a rhenium-complexed alkyne, followed by the formation of a rhenium carbene and elimination, was suggested.

Eq. 8.42

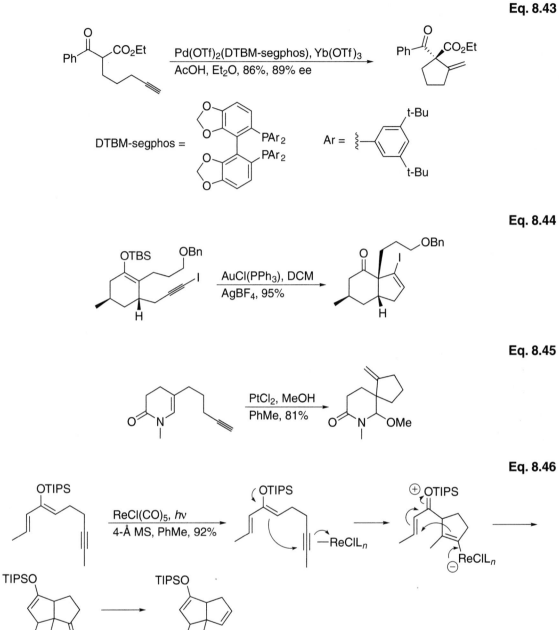

The complexation of 1-alkynylcyclopropanols and cyclobutanols to gold catalysts results in ring-expansion to give cyclobutanones and cyclopentanones, respectively (**Eq. 8.47**).[72]

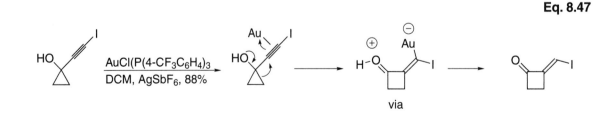

Skeletal rearrangements are observed upon reaction of 1,*n*-enynes (where *n* = 5–7) with gold and platinum catalysts. The gold-catalyzed reactions are particularly facile and some proceed even at –63°C, affording products similar to those observed using platinum at higher temperatures.[73]

Gold(I) complexes are relatively nonnucleophilic and do not usually undergo oxidative addition.[74] Reductive elimination from gold(III) complexes is also a disfavored process.[75] The chemistry of gold complexes is mostly centered around the ability to coordinate with alkynes, thereby activating the alkyne toward nucleophilic addition.

Although a bewildering array of seemingly unrelated products are obtained from 1,*n*-enyne cycloisomerizations using gold and platinum catalysts, most of the reactions can be rationalized with a few mechanisms.[76] For 1,5-enynes, the alkyne coordinates, which initiates a nucleophilic addition–cyclopropane formation to give a gold carbene (**Eq. 8.48**). A carbene is one extreme representation of the bonding, and a carbocation is the other extreme. The true nature of the bonding is unknown and probably varies with substrate. The carbene can undergo a number of different transformations, depending on additives and substitution pattern. A [1,2]-hydride shift–metal-elimination affords bicyclo[3.2.0] ring systems [path (a) in **Eq. 8.48**[77]] and carbon-bond cleavage, migration followed by loss of metal, produces alkylidene cyclopentenes [path (b)]. The carbene intermediate can be intercepted by nucleophiles via cleavage of the cyclopropane ring and the metal removed by hydrolytic cleavage (**Eq. 8.49**). Depending on the site of addition, either a cyclopentene [path (c)] or a cyclohexene may be formed [path (d)]. As a final major mechanism for 1,5-enynes, migration of the carbene carbon, forming a six-membered carbene, followed by elimination is observed in some cases (**Eq. 8.50**).

Eq. 8.48

Eq. 8.49

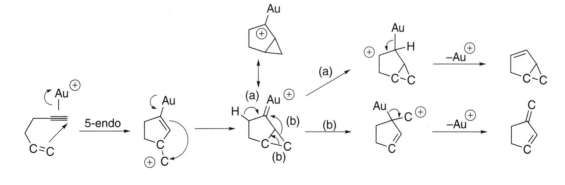

Eq. 8.50

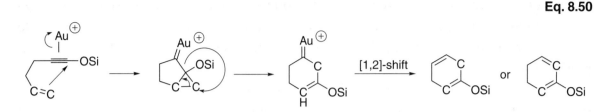

1,6-Enynes have an equally rich chemistry, producing a number of cyclization products that can be rationalized mechanistically. Theoretical studies of platinum-catalyzed 1,6-enyne cyclizations indicate the formation of polarized π-alkyne–platinum complexes having a partial positive charge on the internal *sp*-hybridized carbon. Attack by the alkene (6-endo) results in the formation of a cyclopropanated platinum carbene, followed by a [1,2]-hydride shift that yields a bicyclo[3.1.0]-system[78] or an alkylidene–cyclohexene [**Eq. 8.51**, path (a)].[79] Five-membered products are probably derived from a 5-exo addition of the alkene to the complexed alkyne [path (b)]. The resulting cyclopropanated carbene can rearrange to a sigma-bonded complex by formal migration of CD_2, followed by elimination [path (c)], or to a new carbene by formal migration of CH=Pt, followed by a [1,2]-hydride shift [path (d)]. Two terminal deuterium atoms have been included in the mechanism to show the difference in products formed in paths (c) and (d).[80]

Eq. 8.51

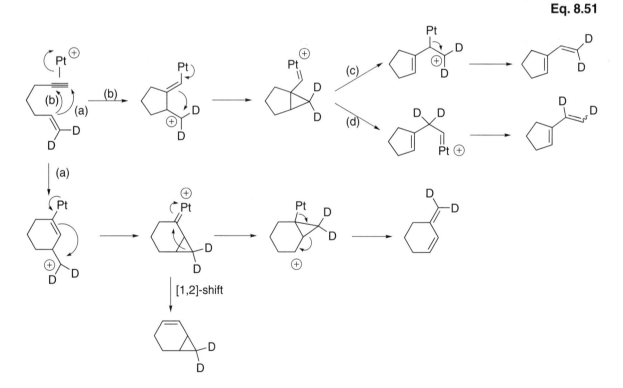

A few examples of ene–yne cyclizations can be found in **Eqs. 8.52,**[81] **8.53,**[82] and **8.54.**[82] Note the complete switch in chemoselectivity by a simple change in stereochemistry of the alkene involved in the initial cyclization (**Eqs. 8.53** and **8.54**).

Eq. 8.52

Eq. 8.53

Eq. 8.54

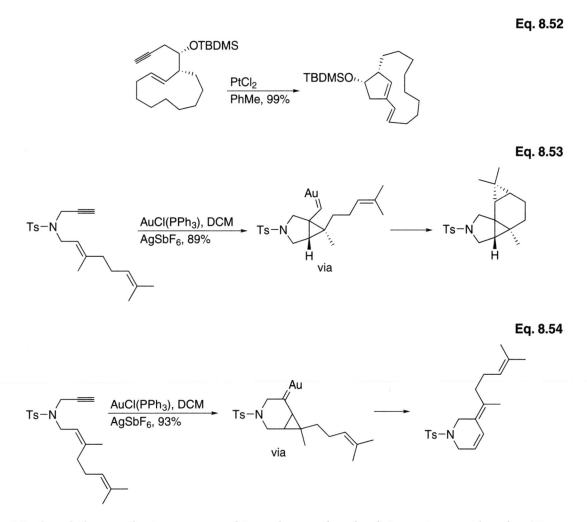

Nucleophiles can be incorporated into the product both in an intermolecular (**Eqs. 8.55 and 8.56**)[82] and intramolecular fashion (**Eq. 8.57**).[83] Electron-rich arenes add to the carbene carbon in a Friedel–Crafts fashion to form highly functionalized compounds (**Eq. 8.58**).[84] Gold, platinum, palladium, rhodium, and iridium complexes catalyze the formation of alkynyl-substituted furans to give phenols.

Eq. 8.55

Eq. 8.56

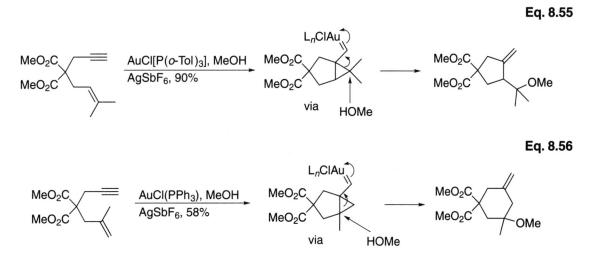

Eq. 8.57

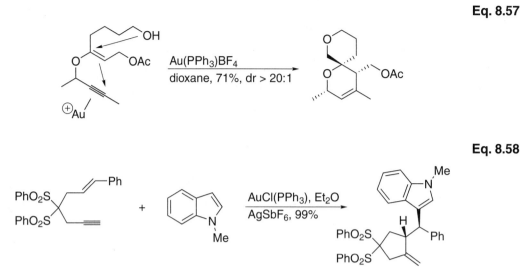

Eq. 8.58

Not only hydrides, but acyloxy groups can also undergo [1,2]-migrations.[85] Two mechanistic possibilities have been suggested, differing in the timing of the [1,2]-acyloxy shift (**Eq. 8.59**).[86] For example, for $n = 1$ in **Eq. 8.59**, the migration of the acyloxy group may be initiated before or after cyclization. The reactions are terminated as described previously with the net effect of a [1,2]-shift of the ester group. For $n = 2$, the migration occurs in the bicyclo[4.1.0] case and, for $n = 0$, cyclopentadienes are formed.

Eq. 8.59

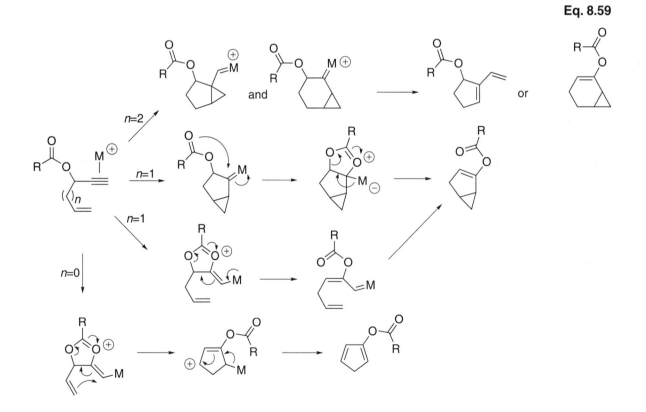

The cyclization of 1,4-enyne esters is a variation of the palladium-catalyzed Rautenstrauch rearrangement.[87] Very high transfer of chirality is observed (**Eq. 8.60**).[88]

In addition to gold and platinum, ruthenium, iridium, and rhodium complexes catalyze cyclopropanations of alkenes with alkynyl aldimines, iminoethers, esters, and amides. A metal–carbene is a likely intermediate in these reactions (**Eqs. 8.61**[89] and **8.62**[90]). An example of an enyne cyclization in total synthesis is shown in **Eq. 8.63**.[91]

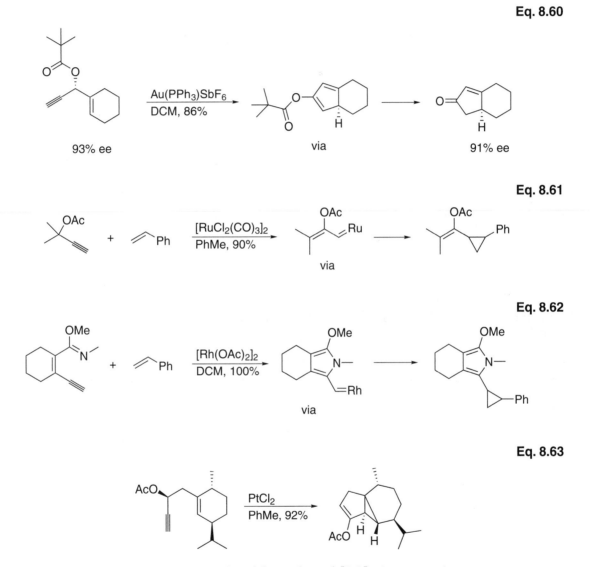

Eq. 8.60

Eq. 8.61

Eq. 8.62

Eq. 8.63

Propargylic esters participate in gold-catalyzed [3,3]-sigmatropic rearrangements to give allenic esters. Substrates having an alkene (**Eq. 8.64**)[92] or an arene (**Eq. 8.65**)[93] undergo further gold-catalyzed reactions, affording bicyclic products.

Eq. 8.64

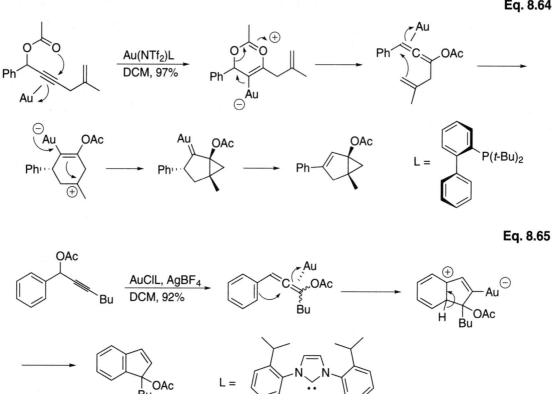

Eq. 8.65

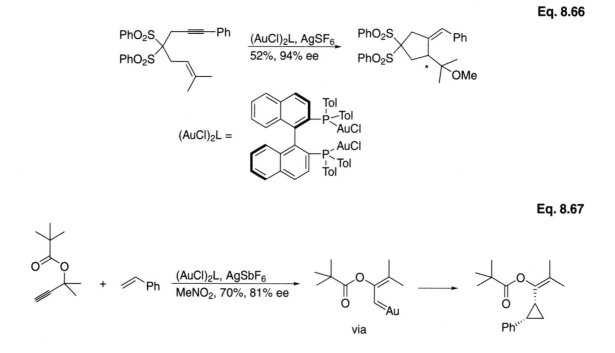

Asymmetric reactions using gold catalysts are hampered by the coordination geometry of gold(I) complexes. Gold(I) complexes are linear (180° angle), so the influence of a chiral ligand on a coordinated substrate is diminished. However, some asymmetric reactions have been developed, such as the cyclizations of enynes in the presence of an alcohol (**Eq. 8.66**)[94] and intermolecular cyclopropanations (**Eq. 8.67**).[95]

Eq. 8.66

Eq. 8.67

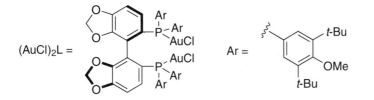

Late transition metals catalyze interesting cyclization–migration reactions of a variety of ortho-substituted alkynylarenes (**Eqs. 8.68,**[96] **8.69,**[97] and **8.70**[98]). The mechanisms of these reactions are not yet fully understood.

Eq. 8.68

Eq. 8.69

Eq. 8.70

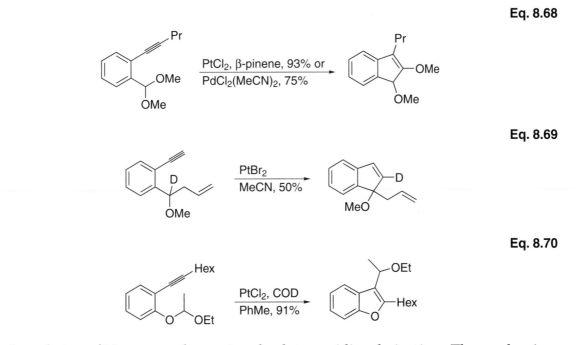

Annulation of *N*-propargyl enamines leads to pyridine derivatives. The mechanism is proposed to proceed by initial formation of an alkylidene–rhodium complex, nucleophilic addition, and rearrangement (**Eq. 8.71**).[99] An alkylidene–ruthenium complex is proposed as an intermediate from cyclizations of 2-(2′-iodoalkenyl)ethynylarenes. The iodine migrates during the reaction (**Eq. 8.72**).[100] Allenylidene–ruthenium complexes formed from propargylic alcohols react with a variety of nucleophiles to form substitution products (**Eqs. 8.73**[101] and **8.74**[102]) or annulated compounds (**Eq. 8.75**[103]). An allenyldiene–ruthenium complex having two ruthenium atoms bridged by a thiolate has been characterized by X-ray crystallography.

Eq. 8.71

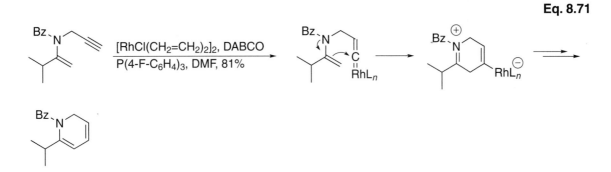

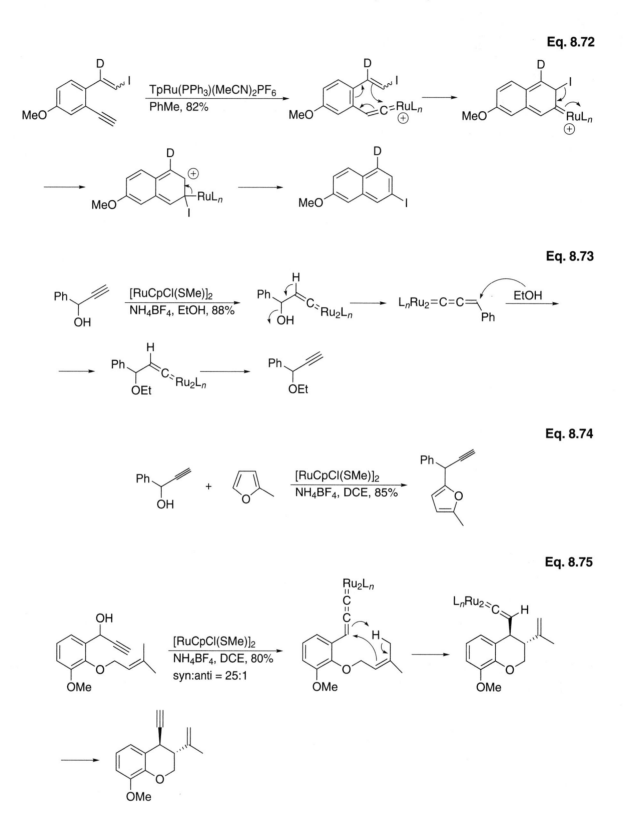

Eq. 8.72

Eq. 8.73

Eq. 8.74

Eq. 8.75

8.3 Stable Alkyne Complexes

a. As Alkyne Protecting Groups

In contrast to most other metals, dicobalt octacarbonyl forms very stable complexes with alkynes, which act as four-electron, bridging ligands, perpendicular to the Co–Co bond (**Eq. 8.76**).[104] This complexation effectively reduces the reactivity of the alkyne to the extent that it can be used to protect alkynes from reduction or hydroboration (**Eq. 8.77**).[105] The alkyne can be regenerated by mild oxidative removal of the cobalt, using cerium ammonium nitrate, trialkylamine oxides, or dimethyl sulphoxide, or by treatment with tetrabutylammonium fluoride in THF (3h, –10°C).[106] Reductive removal of the metal using tributyltin hydride affords the corresponding alkene (**Eq. 8.78**[107]).[108] Alkenyl silanes are formed upon reductive decomplexation using triethylsilane (**Eq. 8.79**).[109] Addition of 1,2-bis(trimethylsilyl)ethyne inhibits alkene isomerization and alcohol silylation by removing cobalt intermediates from the reaction mixture.

Eq. 8.76

Eq. 8.77

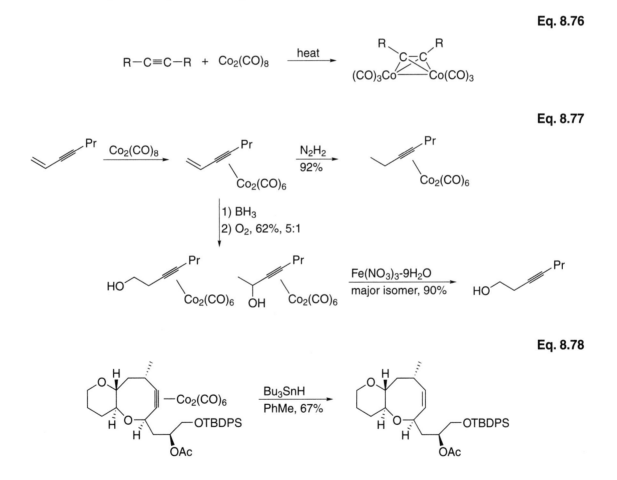

Eq. 8.78

Eq. 8.79

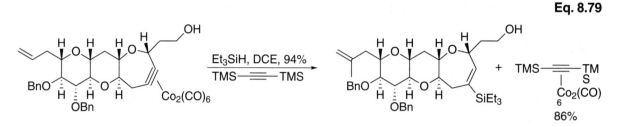

Complexation of an alkyne to cobalt carbonyl also distorts its geometry towards that of an alkene with an approximate angle of 140°![110] This distortion away from linearity has been used to facilitate otherwise reluctant reactions, such as ring-closing metathesis (**Eq. 8.80**)[111] and [4+2] cycloadditions (**Eq. 8.81**).[112]

Eq. 8.80

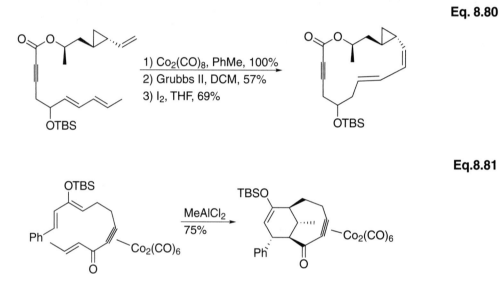

Eq.8.81

The complexation of alkynes to cobalt also stabilizes positive charge in the propargylic position, permitting synthetically useful reactions with propargyl cations (the Nicholas reaction) without the production of allenic byproducts (**Eq. 8.82**).[113] The cobalt-stabilized propargyl cations can be made from a variety of precursors, including propargyl alcohols, ethers, epoxides, and even enynes by treatment with a Lewis acid. These undergo clean reactions with a variety of nucleophiles[114] to give the propargyl substitution product.

Eq. 8.82

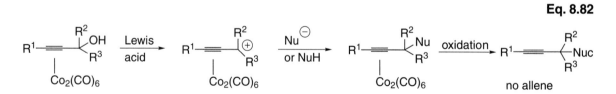

The distortion of the alkyne away from linearity allows the formation of relatively small cyclic alkynes via the Nicholas reaction (**Eq. 8.83**[115]). This process has also found extensive use in the synthesis of ene–diyne antitumor agents (**Eq. 8.84**),[116] and to promote more complex cyclizations (**Eqs. 8.85,**[117,118] **8.86,**[119] **8.87,**[120] and **8.88**[121]).

Eq. 8.83

Eq. 8.84

Eq. 8.85

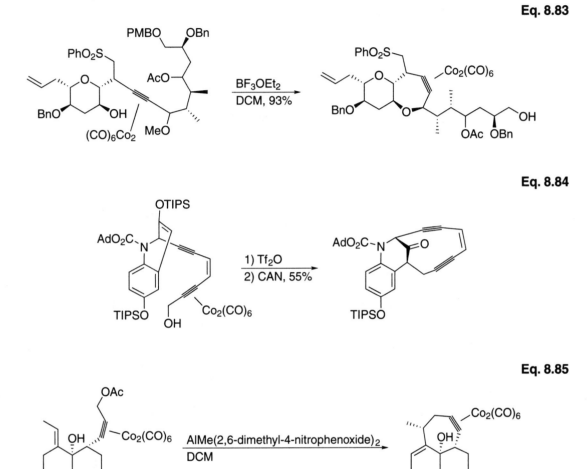

Eq. 8.86

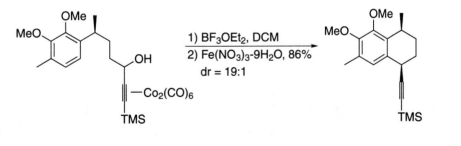

Eq. 8.87

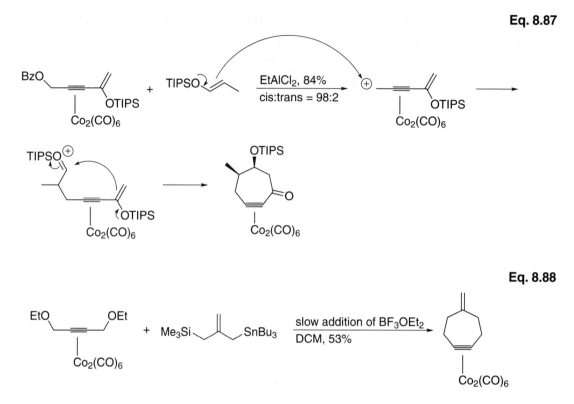

Eq. 8.88

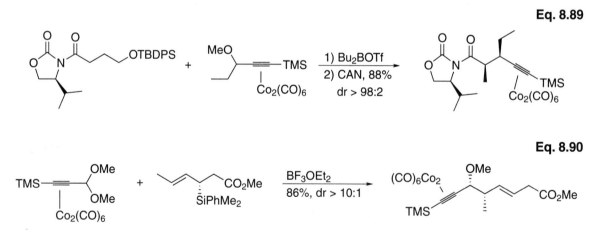

With chiral boron enolates, high enantiomeric excesses can be obtained (**Eq. 8.89**[122]).[123] Cobalt-complexed propargylic acetals can be utilized in asymmetric reactions with allyl silanes (**Eq. 8.90**).[124] Little or no diastereoselectivity was observed for the uncomplexed alkyne.[125]

Eq. 8.89

Eq. 8.90

Another use of alkyne–cobalt complexes is in the stereoselective aldol coupling of alkynyl aldehydes (**Eq. 8.91**[126]). Although free alkynyl aldehydes undergo aldol reactions with silyl enol ethers with little stereoselectivity, the complexed aldehydes react with high syn selectivity.[127] Very high and opposite syn:anti selectivity is observed from reactions of ketene O,S-acetals with complexed and uncomplexed aldehydes

(**Eq. 8.92**).[128] With optically active allylboranes, very high enantioselectivity as well as diastereoselectivity is observed and the cobalt can be readily removed by mild oxidation, giving the aldol product in good yield (**Eq. 8.93**).[129] A highly enantioselective alkylation using diethylzinc in the presence of a chiral ligand was used in a synthesis of (+)-incrustoporin (**Eq. 8.94**).[130]

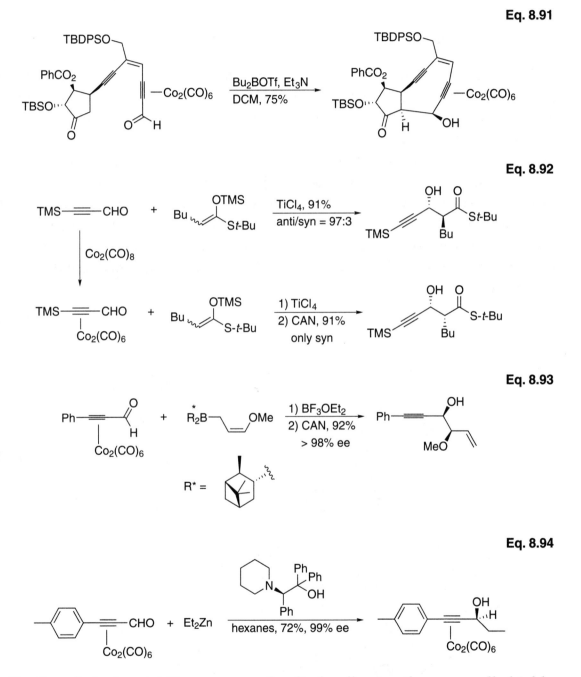

Eq. 8.91

Eq. 8.92

Eq. 8.93

Eq. 8.94

Finally, cobalt also stabilizes propargyl radicals, allowing these normally highly reactive species to be used in synthesis (**Eq. 8.95**).[131]

Eq. 8.95

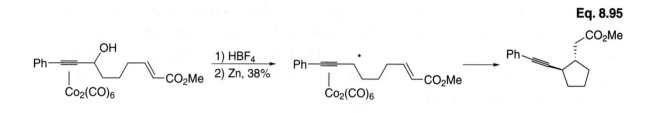

b. The Pauson–Khand Reaction[132,133]

Although dicobalt complexes of alkynes are quite stable to a wide variety of electrophilic and nucleophilic reagents, when they are heated in the presence of an alkene, an interesting and useful "2+2+1" cycloaddition (the Pauson–Khand reaction) occurs, producing cyclopentenones (**Eq. 8.96**).[134] The reaction joins an alkyne, an alkene, and carbon monoxide in a regioselective manner, tending to place alkene substituents next to the CO and large alkyne substituents adjacent to the CO, although there are many exceptions. The mechanism of this process has been studied in some detail,[135] and a likely one, consistent with the products formed, is shown in **Eq. 8.97** and involves the (by now) expected loss of CO to generate a vacant coordination site, coordination of the alkene,[136] insertion of the alkene into a Co–C bond, insertion of CO, and reductive elimination.

Eq. 8.96

Eq. 8.97

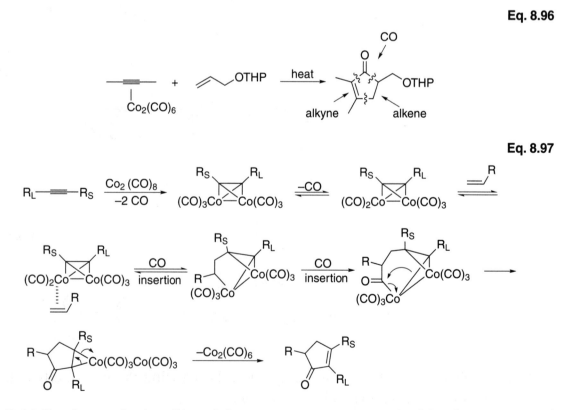

Initially, the synthetic utility of the process was compromised by the necessity of severe conditions and the resulting low yields. However, the reaction can be dramatically accelerated by the addition of Lewis base promoters, such as tertiary amine oxides,[137] amines,[138] thioethers,[139] alcohols, ethers, water, and phosphine oxides, and the yields increased by immobilizing the system on polymers or on silica gel.[140] Calculations suggest that the Lewis base inhibits the reversibility of the alkene insertion step.[141]

Although many intermolecular Pauson–Khand reactions have been reported (**Eq. 8.98**[142]), intramolecular versions are more synthetically interesting, and have been used to synthesize a variety of bicyclic systems (**Eqs. 8.99**[143] and **8.100**[144]). With chiral substrates, high diastereoselectivity can be achieved (**Eqs. 8.101,**[145] **8.102,**[146] and **8.103**[147]). High diastereoselectivity was also observed in some cases with a chiral auxilliary on the alkyne (chiral alkoxyalkyne or chiral propiolic ester).[148] For a very limited range of reactants (i.e., terminal alkynes, norbornene, or norbornadiene), fair to excellent enantiomeric excess could be obtained utilizing a chiral phosphine ligand on the cobalt.[149,150]

Eq. 8.98

Eq. 8.99

Eq. 8.100

Eq. 8.101

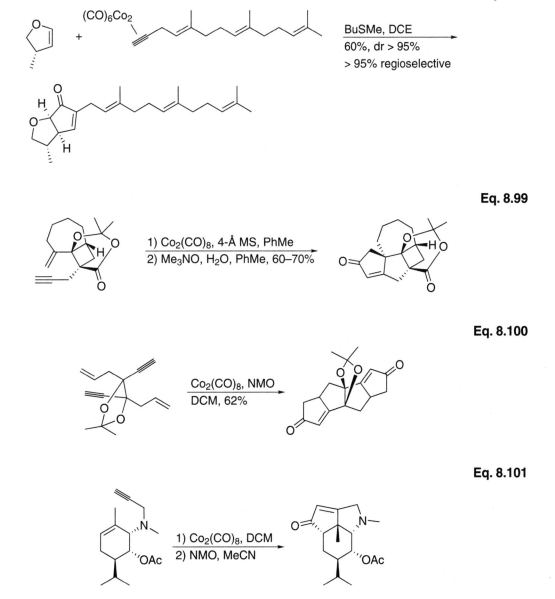

Eq. 8.102

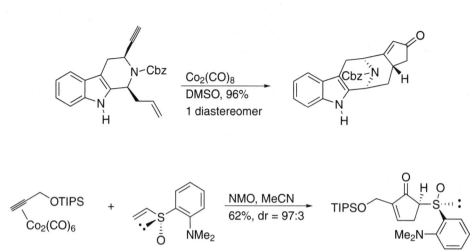

Eq. 8.103

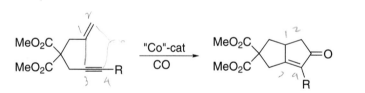

All of the above reactions rely on the use of stoichiometric amounts of dicobalt octacarbonyl. Recently a number of catalytic processes involving the use of catalytic amounts of $Co_2(CO)_8$ under high-intensity visible light irradiation,[151] $Co_2(CO)_8$ in the presence of triphenyl phosphite[152] or tetramethylthiourea and 1 atm of CO pressure,[153] $Co_2(PPh_3)(CO)_7$,[154] or $Co_4(CO)_{12}$ under 10 atm of CO pressure[155] have been developed.[156] All have been carried out on a limited range of substrates, usually that shown in **Eq. 8.104**, and all require carbon monoxide, but efficient catalytic systems for these very interesting transformations are indeed available.

Eq. 8.104

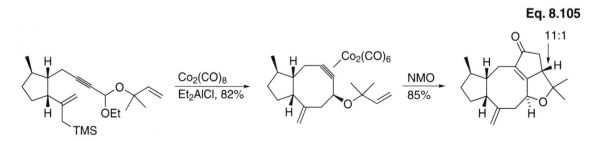

Since stable alkyne–cobalt complexes are involved in the Pauson–Khand reaction, the ability of cobalt complexation to stabilize propargyl cations can be utilized to synthesize precursors to the Pauson–Khand reaction, greatly expanding the scope of the process. Allylsilanes have been used as nucleophiles (**Eq. 8.105**).[157] By taking advantage of the improved reaction conditions (i.e., R_3NO, SiO_2) mentioned earlier, good yields of complex products can be obtained. Homopropargylic acetals can also be used. An example utilizing allylboranes as the nucleophile is shown in **Eq. 8.106**.[158] Note the high stereoselectivity of the allylborane reaction.

Eq. 8.105

Eq. 8.106

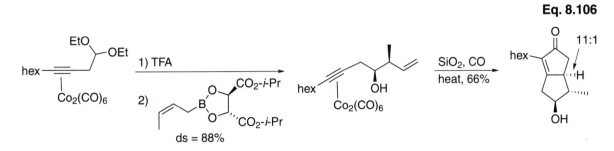

Although intramolecular reactions tend to be quite regioselective, intermolecular reactions of disubstituted olefins often give mixtures of regioisomers. However, regiochemical control can often be regained by having a ligand—usually a heteroatom—in the homoallylic position of the alkene (**Eq. 8.107**).[159] Although a substantial amount of experimental data is available, this process is not completely understood.[160] In addition to controlling the regioselectivity, heteroatom ligands significantly accelerate the reactions (**Eq. 8.108**).[161] Regio- and diastereoselective intermolecular reactions of chiral cyclopropenes can be achieved (**Eq. 8.109**).[162] Regioselective reactions are also observed using electron-deficient alkynes (**Eq. 8.110**).[163] The alkyne polarization was a major factor determining the regioselectivity when electron-deficient alkynes were employed.[164]

Eq. 8.107

Eq. 8.108

Eq. 8.109

Eq. 8.110

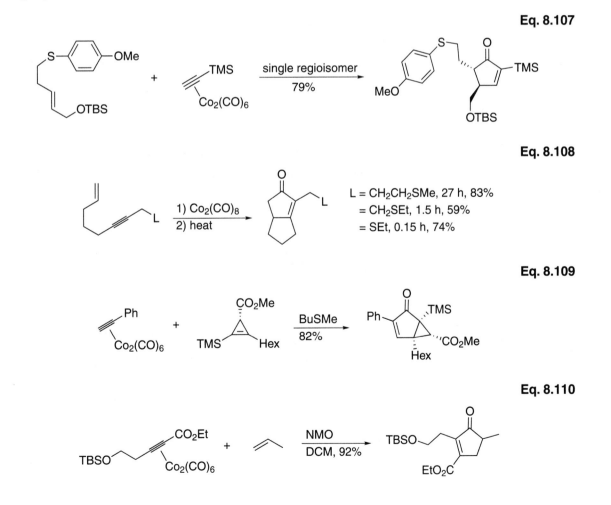

Cobalt carbonyls were initially used in Pauson–Khand reactions, but zirconium, titanium (**Eq. 8.111**),[165] nickel, palladium,[166] iron, iridium (**Eq. 8.112**),[167] molybdenum, tungsten, rhodium,[168] and ruthenium[169,170] catalysts have more recently been developed. Aldehydes can be used as the source of carbon monoxide by an initial metal-catalyzed decarbonylation.[171]

Eq. 8.111

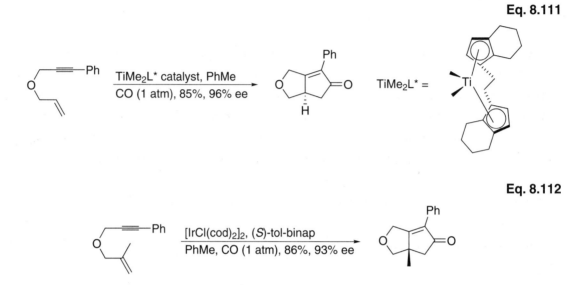

Eq. 8.112

Allene–ynes are interesting substrates for Pauson–Khand type reactions.[172] A variety of transition metal complexes can be used to mediate or catalyze inter- and intramolecular reactions. Both alkenes of the allenic system can participate and the outcome depends on the substituents on the allene and the catalyst employed. A complete reversal of double bond selectivity can be realized by switching catalyst (**Eq. 8.113**)[173]. A few example of the allenic Pauson–Khand reaction are shown in **Eqs. 8.114**[174] and **8.115**.[175] Allene–enes (**Eq. 8.116**),[176] 1,3-diene–ynes (**Eq. 8.117**)[177] and 1,3-diene–allenes (**Eq. 8.118**)[178] undergo rhodium-complex-catalyzed Pauson–Khand-type reactions.

Eq. 8.113

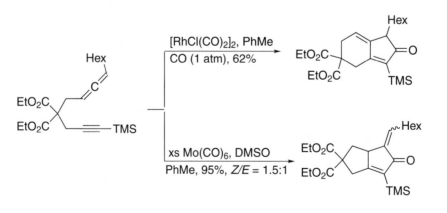

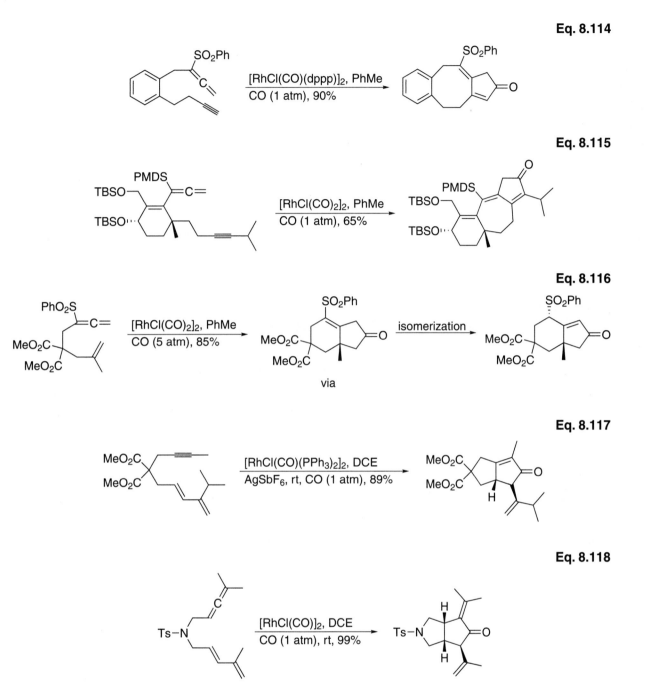

The Pauson–Khand reaction is not limited to carbon–carbon double bonds. Imines,[179] carbodiimides (**Eq. 8.119**),[180] ketones,[181] and aldehydes[182] also participate in cycloaddition with alkynes and carbon monoxide to give either α,β-unsaturated lactams or lactones. Molybdenum hexacarbonyl mediates related reactions, forming α-methylene-γ-butyrolactones from allene-tethered ketones or aldehydes via carbon monoxide insertion (**Eq. 8.120**).[183]

Eq. 8.119

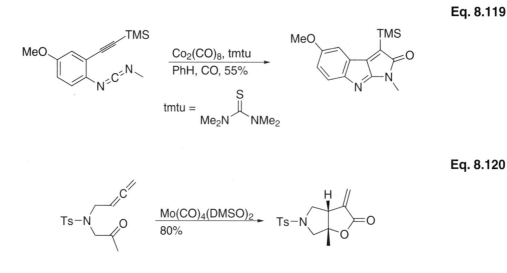

Eq. 8.120

8.4 Metal-Catalyzed Cyclooligomerization of Alkynes

Several transition metals form metallacyclopentadienes from oxidative cyclization with two equivalents of an alkyne (or two different alkynes). These intermediates readily undergo an insertion–reductive elimination sequence with alkynes, alkenes, allenes, nitriles, and isocyanates to give highly functionalized products (**Figure 8.2**).

Although complexes of many transition metals, such as nickel, iridium, rhodium, ruthenium, catalyze the cyclotrimerization of alkynes to arenes, $CpCo(CO)_2$ is among the most efficient. The mechanism of the process has been studied extensively[184] and is shown in **Eq. 8.121**. It involves loss of CO from the catalyst (hence the requirement of relatively high temperatures to generate vacant coordination sites), followed by coor-

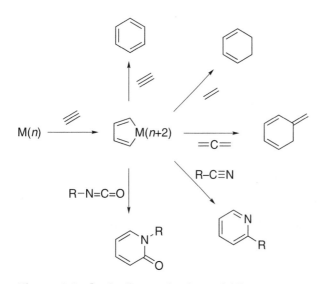

Figure 8.2. Cyclooligomerizations of Alkynes

dination of two alkynes, and formation of the metallacyclopentadiene (a "reductive" coupling of the alkynes and concomitant "oxidation" of the metal). This unsaturated metallacyclopentadiene can then either coordinate and insert another alkyne, and undergo a reductive elimination to produce the arene and regenerate the catalytically active species, or it can undergo a [4+2] cycloaddition, achieving the same result.

Eq. 8.121

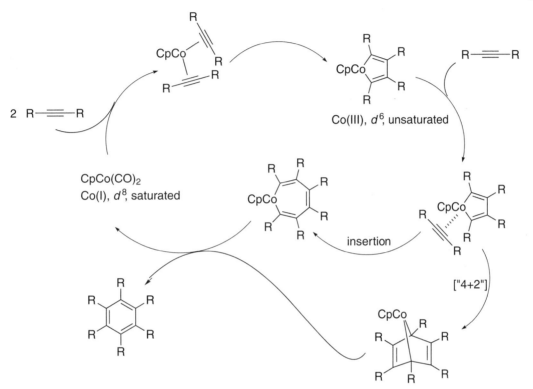

Initially, the process was of little synthetic interest, since it was restricted to symmetrical, internal alkynes. Terminal alkynes gave all possible regioisomers of the trisubstituted benzenes, and "crossed" cyclotrimerization of two different alkynes gave all possible products, with no control.

An elegant solution to this latter problem and the basis of most synthetically useful applications of this methodology was to use a diyne as one component, and a huge alkyne bis-trimethylsilylethyne, as the other.[185] The diyne ensures that the metallacyclopentadiene is formed only from it, since incorporation of the second alkyne unit becomes intramolecular. The bulk of the TMS groups prevents self-trimerization of this member. These two components cleanly cocyclotrimerize to give complex organic molecules in excellent yield (**Eq. 8.122**).[186] Using 1,5-diynes and bis-trimethylsilylethyne affords benzocyclobutenes. As a bonus, these are thermally cleaved to give dienes, which undergo Diels–Alder reactions to form polycyclic compounds. When the dienophile is built into the diyne fragment, complex products are formed expeditiously, as shown by the very direct synthesis of estrone (**Eq. 8.123**)[187] by this methodology. Facile cyclotrimerizations are observed when all three alkynes are contained in the same molecule (**Eq. 8.124**).[188]

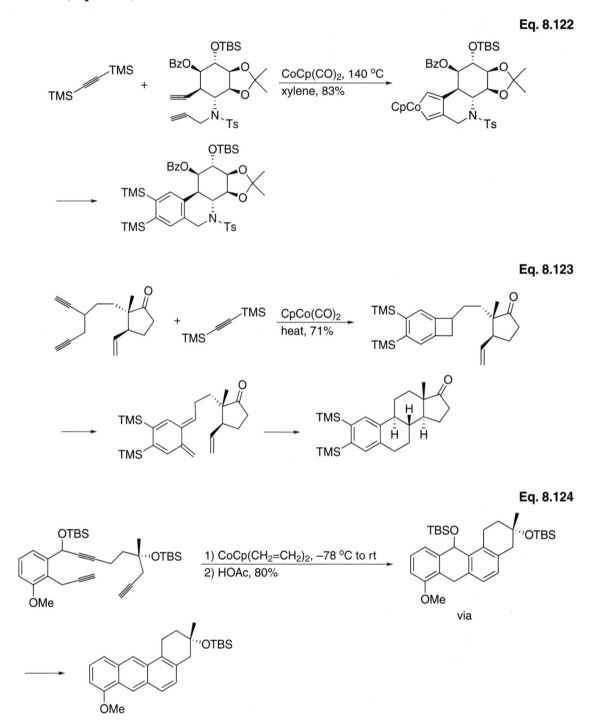

Eq. 8.122

Eq. 8.123

Eq. 8.124

Although CoCp(CO)$_2$ is by far the most extensively studied alkyne cyclotrimerization system catalyst, many other metals will effect the reaction efficiently. Wilkinson's catalyst, (Ph$_3$P)$_3$RhCl, cocyclotrimerizes diynes with alkynes (**Eq. 8.125**),[189] including acetylene itself (**Eq. 8.126**)[190] and asymmetric reactions lead to axially chiral compounds

(**Eq. 8.127**),[191] as do nickel(0) complexes (**Eq. 8.128**[192]).[193] The mode of intermolecular cyclization was modulated by the choice of ligand in iridium(I)-catalyzed reactions (**Eq. 8.129**).[194] Ruthenium(II) complexes can be used as the catalysts (**Eq. 8.130**).[195] Palladium(0) complexes catalyze the cyclotriomerization of alkynes (**Eq. 8.131**).[196]

Eq. 8.125

Eq. 8.126

Eq. 8.127

Eq. 8.128

Eq. 8.129

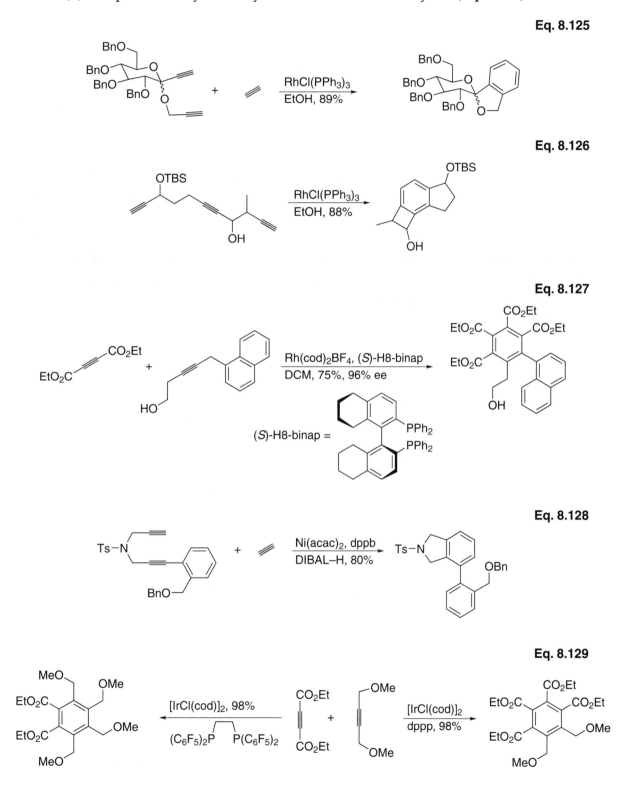

Eq. 8.130

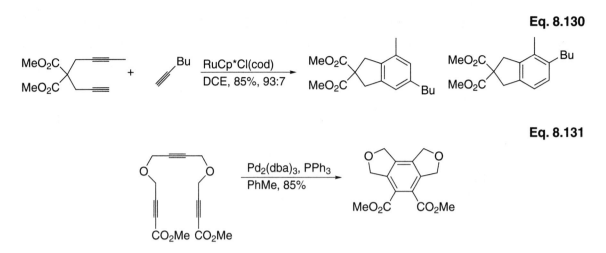

Eq. 8.131

The nitrile group is another triple-bonded species that does not self-trimerize, but can be cotrimerized with alkynes to give pyridines.[197] As with simple alkyne cyclotrimerization, unsymmetric alkynes give all possible regioisomers. However, cocyclotrimerization with a diyne proceeds cleanly to a single pyridine (**Eq. 8.132**[198]).[199] The process also works for simple alkynes (**Eqs. 8.133**,[200,201] **8.134**,[202] and **8.135**[203]). Note the asymmetric induction observed in the latter example. Ruthenium complexes also catalyze this reaction (**Eq. 8.136**).[204]

Eq. 8.132

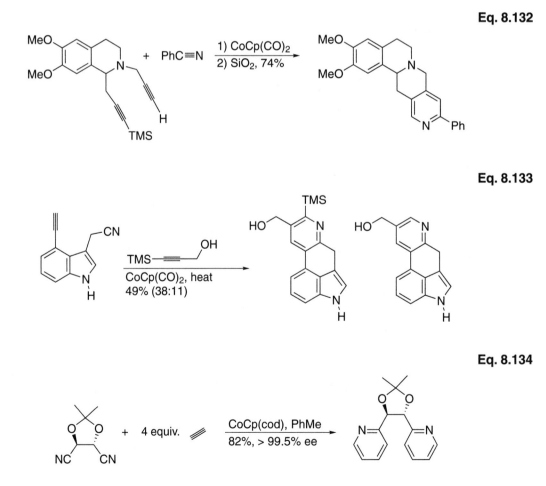

Eq. 8.133

Eq. 8.134

Eq. 8.135

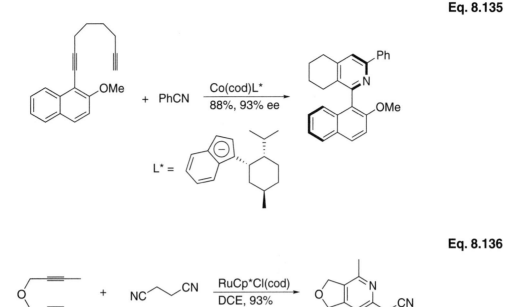

Eq. 8.136

Alkenes will cocyclotrimerize with alkynes, provided homotrimerization of the alkyne can be suppressed. This can be achieved by preforming the metallacyclopentadiene, and then adding the alkene (**Eq. 8.137**),[205] or by carrying the reaction out partially[206] (**Eq. 8.138**)[207] or completely (**Eq. 8.139**)[208] intramolecularly.[209] A cobalt-complexed diene is isolated from these reactions. The metal can easily be removed by oxidation with $Fe(NO)_3$ or by using SiO_2. Allenes undergo cyclotrimerization reactions with alkynes (**Eq. 8.140**).[210]

Eq. 8.137

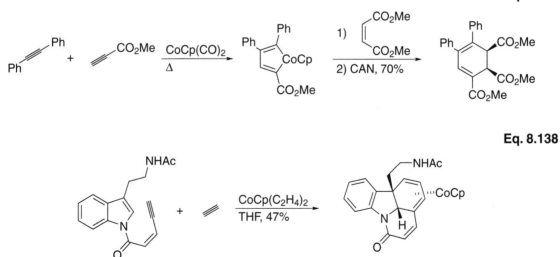

Eq. 8.138

Eq. 8.139

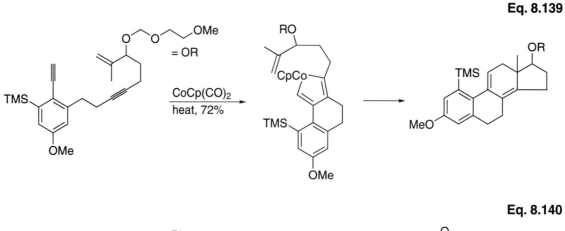

Eq. 8.140

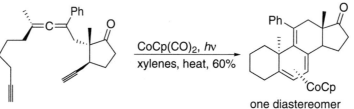

one diastereomer

Ruthenium- and nickel-complex-catalyzed variations on the alkyne–alkyne–alkene cyclotrimerization are shown in **Eqs. 8.141**[211] and **8.142**.[212] Zinc metal is used in **Eq. 8.142** to reduce nickel(II) to nickel(0). Palladium(0) complexes catalyze the cocyclotrimerization of enynes with diynes (**Eq. 8.143**).[213,214] A better yield was obtained using PPh$_3$ and P(o-tol)$_3$. The mechanism for this reaction is unclear.[215]

Eq. 8.141

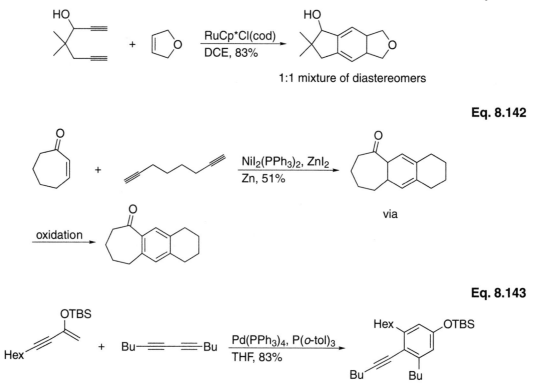

1:1 mixture of diastereomers

Eq. 8.142

Eq. 8.143

Isocyanates can fulfill the role of the double-bonded species in these cocyclotrimerizations, resulting in the production of pyridones (**Eqs. 8.144**[216] and **8.145**[217]).[218] Stable metallacycles derived from one alkyne and one isocyanate can be obtained from nickel(0) complexes (**Eq. 8.146**).[219] Depending on the stoichiometry, another alkyne can then insert to give a pyridone, or another isocyanate can insert to give a pyrimidine. This metallacycle can also react with CO or acids to give imides or conjugated amides. A number of compounds having heteroatom–carbon double bonds participate in ruthenium-catalyzed cyclotrimerizations (**Eq. 8.147**).[220]

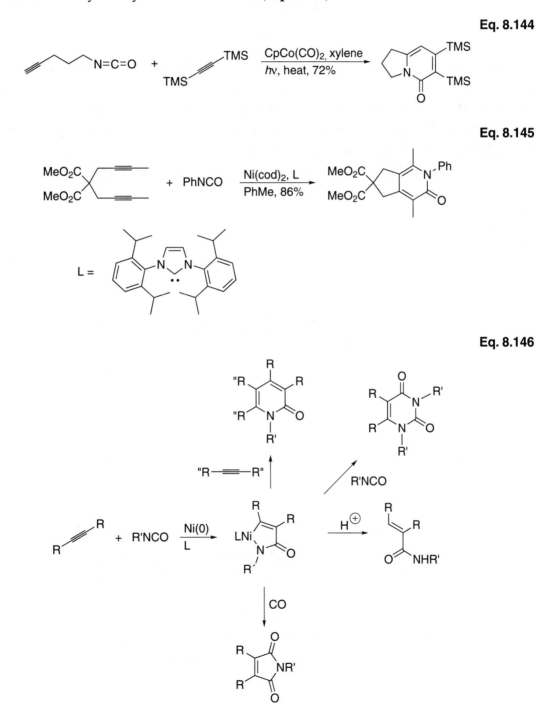

Eq. 8.144

Eq. 8.145

Eq. 8.146

Eq. 8.147

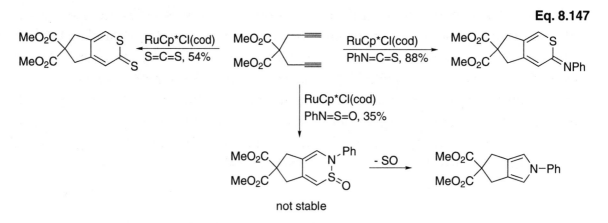

Nickel(0) complexes also "reductively couple" alkynes with carbon dioxide to give stable metallacycles. These can insert another alkyne, a molecule of carbon monoxide, or undergo protolytic cleavage (**Eq. 8.148**).[221] Catalytic reactions are also feasible (**Eq. 8.149**).[222] This process has not yet been used to synthesize complex molecules. Nickel(0) complexes mediate an alkyne–alkene–carbon dioxide coupling, probably via the formation of a metallcyclopentene intermediate (**Eq. 8.150**).[223]

Eq. 8.148

Eq. 8.149

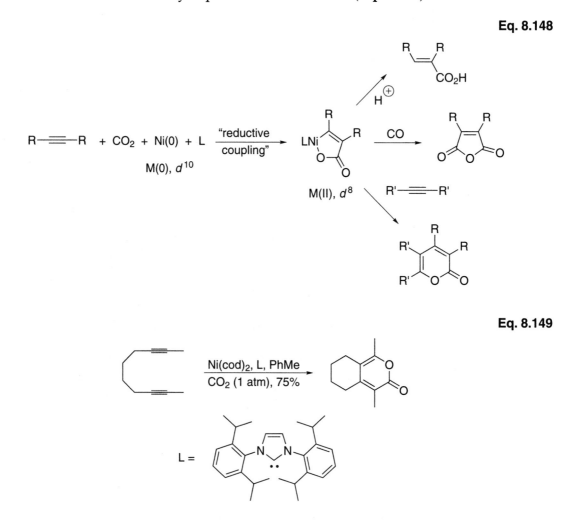

Eq. 8.150

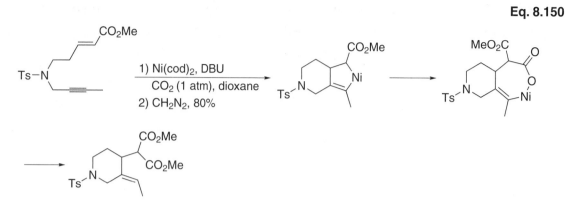

Alkynes react with many transition metals in the presence of carbon monoxide to produce compounds containing both carbon monoxide and alkyne-derived fragments. In addition, a profusion of organometallic complexes containing ligands constructed of alkyne-derived fragments and carbon monoxide are often obtained.[224] Reactions of this type are usually quite complex, and few have been controlled sufficiently to be of use in the synthesis of organic compounds. Quinones have long been known to be one of the products of the reaction of alkynes with carbon monoxide, and evidence for the intermediacy of maleoyl complexes was strong (**Eq. 8.151**). These processes have found little use in organic synthesis because a myriad of other organic and organometallic products formed as well. However, the reactions are potentially useful and will likely see future development.

Eq. 8.151

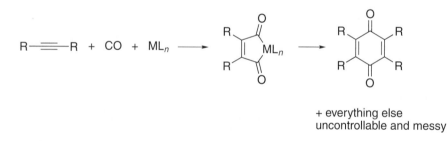

+ everything else
uncontrollable and messy

By devising an efficient synthesis of maleoyl and phthaloyl complexes, and permitting these discrete complexes to react with alkynes, quinones can be made efficiently and in high yield. The most efficient phthaloyl complex is that of cobalt, prepared by the reaction of benzocyclobutanediones with $CoCl(PPh_3)_3$.[225] The initially formed bisphosphine–cobalt(III) complex is unreactive toward alkynes, since a labile site axial to the phthaloyl plane is required for alkyne complexation. However, treatment with silver(I) to produce the unsaturated cationic complex[226] or better, with dimethylglyoxime in pyridine to produce an octahedral complex with a labile cis pyridine ligand (X-ray), leads to active phthaloyl complexes that are efficiently converted to naphthoquinones upon reaction with a variety of alkynes (**Eq. 8.152**). In a similar manner, benzoquinones are efficiently synthesized from cyclobutenediones (**Eqs. 8.153**[227] and

8.154[228]). With unsymmetric systems, mixtures of regioisomers are often obtained, but significantly better selectivity is observed in the presence of a Lewis acid.[229]

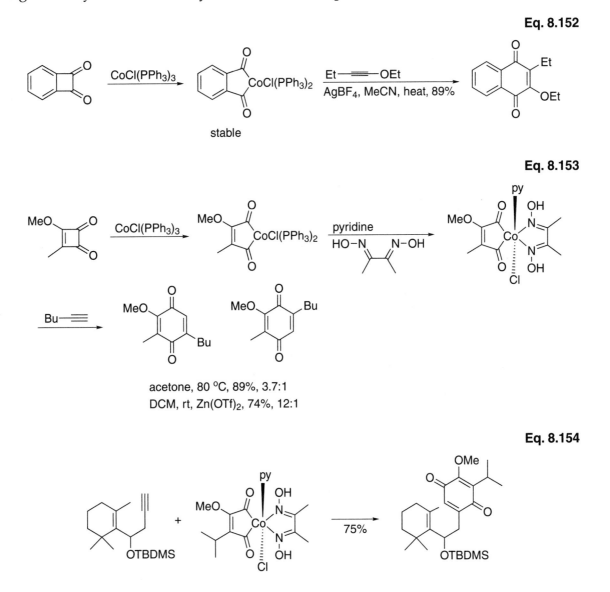

Eq. 8.152

stable

Eq. 8.153

acetone, 80 °C, 89%, 3.7:1
DCM, rt, Zn(OTf)$_2$, 74%, 12:1

Eq. 8.154

75%

8.5 Zirconium- and Titanium-Complex-Mediated Reactions[230]

All of the alkyne chemistry discussed above involves late transition metals, with their normal manifold of reactions. However, the early transition metals also have a rich chemistry with alkynes (see, for example, Chapter 4). Hexasubstituted arenes are

available by stepwise zirconium-assisted cyclotrimerization of three different alkynes (**Eq. 8.155**).[231] This stepwise zirconium-complex-mediated methodology can also be applied to make pyridines (**Eq. 8.156**).[232]

Eq. 8.155

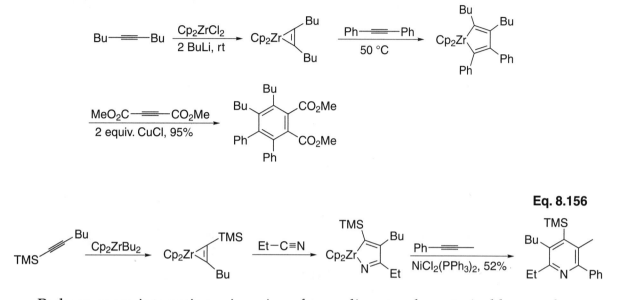

Eq. 8.156

Perhaps more interesting, zirconium forms discrete, characterizable complexes of *benzyne*, produced by the treatment of aryllithium reagents with $Cp_2Zr(Me)Cl$, followed by elimination of methane (**Eq. 8.157**). These reactive benzyne complexes undergo a range of insertion reactions, which ultimately result in functionalization of the arene ring. When unsymmetric aryne complexes are used, these insertions are not regiospecific, but insertion into the sterically less-hindered Zr–C bond is favored. Thus, isonitriles insert to give azazirconacyclopentadienes that can be cleaved to aryl ketones or iodoaryl ketones (**Eq. 8.158**).[233] Alkynes insert to give zirconacyclopentadienes, which are converted to benzothiophenes by treatment with sulfur dichloride (**Eq. 8.159**).[234] Alkenes also insert to give zirconazacyclopentenes, which can be converted to a variety of substituted systems by cleavage with electrophiles (**Eq. 8.160**).[235]

Eq. 8.157

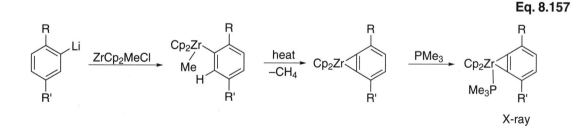

X-ray

Eq. 8.158

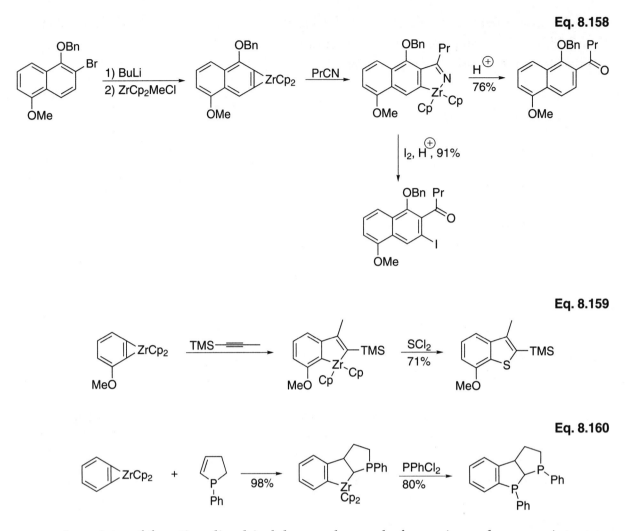

Eq. 8.159

Eq. 8.160

A variety of functionalized indoles can be made from zirconabenzyne intermediates (**Eq. 8.161**)[236] utilizing similar insertion chemistry. An intramolecular alkene insertion was utilized in a synthesis of a CC-1065 analog (**Eq. 8.162**).[237]

Eq. 8.161

Eq. 8.162

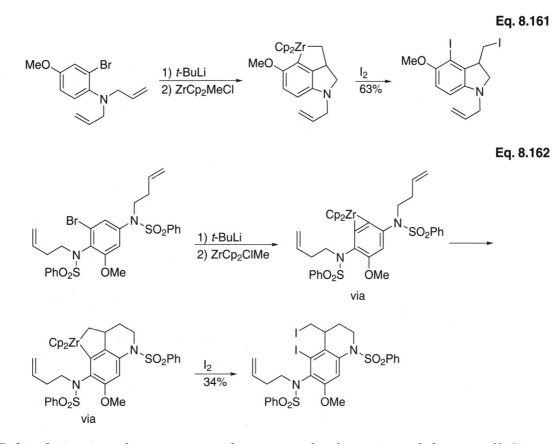

Related titanium–benzyne complexes can also be prepared from aryllithium reagents and TiCp$_2$ClMe and further reacted with alkynes, alkenes, ketones, and nitriles (**Eq. 8.163**).[238]

Eq. 8.163

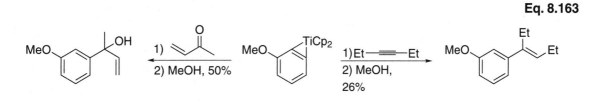

References

(1) a) Reger, D. L.; Belmore, K. A.; Mintz, E.; McElligot, P. J. *Organometallics* **1984**, *3*, 134. b) Reger, D. L.; Mintz, E. *Organometallics* **1984**, *3*, 1759.

(2) For reviews, see a) Alonso, F.; Beletskaya, I. P.; Yus, M. *Chem. Rev.* **2004**, *104*, 3079. b) Zeni, G.; Larock, R. C. *Chem. Rev.* **2004**, *104*, 2285. c) Nakamura, I.; Yamamoto, Y. *Chem. Rev.* **2004**, *104*, 2127.

(3) For a review, see Müller, T. E.; Beller, M. *Chem. Rev.* **1988**, *98*, 675.

(4) Utimoto, K. *Pure Appl. Chem.* **1983**, *55*, 1845.

(5) In a synthesis of furoquinocin D. Saito, T.; Morimoto, M.; Akiyama, C.; Matsumoto, T.; Suzuki, K. *J. Am. Chem. Soc.* **1995**, *117*, 10757.

(6) In a synthesis of peridinin. Furuichi, N.; Hara, H.; Osaki, T.; Nakano, M.; Mori, H.; Katsumura, S. *J. Org. Chem.* **2004**, *69*, 7949.

(7) In a synthesis of (+)-broussonetine G. Trost, B. M.; Horne, D. B.; Woltering, M. J. *Chem.—Eur. J.* **2006**, *12*, 6607.

(8) a) Consorti, C. S.; Ebeling, G.; Dupont, J. *Tetrahedron Lett.* **2002**, *43*, 753. b) See also Norton, J. R.; Shenton, K. E.; Schwartz, *Tetrahedron Lett.* **1975**, 51.

(9) a) Kondo, Y.; Shiga, F.; Murata, N.; Sakamoto, T.; Yamanaka, H. *Tetrahedron* **1994**, *50*, 11803. b) In a synthesis of coumestrol. Hiroya, K.; Suzuki, N.; Yasuhara, A.; Egawa, Y.; Kasano, A.; Sakamoto, T. *J. Chem. Soc., Perkin Trans. 1* **2000**, 4339.

(10) Arcadi, A.; Cacchi, S.; Del Rosario, M.; Fabrizi, G.; Marinelli, F. *J. Org. Chem.* **1996**, *61*, 9280.

(11) Asao, N.; Nogami, T.; Takahashi, K.; Yamamoto, Y. *J. Am. Chem. Soc.* **2002**, *124*, 764.

(12) Miyakoshi, N.; Aburano, D.; Mukai, C. *J. Org. Chem.* **2005**, *70*, 6045.

(13) In a synthesis of SB 242784. Yu, M. S.; Lopez de Leon, L.; McGuire, M. A.; Botha, G. *Tetrahedron Lett.* **1998**, *39*, 9347.

(14) For a review, see Battistuzzi, G.; Cacchi, S.; Fabrizi, G. *Eur. J. Org. Chem.* **2002**, 2671.

(15) For related iridium-complex-catalyzed reactions forming indoles, benzo[*b*]furans, isocoumarins, and isochromenes, see Li, X.; Chianese, A. R.; Vogel, T.; Crabtree, R. H. *Org. Lett.* **2005**, *7*, 5437.

(16) Kondo, Y.; Shiga, F.; Murata, N.; Sakamoto, T.; Yamanaka, H. *Tetrahedron* **1994**, *50*, 11803.

(17) Flynn, B. L.; Hamel, E.; Jung, M. K. *J. Med. Chem.* **2002**, *45*, 2670.

(18) Arcadi, A.; Cacchi, S.; Marinelli, F. *Tetrahedron Lett.* **1992**, *33*, 3915.

(19) Wolf, L. B.; Tjen, K. C. M. F.; ten Brink, H. T.; Blaauw, R. H.; Hiemstra, H.; Schoemaker, H. E.; Rutjes, F. P. J. T. *Adv. Synth. Cat.* **2002**, *344*, 70.

(20) Yu, Y.; Stephenson, G. A.; Mitchell, D. *Tetrahedron Lett.* **2006**, *47*, 3811.

(21) Dai, G.; Larock, R. C. *Org. Lett.* **2001**, *3*, 4035.

(22) Dai, G.; Larock, R. C. *J. Org. Chem.* **2002**, *67*, 7042.

(23) Huang, Q.; Larock, R. C. *J. Org. Chem.* **2003**, *68*, 980.

(24) For an example, see Shimada, T.; Yamamoto, Y. *J. Am. Chem. Soc.* **2002**, *124*, 12670.

(25) For reviews, see a) Bruneau, C.; Dixneuf, P. H. *Angew. Chem. Int. Ed.* **2006**, *45*, 2176. b) Beller, M.; Seayad, J.; Tillack, A.; Jiao, H. *Angew. Chem. Int. Ed.* **2004**, *43*, 3368.

(26) For mechanistic discussions, see Alonso, F.; Beletskaya, I. P.; Yus, M. *Chem. Rev.* **2004**, *104*, 3079.

(27) Doucet, H.; Martin-Vaca, B.; Bruneau, C.; Dixneuf, P. H. *J. Org. Chem.* **1995**, *60*, 7247.

(28) Hua, R.; Tian, X. *J. Org. Chem.* **2004**, *69*, 5782.

(29) Mitsudo, T.-a.; Hori, Y.; Yamakawa, Y.; Watanabe, Y. *J. Org. Chem.* **1987**, *52*, 2230.

(30) Reactions forming indoles, benzo[*b*]furans, and enol lactones via alkenylidene intermediates. Trost, B. M.; McClory, A. *Angew. Chem. Int. Ed.* **2007**, *46*, 2047.

(31) Trost, B. M.; Rhee, Y. H. *J. Am. Chem. Soc.* **2002**, *124*, 2528.

(32) For a review, see Trost B. M.; Fredricksen, M. U.; Rudd, M. T.; *Angew. Chem. Int. Ed.* **2005**, *44*, 6630.

(33) In a synthesis of rosefuran. Trost, B. M.; Flygare, J. A. *J. Org. Chem.* **1994**, *59*, 1078.

(34) Mamane, V.; Hannen, P.; Fürstner, A. *Chem.—Eur. J.* **2004**, *10*, 4556.

(35) Seregin, I. V.; Gevorgyan, V. *J. Am. Chem. Soc.* **2006**, *128*, 12050.

(36) Bäckvall, J.-E.; Nilsson, Y. I. M.; Gatti, R. P. G. *Organometallics* **1995**, *14*, 4242.

(37) Li, J.-H.; Tang, S.; Xie, Y.-X. *J. Org. Chem.* **2005**, *70*, 477.

(38) a) Wang, Z.; Lu, X. *Tetrahedron Lett.* **1997**, *38*, 5213. b) For a review, see Lu, X.; Zhu, G.; Wang, Z.; Ma, S.; Ji, J.; Zhang, Z. *Pure Appl. Chem.* **1997**, *69*, 553.

(39) Ma, S.; Wu, B.; Jiang, X.; Zhao, S. *J. Org. Chem.* **2005**, *70*, 2568.

(40) Ma, S.; Wu, B.; Jiang, X. *J. Org. Chem.* **2005**, *70*, 2588.

(41) In a synthesis of tetrahydrodicranenone. B. Trost, B. M.; Pinkerton, A. B. *J. Org. Chem.* **2002**, *66*, 7714.

(42) Trost, B. M.; Pinkerton, A. B. *J. Am. Chem. Soc.* **2000**, *122*, 8081.

(43) Tong, X.; Li, D.; Zhang, Z.; Zhang, X. *J. Am. Chem. Soc.* **2004**, *126*, 7601.

(44) Kokubo, K.; Matsumasa, K.; Miura, M.; Nomura, M. *J. Org. Chem.* **1996**, *61*, 6941.

(45) Hua, R.; Shimada, S.; Tanaka, M. *J. Am. Chem. Soc.* **1998**, *120*, 12365.

(46) Hua, R.; Onozawa, S.-y.; Tanaka, M. *Chem.—Eur. J.* **2005**, *11*, 3621.

(47) For a review of gold-catalyzed hydroaminations of carbon–carbon multiple bonds, see Widenhoefer, R. A.; Han, X. *Eur. J. Org. Chem.* **2006**, *20*, 4555.

(48) Antoniotti, S.; Genin, E.; Michelet, V.; Genet, J.-P. *J. Am. Chem. Soc.* **2005**, *127*, 9976.

(49) In a synthesis of solenopsin A. Fukuda, Y.; Utimoto, K. *Synthesis* **1991**, 975.

(50) Genin, E.; Toullec, P. Y.; Antoniotti, S.; Brancour, C.; Genet, J.-P. Michelet, M. *J. Am. Chem.* **2006**, *128*, 3112.

(51) a) In a synthesis of (+)-ochromycinone and (+)-rubiginone B_2. Sato, K.; Asao, N.; Yamamoto, Y. *J. Org. Chem.* **2005**, *70*, 8977. b) See also Asao, N.; Nogami, T.; Lee, S. Yamamoto, Y. *J. Am. Chem. Soc.* **2003**, *125*, 10921.

(52) a) Gold-catalyzed: Yao, T.; Zhang, X.; Larock, R. C. *J. Am. Chem. Soc.* **2004**, *126*, 11164. b) Yao, T.; Zhang, X.; Larock, R. C. *J. Org. Chem.* **2005**, *70*, 7679. c) Platinum-catalyzed: Oh, C. H.; Reddy, V. R.; Kim, A.;

Rhim, C. Y. *Tetrahedron Lett.* **2006**, *47*, 5307. d) Copper-catalyzed: Patil, N. T.; Wu, H.; Yamamoto, Y. *J. Org. Chem.* **2005**, *70*, 4531.

(53) Binder, J. T.; Crone, B.; Kirsch, S. F.; Liebert, C.; Manz, H. *Eur. J. Org. Chem.* **2007**, 1636.

(54) For reviews, see a) Bandini, M.; Emer, E.; Tommasi, S.; Umani-Ronchi, A. *Eur. J. Org. Chem.* **2006**, 3527. b) Nevado, C.; Echavarren, A. M. *Synthesis* **2005**, 167.

(55) Tunge, J. A.; Foresse, L. N. *Organometallics* **2005**, *24*, 6440.

(56) Soriano, E.; Marco-Contelles, J. *Organometallics* **2006**, *25*, 4542.

(57) In a synthesis of 6-doexyxclitoriacetal. Khorpheung, P.; Tummatorn, J.; Petsom, A.; Taylor, R. J. K.; Roengsumran, S. *Tetrahedron Lett.* **2006**, *47*, 5989.

(58) Ferrer, C.; Amjis, C. H. M.; Echavarren, A. M. *Chem.—Eur. J.* **2007**, *13*, 1358.

(59) Hashmi, A. S. K.; Blanco, M. C. *Eur. J. Org. Chem.* **2006**, 4340.

(60) Hashmi, A. S. K.; Frost, T. M.; Bats, J. W. *J. Am. Chem. Soc.* **2000**, *122*, 11553.

(61) Hashmi, A. S. K.; Frost, T. M.; Bats, J. W. *Org. Lett.* **2001**, *3*, 3769.

(62) Martin-Matute, B.; Nevado, C.; Cardenas, D. J.; Echvarren, A. M. *J. Am. Chem. Soc.* **2003**, *125*, 5757.

(63) For a theoretical study of gold-catalyzed reactions, see Rabaa, H.; Engels, B.; Hupp, T.; Hashmi, A. S. K. *Int. J. Quantum Chem.* **2007**, *107*, 359.

(64) Hashmi, A. S. K.; Rudolph, M.; Weyerauch, J. P.; Woelfle, M.; Frey, W.; Bats, J. W. *Angew. Chem. Int. Ed.* **2005**, *44*, 2788.

(65) Kennedy-Smith, J. J.; Staben, S. T.; Toste, F. D. *J. Am. Chem. Soc.* **2004**, *126*, 4526.

(66) For a nickel-complex-catalyzed reaction, see Gao, Q.; Zheng, B.-F.; Li, J.-H.; Yang, D. *Org. Lett.* **2005**, *7*, 2185.

(67) For a rhenium-complex-catalyzed reaction, see Kuninobu, Y.; Kawata, A.; Takai, K. *Org. Lett.* **2005**, *7*, 4823.

(68) Corkey, B. K.; Toste, F. D. *J. Am. Chem. Soc.* **2006**, *127*, 17168.

(69) In a synthesis of (+)-lycopladine A. Staben, S. T.; Kennedy-Smith, J. J.; Huang, D.; Corkney, B. K.; LaLonde, R. L.; Toste, D. F. *Angew. Chem. Int. Ed.* **2006**, *45*, 5991.

(70) Harrison, T. J.; Patrick, B. O.; Dake, G. R. *Org. Lett.* **2007**, *9*, 367.

(71) Kusama. H.; Yamabe, H.; Onizawa, Y.; Hoshino, T.; Iwasawa, N. *Angew. Chem. Int. Ed.* **2005**, *44*, 468.

(72) Markham, J. P.; Staben, S. T.; Toste, D. F. *J. Am. Chem. Soc.* **2005**, *127*, 9708.

(73) For reviews on gold- and platinum-catalyzed reactions, see a) Fürstner, A.; Davies, P. W. *Angew. Chem. Int. Ed.* **2007**, *46*, 3410. b) Gorin, D. J.; Toste, F. D. *Nature* **2007**, 395. c) Jiminez-Nunea, E.; Echavarren, A. M. *Chem. Commun.* **2007**, 333. d) Zhang, L.; Sun, J.; Kozmin, S. A. *Adv. Synth. Catal.* **2006**, *348*, 2271. e) Hashmi, A. S. K.; Hutchings, G. J. *Angew. Chem. Int. Ed.* **2006**, *45*, 7896. f) Ma, S.; Yu, S.; Gu, Z. *Angew. Chem. Int. Ed.* **2006**, *45*, 200.

(74) Nakanishi, W.; Yamanaka, M.; Nakamura, E. *J. Am. Chem. Soc.* **2005**, *127*, 1446.

(75) For a theoretical study, see Komiya, S.; Albright, T. A.; Hoffmann, R.; Kochi, J. K. *J. Am. Chem. Soc.* **1976**, *98*, 7255.

(76) For a review of the mechanistic aspects, see Lloyd-Jones, G. C. *Org. Biomolec. Chem.* **2003**, *1*, 215.

(77) For a theoretical study using platinum, see Soriano, E.; Marco-Contelles, J. *J. Org. Chem.* **2005**, *70*, 9345.

(78) For an iridium-complex-catalyzed reaction, see Shibata, T.; Kobayashi, Y.; Maekawa, S.; Toshida, N.; Takagi, K. *Tetrahedron* **2006**, *61*, 9018.

(79) Soriano, E.; Ballesteros, P.; Marco-Contelles, J. *J. Org. Chem.* **2004**, *69*, 8018.

(80) For reactions with deuterium- and 13C-labelled substrates, see a) Nieto-Oberhuber, C.; Munoz, M. P.; Lopez, S.; Jiminez-Nunes, E.; Nevado, C.; Herrero-Gomez, Raducan, M.; Echavarren, A. M. *Chem.—Eur. J.* **2006**, *12*, 1677. b) Chatani, N.; Furukawa, N.; Sakurai, H.; Murai, S. *Organometallics* **1996**, *15*, 901. c) Oi, S.; Tsukamoto, I.; Miyano, S.; Inoue, Y. *Organometallics* **2001**, *20*, 3704.

(81) In a synthesis of roseophilin. Trost, B. M.; Doherty, G. A. *J. Am. Chem. Soc.* **2000**, *122*, 3801.

(82) Nieto-Oberhuber, C.; Lopez, S.; Munoz, M. P.; Jiminez-Nunes, E.; Bunuel, E.; Cardenas, D. J.; Echavarren, A. M. *Chem.—Eur. J.* **2006**, *12*, 1694.

(83) Sherry, B. D.; Maus, L.; Laforteza, B. N.; Toste, D. F. *J. Am. Chem. Soc.* **2006**, *128*, 8132.

(84) Toullec, P. Y.; Genin, E.; Leseurre, L.; Genet, J.-P.; Michelet, V. *Angew. Chem. Int. Ed.* **2006**, *45*, 7427.

(85) For a review, see Marco-Contelles, J.; Soriano, E. *Chem.—Eur. J.* **2007**, *13*, 1350.

(86) Soriano, E.; Ballesteros, P.; Marco-Contelles, J. *Organometallics* **2005**, *24*, 3182.

(87) Rautenstrauch, V. *J. Org. Chem.* **1984**, *49*, 950

(88) Shi, X.; Gorin, D. J.; Toste, F. D. *J. Am. Chem. Soc.* **2005**, *127*, 5802.

(89) Miki, K.; Ohe, K.; Uemura, S. *J. Org. Chem.* **2003**, *68*, 8505.

(90) Nishino, F.; Miki, K.; Kato, Y.; Ohe, K.; Uemura, S. *Org. Lett.* **2003**, *5*, 2615.

(91) In a synthesis of (–)-α-cubebene. Fürstner, A.; Hannen, P. *Chem.—Eur. J.* **2006**, *12*, 3006.

(92) Buzas, A.; Gagosz, F. *J. Am. Chem. Soc.* **2006**, *128*, 12614.

(93) Marion, N.; Diez-Gonzalez, S.; de Fremont, P.; Noble, A. P.; Nolan, S. P. *Angew. Chem. Int. Ed.* **2006**, *45*, 3647.

(94) Munoz, M. P.; Adrio, J.; Carretero, J. C.; Echavarren, A. M. *Organometallics* **2005**, *24*, 1293.

(95) Johansson, M. J.; Gorin, D. J.; Staben, S. T.; Toste, F. D. *J. Am. Chem. Soc.* **2005**, *127*, 18002.

(96) Nakamura, I.; Bajracharya, G. B.; Wu, H.; Oishi, K.; Mitzushima, Y.; Gridnev, I. D.; Yamamoto, Y. *J. Am. Chem. Soc.* **2004**, *126*, 15423.

(97) Bajracharya, G. B.; Pahadi, N. K.; Gridnev, I. D.; Yamamoto, Y. *J. Org. Chem.* **2006**, *71*, 6204.

(98) Nakamura, I.; Mizushima, Y.; Yamamoto, Y. *J. Am. Chem. Soc.* **2005**, *127*, 15022.

(99) Kim, H.; Lee, C. *J. Am. Chem. Soc.* **2006**, *128*, 6336.

(100) Shen, H.-C.; Pal, S.; Lian, J.-J.; Liu, R.-S. *J. Am. Chem. Soc.* **2003**, *125*, 15762.

(101) Nishibayashi, Y.; Wakiji, I.; Hidai, M.; Uemura, S. *J. Am. Chem. Soc.* **2000**, *122*, 11019.

(102) Inada, Y.; Yoshikawa, M.; Milton, M. D.; Nishiba-yashi, Y.; Uemura, S. *Eur. J. Org. Chem.* **2006**, 881.

(103) Nishibayashi, Y.; Inada, Y.; Hidai, M.; Uemura, S. *J. Am. Chem. Soc.* **2003**, *125*, 6060.

(104) Dickson, R. S.; Fraser, P. J. *Adv. Organomet. Chem.* **1974**, *12*, 323.

(105) Nicholas, K. M.; Pettit, R. *Tetrahedron Lett.* **1971**, 3475.

(106) Jones, G. B.; Wright, J. M.; Rush, T. M.; Plourok, G. W.; Kelton, T. F.; Mathews, J. E.; Huber, R. S.; Davidson, J. P. *J. Org. Chem.* **1997**, *62*, 9379.

(107) In a synthesis of the HIJK-ring fragment of cigua-toxin. Liu, T.-Z.; Isobe, M. *Tetrahedron* **2000**, *56*, 5391.

(108) Hosokawa, S.; Isobe, M. *Tetrahedron Lett.* **1998**, *39*, 2609.

(109) Kira, K.; Tanda, H.; Hamajima, A.; Baba, T.; Takai, S.; Isobe, M. *Tetrahedron* **2002**, *58*, 6485.

(110) Dickson, R. S.; Fraser, P. J. *Adv. Organomet. Chem.* **1974**, *12*, 323.

(111) In a synthesis of cycloproparadiciol. Yang, Z.-Q.; Danishefsky, S. J. *J. Am. Chem. Soc.* **2003**, *125*, 9602.

(112) Iwasawa, N.; Inaba, K.; Nakayama, S.; Aoki, M. *Angew. Chem. Int. Ed.* **2005**, *44*, 7447.

(113) For reviews, see a) Diaz, D. D.; Betancort, J. M.; Martin, V. S. *Synlett* **2007**, 343. b) Teobald, B. J. *Tetrahedron* **2002**, *58*, 4133. c) Caffyn, A.; Nicholas, K. M. Transition Metal Alkyne Complexes. Transition Metal Stabilized Propargyl Systems. In *Comprehensive Organometallic Chemistry II*; Abel, E. W., Stone, F. G. A., Wilkinson, G., Eds.; Pergamon: Oxford, U.K., 1995; Vol. 12, pp 685–702.

(114) For an extensive study of the reactions of cobalt-stabilized propargyl cations with a broad range of nucleophiles, see Kuhn, O.; Raw, D.; Mayr, H. *J. Am. Chem. Soc.* **1998**, *120*, 900.

(115) In a synthesis of the JKLM-ring fragment of cigua-toxin. Baba, T.; Huang, G.; Isobe, M. *Tetrahedron* **2003**, *59*, 6851.

(116) For a review of these classes of compounds, including a discussion of the Nicholas reaction in this context, see a) Nicolaou, K. C.; Dai, W. M. *Angew. Chem., Int. Ed. Engl.* **1991**, *30*, 1387. b) Magnus, P.; Pitterna, T. *J. Chem. Soc., Chem. Commun.* **1991**, 541. c) Magnus, P.; Miknis, G. F.; Press, N. J.; Grandjean, D.; Taylor, G. M.; Harling, J. *J. Am. Chem. Soc.* **1997**, *119*, 6739. d) Magnus, P.; Eisenbeis, S. A.; Fourhurst, R. H.; Ikadis, T.; Magnus, T. A.; Parry, D. *J. Am. Chem. Soc.* **1997**, *119*, 5591.

(117) In a synthesis of ingenol. Tanino, K.; Onuki, K.; Asano, K.; Miyashita, M.; Nakamura, T.; Takahashi, Y.; Kuwajima, I. *J. Am. Chem. Soc.* **2003**, *125*, 1498.

(118) Nakamura, T.; Matsui, T.; Tamino, K.; Kuwajima, I. *J. Org. Chem.* **1997**, *62*, 3032.

(119) In a synthesis of pseudopterosin G aglycon dimethyl ether. LeBrazidec, J.-Y.; Kocienski, P. J.; Connolly, J. D.; Muir, K. W. *J. Chem. Soc., Perkin Trans. 1* **1998**, 2475.

(120) Tanino, K.; Kondo, F.; Shimizu, T.; Miayshita, M. *Org. Lett.* **2002**, *4*, 2217.

(121) Patel, M. M.; Green, J. R. *Chem. Commun.* **1999**, 509.

(122) In a synthesis of PS-5. Jacobi. P. A.; Murphree, S.; Rupprecht, F.; Zheng, W. *J. Org. Chem.* **1996**, *61*, 2413.

(123) a) Jacobi, P. A.; Buddu, S. C.; Fry, M. D.; Rajeswari, S. *J. Org. Chem.* **1997**, *62*, 2894. b) Jacobi, P. A.; Herradina, P. *Tetrahedron Lett.* **1997**, *38*, 6621.

(124) In a synthesis of cystothiazoles. Shao, J.; Panek, J. S. *Org. Lett.* **2004**, *6*, 3083.

(125) Sui, M.; Panek, J. S. *Org. Lett.* **2001**, *3*, 2439.

(126) In a synthesis of the core of neocarzinostatin and kedarcidin chromophores. Caddick, S.; Delisser, V. M. *Tetrahedron Lett.* **1997**, *38*, 2355.

(127) For a review, see Mukai, C.; Hanaoka, M. *Synlett* **1996**, 11.

(128) In a synthesis of blastomycinone. Mukai, C.; Kataoka, O.; Hanaoka, M. *J. Org. Chem.* **1993**, *58*, 2946.

(129) Ganesh, P.; Nicholas, K. M. *J. Org. Chem.* **1997**, *62*, 1737.

(130) Fontes, M.; Verdaguer, X.; Sola, L.; Vidal-Ferran, A.; Reddy, K. S.; Riera, A.; Pericas, M. A. *Org. Lett.* **2002**, *4*, 2381.

(131) Salazar, K. L.; Khan, M. A.; Nicholas, K. M. *J. Am. Chem. Soc.* **1997**, *119*, 9053.

(132) For reviews, see a) Gibson, S. E.; Mainolfi, N. *Angew. Chem. Int. Ed.* **2005**, *44*, 3022. b) Blanco-Urgoiti, J.; Anorbe, L.; Perez-Serrano, L.; Dominguez, G.; Perez-Castells, J. *Chem. Soc. Rev.* **2004**, *33*, 32. c) Brummond, K. M.; Kent, J. L. *Tetrahedron* **2000**, *56*, 3263. d) Gao, O.; Schmalz, H.-O. *Angew. Chem. Int. Ed.* **1998**, *37*, 911. e) Chung, Y. K. *Coord. Chem. Rev.* **1999**, *188*, 297. f) Schore, N.E. *Org. React.* **1991**, *40*. 1.

(133) For a review of the side reactions in Pauson–Khand reactions, see Bonaga, L. V. R.; Krafft, M. E. *Tetrahedron* **2004**, *60*, 9795.

(134) Billington, D. C.; Pauson, P. L. *Organometallics* **1982**, *1*, 1560.

(135) Gimbert, Y.; Lesage, D.; Milet, A.; Fournier, F.; Greene, A. E.; Tabet, J.-C. *Org. Lett.* **2003**, *5*, 4073.

(136) This type of intermediate has been isolated. Banide, E. V.; Müller-Bunz, H.; Manning, A. R.; Evans, P.; McGlinchey, M. J. *Angew. Chem. Int. Ed.* **2007**, *46*, 2907.

(137) Shambayati, S.; Crowe, W. E.; Schreiber, S. L. *Tetrahedron Lett.* **1990**, *31*, 5289.

(138) Rajesh, T.; Periasamy, M. *Tetrahedron Lett.* **1998**, *39*, 117.

(139) Brown, J. A.; Irvine, S.; Kerr, W. J.; Pearson, C. M. *Org. Biomolec. Chem.* **2005**, *3*, 2396.

(140) a) Smit, W. D.; Kireev, S. L.; Nefedov, O. M.; Tarasov, V. A. *Tetrahedron Lett.* **1989**, *30*, 4021. b) Becker, D. P.; Flynn, T. *Tetrahedron Lett.* **1993**, *34*, 2087.

(141) Perez del Valle, C.; Milet, A.; Gimbert, Y.; Greene, A. E. *Angew. Chem. Int. Ed.* **2005**, *44*, 5717.

(142) In a synthesis of terpesacin. Chan, J.; Jamison, T. F. *J. Am. Chem. Soc.* **2004**, *126*, 10682.

(143) In an approach to ingenol. Winkler, J. D.; Lee, E. C. Y.; Nevels, L. I. *Org. Lett.* **2005**, *7*, 1489.

(144) a) Van Ornum, S. G.; Cook, J. M. *Tetrahedron Lett.* **1997**, *38*, 3657. b) Van Ornum, S. G.; Bruendel, M. M.; Cooke, J. M. *Tetrahedron Lett.* **1998**, *37*, 6649.

(145) In a synthesis of (–)-dendrobine. Cassayre, J.; Zard, S. Z. *J. Organomet. Chem.* **2001**, *624*, 316.

(146) In a synthesis of (–)-alostonerine. Miller, K. A.; Martin, S. F. *Org. Lett.* **2007**, *9*, 1113.

(147) In a synthesis of (–)-pentenomycin I. Rivero, M. R.; Alonso, I.; Carretero, J. C. *Chem.—Eur. J.* **2004**, *10*, 5443.

(148) a) Bernardes, V.; Kam, N.; Riera, A.; Moyano, A.; Pericas, M. A.; Green, A. E. *J. Org. Chem.* **1995**, *60*, 6670. b) Fonquerna, S.; Moyano, A.; Pericas, M. A.; Riera, A. *Tetrahedron* **1995**, *51*, 4639.

(149) Hay, A. M.; Kerr, W. J.; Kirk, G. G.; Middlemiss, D. *Organometallics* **1995**, *14*, 4986.

(150) Verdaguer, X.; Lledo, A.; Lopez-Mosqera, C.; Maestro, M. A.; Pericas, M. A.; Riera, A. *J. Org. Chem.* **2004**, *69*, 8053.

(151) Pagendorf, B. L.; Livinghouse, T. *J. Am. Chem. Soc.* **1996**, *118*, 2285.

(152) Jeong, N.; Hwang, S. H.; Lee, Y.; Chung, Y. K. *J. Am. Chem. Soc.* **1994**, *116*, 3159.

(153) Tang, Y.; Deng, L.; Zhang, Y.; Dong, G.; Chen, J.; Yang, Z. *Org. Lett.* **2005**, *7*, 593.

(154) Gibson, S. E.; Johnstone, C.; Stevenazzi, A. *Tetrahedron* **2002**, *58*, 4937.

(155) Kim, J. W.; Chung, Y. K. *Synthesis* **1998**, 142.

(156) For a review on catalytic reactions, see a) Shibata, T. *Adv. Synth. Catal.* **2006**, *348*, 2328. b) Gibson, S. E.; Stevenazzi, A. *Angew. Chem. Int. Ed.* **2003**, *42*, 1800.

(157) Jameson, T. F.; Shambayati, S.; Crowe, W. E.; Schreiber, S. L. *J. Am. Chem. Soc.* **1997**, *119*, 4353.

(158) Roush, W. R.; Park, J.-C. *Tetrahedron Lett.* **1991**, *32*, 6285.

(159) Krafft, M. E.; Juliano, C. A.; Scott, I. L.; Wright, C.; McEachin, M. P. *J. Am. Chem. Soc.* **1991**, *113*, 1693.

(160) Krafft, M. E.; Juliano, C. A. *J. Org. Chem.* **1992**, *57*, 5106.

(161) Krafft, M. E.; Scott, I. L.; Romulo, R. H.; Feibelmann, S.; Van Pelt, C. E. *J. Am. Chem. Soc.* **1993**, *115*, 7199.

(162) Pallerla, M. K.; Fox, J. M. *Org. Lett.* **2005**, *7*, 3593.

(163) In a synthesis of asteriscanolide. Krafft, M. E.; Cheung, Y. Y.; Abboud, K. A. *J. Org. Chem.* **2001**, *66*, 7443.

(164) de Bruin, T. J. M.; Michel, C.; Vekey, K.; Greene, A. E.; Gimbert, Y.; Milet, A. *J. Organomet. Chem.* **2006**, *691*, 4281.

(165) Hicks, F. A.; Buchwald, S. L. *J. Am. Chem. Soc.* **1999**, *121*, 7026.

(166) Tang, Y.; Deng, L.; Zhang, Y.; Dong, G.; Chen, J.; Yang, Z. *Org. Lett.* **2005**, *7*, 1657.

(167) Shibata, T.; Toshida, N.; Yamasaki, M. Maekawa, S.; Takagi, K. *Tetrahedron* **2005**, *61*, 9974.

(168) Koga, Y.; Kobayashi, T.; Narasaka, K. *Chem. Lett.* **1998**, 249.

(169) Morimoto, T.; Chatani, N.; Fukumoto, Y.; Murai, S. *J. Org. Chem.* **1997**, *62*, 3762.

(170) Kondo, T.; Suzuki, N.; Okada, T.; Mitsudo, T. *J. Am. Chem. Soc.* **1997**, *119*, 6187.

(171) For rhodium-catalyzed reactions, see a) Morimoto, T.; Fuji, K.; Tsutsumi, K.; Kakiuchi, K. *J. Am. Chem. Soc.* **2002**, *124*, 3806. b) Shibata, T.; Toshida, N.; Takagi, K. *J. Org. Chem.* **2002**, *67*, 7446. c) For iridium-catalyzed reactions, see Kwong, F. Y.; Lee, H. W.; Lam, W. H.; Qui, L.; Chan, A. S. C. *Tetrahedron: Asymmetry* **2006**, *17*, 1238.

(172) For a review, see Alcaide, B.; Almendros, P. *Eur. J. Org. Chem.* **2004**, 3377.

(173) Brummond, K. M.; Chen, H.; Fisher, K. D.; Kerekes, A. D.; Rickards, B.; Sill, P. C.; Geib, S. J. *Org. Lett.* **2002**, *4*, 1931.

(174) Mukai, C.; Hirose, T.; Teramoto, S.; Kitagaki, S. *Tetrahedron* **2005**, *61*, 10983.

(175) In a synthesis of the core of guanacasterpene A. Brummond, K. M.; Gao, D. *Org. Lett.* **2003**, *5*, 3491.

(176) Inagaki, F.; Mukai, C. *Org. Lett.* **2006**, *8*, 1217.

(177) Wender, P. A.; Deschamps, N. M.; Gamber, G. G. *Angew. Chem. Int. Ed.* **2003**, *42*, 1853.

(178) Wender, P. A.; Croatt, M. P.; Deschamps, N. M. *Angew. Chem. Int. Ed.* **2006**, *45*, 2459.

(179) Chatani, N.; Morimoto, T.; Kamitani, A.; Fukumoto, Y.; Murai, S. *J. Organomet. Chem.* **1999**, *579*, 177.

(180) In a synthesis of physostigmine. Mukai, C.; Yoshida, T.; Sorimachi, M.; Odani, A. *Org. Lett.* **2006**, *8*, 83.

(181) Kablaoui, N. M.; Hicks, F. A.; Buchwald, S. L.; *J. Am. Chem. Soc.* **1997**, *119*, 4424.

(182) Adrio, J.; Carretero, J. C. *J. Am. Chem. Soc.* **2007**, *129*, 778.

(183) Yu, C.-M.; Hong, Y.-T.; Lee, J.-H. *J. Org. Chem.* **2004**, *69*, 8506.

(184) a) Wakatsuki, Y.; Kuramitsu, T.; Yamazaki, H. *Tetrahedron Lett.* **1974**, *15*, 4549. b) McAllister, D. R.; Bercaw, J. E.; Bergman, R. G. *J. Am. Chem. Soc.* **1977**, *99*, 1666.

(185) For reviews, see a) Malacria, M. *Chem. Rev.* **1996**, *96*, 289. b) Ojima, I.; Tzamarcoudaki, M.; Li, Z.; Donovan, R. J. *Chem. Rev.* **1996**, *96*, 635. c) Vollhardt, K. P. C. *Acc. Chem. Res.* **1977**, *10*, 1.

(186) In a synthesis of the deoxygenated pancratistatin core. Moser, M.; Sun, X.; Hudlicky, T. *Org. Lett.* **2005**, *7*, 5669.

(187) Funk, R. L.; Vollhardt, K. P. C. *J. Am. Chem. Soc.* **1980**, *102*, 5253.

(188) In a synthesis of (–)-8-O-methyltetrangomycin. Kesenheimer, C.; Groth, U. *Org. Lett.* **2006**, *8*, 2507.

(189) In a synthesis of viridin. Anderson, E. A.; Alexanian, E. J.; Sorensen, E. J. *Angew. Chem. Int. Ed.* **2004**, *43*, 1998.

(190) McDonald, F. E.; Zhu, H. Y. H.; Holmquist, C. R. *J. Am. Chem. Soc.* **1995**, *117*, 6605.

(191) Tanaka, K.; Nishida, G.; Ogino, M.; Hirano, M.; Noguchi, K. *Org. Lett.* **2005**, *7*, 3119.

(192) Sato, Y.; Ohashi, K.; Mori, M. *Tetrahedron Lett.* **1999**, *40*, 5231.

(193) Sato, Y.; Nishimata, T.; Mori, M. *Heterocycles* **1997**, *44*, 443.

(194) Takeguchi, R.; Nakaya, Y. *Org. Lett.* **2003**, *5*, 3659.

(195) Yamamoto, Y.; Arakawa, T.; Ogawa, R.; Itoh, K. *J. Am. Chem. Soc.* **2003**, *125*, 12143.

(196) Yamamoto, Y.; Nagata, A.; Itoh, K. *Tetrahedron Lett.* **1999**, *40*, 5053.

(197) For reviews, see a) Heller, B.; Hapke, M. *Chem. Soc. Rev.* **2007**, *36*, 1085. b) Varela, J. A.; Saa, C. *Chem. Rev.* **2003**, *103*, 3787.

(198) Hillard, R. L., III; Parnell, C. A.; Volhardt, K. P. C. *Tetrahedron* **1983**, *39*, 905.

(199) a) Naiman, A.; Vollhardt, K. P. C. *Angew. Chem., Int. Ed. Engl.* **1977**, *16*, 708. b) For a review, see Bonnemann, H.; Brijoux, W. *Adv. Heterocycl. Chem.* **1990**, *48*, 177.

(200) Saa, C.; Crotts, D. D.; Hsu, G.; Vollhardt, K. P. C. *Synlett* **1994**, 487.

(201) a) Boese, R.; Harvey, D. F.; Malaska, M. J.; Vollhardt, K. P. C. *J. Am. Chem. Soc.* **1994**, *116*, 11153. b) Boese, R.; van Sickle, A. P.; Vollhardt, K. P. C. *Synthesis* **1994**, 1374.

(202) Heller, B.; Sundermann, B.; Fischer, C.; You, J.; Chen. W.; Drexler, H.-J.; Knochel, P.; Bonrath, W.; Gutnov, A. *J. Org. Chem.* **2003**, *68*, 9221.

(203) Gutnov, A.; Heller, B.; Fischer, C.; Drexler, H.-J.; Spannenberg, A.; Sundermann, B.; Sundermann, C. *Angew. Chem. Int. Ed.* **2004**, *43*, 3795.

(204) Yamamoto, Y.; Kinpara, K.; Ogawa, R.; Nishiyama, H.; Itoh, K. *Chem.—Eur. J.* **2006**, *12*, 5618.

(205) Wakatsuki, Y.; Yamazaki, H. *J. Organomet. Chem.* **1977**, *139*, 169.

(206) Grotjahn, D. B.; Vollhardt, K. P. C. *Synthesis* **1993**, 579.

(207) Eichberg, M. J.; Dorta, R. L.; Lamottke, K.; Vollhardt, K. P. C. *Org. Lett.* **2000**, *2*, 2479.

(208) Butenschon, H.; Winkler, M.; Vollhardt, K. P. C. *J. Chem. Soc., Chem. Commun.* **1986**, 388.

(209) Cammack, J. K.; Jalisatgi, S.; Matzger, A. J.; Negron, A.; Vollhardt, K. P. C. *J. Org. Chem.* **1996**, *61*, 2699.

(210) Petit, M.; Aubert, C.; Malacria, M. *Tetrahedron* **2006**, *62*, 10582.

(211) Yamamoto, Y.; Kitahara, H.; Ogawa, R.; Kawaguchi, H.; Tatsumi, K.; Itoh, K. *J. Am. Chem. Soc.* **2000**, *122*, 4310.

(212) Sambaiah, T.; Li, L.-P.; Huang, D.-J.; Lin, C.-H.; Rayabarapu, D. K.; Cheng, C.-H. *J. Org. Chem.* **1999**, *64*, 3663.

(213) a) Weibel, D.; Gevorgyan, V.; Yamamoto, Y. *J. Org. Chem.* **1998**, *63*, 1217. b) Gevorgyan, V.; Quan, L. G.; Yamamoto, Y. *J. Org. Chem.* **1998**, *63*, 1244.

(214) For a review, see Rubin, M.; Sromek, A. W.; Gevorgyan, V. *Synlett* **2003**, 2265.

(215) For a discussion of the mechanism, see Rubina, M.; Conley, M.; Gevorgyan, V. *J. Am. Chem. Soc.* **2006**, *128*, 5818.

(216) a) Earl, R. A.; Vollhardt, K. P. C. *J. Org. Chem.* **1984**, *49*, 4786. b) Earl, R. A.; Vollhardt, K. P. C. *J. Am. Chem. Soc.* **1983**, *105*, 6991.

(217) Duong, H. A.; Cross, M. J.; Louie, J. *J. Am. Chem. Soc.* **2004**, *126*, 11438.

(218) For rhodium-complex-catalyzed reactions, see Kondo, T.; Nomura, M.; Ura, Y.; Wada, K.; Mitsudo, T.-a. *Tetrahedron Lett.* **2006**, *47*, 7107.

(219) a) Hoberg, H.; Oster, B. W. *J. Organomet. Chem.* **1982**, *234*, C35. b) Hoberg, H.; Oster, B. W. *J. Organomet. Chem.* **1983**, *252*, 359.

(220) Yamamoto, Y.; Kinpara, K.; Saigoku, T.; Takagishi, H.; Okuda, S.; Nishiyama, H.; Itoh, K. *J. Am. Chem. Soc.* **2005**, *127*, 605.

(221) a) Hoberg, H.; Schaefer, P.; Burkhart, G.; Krüger, C.; Ramao, M.J. *J. Organomet. Chem.* **1984**, *266*, 203.

b) Hoberg, H.; Apotecher, B. *J. Organomet. Chem.* **1984**, *270*, C15.

(222) a) Louie, J.; Gibby, J. E.; Farnworth, M. V.; Tekavec, T. N. *J. Am. Chem. Soc.* **2002**, *124*, 15188. b) Tsuda, T.; Morikawa, S.; Sumiya, R.; Saegusa, T. *J. Org. Chem.* **1988**, *53*, 3140.

(223) Takimoto, M.; Mizuno, T.; Mori, M.; Sato, Y. *Tetrahedron* **2006**, *62*, 7589.

(224) a) Pino, P.; Braca, G. Carbon Monoxide Addition to Acetylenic Substrates. In *Organic Synthesis via Metal Carbonyls*; Wender, I., Pino, P., Eds.; Wiley: New York, 1977; Vol. 2, pp 420–516. b) Hubel, W. Organometallic Derivatives from Metal Carbonyls and Acetylene Compounds. In *Organic Synthesis via Metal Carbonyls*; Wender, I., Pino, P., Eds.; Wiley: New York, 1968; Vol. 1, pp 273–340.

(225) Liebeskind, L. S.; Baysdon, S. L.; South, M. S.; Blount, J. F. *J. Organomet. Chem.* **1980**, *202*, C73.

(226) Liebeskind, L. S.; Baysdon, S. L.; South, M. S.; Iyer, S.; Leeds, J. P. *Tetrahedron* **1985**, *41*, 5839.

(227) Liebeskind, L. S.; Jewell, C. F. *J. Organomet. Chem.* **1985**, *285*, 305.

(228) In a synthesis of royleanone. Liebeskind, L. S.; Chidambaram, R.; Nimkar, S.; Liotta, D. *Tetrahedron Lett.* **1990**, *31*, 3723.

(229) Iyer, S.; Liebeskind, L. S. *J. Am. Chem. Soc.* **1987**, *109*, 2759.

(230) For reviews, see a) Buchwald, S. L.; Nielsen, R .B. *Chem. Rev.* **1988**, *88*, 1047. b) Buchwald, S. L.; Broene, R. D. Transition Metal Alkyne Complexes—Zirconium-Benzyne Complexes. In *Comprehensive Organometallic Chemistry II*; Abel, E. W., Stone, F. G. A., Wilkinson, G., Eds.; Pergamon: Oxford, U.K., 1995; Vol. 12, pp 771–784.

(231) Takahashi, T.; Xi, Z.; Yamazaki, A.; Liu, Y.; Nakajima, K.; Kotora, M. *J. Am. Chem. Soc.* **1998**, *120*, 1672.

(232) Takahashi, T.; Tsai, F.-Y.; Li, Y.; Wang, H.; Kondo, Y.; Yamanaka, M.; Nakajima, K. Kotora, M. *J. Am. Chem. Soc.* **2002**, *124*, 5059.

(233) Buchwald, S. L.; King, S. M. *J. Am. Chem. Soc.* **1991**, *113*, 258.

(234) Buchwald, S. L.; Fang, Q. *J. Org. Chem.* **1989**, *54*, 2793.

(235) Cuny, G. D.; Gutierrez, A.; Buchwald, S. L. *Organometallics* **1991**, *10*, 537.

(236) a) Tidwell, J. H.; Senn, D. R.; Buchwald, S. L. *J. Am. Chem. Soc.* **1991**, *113*, 4685. b) Tidwell, J. H.; Buchwald, S. L. *J. Am. Chem. Soc.* **1994**, *116*, 11797. c) Tidwell, J. H.; Peat, A. J.; Buchwald, S. L. *J. Org. Chem.* **1994**, *59*, 7164.

(237) Tietze, L. F.; Looft, J.; Feuerstein, T. *Eur. J. Org. Chem.* **2003**, *15*, 2749.

(238) Campora, J.; Buchwald, S. L.; *Organometallics* **1993**, *12*, 4182.

Synthetic Applications of η^3-Allyl Transition Metal Complexes

9

9.1 Introduction

Although η^3-allyl complexes are known for virtually all of the transition metals, relatively few have found use in organic synthesis. However, those that have, mainly those of Pd, have broad utility. η^3-Allyl metal complexes can be made from a wide range of organic substrates in a variety of ways (**Figure 9.1**). These include (1) oxidative addition of allylic substrates to metal(0) or (I) complexes, (2) reaction of main group allyl

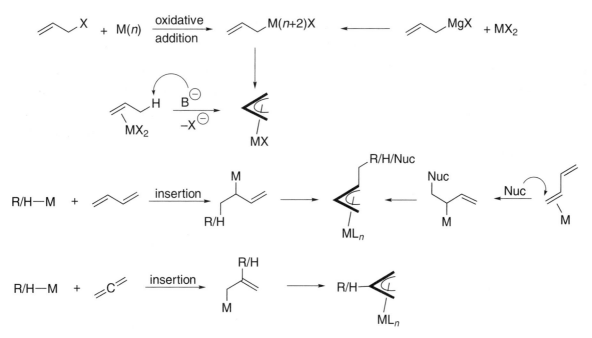

Figure 9.1. Preparation of η^3-Allyl Metal Complexes

metal complexes with transition metals (transmetallation), (3) nucleophilic attack on a 1,3-diene–metal complex, (4) insertion of 1,2- and 1,3-dienes into a metal hydride (or metal alkyl), (5) acidic cleavage of complexed allylic ethers, and (6) allylic proton abstraction from a π-alkene complex.

Although many η^3-allyl complexes are stable and isolable, they also are quite reactive under appropriate conditions. The scope of the reactions in which these complexes participate is presented in the following sections.

9.2 Transition-Metal-Catalyzed Telomerization of 1,3-Dienes[1]

Treatment of 1,3-dienes with nucleophiles in the presence of palladium(0) catalysts results in the production of functionalized octadienes, made by the joining of two 1,3-dienes with incorporation of the nucleophile (**Eq. 9.1**).

Eq. 9.1

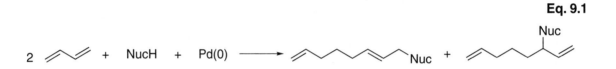

The mechanism of this process has not been closely studied, but is thought to involve the complexation of two dienes to the Pd(0), followed by the (by now) familiar "reductive dimerization" of the diene units (actually an oxidative addition of two dienes to the metal), thus joining them and generating a bis-η^3-allylpalladium species (**Eq. 9.2**). As we shall soon see, η^3-allylpalladium complexes are generally subject to nucleophilic attack, usually at the less-substituted terminus. Thus, nucleophilic attack produces an anionic π-alkene–η^3–η^1-allyl complex, which undergoes proton transfer to the metal, followed by reductive elimination to produce the diene "telomer" (an imprecise term used here to denote the combination of two diene units with a nucleophile). This procedure offers a convenient way to assemble functionalized chains (**Eq. 9.3**[2]),[3] but the intermolecular version has found only modest use in complex molecule synthesis.

Eq. 9.2

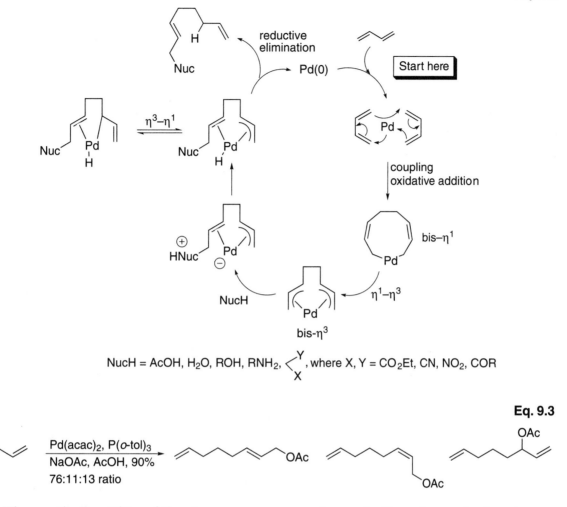

NucH = AcOH, H₂O, ROH, RNH₂, $\overset{Y}{\underset{X}{<}}$, where X, Y = CO₂Et, CN, NO₂, COR

Eq. 9.3

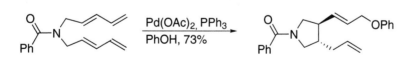

Pd(acac)₂, P(o-tol)₃
NaOAc, AcOH, 90%
76:11:13 ratio

The synthetic utility of the above process was dramatically enhanced when it was applied in an intramolecular fashion (**Eq. 9.4**),[4] thus assembling five-membered rings with appended functionality quickly and efficiently. Intramolecular trapping is also efficient (**Eq. 9.5**).[5]

Eq. 9.4

Pd(OAc)₂, PPh₃
PhOH, 73%

Eq. 9.5

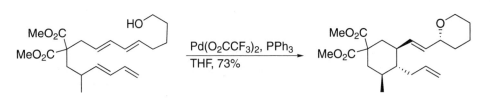

Provided the substrate has appropriately situated β-hydrogens, this cyclization can take place without nucleophilic attack, thus producing trienes instead of dienes (**Eq. 9.6**).[6] This process is closely related to the eneyne cyclizations that proceed by "hydrometallation" (Chapter 3), as well as to a number of other cyclization reactions catalyzed by low-valent transition metals.

Eq. 9.6

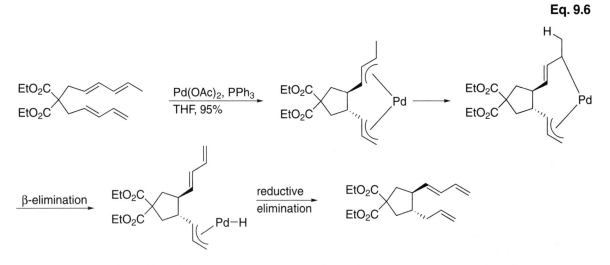

Most closely related are the iron(0)-catalyzed coupling reactions of trienes, as exemplified by **Eqs. 9.7**,[7] **9.8**,[8] and **9.9**.[9] Again, the mechanism has not been studied, but is likely to be similar to that in **Eq. 9.6**. A related rhodium-catalyzed reaction has also been described (**Eq. 9.10**).[10]

Eq. 9.7

Eq. 9.8

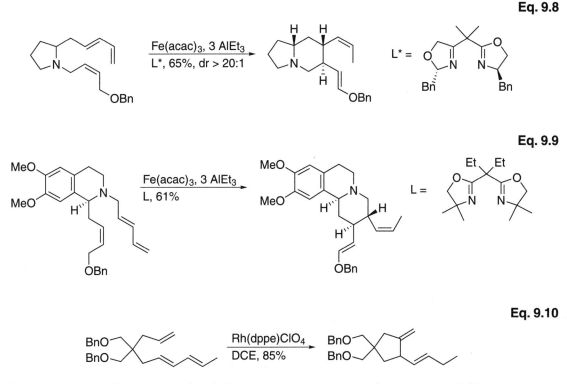

Eq. 9.9

Eq. 9.10

Perhaps the earliest example of this type of process is the commercially important nickel(0)-catalyzed cyclooligomerization of 1,3-dienes that produces a wide variety of compounds depending on conditions (**Figure 9.2**).[11]

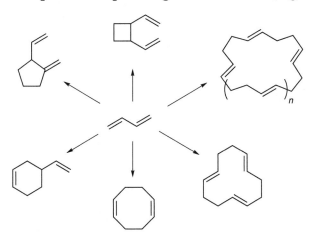

Figure 9.2. Ni(0)-Catalyzed Cyclooligomerization of Butadiene

Whatever the ultimate product, they all derive from an initial dimerization of the diene to form the bis-η^3-allylnickel complex (much as in Eq. 9.2 with Pd). Insertion of an additional diene leads to the cyclic trimers, while reductive elimination from one of the several bis-η^1-allyl species produces cyclooctadienes, vinylcyclohexanes, or divinylcyclobutanes (**Eq. 9.11**). Alternatively, β-elimination–reductive elimination affords trienes that can undergo further reactions with nickel(0).

Eq. 9.11

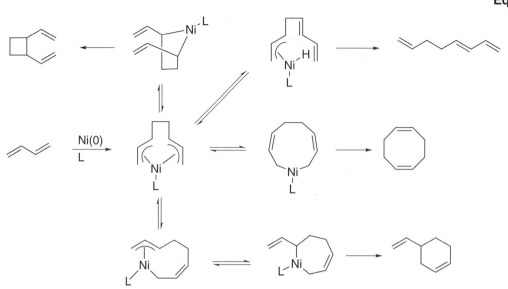

Although this chemistry is very well developed with simple substrates, its real synthetic utility for complex systems resides in intramolecular reactions directed toward forming eight-membered rings, such as in the synthesis of (+)-astericanolide (**Eq. 9.12**).[12] Alkynes are also efficient partners in a similar process for forming bicyclic compounds (**Eq. 9.13**).[13] The types of processes shown in **Eqs. 9.12** and **9.13** are covered in more detail in Chapter 7.

Eq. 9.12

Eq. 9.13

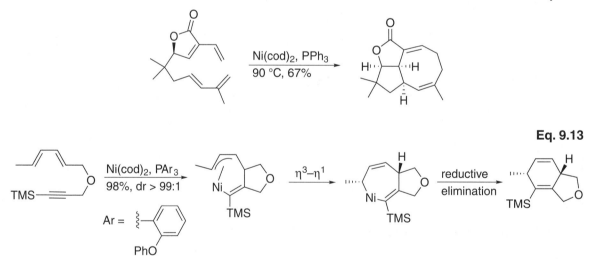

9.3 Palladium-Catalyzed Reactions of Allylic Substrates[14]

a. General[15]

Palladium complexes catalyze a wide variety of synthetically useful reactions of allylic substrates that proceed through η³-allyl intermediates (**Figure 9.3**). Whatever the ulti-

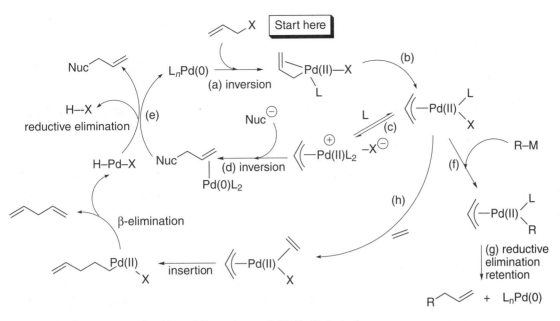

Figure 9.3. Palladium-Catalyzed Reactions of Allylic Substrates

mate product, they all involve the same manifold of familiar steps, and most involve nucleophilic attack on the η^3-allylpalladium complex. Palladium initially coordinates to the alkene of the allylic substrate. This is followed by oxidative addition of the allylic substrate to the palladium(0) catalyst [path (a) in **Figure 9.3**], a process that goes with clean inversion to initially give the η^1-allyl complex, which is rarely detected and is in equilibrium with the η^3-allyl complex [path (b)]. In contrast, the η^3-allyl complexes are quite stable, yellow solids that can easily be isolated if desired. However, in the presence of excess ligand, low equilibrium concentrations of a cationic η^3-allyl complex [path (c)] are generated. This is very reactive towards a wide range of nucleophiles, and undergoes nucleophilic attack from the face opposite the metal [path (d)] (inversion) to produce the allylated nucleophile, and to regenerate palladium(0) [path (e)] to re-enter the catalytic cycle. The net stereochemistry of this allylic substitution is retention, which is the result of two inversions. The regiochemistry with unsymmetric allylic substrates generally favors attack at the less-substituted allyl terminus, but depends somewhat on the nucleophile, the metal (see below), the cation,[16] as well as the ligands.[17]

η^3-Allyl complexes undergo transmetallation with main group organometallics, generating an η^3-allyl–η^1-alkylpalladium(II) complex [path (f)], which undergoes reductive elimination [path (g)] to produce the alkylated allylic compound. Since, in this case, the nucleophile (R of R–M) is delivered first to the metal and then to the η^3-allyl group from the *same* face as the metal, this step occurs with retention, and the overall alkylation goes with net inversion. η^3-Allyl complexes can also insert alkenes [path (h)] (probably via their η^1-isomers), producing 1,5-dienes after β-hydrogen elimination. All of these processes have been applied to the synthesis of complex molecules and are discussed in detail below.

A number of palladium(0) complexes and palladium(II) precatalysts can be used, and the choice of catalyst is often unimportant. The most frequently utilized catalyst is $(Ph_3P)_4Pd$, which is commercially available. However, commercial material is often of widely varying activity and it is best to prepare one's own. η^3-Allylpalladium(II) complexes are themselves catalysts, and because of their ease in preparation and handling, they are often used. A very convenient catalyst results from the treatment of the air-stable, easily prepared and handled $Pd(dba)_2$ (dba = PhCH=CH–CO–CH=CHPh, dibenzylidene acetone) with varying amounts of triphenylphosphine, generating PdL_n in situ. Finally, a variety of palladium(II) salts are readily reduced in the presence of substrate and ligand, and are often efficient precatalysts for the process.

b. Allylic Alkylation[18,19]

Alkylation of allylic substrates by stabilized carbanions (i.e., the "Tsuji–Trost reaction") is one of the synthetically most useful reactions catalyzed by palladium(0) complexes (**Eq. 9.14**). A wide range of allylic substrates undergo this reaction with a number of carbanions, making this an important process for the formation of carbon–carbon bonds. The reaction is very stereoselective, and proceeds with overall retention of configuration, the result of two inversions.[20] The reaction is also quite regioselective, with attack at the less-substituted terminus of the η^3-allyl intermediate favored, regardless of the initial position of the allylic leaving group. However, the regioselectivity depends on the substrate, nucleophile, catalyst, and conditions. For example, the more-substituted terminus is favored when using tricyclohexylphosphine as the ligand.[21]

Eq. 9.14

$$X \overset{\ominus}{\frown} Y \ + \ R \diagdown Z \ \xrightarrow{PdL_4} \ R \diagdown \overset{Y}{\underset{X}{\diagup}}$$

$Z = Br, Cl, OAc, -\overset{O}{\overset{\|}{C}}OR, O\overset{O}{\overset{\|}{P}}(OEt)_2, O-\overset{O}{\overset{\|}{S}}-R, OPh, OH, R_3\overset{\oplus}{N}, NO_2, SO_2Ph, CN,$

$X, Y = CO_2R, COR, SO_2Ph, CN, NO_2$

This process has found extensive use in synthesis over the years, and the literature abounds with applications to highly complex systems, so the methodology is well established and reliable.[22,23,24] Examples are shown in **Eqs. 9.15**,[25] **9.16**,[26] and **9.17**.[27]

Eq. 9.15

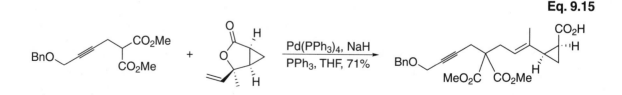

Eq. 9.16

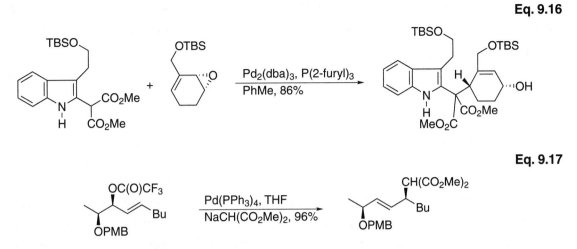

Eq. 9.17

Intramolecular versions of this process are efficient,[28] and have been used to make rings from three to eleven members, as well as macrocycles.[29,30,31] Examples are shown in **Eqs. 9.18,**[32] **9.19,**[33] and **9.20.**[34] Reactions of allylic carbonates and epoxides do not require the addition of a base to generate a carbanion since a base is generated in the formation of the η^3-allyl intermediate.

Eq. 9.18

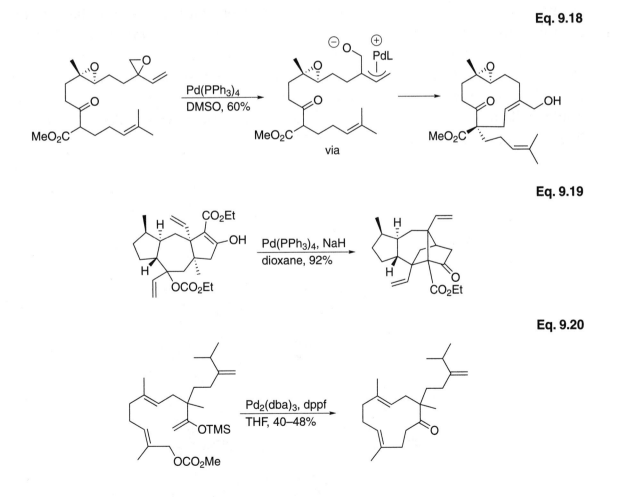

Eq. 9.19

Eq. 9.20

Because of the difference in reactivity of various allylic functional groups (e.g., Cl > OCO₂R > OAc >> OH), high regioselectivity can be achieved with bis-allylic substrates, in addition to high stereoselectivity (**Eq. 9.21**).[35]

Eq. 9.21

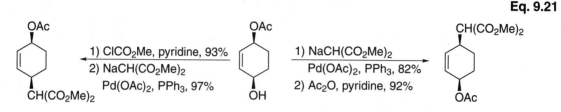

The development of procedures to induce asymmetry into palladium-catalyzed allylic reactions has dramatically enhanced the synthetic utility of this process.[36] The situation for asymmetric induction in these systems is complex, and the requirements differ with different classes of substrates. The most difficult substrates are chiral, racemic, unsymmetrically 1,3-disubstituted allylic compounds (**Eq. 9.22**). Since oxidative addition goes with clean inversion, the reaction of optically active palladium(0) complexes with chiral, racemic allylic acetates will lead to *two* diastereoisomeric η³-allylpalladium complexes. Since nucleophilic attack also occurs with clean inversion, to achieve high asymmetric induction, one of the two diastereoisomers must react substantially faster than the other, *and the less reactive diastereoisomer must be able to isomerize to the more reactive one at a rate that exceeds nucleophilic attack.* (This is the same situation observed for asymmetric hydrogenation in Chapter 4.) With acyclic systems, this isomerization is readily accomplished via a π → σ rearrangement, followed by rotation about the σ-alkylpalladium bond, followed by σ → π rearrangement, which effectively results in enantioface exchange (**Eqs. 9.22** and **9.23**).[37]

Eq. 9.22

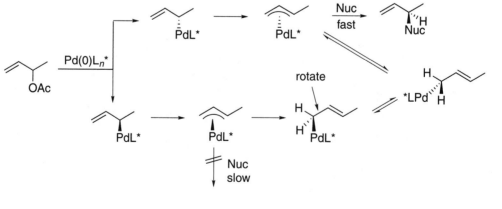

Eq. 9.23

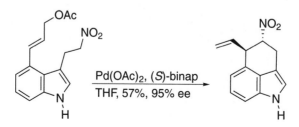

This $\pi \rightarrow \sigma \rightarrow \pi$ isomerization route is *not* available to cyclic systems, since the ring prevents rotation about the palladium–carbon σ-bond. However, asymmetric induction in the allylic alkylation of racemic chiral cyclic substrates has been achieved. In this case, diastereoface exchange is likely to occur by attack of the palladium *catalyst* $[Pd(0)L_n]$ on an η^3-allylpalladium complex (**Eq. 9.24**). This also provides a mechanism for the *erosion* of selectivity when optically active substrates are used (e.g., "racemization"), as has sometimes been observed when relatively high concentrations of catalyst are used.[38]

Eq. 9.24

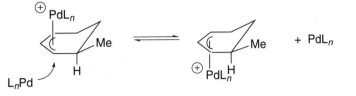

Enantioselective leaving group discrimination is perhaps equally difficult and uncommon. In a synthesis of sphingofungins E and F, a chiral palladium complex catalyzed an asymmetric allylic alkylation of a geminal diacetate. This reaction is remarkable in that it differentiates between the two potential leaving groups and the enantiotopic faces of the enolate nucleophile (**Eq. 9.25**).[39] A differentiation of the prochiral faces of the nucleophiles was used in a double allylic alkylation in a synthesis of (–)-huperzine A (**Eq. 9.26**).[40]

Eq. 9.25

Eq. 9.26

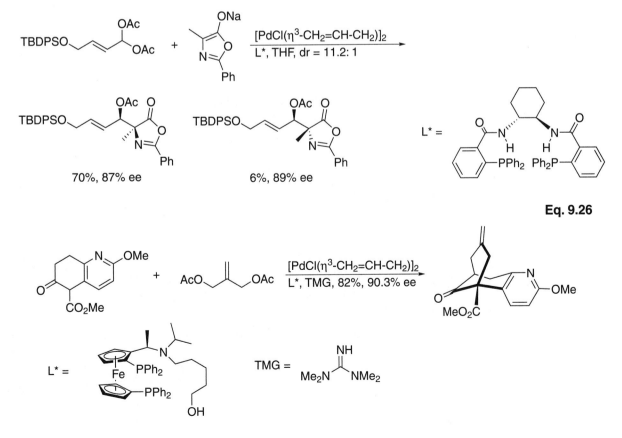

Considerably less challenging (and thus much more extensively studied) are symmetrical allylic substrates in which "only" discrimination between the two enantiotopic termini of the η^3-allyl complex is required.[41] A very large number of ligands have been (and continue to be) developed for this class of substrates and most of them result in high asymmetric induction[42] (**Eq. 9.27**).[43]

Eq. 9.27

Perhaps the most useful class of asymmetric η^3-allylpalladium reactions involves catalytic desymmetrizations of meso substrates (**Eq. 9.28**).[44] In these cases, the catalyst discriminates between the two enantiotopic allylic leaving groups, resulting again in high asymmetric induction. In many synthetic applications, the second leaving group is displaced in a subsequent step, resulting in rapid construction of cyclic systems (**Eq. 9.29**).[45] A particularly impressive example utilized two palladium-catalyzed allylic displacements followed by a Heck alkene arylation (Chapter 4) (**Eq. 9.30**).[46,47]

Eq. 9.28

Eq. 9.29

Eq. 9.30

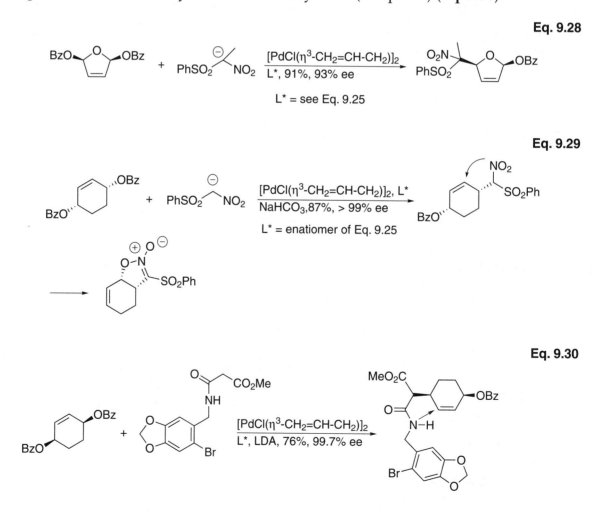

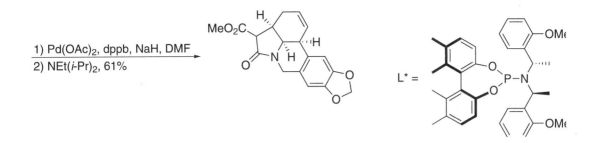

1) Pd(OAc)$_2$, dppb, NaH, DMF
2) NEt(*i*-Pr)$_2$, 61%

c. Allylic Alkylation by Transmetallation

Transmetallation has been used much less frequently in allylic systems than with aryl or vinyl systems, for several practical reasons. Allylic acetates are the most attractive class of allylic substrates, because of their ready availability from the corresponding alcohols, but they are substantially less reactive towards oxidative addition to Pd(0) than are allylic halides, and phosphine ligands are required to promote this process. Acetate coordinates quite strongly to palladium and this, along with the presence of excess phosphine, slows down the crucial transmetallation step, already the rate-limiting step. Finally, in contrast to dialkylpalladium(II) complexes, σ-alkyl–η^3-allylpalladium(II) complexes undergo reductive elimination only slowly, compromising this final step in the catalytic cycle (**Eq. 9.31**). It is thus not surprising that the alkylation of allylic acetates via transmetallation has been slow to develop. However, under appropriate reaction conditions [i.e., polar solvents such as DMF, "ligandless" (no phosphine) catalysts such as Pd(dba)$_2$ or PdCl$_2$(MeCN)$_2$, and excess LiCl to facilitate transmetallation], allylic acetates can be alkylated by a variety of aryl- and alkenyltin reagents. Coupling occurs at the less-substituted terminus of the allylic system with clean inversion (i.e., inversion in the oxidative addition step, and retention in the transmetallation–reductive elimination step), and the geometry of the alkene in both the allylic substrate[48] and alkenyltin partner is maintained (**Eq. 9.32**).[49] An impressive example of this reaction in a synthesis of azaspiracid-1 is shown in **Eq. 9.33**.[50]

Eq. 9.31

Eq. 9.32

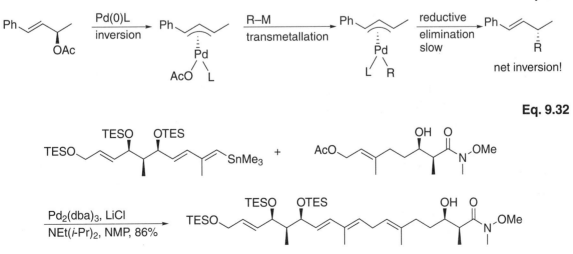

Eq. 9.33

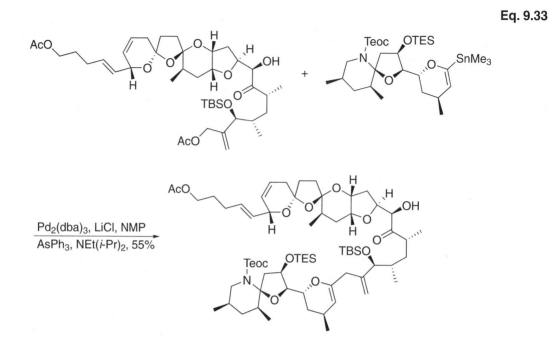

Allylic epoxides and carbonates[51] are better substrates for this process, and are alkylated by a variety of aryl- and alkenyltin reagents (**Eqs. 9.34**[52] and **9.35**[53]). Acetylenic, benzylic, or allylic tin compounds fail to couple. Allylic chlorides are the most reactive of the substrates, undergoing this coupling reaction under "normal" reaction conditions (i.e., phosphine present) in excellent yield with highly functionalized substrates[54] (**Eq. 9.36**[55]). The process is restricted neither to allylic chlorides nor to organostannanes (**Eq. 9.37**).[56]

Eq. 9.34

Eq. 9.35

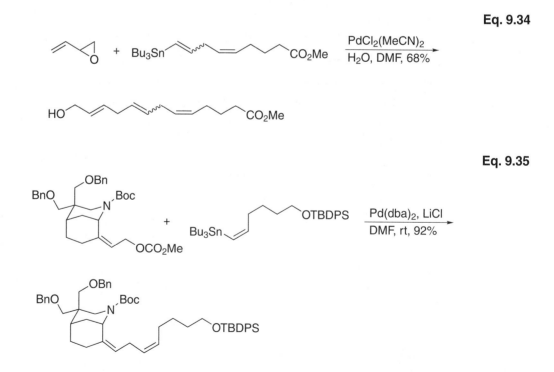

Eq. 9.36

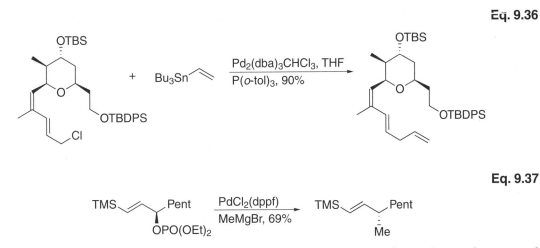

Eq. 9.37

Nucleophilic addition of ester enolates or anions of hindered nitriles to the central carbon of η^3-allylpalladium complexes is observed in some cases. A cyclopropane is formed in this case via a palladacyclobutane.[57,58] This mode of addition is favored by the use of strong donor ligands and α-branched nucleophiles. An example is shown in **Eq. 9.38** wherein the ester enolate is generated by a decarbonylation–intermolecular Michael addition.[59]

Eq. 9.38

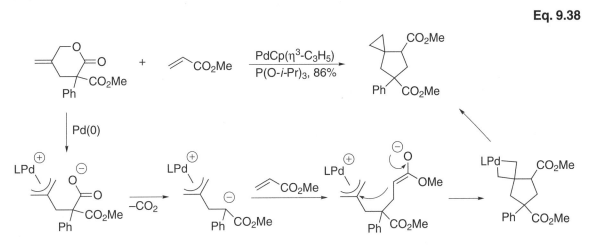

d. Allylic Amination,[60] Alkoxylation, Reduction, and Deprotonation

The reactions of nitrogen nucleophiles with allylic substrates are among the most useful for organic synthesis, and both 1° and 2° amines (but *not* NH$_3$) efficiently aminate allylic substrates, as do anions of amides and imides, in addition to azide. The stereo- and regioselectivity of allylic amination parallels that of allylic alkylation, and the range of reactive substrates is comparable. The nitrogen almost invariably ends up at the less-hindered terminus of the allylic system, although that may not be the initial site of attack, since Pd(0) complexes catalyze the allylic transposition of allylic amines.[61]

Intermolecular aminations are efficient and a wide range of functionality is tolerated. The process is used extensively in the synthesis of carbocyclic nucleoside analogs[62] (**Eq. 9.39**)[63] and in other applications (**Eq. 9.40**[64]). A large-scale reaction was used in a synthesis of an anti-MRSA β-methylcarbapenem (**Eq. 9.41**).[65]

Eq. 9.39

Eq. 9.40

Eq. 9.41

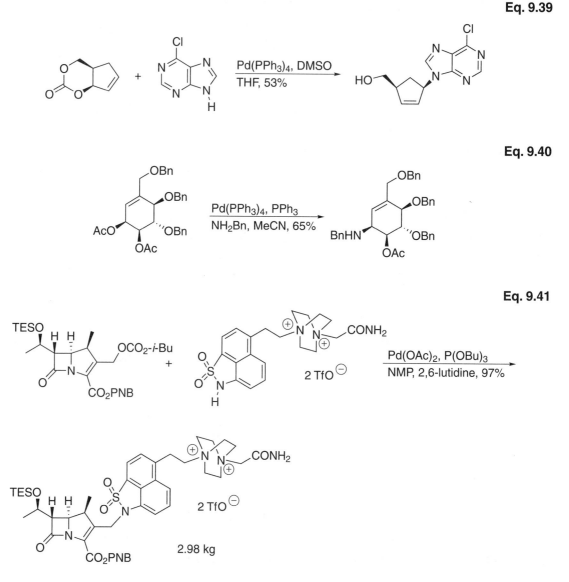

Asymmetric induction in the palladium-catalyzed allylic amination process is also efficient, with very much the same classes of substrates as utilized in asymmetric allylic alkylation, particularly symmetrically 1,3-disubstituted substrates (**Eqs. 9.42**[66] and **9.43**).[67] It is most extensively used for the desymmetrization of meso-allylic substrates, and is efficient for the synthesis of highly functionalized cyclohexenes (**Eq. 9.44**).[68]

Eq. 9.42

Eq. 9.43

Eq. 9.44

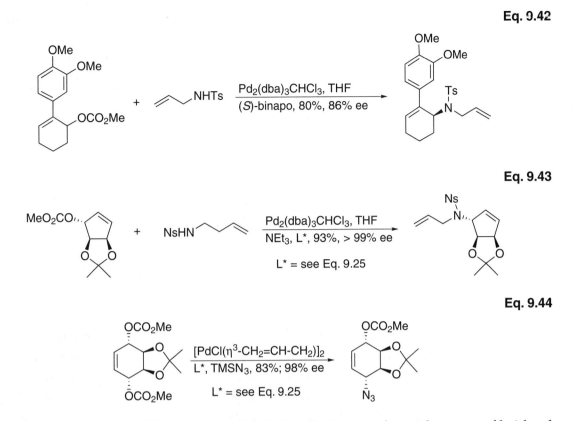

Intramolecular aminations are particularly effective, and a wide range of bridged (**Eq. 9.45**),[69] fused (**Eq. 9.46**),[70] spiro (**Eq. 9.47**),[71] and macrocyclic systems (**Eq. 9.48**)[72] can be made efficiently. The amination of the η^3-allyl intermediate in **Eq. 9.46** is unusual in that it occurs from the same face as the metal (an overall inversion). In this case, the nucleophile cannot easily reach the opposite side of the ring.

Eq. 9.45

Eq. 9.46

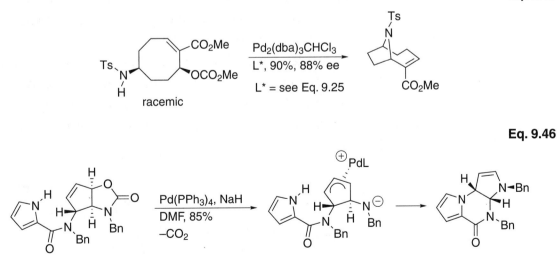

Eq. 9.47

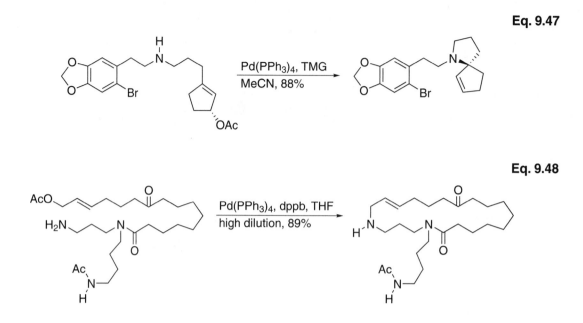

Eq. 9.48

Although carbon and nitrogen nucleophiles have been most extensively used in palladium-catalyzed allylic substitution processes, a range of oxygen nucleophiles can also participate. Phenols attack allylic epoxides[73] and carbonates[74,75] in the presence of palladium(0) catalysts and, in the presence of chiral ligands, high asymmetric induction is observed (**Eqs. 9.49**[76] and **9.50**[77]). Since a racemic mixture of the carbonate was used in the last example, an isomerization step must be involved at some point in the mechanism. Two suggested possibilities are anti-addition of a second equivalent of the palladium catalyst and the formation of a σ-palladium complex, followed by reformation of the η³-complex from the opposite face. Regardless of the mechanism, the nucleophilic addition must be much faster for one of the complexes. Glycal acetates were coupled to the anomeric OH group of another carbohydrate (**Eq. 9.51**).[78] Glycosylation reactions (**Eq. 9.52**),[79] and intramolecular alkoxylations were efficient[80,81] (**Eq. 9.53**).[82] In the latter example, an acceleration of the reaction rate was observed in the presence of trimethyltin chloride. A tin alkoxide is probably formed and these intermediates have been shown to be very reactive nucleophiles toward η³-allyl complexes.[83] Although hydroxide (or water) has not been used in this process, triphenylsilanol is an efficient surrogate (**Eq. 9.54**).[84,85] Palladium(0) complexes even catalyze the O-alkylation of enolates (**Eq. 9.55**).[86]

Eq. 9.49

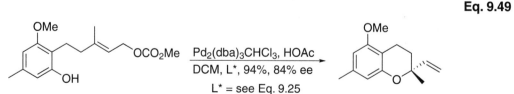

Eq. 9.50

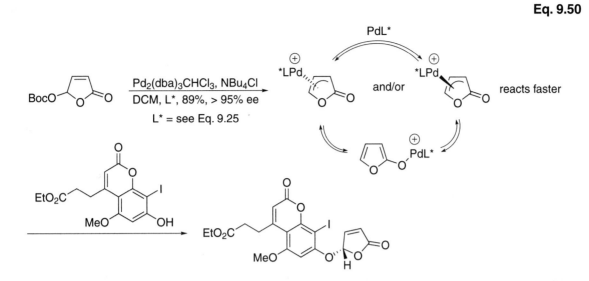

Eq. 9.51

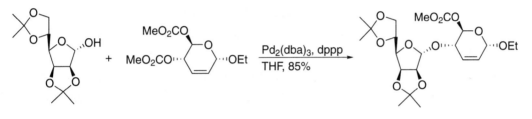

Eq. 9.52

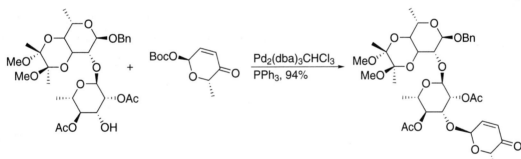

Eq. 9.53

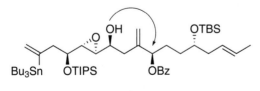

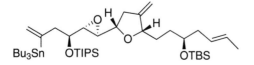

Eq. 9.54

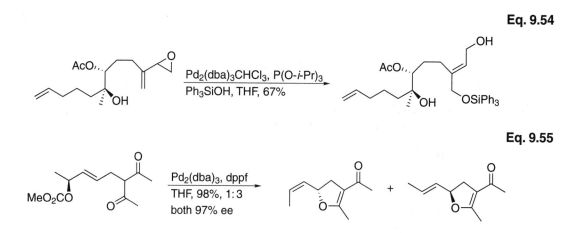

Eq. 9.55

Acetate is rarely used as a nucleophile since it is most often the leaving group in palladium-catalyzed allylic substitution reactions. However, carboxylates are capable of attacking η³-allylpalladium complexes and, in contrast to most nucleophiles, either can directly attack the η³-allyl ligand from the face opposite the metal, resulting in inversion in this step, or can attack the metal, resulting in retention in this step. Attack by acetate plays a major role in a useful variant involving palladium(II)-catalyzed bis-acetoxylation or chloroacetoxylation of dienes (**Eq. 9.56**). The process begins with a palladium(II)-assisted acetoxylation of one of the two double bonds of the complexed diene, generating an η³-allylpalladium(II) complex. In the absence of added chlorides, the second acetate is delivered from the metal, leading to the trans diacetate. Added chloride blocks this coordination site on the metal and leads to nucleophilic attack by uncomplexed acetate from the face opposite the metal, giving the corresponding cis-diacetate and palladium(0).[87] The reoxidation of palladium(0) to palladium(II) can be carried out using a number of oxidants, including benzoquinone–MnO$_2$.[88] Because the products of this reaction are allylic acetates, and because of the wide variety of palladium(0)-catalyzed reactions of allylic acetates, this is a useful process that has been extensively studied.

Eq. 9.56

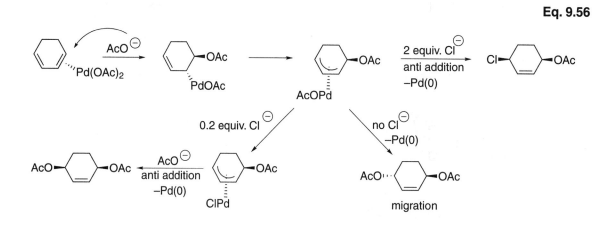

This chemistry has been used to develop quite efficient and stereoselective approaches to fused tetrahydrofurans and tetrahydropyrans (**Eq. 9.57**).[89] In this case, the tethered alcohol group is the nucleophile, which initially attacks the π-alkene complex, giving the trans η³-allyl complex. Inter- or intramolecular attack of acetate, chloride, or alkoxide on this η³-allyl complex leads to the observed products. Other applications of this very useful process are presented later in this chapter.

Eq. 9.57

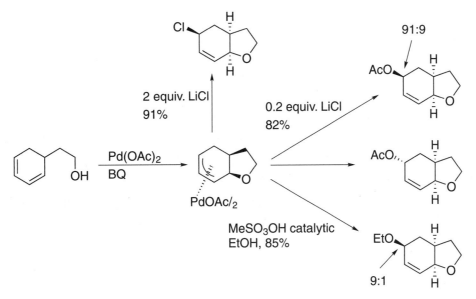

Allylation of N- and some C-nucleophiles[90] is not limited to allylic esters, halides, and carbonates. Allylic alcohols, for example, can also be used in the presence of a Lewis acid promoter.[91] The mechanism is likely related to that of the more commonly used substrates via the formation of an η³-allylpalladium intermediate. However, the mechanism for the formation of the η³-allylpalladium intermediate is more complex and not clearly understood. Examples of reactions of allylic alcohols are shown in **Eqs. 9.58**[92] and **9.59**.[93]

Eq. 9.58

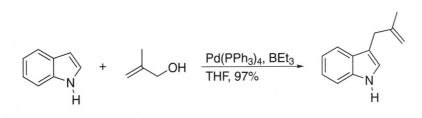

Eq. 9.59

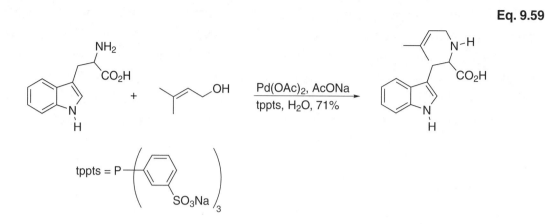

Palladium(0) complexes catalyze a number of other useful reactions of allylic substrates. Allylic esters are easily reduced to alkenes with clean inversion in the presence of a hydride source, which attacks the η³-allylpalladium intermediate (**Eq. 9.60**).[94] Ammonium formates are particularly effective in complex systems[95] (**Eqs. 9.61**[96] and **9.62**[97]). With chiral ligands, asymmetry can be induced[98] (**Eq. 9.63**).[99] Allylic sulfones can be reduced to alkenes in a similar fashion (**Eq. 9.64**).[100] Allylic formates undergo intramolecular reduction, with formate acting as the hydride source (**Eq. 9.65**).[101] In these cases, the hydride is delivered from the same face as the palladium.

Eq. 9.60

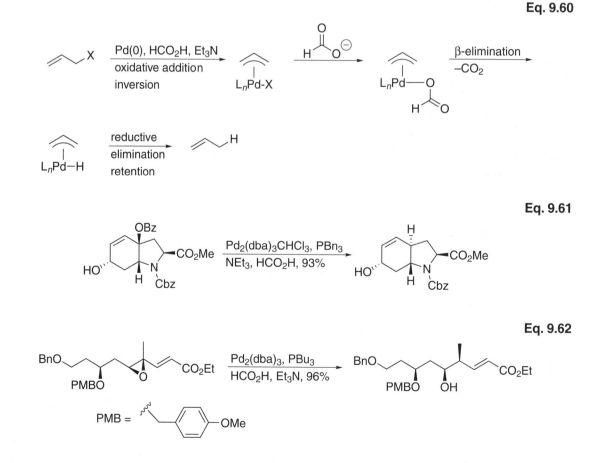

Eq. 9.61

Eq. 9.62

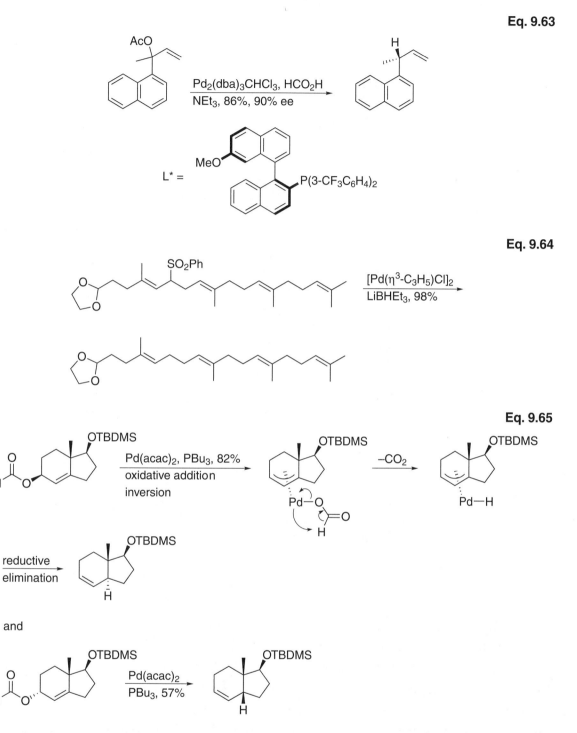

Eq. 9.63

Eq. 9.64

Eq. 9.65

In the absence of nucleophiles, palladium(0) complexes, particularly the one derived from treating palladium acetate with one equivalent of tributylphosphine,[102] efficiently form dienes from allylic carbonates. The process occurs by oxidative addition with inversion, followed by syn elimination. In sterically biased systems, high regioselectivity is observed (**Eq. 9.66**).[103] Decent asymmetric induction has been observed in appropriate symmetrical systems (**Eq. 9.67**).[104] In the presence of base, an anti elimination is possible.[105]

Eq. 9.66

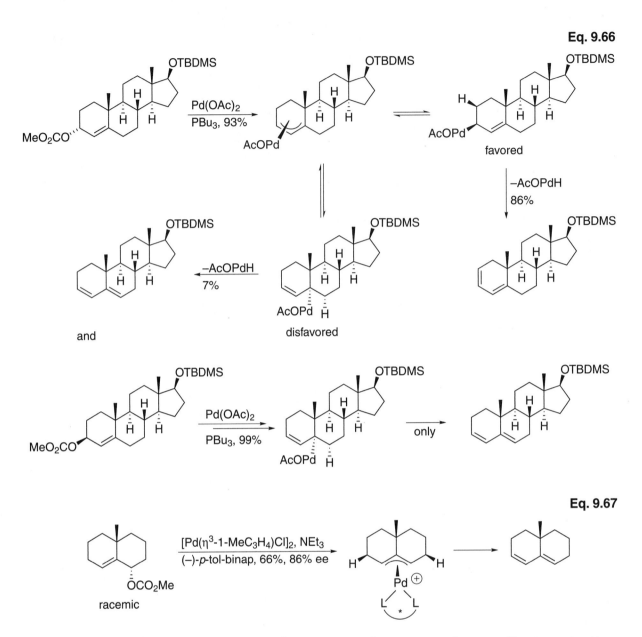

Eq. 9.67

Finally, allylic carbonates, carbamates, and esters make excellent protecting groups[106] for carboxylic acids, alcohols, and amines, because they are stable to a wide range of reaction conditions, and can be exclusively removed in the presence of other labile protecting groups. In these cases, a nucleophile must be present to regenerate the catalyst from the initially formed η³-allyl complex (**Eq. 9.68**). This system can be used with quite complex substrates (**Eqs. 9.69**[107] and **9.70**).[108] The most spectacular example of the utility of this protecting group strategy was the removal of 104 allylic protecting groups from NH₂ and phosphate moieties in a 60-mer oligonucleotide in almost 100% overall yield in a *single* palladium-catalyzed reaction.[109]

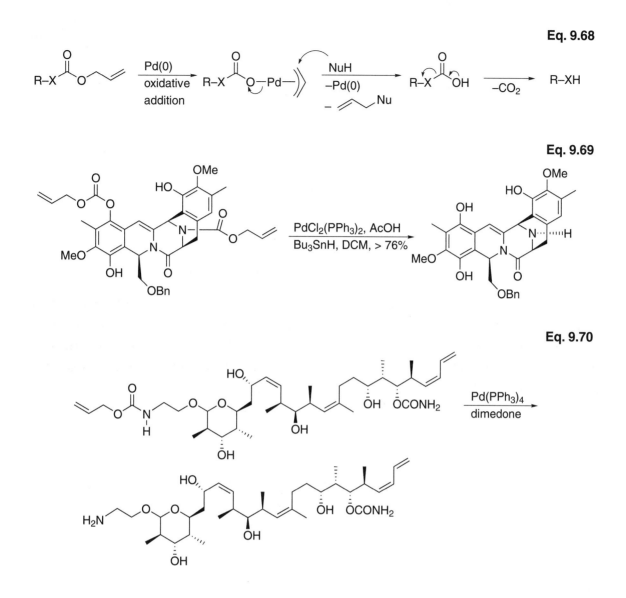

Eq. 9.68

Eq. 9.69

Eq. 9.70

e. Insertion Processes Involving η³-Allylpalladium Complexes

Alkenes, alkynes, and carbon monoxide insert into η³-allylpalladium complexes (perhaps via the η¹-isomer), generating σ-alkylpalladium complexes that can undergo the very rich chemistry described in Chapter 4. A computational study indicated that the alkene inserts directly into the η³-allylpalladium complex.[110] When carried out in an intramolecular manner, efficient cyclization processes can be developed (**Eq. 9.71**).[111] These reactions are synthetically very powerful, and demonstrate the real potential for the use of transition metals in organic synthesis, since often even the substrates are synthesized using transition metals.

Eq. 9.71

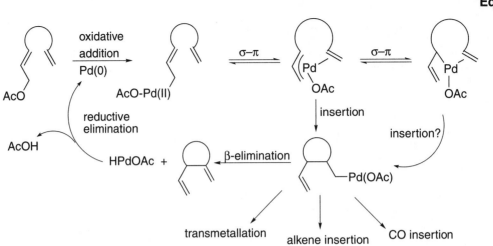

In **Eq. 9.72**, for example, the 1-acetoxy-4-chlorobut-2-ene was synthesized by the Pd(OAc)₂-catalyzed chloroacetoxylation of dienes and the diallylic bis-sulfone prepared by the Pd-catalyzed allylic alkylation of allylic halides. Treatment of this compound with Pd(dba)₂ in acetic acid results in efficient cyclization.[112] The reaction is quite general, and tolerates a variety of functional groups (**Eq. 9.73**).[113]

Eq. 9.72

Eq. 9.73

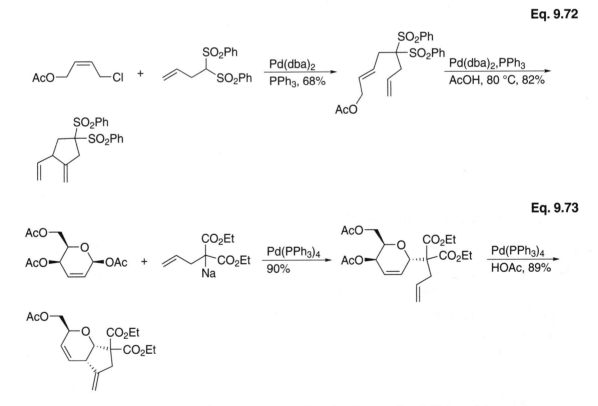

When additional unsaturation is appropriately situated, additional insertions can occur, resulting in the formation of several rings in a single efficient process (**Eqs. 9.74**[114] and **9.75**[115]). Dienes also insert into η³-allylpalladium complexes, producing another η³-allyl complex, which can undergo nucleophilic attack by acetate to form diene acetates (**Eq. 9.76**).[116]

Eq. 9.74

Eq. 9.75

Eq. 9.76

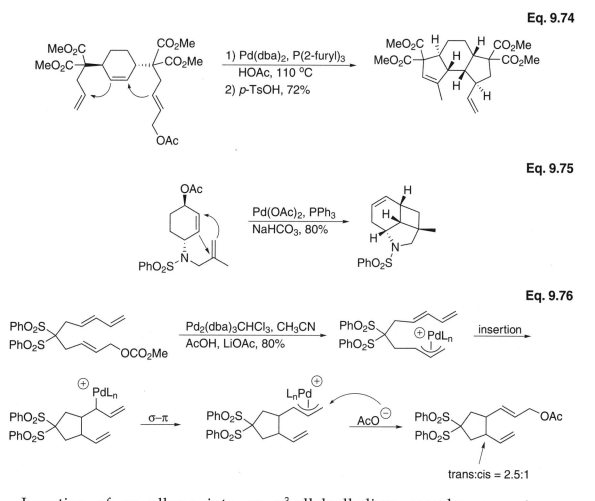

trans:cis = 2.5:1

Insertion of an alkene into an η^3-allylpalladium complex generates an η^1-alkylpalladium complex, which itself is very reactive towards the insertion of carbon monoxide (Chapter 4). By carrying out these cyclizations under an atmosphere of carbon monoxide, the η^1-alkylpalladium intermediate can be intercepted, making an η^1-acylpalladium complex that can undergo further insertion. With appropriate adjustment of the substrate structure and reaction conditions, impressive polycyclizations can be achieved (**Eqs. 9.77**[117] and **9.78**[118]). The process can also be truncated by transmetallation (**Eq. 9.79**).[119]

Eq. 9.77

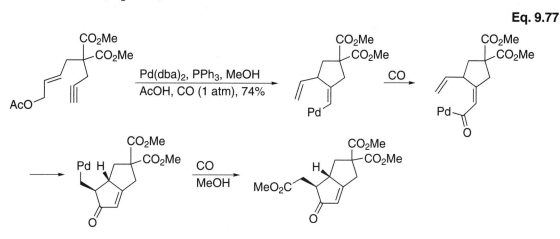

Eq. 9.78

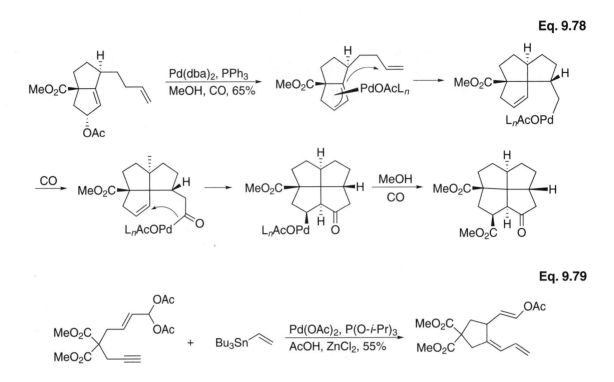

Eq. 9.79

f. Palladium(0)-Catalyzed Cycloaddition Reactions via Trimethylenemethane Intermediates

Although stable trimethylenemethane complexes of transition metals have been known for a long time, they have found little use in synthesis because of their lack of reactivity. However, 1,1'-bifunctional allylic compounds having an allylic acetate and an allylic silane undergo reaction with palladium(0) complexes to produce unstable, uncharacterized intermediates that react as if they were zwitterionic trimethylenemethane complexes, undergoing [3+2] cycloadditions to a range of electron-deficient alkenes (**Eq. 9.80**).[120] The reaction is thought to involve oxidative addition of the allylic acetate to the Pd(0) complex, generating a cationic η³-allylpalladium complex. Displacement of the remaining allyl silane group, perhaps by acetate, produces a zwitterionic trimethylenemethane complex, the anionic end of which attacks the β-position of a conjugated enone, generating an α-enolate, which in turn attacks a terminus of the electrophilic cationic η³-allylpalladium complex.

Eq. 9.80

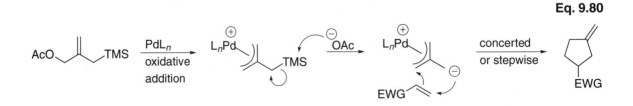

It is unknown if this cycloaddition is concerted; however, since the stereochemistry of the alkene partners is maintained in the product, ring closure must occur faster than rotation about an incipient single bond. In addition, the process can be quite diastereoselective (**Eqs. 9.81**[121] and **9.82**).[122] With unsymmetrical bifunctional allylic acetates, regioselective coupling occurs with coupling at the more-substituted terminus, regardless of the electronic nature of the functional group and the initial position of the acetate and silyl leaving groups (**Eq. 9.83**).[123] This indicates that equilibration of the three termini of the trimethylmethane fragment must occur more rapidly than does coupling. With more extended acceptors such as pyrones or tropone, [3+2], [3+4], or [3+6] cycloadditions may occur, depending on the substituents on the acceptor (**Eqs. 9.84**[124] and **9.85**[125]). Intramolecular versions of this reaction are quite efficient for the production of highly functionalized polycyclic compounds[126] (**Eq. 9.86**).[127] Asymmetric reactions have also been realized (**Eq. 9.87**).[128]

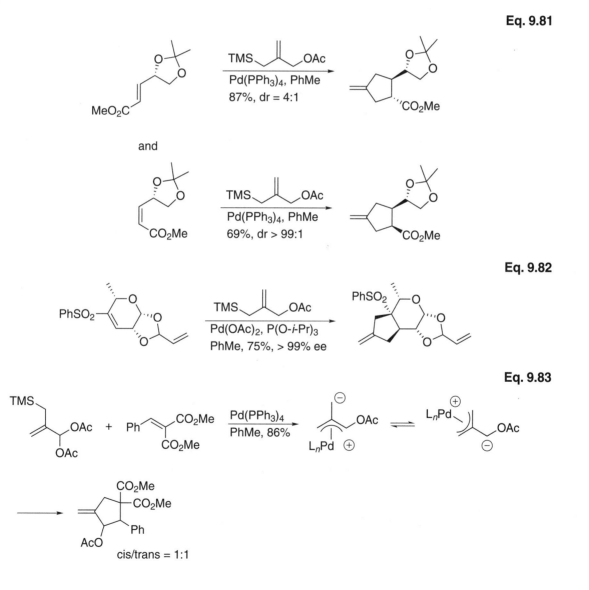

Eq. 9.81

and

Eq. 9.82

Eq. 9.83

cis/trans = 1:1

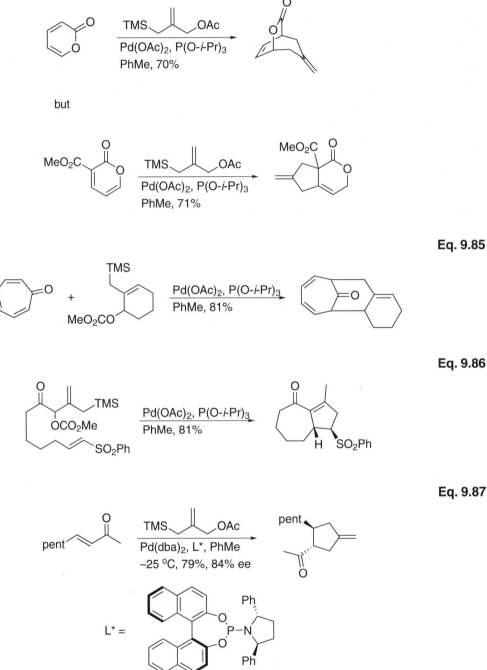

Eq. 9.84

but

Eq. 9.85

Eq. 9.86

Eq. 9.87

Carbon–oxygen and carbon–nitrogen double bonds participate in cycloaddition reactions with trimethylenemethane–palladium complexes. An example of a cycloaddition of carbon dioxide to give a lactone is shown in **Eq. 9.88**.[129] Aldehydes also participate in [3+2] cycloaddition reactions with bifunctional allylic acetates in the presence of palladium(0) catalysts, forming five-membered oxygen heterocycles[130] (**Eq. 9.89**).[131]

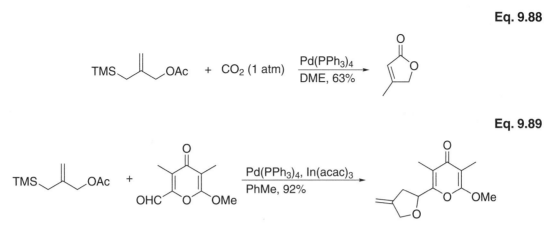

Eq. 9.88

Eq. 9.89

With unsymmetrical trimethylenemethane intermediates, mixtures of regioisomers are obtained, in contrast to the observations with conjugated enones. This is rationalized by asserting that cyclization competes effectively with rearrangement of the initially formed trimethylenemethane complexes. Tin co-catalysts dramatically improve the yields and regioselectivity of the process, perhaps by acting as a Lewis acid to increase the reactivity of the aldehyde. A more profound effect is noted when an In(III) co-catalyst is used with conjugated enones as acceptors. This additive completely changes the regioselectivity of this reaction, from 1,4 (i.e., addition to the alkene) to 1,2 (i.e., addition to the carbonyl group) (**Eq. 9.90**).[132] DIBAL is used in this reaction to reduce palladium(II) to palladium(0). This change is thought to be the result of coordination of the highly electropositive In(III) to the carbonyl oxygen, thus enhancing reaction at this site over attack of an alkene. Trimethylenemethane complexes even react with aziridines to give [3+3] cycloaddition products (**Eq. 9.91**).[133]

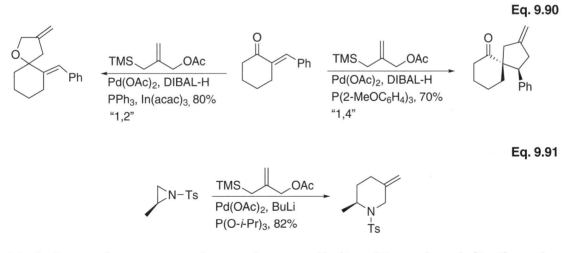

Eq. 9.90

Eq. 9.91

Methylenecyclopropanes also undergo palladium(0)-catalyzed [3+2] cycloaddition reactions—again, most likely through trimethylenemethane intermediates (**Figure 9.4**)[134] formed by "oxidative addition" into the activated cyclopropane carbon–carbon bond. Nickel(0) complexes catalyze similar processes. Although most of the

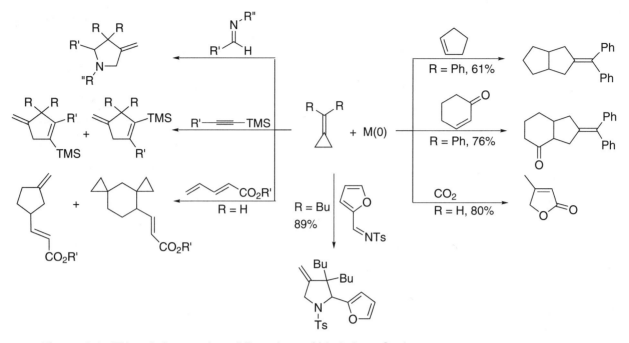

Figure 9.4. "Trimethylenemethane" Reactions of Methylene Cyclopropenes

studies of this system have been carried out on relatively simple substrates, a few, more complex systems have been examined (**Eq. 9.92**).[135]

Eq. 9.92

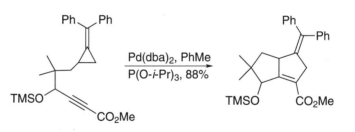

With diastereoisomerically pure methylenecyclopropanes, the reaction is stereospecific and proceeds with retention of configuration at the chiral carbon center at which reaction occurred (**Eq. 9.93**).[136] With alkene rather than alkyne acceptors, the reaction is stereospecific (i.e., retention) with respect to the pre-existing cyclopropane stereocenter, but the alkene geometry is not retained (**Eq. 9.94**).[137]

Eq. 9.93

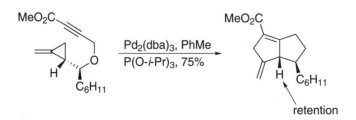

Eq. 9.94

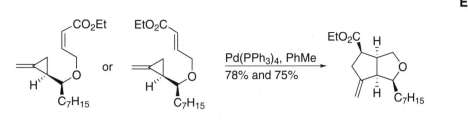

g. Reaction of Propargylic Substrates

Oxidative addition of palladium(0) complexes to propargylic substrates affords η^1-propargyl complexes that may rearrange to η^3-propargyl complexes,[138] depending on ligands and substituents present (**Eq. 9.95**). As a result, palladium-catalyzed nucleophilic addition to propargylic substrates often results in the formation of mixtures of propargylic and allenic product.[139] The regioselectivity of the addition of organozinc reagents can be controlled by the ligand employed (**Eq. 9.96**).[140] Propargylic carbonates have been used in synthesis and an example is shown in **Eq. 9.97**.[141]

Eq. 9.95

Eq. 9.96

Eq. 9.97

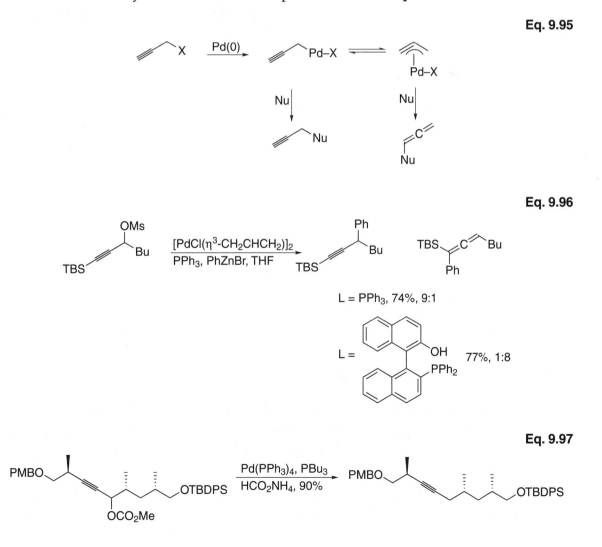

L = PPh₃, 74%, 9:1

L =   77%, 1:8

9.4 η^3-Allyl Complexes of Metals Other than Palladium

a. Molybdenum, Tungsten, Ruthenium, Rhodium, and Iridium Catalysis

Molybdenum hexacarbonyl catalyzes the alkylation of allylic acetates by stabilized carbanions. However, the regioselectivity is quite different from that observed with palladium(0).[142] Whereas palladium directs the nucleophile to the less-substituted terminus of the η^3-allyl system, molybdenum leads to reaction at the more-substituted end when malonate is the attacking anion, and the less-substituted end when more sterically demanding nucleophiles are used. The complex $Mo(RNC)_4(CO)_2$ is a more efficient catalyst than $Mo(CO)_6$, but, in this case, attack at the less-substituted allylic terminus predominates.[143] With sterically bulky nucleophiles in the presence of chiral diamine ligands, $Mo(CO)_3(EtCN)_3$ catalyzes allylic alkylation at the more-substituted terminus with high ee.[144,145]

The molybdenum-complex-catalyzed reaction proceeds with overall net retention. However, in contrast to the palladium-catalyzed reactions involving two inversions, the overall retention in molybdenum-complex-catalyzed reactions is the result of two retention steps for acyclic systems (**Eq. 9.98**).[146] A synthetic application of this catalyst system is shown in **Eq. 9.99**.[147]

Eq. 9.98

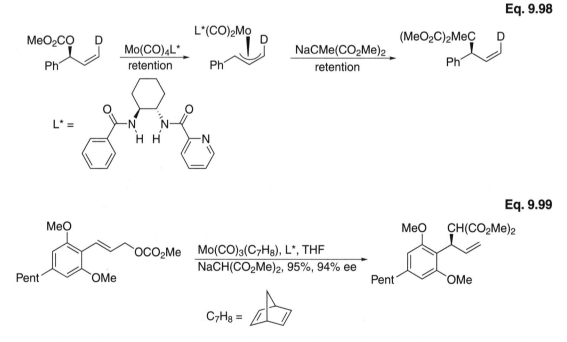

Eq. 9.99

In a similar fashion, tungsten hexacarbonyl and related molybdenum(0) complexes also catalyze allylic alkylations. Although tungsten catalysts usually result in attack at the more-substituted terminus,[148] regardless of the steric bulk of the nucleophile, exceptions do exist.[149]

Rhodium(I),[150] ruthenium(0 and II),[151,152] and iridium(I)[153,154] complexes catalyze the allylic alkylation of allylic carbonates and acetates at the *more*-substituted position.

High enantiomeric excess in the presence of chiral ligands can be obtained.[155,156,157] Although a significant number of papers have been published, only a handful of applications in organic synthesis have appeared, utilizing rhodium, ruthenium, and iridium catalysts in allylic substitution reactions. Three examples are shown in **Eqs. 9.100**,[158] **9.101**,[159] and **9.102**.[160] Copper alkoxides and enolates are much better nucleophiles in these reactions.

Eq. 9.100

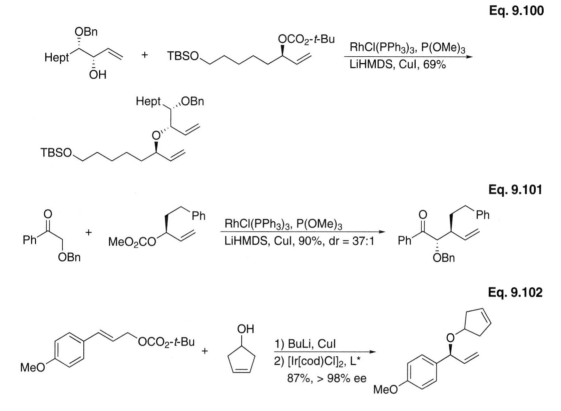

Eq. 9.101

Eq. 9.102

Further complicating the regioselectivity in transition-metal-catalyzed allylic alkylation reactions is that nucleophiles, for some catalyst systems, add to the carbon bearing the leaving group, regardless of substitution. This "memory effect" is perhaps the result of the formation of an η¹-allyl complex and not an η³-allyl complex (**Eq. 9.103**).[161] The issue of regioselectivity in these processes is not yet resolved. Reactions of preformed η³-allyl metal complexes are discussed in the following sections.

Eq. 9.103

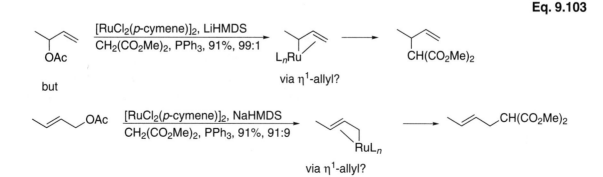

b. η³-Allylmolybdenum and Tungsten Complexes

Preformed η³-allylmolybdenum complexes can be used both to activate adjacent functional groups and to control stereochemistry. For example, the acetyl complex in **Eq. 9.104** undergoes facile aldol condensation with benzaldehyde, followed by reduction by $NaBH_4$ from the face opposite the metal. Replacement of the relatively inert CO ligands by nitrosyl (formally NO⁺) and chloride produces a more labile complex that slowly condenses with benzaldehyde to give the 1,3-diol in modest yield after hydrolysis.[162] The neutral η³-allylmolybdenum complexes are quite robust. For example, the Horner–Wadsworth–Emmons reaction and osmium tetroxide oxidation of the thus formed alkene can be performed. Reaction of functional groups adjacent to the η³-allylmolybdenum moiety occurs with high stereoselectivity. After completed side-chain manipulations, the complex can be activated toward nucleophilic addition by a CO to NO exchange (**Eq. 9.105**).[163]

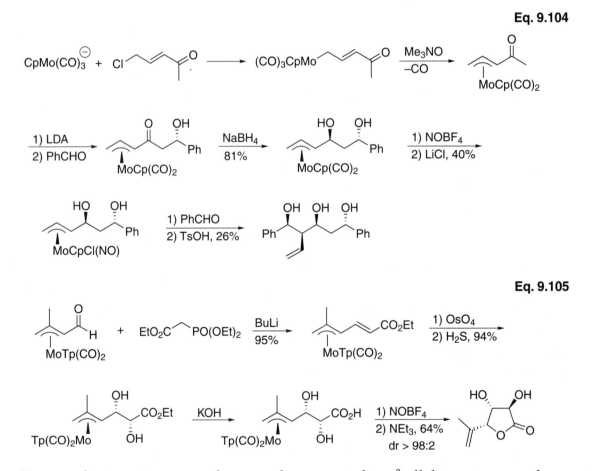

Eq. 9.104

Eq. 9.105

Propargylic tungsten σ-complexes can be converted to η³-allyltungsten complexes via inter- or intramolecular alkoxycarbonylations. The resulting complexes react with aldehydes to give lactones and this reaction has been used in organic synthesis (**Eqs. 9.106**[164] and **9.107**[165]).

Eq. 9.106

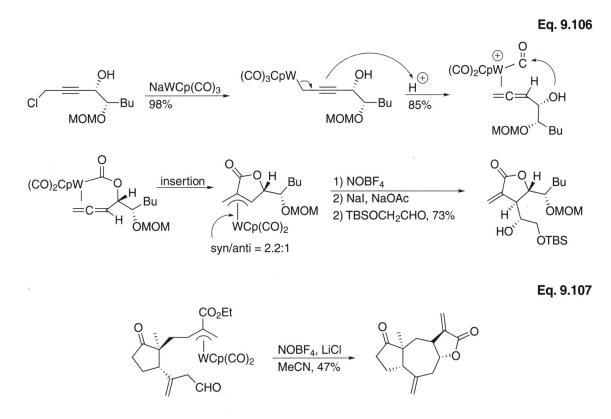

Eq. 9.107

The η³-allylmolybdenum group can be utilized to direct the facial selectivity of the addition of nucleophiles to cyclic complexes. For example, abstraction of a hydride or a methoxy group from positions adjacent to the allylic functionality yields a cationic diene complex. Regio- and stereoselective addition of a nucleophile restores the η³-allylmolybdenum complex. This type of reaction was used in an elegant synthesis of (−)-andrachcinidine (**Eq. 9.108**).[166]

Eq. 9.108

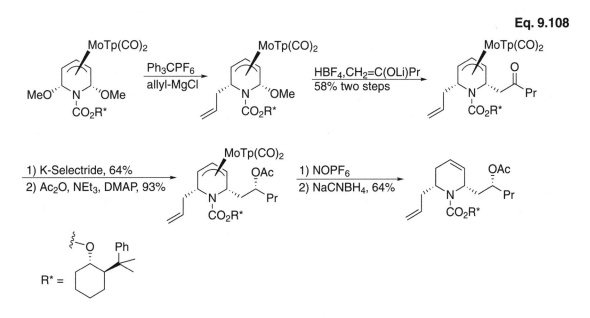

Neutral, optically active η³-pyranylmolybdenum participates in [5+2] and [5+3] cycloadditions. Again, the reaction occurs from the face opposite the metal (**Eqs. 9.109**[167] and **9.110**[168]).

Eq. 9.109

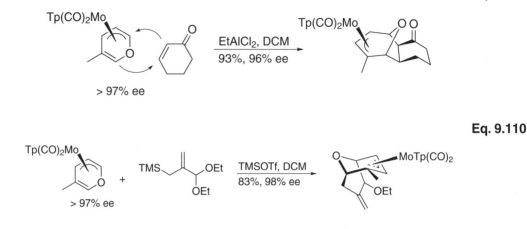

Eq. 9.110

c. η³-Allyliron Complexes

Cationic η³-allyliron tricarbonyl complexes undergo nucleophilic attack by stabilized carbanions in much the same way as η³-allylpalladium complexes, and Na[Fe(CO)$_3$(NO)] and [Bu$_4$N][Fe(CO)$_3$(NO)] catalyze the allylic alkylation of allylic chlorides, carbonates, and acetates.[169,170] However, perhaps because of the efficiency of the palladium-catalyzed processes, this related iron chemistry has found little use in organic synthesis.

In contrast to catalytic systems, η³-allyliron complexes preformed from γ-alkoxy or acetoxy enones are becoming quite useful in organic synthesis.[171] This utility relies on the ability of iron to stereoselectively complex a single prochiral face of the alkene, and to direct nucleophilic attack to the face opposite the metal, thereby leading to highly stereoselective processes. Early studies centered on unsaturated lactams and indicated that the γ-alkoxy group directed complexation primarily to the same face it occupied (**Eq. 9.111**).[172] The diastereoisomers could be separated, treated with allylsilane and BF$_3$OEt$_2$, and decomplexed to give a single diastereoisomer of allylated product resulting from exclusive attack from the face opposite the metal. Acyclic γ-alkoxyenones show similar selectivity and reactivity, and have been more often utilized in syntheses (**Eq. 9.112**).[173]

Eq. 9.111

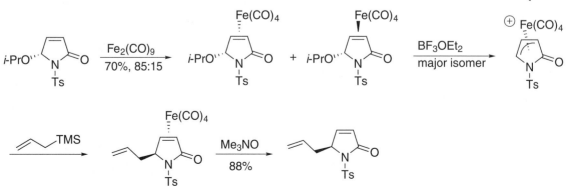

Eq. 9.112

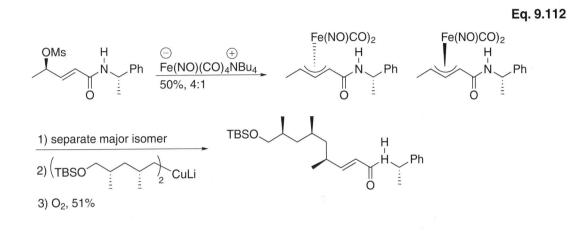

d. η³-Allylcobalt Complexes

The cobalt carbonyl anion, $[Co(CO)_4]^-$, is a weak base but modest nucleophile that reacts with allylic halides to produce η³-allylcobalt carbonyl complexes (**Eq. 9.113**).[174] These are relatively unstable, deep red oils, and little synthetic use for them has been found. When treated with methyl iodide, the $[Co(CO)_4]^-$ anion reacts to form the η¹-complex, which readily inserts CO to form the η¹-acyl complex. Treatment of this complex with butadiene results in another insertion, producing the β-acyl–η³-allylcobalt complex (**Eq. 9.114**). Treatment with a base abstracts the quite acidic α-proton, producing the acyl diene, and regenerating the cobalt carbonyl anion.[175] In contrast, treatment with a stabilized carbanion results in nucleophilic attack at the unsubstituted end of the η³-allyl group, resulting in an overall alkylation–acylation of the 1,3-diene.[176] Neither of these processes has been utilized with complex substrates.

Eq. 9.113

$$Co_2(CO)_8 \ + \ Na/Hg \ \longrightarrow \ 2 \ Na \ Co(CO)_4 \ \xrightarrow{\text{allyl-Br}} \ \text{(–Co(CO)}_3$$

$$M(0), \ d^9 \qquad\qquad\qquad M(-1), \ d^{10}$$

Eq. 9.114

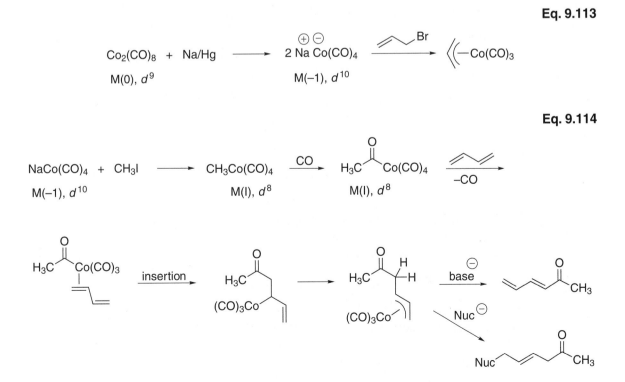

e. η^3-Allylnickel Complexes[177]

η^3-Allylnickel halides are generated in high yield by the reaction of allylic halides with nickel(0) complexes, usually nickel carbonyl or *bis*(cyclooctadiene)nickel, in nonpolar solvents such as benzene (**Eq. 9.115**). This reaction tolerates a range of functional groups in the allyl chain, and allows the preparation of a number of synthetically useful functionalized complexes. η^3-Allylnickel halides are not directly accessible by allylic deprotonation of alkene complexes (see **Figure 9.1**), in contrast to the corresponding palladium complexes, at least in part because nickel(II)–alkene complexes are virtually unknown, allowing no pathway for the activation of the allylic position for proton removal. η^3-Allylnickel halide complexes are deep red to red-brown, crystalline solids, which are quite air-sensitive in solution, but are stable in the absence of air.

Eq. 9.115

Although η^3-allylpalladium halides are subject to nucleophilic attack, η^3-allylnickel halides are not, in most cases. Instead, they behave, at least superficially, as if they were nucleophiles themselves, reacting with organic halides, aldehydes, and ketones to transfer the allylic group. However, these reactions are radical-chain processes rather than nucleophilic reactions, and the chemistry of η^3-allylnickel halides is drastically different from that of the corresponding palladium complexes. The process is initiated by light or added reducing agents and completely inhibited by less than one mole percent of *m*-dinitrobenzene, an efficient radical anion scavenger. Chiral secondary halides racemize upon allylation, implying the intermediacy of free carbon-centered radicals, but alkenyl halides maintain their stereochemistry, implying an absence of free carbon-centered radicals.[178]

The best-established and most widely used reaction of η^3-allylnickel halide complexes is their reaction with organic halides to replace the halogen with the allyl group (**Eq. 9.116**). This reaction proceeds only in polar coordinating solvents, such as DMF, HMPA, or *N*-methylpyrrolidone.

Eq. 9.116

R = alkyl, aryl, alkenyl, benzyl
tolerates: OH, NH_2, CO_2R, CO_2H, CHO, COR, CN
rate of reaction: I > Br > Cl

Aryl, alkenyl, primary, and secondary alkyl bromides and iodides react in high yields, with aryl and alkenyl halides being considerably more reactive than the alkyl halides. Chlorides react much more slowly than bromides or iodides. This reaction tolerates a wide variety of functional groups, including hydroxyl, ester, amide, and nitrile. These complexes react with bromides in preference to chlorides in the same molecule, and tolerate ketones and aldehydes in some instances. With unsymmetric η³-allyl groups, coupling occurs exclusively at the less-substituted terminus, in contrast to most main group allyl organometallics. This property was used to an advantage in the synthesis of (+)-cerulenine (**Eq. 9.117**)[179] and lavanduquinocin (**Eq. 9.118**).[180]

Eq. 9.117

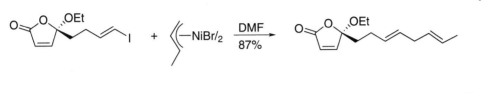

Eq. 9.118

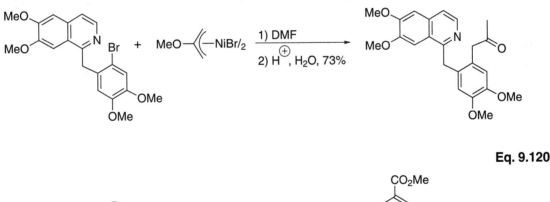

The ability of η³-allylnickel halides to react with aryl halides under very mild conditions and to tolerate a wide range of functional groups permits the introduction of allyl groups into a very broad array of substrates (**Eqs. 9.119**[181] and **9.120**[182]).

Eq. 9.119

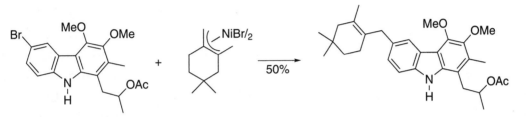

Eq. 9.120

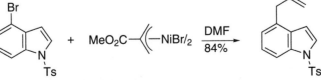

In contrast to reactions in apolar solvents such as benzene, which yield η³-allylnickel halide dimers, allylic halides undergo clean homocoupling at the less-substituted terminus when treated with nickel carbonyl in polar solvents such as DMF or THF (**Eq. 9.121**).[183]

Eq. 9.121

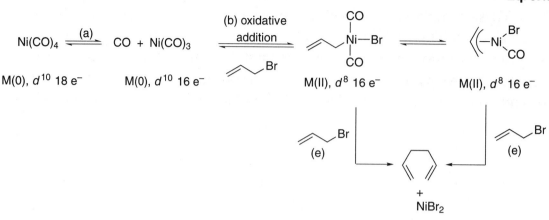

The reaction is reasonably well-understood, and involves dissociation of one CO from the coordinatively saturated, very labile, nickel carbonyl [path (a) in **Eq. 9.122**], followed by oxidative addition of the allylic halide to the unsaturated Ni(CO)₃ fragment [path (b)], giving ultimately the η³-allylnickel carbonyl intermediate [path (c)]. This complex has been detected spectroscopically (v_{CO} 2060 cm⁻¹) and can be generated independently by treating the η³-allylnickel bromide dimer with CO. In DMF, however, both the dimer and the η³-allylnickel carbonyl bromide monomer react quickly with excess allylic bromide to give the coupled product, biallyl. All steps except the last one appear to be reversible, since simple treatment of the η³-allylnickel bromide dimer with carbon monoxide in DMF results in efficient coupling, and both allyl bromide and Ni(CO)₄ can be detected in solution. Under high dilution conditions, α,ω-bis-allylic halides are cyclized (**Eq. 9.122**).[184] In this way, simple 12-, 14-, and 18-membered rings,[185] humulene,[186] and macrocyclic lactones[187] have been prepared. This potentially useful coupling has been utilized very little, primarily because nickel carbonyl is a volatile (bp 43°C), highly toxic, colorless liquid that is difficult to handle and to dispose of.

Eq. 9.122

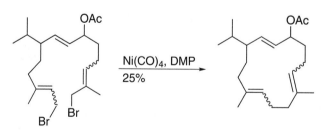

Although allylic halides are among the most reactive toward these complexes, coupling normally gives all possible products because of rapid exchange of the η³-allyl ligand with the allylic halide. If the two allyl groups are somewhat different electronically, selective cross coupling is sometimes observed in reasonable yield.[188] The stereochemistry of the alkene in the η³-allyl system is lost in these allyl transfer reactions, but the stereochemistry of the alkene in alkenyl halide substrates is normally maintained.

Conjugated enones react with nickel(0) complexes in the presence of trimethylsilyl chloride to produce 1-trimethylsilyloxy-η³-allylnickel complexes. These undergo typical coupling reactions with organic halides to produce silylenol ethers (**Eq. 9.123**).[189] This product is one that would arise by nucleophilic (R⁻) addition to the β-position of the starting enone. However, by complexation to nickel, the normal reactivity patterns are reversed, and the β-alkyl group is introduced as an electrophile (RX = "R⁺"). By using η³-allylnickel complexes as intermediates, the conjugate addition of alkyltin reagents to conjugated enones can be catalyzed. Although these two processes are related, they differ somewhat mechanistically (**Eq. 9.124**).[190]

Eq. 9.123

Eq. 9.124

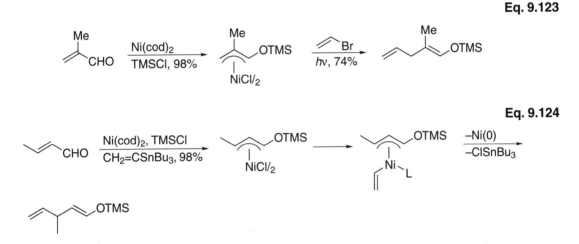

Although η³-allylnickel halides react with organic halides in preference to aldehydes or ketones, under slightly more vigorous conditions (50°C versus 20°C) carbonyl compounds do react to produce homoallylic alcohols.[191] α-Diketones are the most reactive substrates, producing α-ketohomoallylic alcohols. Aldehydes and cyclic ketones, including cholestanone, progesterone, and 5-α-androstane-3,17-dione, react well, but simple aliphatic and α,β-unsaturated ketones react only sluggishly. Again, reaction occurs at the less-substituted end of the allyl group, in contrast to main group allyl organometallics. The reaction of η³-(2-carboethoxyallyl)nickel bromide with aldehydes and ketones produces α-methylene-γ-butyrolactones (**Eq. 9.125**). Finally, nickel catalyzes an asymmetric reductive cyclization of conjugated dienes having a pendant aldehyde. This reaction probably involves the formation of an η³-allylnickel complex via hydro-nickelation of the diene, followed by cyclization (**Eq. 9.126**).[192]

Eq. 9.125

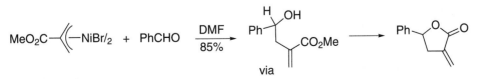

Eq. 9.126

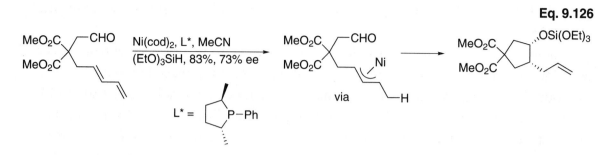

In a process that must involve η³-allylnickel complex intermediates and oxidative addition–transmetallation–reductive elimination cycles, nickel(II) phosphine complexes catalyze the alkylation of allylic acetates by aryl[193] and alkenyl boronates[194] (**Eq. 9.127**).[195] These processes bear a striking resemblance to η³-allylpalladium chemistry discussed above, and are likely to increase the scrutiny of nickel complexes to catalyze other processes thought to be the exclusive domain of palladium chemistry, such as allylic amination (**Eq. 9.128**).[196] The nickel-catalyzed asymmetric allylic amination shown in **Eq. 9.129** is a good example of the increasing attention to this transition metal.[197]

Eq. 9.127

Eq. 9.128

Eq. 9.129

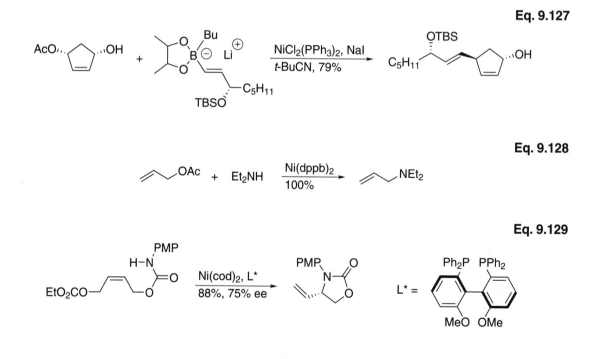

References

(1) For reviews, see a) Tsuji, J. *Acc. Chem. Res.* **1973**, *6*, 8. b) Tsuji, J. *Pure Appl. Chem.* **1981**, *53*, 2371. c) Tsuji, J. *Pure Appl. Chem.* **1982**, *54*, 197. d) Takacs, J. M. Transition Metal Allyl Complexes: Telomerization of Dienes. In *Comprehensive Organometallic Chemistry*; Wilkinson, G., Stone, F. G. A., Abel, E. W., Eds.; Pergamon: New York, 1982; Vol. 6, pp 785–797.

(2) Rodriguez, A.; Nomen, M.; Spur, B. W.; Godfroid, J.-J.; Lee, T. K. *Eur. J. Org. Chem.* **2000**, 2991.

(3) Takahashi, T.; Minami, I.; Tsuji, J. *Tetrahedron Lett.* **1981**, *22*, 2651.

(4) a) Takacs, J. M.; Zhu, J. *J. Org. Chem.* **1989**, *54*, 5193. b) Takacs, J. M.; Chandramouli, S. *Organometallics* **1990**, *9*, 2877. c) Takacs, J. M.; Zhu, J. *Tetrahedron Lett.* **1990**, *31*, 1117.

(5) Takacs, J. M.; Chandramouli, S.V. *J. Org. Chem.* **1993**, *58*, 7315.

(6) Takacs, J. M.; Zhu, J.; Chandramouli, S. *J. Am. Chem. Soc.* **1992**, *114*, 773.

(7) In a synthesis of (–)-gibboside. Takacs, J. M.; Vayalakkada, S.; Mehrman, S. J.; Kingsbury, C. L. *Tetrahedron Lett.* **2002**, *43*, 8417.

(8) Takacs, J. M.; Weidner, J. J.; Takacs, B. E. *Tetrahedron Lett.* **1993**, *34*, 6219.

(9) In a synthesis of a homolog of (–)-protoemetinol. Takacs, J.M.; Boito, S.C. *Tetrahedron Lett.* **1995**, *36*, 2941.

(10) Sato, Y.; Oonishi, Y.; Mori, M. *Organometallics* **2003**, *22*, 30.

(11) Jolly, P. W. Nickel-Catalyzed Oligomerization of Alkenes and Related Reactions. In *Comprehensive Organometallic Chemistry*; Wilkinson, G., Stone, F. G. A., Abel, E. W., Eds.; Pergamon: New York, 1982; Vol. 6, pp 615–648.

(12) a) Wender, P. A.; Ihle, N. C.; Corriea, C. R. D. *J. Am. Chem. Soc.* **1988**, *110*, 5904. b) For related cycloadditions, see Wender, P. M.; Nuss, J. M.; Smith, D. B.; Swarez-Sobrino, A.; Vågberg, J.; DeCosta, D.; Bordner, J. *J. Org. Chem.* **1997**, *62*, 4908.

(13) Wender, P. A.; Jenkins, T. E. *J. Am. Chem. Soc.* **1989**, *118*, 6432.

(14) For a review, see Hyland, C. *Tetrahedron* **2005**, *61*, 3457.

(15) a) For a review on selectivity in these reactions, see Frost, C. G.; Howarth, J.; Williams, J. M. J. *Tetrahedron: Asymmetry* **1992**, *3*, 1089. b) For a general review, see Harrington, P. M. Transition Metal Allyl Complexes: Pd, W, Mo-Assisted Nucleophilic Attack. In *Comprehensive Organometallic Chemistry II*; Abel, E. W., Stone, F. G. A., Wilkinson, G., Eds.; Pergamon: Oxford, U.K., 1995; Vol. 12, pp 797–904.

(16) a) Trost, B. M.; Bant, R. C. *J. Am. Chem. Soc.* **1998**, *120*, 70. b) Kawatsura, M.; Uozumi, Y.; Hayashi, T. *Chem. Commun.* **1998**, 217.

(17) Hayashi, T.; Kawatsura, M.; Uozumi, Y. *J. Am. Chem. Soc.* **1998**, *120*, 1681.

(18) For old reviews, see a) Trost, B. M.; Verhoeven, T. R. *J. Am. Chem. Soc.* **1980**, *102*, 4730. b) Yamamoto, T.; Saito, O.; Yamamoto, A. *J. Am. Chem. Soc.* **1981**, *103*, 5600. c) Trost, B. M. *Acc. Chem. Res.* **1980**, *13*, 385.

(19) For a review using ketone enolates, see Braun, M.; Meier, T. *Angew. Chem. Int. Ed.* **2006**, *45*, 6952.

(20) a) Hayashi, T.; Konishi, M.; Kumada, M. *J. Chem. Soc., Chem. Commun.* **1984**, 107. b) Hayashi, T.; Hagihara, T.; Konishi, M.; Kumada, M. *J. Am. Chem. Soc.* **1983**, *105*, 7767. c) Leutenegger, V.; Umbricht, G.; Fahrni, C. von Matt, P.; Pfaltz, A. *Tetrahedron* **1992**, *48*, 2143.

(21) Blacker, A. J.; Clarke, M. I.; Loft, M. S.; Williams, J. M. J. *Org. Lett.* **1999**, *1*, 1969.

(22) Trost, B. M.; Ceschi, M. A.; König, B. *Angew. Chem., Int. Ed. Engl.* **1997**, *36*, 1486.

(23) Naz, N.; Al-Tey, T.H.; Al-Abed, Y.; Voelter, W.; Fikes, R.; Hiller, W. *J. Org. Chem.* **1996**, *61*, 3230.

(24) Braun, M.; Onuma, H.; Arinaga, Y. *Chem. Lett.* **1995**, 1099.

(25) In a synthesis of tremulenediol A and tremulenolide A. Ashfeld, B. L.; Martin, S. F. *Org. Lett.* **2005**, *7*, 4535.

(26) In a synthesis of (–)-strychnine. Kaburagi, Y.; Tokuyama, H.; Fukuyama, T. *J. Am. Chem. Soc.* **2004**, *126*, 10246.

(27) Maezaki, N.; Hirose, Y.; Tanaka, T. *Org. Lett.* **2004**, *6*, 2177.

(28) For a review, see Heumann, A.; Regher, M. *Tetrahedron* **1995**, *51*, 975.

(29) Roland, S.; Durand, J. O.; Savignac, M.; Genet, J. P. *Tetrahedron Lett.* **1995**, *36*, 3007.

(30) Fürstner, A.; Weintritt, H. *J. Am. Chem. Soc.* **1998**, *120*, 2817.

(31) a) Boeckman, R. K.; Shair, M. D.; Vargas, J. R.; Stoltz, L. A. *J. Org. Chem.* **1993**, *58*, 1295. b) Michelet, V.; Besner, I.; Genet, J. P. *Synlett* **1996**, 215.

(32) In a synthesis of vibsanin F. Yuasa, H.; Makado, G.; Fukuyama, Y. *Tetrahedron Lett.* **2003**, *44*, 6235.

(33) In a synthesis of sordaricin. Kitamura, M.; Chiba, S.; Narasaki, K. *Chem. Lett.* **2004**, *33*, 942.

(34) In a synthesis of δ-araneosene. Hu, T.; Corey, E. J. *Org. Lett.* **2002**, *4*, 2441.

(35) a) Schink, H. E.; Bäckvall, J.-E. *J. Org. Chem.* **1992**, *57*, 1588. b) Bäckvall, J.-E.; Gatti, R.; Shink, H. E. *Synthesis* **1993**, 343.

(36) For reviews, see a) Trost, B. M. *J. Org. Chem.* **2004**, *69*, 5813. b) Trost, B. M.; Crawley, M. L. *Chem. Rev.* **2003**, *103*, 2921. c) Helmchen, G.; Kudis, S.; Sennhenn, P.; Steinhaugher, H. *Pure Appl. Chem.* **1997**, *69*, 513. d) Trost, B. M.; Van Vranken, D. L. *Chem. Rev.* **1996**, *96*, 395. e) Trost, B. M. *Acc. Chem. Res.* **1996**, *29*, 355. f) Williams, J. M. J. *Synlett* **1996**, 705.

(37) a) Kardos, N.; Genet, J. P. *Tetrahedron: Asymmetry* **1994**, *5*, 1525. b) See also Pretot, R.; Pfaltz, A. *Angew. Chem. Int. Ed.* **1998**, *37*, 323.

(38) Granberg, K. L.; Bäckvall, J.-E. *J. Am. Chem. Soc.* **1992**, *114*, 6858.

(39) Trost, B. M.; Lee, C. *J. Am. Chem. Soc.* **2001**, *123*, 12191.

(40) For a recent example, see He, X.-C.; Wang, B.; Yu, G.; Bai, D. *Tetrahedron: Asymmetry* **2001**, *12*, 3213.

(41) For a mechanistic study, see Seebach, D.; Devaquat, E.; Ernst, A.; Hayakawa, M.; Kühnle, F. N. M.; Schweizer, W. B.; Weber, B. *Helv. Chim. Acta* **1995**, *78*, 1636.

(42) Baldwin, J. C.; Williams, J. M. J.; Beckett, R. P. *Tetrahedron: Asymmetry* **1995**, *6*, 1515.

(43) Trost, B. M.; Bunt, R. C. *J. Am. Chem. Soc.* **1994**, *116*, 4089.

(44) In a synthesis of C-2-epi-hygromycin A. Trost, B. M.; Dudash, J., Jr.; Dirat, O. *Chem.—Eur. J.* **2002**, 259.

(45) In a synthesis of valienamine. Trost, B. M.; Chipak, L. S.; Lübbers, T. *J. Am. Chem. Soc.* **1998**, *120*, 1732.

(46) Yoshigaki, H.; Satoh, H.; Sato, Y.; Nukui, S.; Shibasaki, M.; Mori, M. *J. Org. Chem.* **1995**, *60*, 2016.

(47) In a synthesis of (+)-γ-lycorane. Chapsal, B. D.; Ojima, I. *Org. Lett.* **2006**, *8*, 1395.

(48) a) Takanashi, S.-i.; Mori, K. *Liebigs Ann. Chem.* **1997**, 825. b) For the original work in this area, see Del Valle, L.; Stille, J. K.; Hegedus, L. S. *J. Org. Chem.* **1990**, *55*, 3019.

(49) In a synthesis of FR182877. Vanderwal, C. D.; Vosburg, D. A.; Weiler, S.; Sorensen, E. J. *J. Am. Chem. Soc.* **2003**, *125*, 5393.

(50) Nicolaou, K. C.; Koftis, T. V.; Vyskocil, S.; Petrovic, G.; Tang, W.; Frederick, M. O.; Chen, D. Y.-K.; Li, Y.; Ling, T; Yamada, Y. M. A. *J. Am. Chem. Soc.* **2006**, *128*, 2859.

(51) Castaño, A. M.; Ruano, M.; Echavarren, A. M. *Tetrahedron Lett.* **1996**, *37*, 6591.

(52) a) White, J. D.; Jensen, M. S. *Synlett* **1996**, 31. b) Echavarren, A. M.; Tueting, D. R.; Stille, J. K. *J. Am. Chem. Soc.* **1988**, *110*, 4039. c) Teuting, D. R.; Echavarren, A. M.; Stille, J. K. *Tetrahedron* **1989**, *45*, 979.

(53) In a synthesis of madangamines. Yoshimura, Y.; Inoue, J.; Yamazaki, N.; Aoyagi, S.; Kibayashi, C. *Tetrahedron Lett.* **2006**, *47*, 3489.

(54) Farina, V.; Baker, S. R.; Benigni, D.A.; Sapino, C., Jr. *Tetrahedron Lett.* **1988**, *29*, 5739.

(55) In a synthesis of (+)-lasonolide A. Yoshimura, T.; Yakushiji, F.; Kondo, S.; Wu, X.; Shindo, M.; Shishido, K. *Org. Lett.* **2006**, *8*, 475.

(56) Urabe, H.; Inami, H.; Sato, F. *J. Chem. Soc., Chem. Commun.* **1993**, 1595.

(57) For examples of stoichiometric reactions, see a) Otte, A. R.; Wilde, A.; Hoffmann, H. M. R. *Angew. Chem., Int. Ed. Engl.* **1994**, *33*, 1280. b) Hoffmann, H. M. R.; Otte, A. R.; Wilde, A.; Menzer, S.; Williams, D. J. *Angew. Chem., Int. Ed. Engl.* **1995**, *34*, 100. c) Hegedus, L. S.; Darlington, W. H.; Russell, C. E. *J. Am. Chem. Soc.* **1980**, *45*, 5193.

(58) For examples of catalytic reactions, see a) Satake, A.; Nakata, T. *J. Am. Chem. Soc.* **1998**, *120*, 10391. b) Satake, A.; Kadohama, H.; Koshino, H.; H.; Nakata, T. *Tetrahedron Lett.* **1999**, *40*, 3597.

(59) Shintani, R.; Park, S.; Hayashi, T. *J. Am. Chem. Soc.* **2007**, *129*, 14866.

(60) For a review on allylic amination, see a) Johannsen, M.; Jørgensen, K. A. *Chem. Rev.* **1998**, *98*, 1689. b) Flegelova, Z.; Patek, M. *J. Org. Chem.* **1996**, *61*, 6735.

(61) For a discussion of this isomerization, see Watson, I. D. G.; Yudin, A. K. *J. Am. Chem. Soc.* **2005**, *127*, 17516.

(62) Kapeller, H.; Marschener, C.; Weissenbacher, M.; Griengl, H. *Tetrahedron* **1998**, *54*, 1439.

(63) In a synthesis of carbovir. Olivo, H. F.; Yu, J. *J. Chem. Soc., Perkin Trans. 1* **1998**, 391.

(64) In a synthesis of valienamine. Kok, S. H.-L.; Lee, C. C.; Shing, T. K. M. *J. Org. Chem.* **2001**, *66*, 7184.

(65) Humphrey, G. R.; Miller, R. A.; Pye, P. J.; Rossen, K.; Reamer, R. A.; Maliakal, A.; Ceglia, S. S.; Grabowski, E. J. J.; Volante, R. P.; Reider, P. J. *J. Am. Chem. Soc.* **1999**, *121*, 11261.

(66) In a synthesis of mesembrane and mesembrine. Mori, M.; Kuroda, S.; Zhang, C.-S.; Sato, Y. *J. Org. Chem.* **1997**, *62*, 3263.

(67) Ovaa, H.; Stragies, R.; van der Marel, G. A.; van Boom, J. H.; Blechert, S. *Chem. Commun.* **2000**, 1501.

(68) In synthesis of (+)-pancratistatin. Trost, B. M.; Pulley, S. R. *J. Am. Chem. Soc.* **1995**, *117*, 10143.

(69) In a synthesis of (–)-anatoxin-a. Trost, B. M.; Oslob, J. D. *J. Am. Chem. Soc.* **1999**, *121*, 3057.

(70) In a synthesis of the B-ring of agelastatin A. Stein, D.; Anderson, T. G.; Chase, C. E.; Koh, Y.; Weinreb, S. M. *J. Am. Chem. Soc.* **1999**, *121*, 9574.

(71) In a synthesis of (–)-cephalotaxin. Tietze, L. F.; Schirok, H. *J. Am. Chem. Soc.* **1999**, *121*, 10264.

(72) In a synthesis of inandenine-12-one. Trost, B. M.; Cossy, J. *J. Am. Chem. Soc.* **1982**, *104*, 6881.

(73) Bradette, T.; Esher, J. L.; Johnson, C. R. *Tetrahedron: Asymmetry* **1995**, *36*, 6251.

(74) Goux, C.; Massacret, M.; Lhoste, P.; Sinow, D. *Organometallics* **1995**, *14*, 4585.

(75) Trost, B. M.; Toste, F. D. *J. Am. Chem. Soc.* **1998**, *120*, 815.

(76) In a synthesis of (+)-clusifoliol. Trost, B. M.; Shen, H. C.; Dong, L.; Surivet, J.-P. *J. Am. Chem. Soc.* **2003**, *125*, 9276.

(77) In a synthesis of aflatoxins. Trost, B. M.; Toste, F. D. *J. Am. Chem. Soc.* **2003**, *125*, 3090.

(78) Sinou, D.; Frappa, I.; Lhoste, P.; Porwanski, S.; Kryczka, B. *Tetrahedron Lett.* **1995**, *36*, 6251.

(79) In a synthesis of the anthrax tetrasaccharide. Guo, H.; O'Doherty, G. A. *Angew. Chem. Int. Ed.* **2007**, *46*, 5206.

(80) Trost, B. M.; Tenaglia, A. *Tetrahedron Lett.* **1988**, *29*, 2974.

(81) Fournier-Nguefack, C.; Lhoste, P.; Sinow, D. *J. Chem. Res., Synop.* **1998**, 105.

(82) In a synthesis toward amphidinolide K. Williams, D. R.; Meyer, K. G. *Org. Lett.* **1999**, *1*, 1303.

(83) Keinan, E.; Sahai, M.; Roth, Z. *J. Org. Chem.* **1985**, *50*, 3558.

(84) Trost, B. M.; Greenspan, P. D.; Geissler, H.; Kim, J. H.; Greeves, N. *Angew. Chem., Int. Ed. Engl.* **1994**, *33*, 2182.

(85) Shimizu, I.; Omura, T. *Chem. Lett.* **1993**, 1759.

(86) a) Hayashi, T.; Yamane, M.; Ohno, A. *J. Org. Chem.* **1997**, *62*, 204. b) Tenaglia, A.; Krammerer, F. *Synlett* **1996**, 576.

(87) a) Bäckvall, J.-E. Metal-Mediated Additions to Conjugated Dienes In *Advances in Metal-Organic Chemistry*; Liebeskind, L. S., Ed.; JAI Press: London, 1989; Vol. 1, pp 135–175. b) Bäckvall, J.-E.; Byström, S. E.; Nordberg, R. E. *J. Org. Chem.* **1984**, *49*, 4619. c) Nyström, J.-E.; Rein, T.; Bäckvall, J.-E. *Org. Synth.* **1989**, *67*, 105.

(88) Verboom, R. C.; Slagt, V. F.; Bäckvall, J.-E. *Chem. Commun.* **2004**, 1282 and references therein.

(89) a) Bäckvall, J.-E.; Andersson, P. G. *J. Am. Chem. Soc.* **1992**, *114*, 6374. b) Itami, K.; Palmgren, A.; Bäckvall, J.-E. *Tetrahedron Lett.* **1998**, *39*, 1223.

(90) For examples of C-nucleophiles, see a) Hou, R.-S.; Wang, H.-M.; Huang, H.-Y.; Chen, L.-C. *Heterocycles* **2005**, *65*, 1917. b) Kimura, M.; Futamata, M.; Mukai, R.; Tamaru, Y. *J. Am. Chem. Soc.* **2005**, *127*, 4592.

(91) For reviews, see a) Muzart, J. *Eur. J. Org. Chem.* **2007**, 3077. b) Tamaru, Y. *Eur. J. Org. Chem.* **2005**, 2647.

(92) Kimura, M.; Futumata, M.; Mukai, R.; Tamaru, Y. *J. Am. Chem. Soc.* **2005**, *127*, 4592.

(93) Yokoyama, Y.; Hikawa, H.; Mitsuhashi, M.; Uyama, A.; Hiroki, Y.; Murakami, Y. *Eur. J. Org. Chem.* **2004**, 1244.

(94) For a review, see Tsuji, J.; Mandai, T. *Synthesis* **1996**, 1.

(95) a) Mandai, T.; Kaihara, Y.; Tsuji, J. *J. Org. Chem.* **1994**, *59*, 5847. b) Lautens, M.; Delanghe, P. H. M. *Angew. Chem., Int. Ed. Engl.* **1994**, *33*, 2448.

(96) In a synthesis of (–)-tuberostemonine. Wipf, P.; Rector, S. R.; Takahashi, H. *J. Am. Chem. Soc.* **2002**, *124*, 14848.

(97) Nagasawa, K.; Shimizu, I.; Nakata, T. *Tetrahedron Lett.* **1996**, *37*, 6881.

(98) a) Hayashi, T.; Iwamura, I.; Naito, M.; Matsumoto, Y.; Nozumi, Y.; Miki, M.; Yanagai, K. *J. Am. Chem. Soc.* **1994**, *116*, 775. b) Hayashi, T.; Kawatsura, M.; Iwamura, H.; Yamura, Y.; Uozumi, Y. *Chem. Comm.* **1996**, 1767.

(99) Kawatsura, M.; Uozumi, Y.; Ogasawara, M.; Hayashi, T. *Tetrahedron* **2000**, *56*, 2247.

(100) In a synthesis of (–)-hippospongic acid A. Trost, B. M.; Machacek, M. R.; Tsui, H. C. *J. Am. Chem. Soc.* **2005**, *127*, 7014.

(101) a) Mandai, T.; Matsumoto, T.; Kawada, M.; Tsuji, J. *J. Org. Chem.* **1992**, *57*, 1326. b) Mandai, T.; Suzuki, S.; Murakami, T.; Fujita, M.; Kawada, M.; Tsuji, J. *Tetrahedron Lett.* **1992**, *33*, 2987.

(102) Mandai, T.; Matsumoto, T.; Tsuji, J. *Tetrahedron Lett.* **1993**, *34*, 2513.

(103) Mandai, T.; Matsumoto, M.; Nakao, Y.; Teramoto, A.; Kawada, M.; Tsuji, J. *Tetrahedron Lett.* **1992**, *33*, 2549.

(104) Shimizu, I.; Matsumoto, Y.; Ono, T.; Satake, A.; Yamamoto, A. *Tetrahedron Lett.* **1996**, *37*, 7115.

(105) Andersson, P. G.; Schab, S. *Organometallics* **1995**, *14*, 1.

(106) For a review, see Guibe, F. *Tetrahedron* **1998**, *54*, 2967.

(107) In a synthesis of (–)-cribostatin 4. Chen. X.; Zhu, J. *Angew. Chem. Int. Ed.* **2007**, *46*, 3962.

(108) In a synthesis of discodermolides. Hang, D. T.; Nerenberg, J. B.; Schreiber, S. L. *J. Am. Chem. Soc.* **1996**, *118*, 11054.

(109) Hayakawa, Y.; Wakabayashi, S.; Kato, H.; Noyori, R. *J. Am. Chem. Soc.* **1990**, *112*, 1691.

(110) Cardenas, D. J.; Alcami, M.; Cossio, F.; Menendez, M.; Echavarren, A. M. *Chem.—Eur. J.* **2003**, *9*, 96.

(111) For reviews, see a) Oppolzer, W. Transition Metal Allyl Complexes: Intramolecular Alkene and Alkyne Insertions. In *Comprehensive Organometallic Chemistry II*; Abel, E. W., Stone, F. G. A., Wilkinson, G., Eds.; Pergamon: Oxford, U.K., 1995; Vol. 12, pp 905–921. b) Oppolzer, W. *Angew. Chem., Int. Ed. Engl.* **1989**, *28*, 38. c) Oppolzer, W. *Pure Appl. Chem.* **1990**, *62*, 1941.

(112) In a synthesis of pentalenolactone E methyl ester. Oppolzer, W.; Xu, J.-Z.; Stone, C. *Helv. Chim. Acta* **1991**, *74*, 465.

(113) Holzapfel, C. W.; Marais, L. *J. Chem. Res., Synop.* **1998**, 60.

(114) Oppolzer, W.; DeVita, R. J. *J. Org. Chem.* **1991**, *56*, 6256.

(115) Grigg, R.; Sridharan, V.; Surkirthalingam, S. *Tetrahedron Lett.* **1991**, *32*, 3855.

(116) Trost, B. M.; Luengo, J. I. *J. Am. Chem. Soc.* **1988**, *110*, 8239.

(117) a) Oppolzer, W.; Bienayme, H.; Genevois-Borella, A. *J. Am. Chem. Soc.* **1991**, *113*, 9660. b) Oppolzer, W.; Xu, J.-Z.; Stone, C. *Helv. Chim. Acta* **1991**, *74*, 465.

(118) Keese, R.; Guidetti-Grept, R.; Herzog, B. *Tetrahedron Lett.* **1992**, *33*, 1207.

(119) Holzapfel, C. W.; Marais, L. *Tetrahedron Lett.* **1998**, *39*, 2179.

(120) For reviews, see a) Trost, B. M. *Angew. Chem., Int. Ed. Engl.* **1986**, *25*. 1. b) Trost, B. M. *Pure Appl. Chem.* **1988**, *60*, 1615. c) Harrington, P. J. Transition Metal Allyl Complexes: Trimethylene Methane Complexes. In *Comprehensive Organometallic Chemistry II*; Abel, E. W., Stone, F. G. A., Wilkinson, G., Eds., Pergamon: Oxford, U.K., 1995; Vol. 12, pp 923–945. d) Lautens, M.; Klute, W.; Tam, W. *Chem. Rev.* **1996**, *96*, 49.

(121) a) Trost, B. M.; Lynch, J.; Renaut, P.; Steinman, D. H. *J. Am. Chem. Soc.* **1986**, *108*, 284. b) Trost, B. M.; Mignani, S. M. *Tetrahedron Lett.* **1986**, *27*, 4137.

(122) Holzapfel, C. W.; van der Merwe, T. L. *Tetrahedron Lett.* **1996**, *37*, 2303.

(123) Trost, B. M.; Nanninga, T. N.; Satoh, T. *J. Am. Chem. Soc.* **1985**, *107*, 721.

(124) Trost, B. M.; Schneider, S. *Angew. Chem.* **1989**, *101*, 215.

(125) Trost, B. M.; Seoane, P. R. *J. Am. Chem. Soc.* **1987**, *109*, 615.

(126) Trost, B. M.; Higuchi, R. L. *J. Am. Chem. Soc.* **1996**, *118*, 10094 and references therein.

(127) Trost, B. M.; Grese, T. A. *J. Org. Chem.* **1992**, *57*, 686.

(128) Trost, B. M.; Stambuli, J. P.; Silverman, S. M.; Schwörer, U. *J. Am. Chem. Soc.* **2006**, *128*, 13328.

(129) Greco, G. E.; Gleason, B. L.; Lowery, T. A.; Kier, M. J.; Hollander, L. B.; Gibbs, S. A.; Worthy, A. D. *Org. Lett.* **2007**, *9*, 3817.

(130) Trost, B. M.; King, S. A. *J. Am. Chem. Soc.* **1990**, *112*, 408.

(131) In a synthesis of streptomyces metabolites. Jacobsen, M. F.; Moses, J. E.; Adlington, R. M.; Baldwin, J. E. *Tetrahedron* **2006**, *62*, 1675.

(132) a) Trost, B. M.; Sharma, S.; Schmidt, T. *J. Am. Chem. Soc.* **1992**, *114*, 7903. b) Trost, B. M.; Sharma, S.; Schmidt, T. *Tetrahedron Lett.* **1993**, *34*, 7183.

(133) Hedley, S. J.; Moran, W. J.; Price, D. A.; Harrity, J. P. A. *J. Org. Chem.* **2003**, *68*, 4286.

(134) a) Binger, P.; Büch, H. M. *Top. Curr. Chem.* **1987**, *135*, 77. b) For stereochemical and mechanistic studies, see Corley, H.; Motherwell, W. B.; Pennell, A. M. K.; Shipman, M.; Slawin, A. M. Z.; Williams, D. J.; Bingh, P.; Stepp, M. *Tetrahedron* **1996**, *52*, 4883.

(135) Lewis, R. T.; Motherwell, W. B.; Shipman, M.; Slawin, A. M. Z.; Williams, D. J. *Tetrahedron* **1995**, *51*, 3285.

(136) Lautens, M.; Ren, Y. *J. Am. Chem. Soc.* **1996**, *118*, 9597.

(137) Lautens, M.; Ren, Y. *J. Am. Chem. Soc.* **1996**, *118*, 10668.

(138) For an X-ray structure, see Tsutsumi, K.; Ogoshi, S.; Nishiguchi, S.; Kurosawa, H. *J. Am. Chem. Soc.* **1998**, *120*, 1938.

(139) For a review of reactions, see Ma, S. *Eur. J. Org. Chem.* **2004**, 1175.

(140) Ma, S.; Wang, G. *Angew. Chem. Int. Ed.* **2003**, *42*, 4215.

(141) In a synthesis of borrelidin. Nagamitsu, T.; Takano, D.; Marumoto, K.; Fukuda, T.; Furuya, K.; Otoguro, K.; Takeda, K.; Kuwajima, I.; Harigaya, Y.; Omura, S. *J. Org. Chem.* **2007**, *72*, 2744.

(142) Trost, B. M.; Hung, M.-H. *J. Am. Chem. Soc.* **1984**, *106*, 6837.

(143) Trost, B. M.; Merlic, C. A. *J. Am. Chem. Soc.* **1990**, *112*, 9590.

(144) Trost, B. M.; Hachiya, I. *J. Am. Chem. Soc.* **1998**, *120*, 1104.

(145) For a review of molybdenum-complex-catalyzed asymmetric allylic alkylations, see Belda, O.; Moberg, C. *Acc. Chem. Res.* **2004**, *37*, 159.

(146) Lloyd-Jones, G. C.; Krska, S. W.; Hughes, D. L.; Gouriou, L.; Bonnet, V. D.; Jack, K.; Sun, Y.; Reamer, R. A. *J. Am. Chem. Soc.* **2004**, *126*, 702.

(147) In a synthesis of (–)-Δ⁹-*trans*-tetrahydrocannabinol. Trost, B. M.; Dogra, K. *Org. Lett.* **2007**, *9*, 861.

(148) Malkov, A. V.; Baxendale, I. R.; Mansfield, D. J.; Kocovsky, P. *J. Chem. Soc., Perkin Trans. 1* **2001**, 1234.

(149) Co, T. T.; Paek, S. W.; Shim, S. C.; Cho, C. S.; Kim, T.-J. Choi, D. W.; Kang, S. O.; Jeong, J. H. *Organometallics* **2003**, *22*, 1475.

(150) Evans, P. A.; Nelson, J. D. *Tetrahedron Lett.* **1998**, *39*, 1729.

(151) Yu, C.-M.; Lee, S.; Hong, Y.-T.; Yoon, S.-K. *Tetrahedron Lett.* **2004**, *45*, 6557.

(152) For a review of ruthenium-catalyzed reactions, see Bruneau, C.; Renaud, J.-L.; Demerseman, B. *Chem.—Eur. J.* **2006**, *12*, 5178.

(153) a) Takeuchi, R.; Kashio, M. *Angew. Chem., Int. Ed. Engl.* **1997**, *36*, 263. b) Takeuchi, R.; Kashio, M. *J. Am. Chem. Soc.* **1998**, *120*, 8647.

(154) For a review, see Helmchen, G., Dahnz, A.; Dübon, P.; Schelwies, M.; Weihofen, R. *Chem. Commun.* **2007**, 675.

(155) Jansson, J. P.; Helmchen, G. *Tetrahedron Lett.* **1997**, *38*, 8025.

(156) Matsushima, Y.; Onitsuka, K.; Kondo, T.; Mitsudo, T.-a.; Takahashi, S. *J. Am. Chem. Soc.* **2001**, *123*, 10405.

(157) Kazmaier, U.; Stolz, D. *Angew. Chem. Int. Ed.* **2006**, *45*, 3072.

(158) In a synthesis of guar acid. Evans, P. A.; Leahy, D. K.; Andrews, W. J.; Uraguchi, D. *Angew. Chem. Int. Ed.* **2004**, *43*, 4788.

(159) Evans, P. A.; Lawler, M. J. *J. Am. Chem. Soc.* **2004**, *126*, 8642.

(160) In a synthesis of (–)-centrolobine. Börsch, M.; Blechert, S. *Chem. Commun.* **2006**, 1968.

(161) Kawatsura, M.; Ata, F.; Hayase, S.; Itoh, T. *Chem. Commun.* **2007**, 4283.

(162) Vong, W.-J.; Peng, S.-M.; Lin, S.-H.; Lin, W.-J.; Liu, R.-S. *J. Am. Chem. Soc.* **1991**, *113*, 573.

(163) Pearson, A. J.; Mesaros, E. F. *Org. Lett.* **2001**, *3*, 2665.

(164) In synthesis of (+)-dihydrocanadensolide. Chen, M.-J.; Narkunan, K.; Liu, R.-S. *J. Org. Chem.* **1999**, *64*, 8311.

(165) In synthesis toward sesquiterpene lactones. Narkunan, K.; Shiu, L.-H.; Liu, R.-S. *Synlett* **2000**, 1300.

(166) Shu, C.; Liebeskind, L. S. *J. Am. Chem. Soc.* **2003**, *125*, 2878.

(167) Yin, J.; Liebeskind, L. S. *J. Am. Chem. Soc.* **1999**, *121*, 5811.

(168) Arryas, R. G.; Liebeskind, L. S. *J. Am. Chem. Soc.* **2003**, *125*, 9026.

(169) Roustan, J. L.; Merour, J. Y.; Houlihan, F. *Tetrahedron Lett.* **1979**, 3721.

(170) Plietker, B. *Angew. Chem. Int. Ed.* **2006**, *45*, 1469.

(171) For reviews, see a) Enders, D.; Jandeleit, B.; von Berg, S. *Synlett* **1997**, 421. b) Speckamp, W. N. *Pure Appl. Chem.* **1996**, *68*, 695. c) For early reports, see Green, J.; Carroll, M. K. *Tetrahedron Lett.* **1991**, *32*, 1141.

(172) Hopman, J. C. P.; Hiemstra, H.; Speckamp, W. N. *J. Chem. Soc., Chem. Commun.* **1995**, 617.

(173) In a synthesis of a fragment of ionomycin. Cooksey, J. P.; Kocienski, P. J.; Li, Y.-F.; Schunk, S.; Snaddon, T. N. *Org. Biomolec. Chem.* **2006**, *4*, 3325.

(174) Heck, R. F.; Breslow, D. S. *J. Am. Chem. Soc.* **1960**, *82*, 750.

(175) Heck, R. F. Organic Syntheses via Alkyl and Acylcobalt Tetracarbonyls. In *Organic Synthesis via Metal Carbonyls*; Wender, P., Pino, P., Eds.; Wiley: New York, 1968; Vol. 1, pp 379–384.

(176) a) Hegedus, L. S.; Inoue, Y. *J. Am. Chem. Soc.* **1982**, *104*, 4917. b) Hegedus, L. S.; Perry, R. J. *J. Org. Chem.* **1984**, *49*, 2570.

(177) For reviews, see a) Semmelhack, M. F. *Org. React.* **1972**, *19*, 115. b) Billington, D. C. *Chem. Soc. Rev.* **1985**, *14*, 93. c) Krysan, D. J. Transition Metal Allyl Complexes π-Allylnickel Halides and Other π-Allyl Complexes Excluding Palladium. In *Comprehensive Organometallic Chemistry II*; Abel, E. W., Stone, F. G. A., Wilkinson, G., Eds.; Pergamon: Oxford, U.K., 1995; Vol. 12, pp 959–978.

(178) Hegedus, L. S.; Thompson, D. H. P. *J. Am. Chem. Soc.* **1985**, *107*, 5663.

(179) Kedar, T. E.; Miller, M. W.; Hegedus, L. S. *J. Org. Chem.* **1996**, *61*, 6121.

(180) Knölker, H.-J.; Fröhner, W. *Tetrahedron Lett.* **1998**, *39*, 2537.

(181) Hegedus, L. S.; Stiverson, R. K. *J. Am. Chem. Soc.* **1974**, *96*, 3250

(182) Hegedus, L. S.; Sestrick, M. R.; Michaelson, E. T.; Harrington, P. J. *J. Org. Chem.* **1989**, *54*, 4141.

(183) Corey, E. J.; Semmelhack, M. F.; Hegedus, L. S. *J. Am. Chem. Soc.* **1968**, *90*, 2416.

(184) In a synthesis of cembrene. Dauben, W. G.; Beasley, G. H.; Broadhurst, M. D.; Muller, B.; Peppard, D. J.; Pesnelle, P.; Suter, C. *J. Am. Chem. Soc.* **1974**, *96*, 4724.

(185) Corey, E. J.; Wat, E. K. W. *J. Am. Chem. Soc.* **1967**, *89*, 2757.

(186) In a synthesis of humulene. Corey, E. J.; Hamanaka, E. *J. Am. Chem. Soc.* **1967**, *89*, 2758.

(187) Corey, E. J.; Kirst, H. A. *J. Am. Chem. Soc.* **1972**, *94*, 667.

(188) Guerrieri, F.; Chinsoli, G. P.; Merzoni, S. *Gazz. Chim. Ital.* **1974**, *104*, 557.

(189) Johnson, J. R.; Tully, P. S.; Mackenzie, P. B.; Sabat, M. *J. Am. Chem. Soc.* **1991**, *113*, 6172.

(190) Grisso, B. A.; Johnson, J. R.; Mackenzie, P. B. *J. Am. Chem. Soc.* **1992**, *114*, 5160.

(191) Hegedus, L. S.; Wagner, S. D.; Waterman, E. L.; Siirala-Hansen, K. *J. Org. Chem.* **1975**, *40*, 593.

(192) a) Sato, Y.; Saito, N.; Mori, M. *J. Org. Chem.* **2002**, *67*, 9310. b) See also Yeh, M.-C.; Liang, J.-H.; Jiang, Y.-L.; Tsai, M.-S. *Tetrahedron* **2003**, *59*, 3409.

(193) Kobayashi, Y.; Takahisa, E.; Usimani, S. B. *Tetrahedron Lett.* **1998**, *39*, 597.

(194) Kobayashi, Y.; Takahisa, E.; Usmani, S. B. *Tetrahedron Lett.* **1998**, *39*, 601.

(195) In a synthesis of Δ^7-PGA$_1$ methyl ester. Kobayashi, Y.; Murugesh, M. G.; Nakano, M.; Takahisa, E.; Usmani, S. B.; Ainai, T. *J. Org. Chem.* **2002**, *67*, 7110.

(196) Bricout, H.; Carpentier, J.-F.; Mortreux, A. *Tetrahedron* **1998**, *54*, 1073.

(197) Berkowitz, D. B.; Maiti, G. *Org. Lett.* **2004**, *6*, 2661.

10

Synthetic Applications of Transition Metal Arene Complexes

10.1 Introduction

Arenes form stable, isolable complexes with a range of transition metals, including Cr, Mo, W, Fe, Ru, Os, and Mn.[1] By far, the most common arene–transition metal complex is the η^6-coordinated type (**Figure 10.1**) in which the entire π-system of the arene is complexed. These η^6-arene complexes have a very rich reaction chemistry, which constitutes the bulk of this chapter. η^2-Arene complexes are much less common, and only those of Os, Re, Mo, and W have found use in organic synthesis. With this type of arene complex, only two carbons of the π-arene system are complexed, in essence deconjugating the arene. The uncomplexed portion of the π-system then displays reactivity common to the residual degree of unsaturation (e.g., 1,3-dienes for Os-complexed benzenes). Both types of arene complexation, η^6 and η^2, result in dramatic alteration in the reactivity of the complexed arene; and therein lies their utility for organic synthesis.

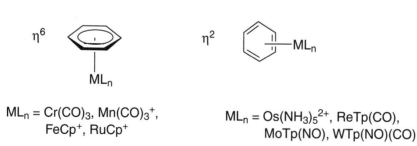

$ML_n = Cr(CO)_3, Mn(CO)_3{}^+,$
$\quad\quad\quad FeCp^+, RuCp^+$

$ML_n = Os(NH_3)_5{}^{2+}, ReTp(CO),$
$\quad\quad\quad MoTp(NO), WTp(NO)(CO)$

Figure 10.1. Modes of Arene Complexation

10.2 Preparation of η^6-Arene Complexes[2]

Arenechromium tricarbonyl complexes are normally prepared by heating $Cr(CO)_6$ in the arene as solvent, although the procedure is complicated by the sublimation of $Cr(CO)_6$.[3] Alternatively, one can start with preformed complexes having labile ligands, such as $Cr(CO)_3(NH_3)_3$, $Cr(CO)_5$(2-picoline), $Cr(CO)_3(MeCN)_3$, or $Cr(CO)_3$(pyridine)$_3$, to avoid this problem (**Eq. 10.1**).[4] Perhaps the most convenient method is to carry out an arene exchange with the $Cr(CO)_3$(naphthalene) complex (**Eq. 10.2**),[5] a procedure that goes under mild conditions and does not require large excesses of arene.

Eq. 10.1

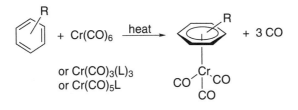

M(0), d^6, 18 e$^-$, saturated

Eq. 10.2

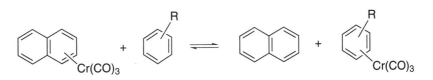

The complexation of arenes to the chromium tricarbonyl fragment is facilitated by electron-donating groups on the arene ring and, in general, electron-rich arenes readily form arenechromium tricarbonyl complexes, while electron-poor arenes react much more slowly or not at all. (The chromium tricarbonyl complex of nitrobenzene is unknown.) In some cases, discrimination between two similar arene rings in the same compound is possible, provided there is sufficient electronic difference between them.

Cationic cyclopentadienyliron–arene complexes are most readily prepared by the treatment of ferrocene with the arene in the presence of aluminum trichloride.[6,7] Although this places some restriction on the arenes to be complexed, a modest range of complexes is available by this procedure (**Eq. 10.3**).

Eq. 10.3

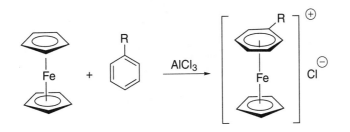

Related ruthenium–arene complexes can be made by the reaction of arenes with labile cyclopentadienyl ruthenium complexes (**Eq. 10.4**).[8] A more versatile method involves the reduction of polymeric $RuCl_3 \cdot xH_2O$ with zinc, followed by the addition of an arene and Cp*H (**Eq. 10.5**).[9] Cationic manganese–arene complexes have also been prepared by ligand exchange processes using $Mn(CO)_5^+$, $Mn(CO)_3(acetone)_3^+$, or $Mn(CO)_3(naphthalene)^+$ (**Eq. 10.6**).[10,11]

Eq. 10.4

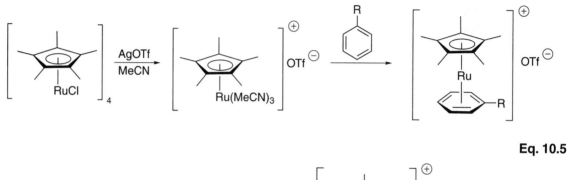

Eq. 10.5

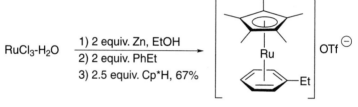

Eq. 10.6

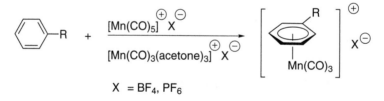

Complexation of arenes has a profound effect on their reaction chemistry (**Figure 10.2**).[12] Relative to the arene ring, the ML_n fragment is net electron-withdrawing, as evidenced by the high dipole moment (5.08 D for benzene chromium tricarbonyl), the

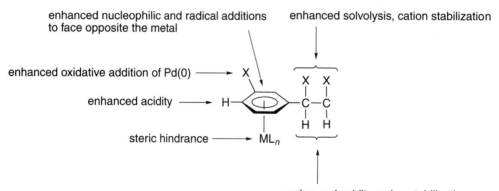

Figure 10.2. Effects of Complexation of Arenes to Metals

increase in acidity of benzoic acid upon complexation [pK_a = 4.77 for the $Cr(CO)_3$–benzoic acid complex vs. 5.75 for free benzoic acid], and the decrease in basicity for $Cr(CO)_3$ complexed with aniline (pK_b = 13.31 vs. 11.70 for aniline itself).

As a consequence, the arene ring becomes activated towards nucleophilic or radical attack, rather than the normal electrophilic attack. In addition, the electron-deficient arene ring is better able to stabilize negative charge, thus both ring and benzylic deprotonation become more favorable. Finally, the metal–ligand fragment completely blocks one face of the arene, and directs incoming reagents to the face opposite the metal. All of these effects have been used to an advantage in organic synthesis.

10.3 Reactions of η^6-Arene–Metal Complexes[13]

a. Nucleophilic Aromatic Substitution of Complexed Aryl Halides

Although simple aryl halides are relatively inert to nucleophilic aromatic substitution, this reaction is greatly facilitated when the aryl halides are complexed to appropriate transition metal fragments. For example, the rate of substitution of chloride by methoxide in η^6-chlorobenzenechromium tricarbonyl approximates that of uncomplexed nitrobenzene,[14] an indication of the electron-withdrawing power of the chromium tricarbonyl fragment. Sodium phenoxide and aniline also readily effect this substitution.[15] (η^6-Fluorobenzene)chromium tricarbonyl is even more reactive, undergoing substitution by a range of nucleophiles, including alkoxides, amines, cyanide,[16] and stabilized carbanions (**Eq. 10.7**).[17]

Eq. 10.7

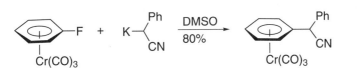

With chromium–arene complexes, these substitution reactions proceed by a two-step process involving nucleophilic attack of the arene ring from the face opposite the metal, forming an anionic η^5-cyclohexadienyl complex, followed by rate-limiting loss of the halide from the endo side of the ring (**Eq. 10.8**).

Eq. 10.8

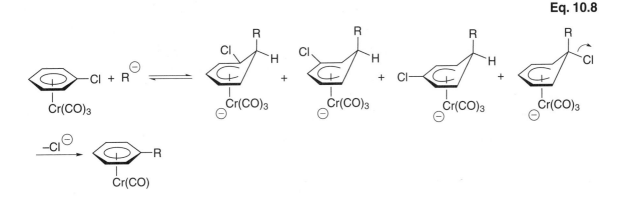

The displacement of halide by carbanions is limited to stabilized carbanions capable of adding reversibly to the complexed arene. More reactive carbanions, such as 2-lithio-1,3-dithiane, attack the complexed arene ortho and meta to the halide, producing η^5-cyclohexadienyl complexes that cannot directly lose chloride, and furthermore, cannot rearrange (equilibrate) to the η^5-cyclohexadienyl complex, which can eliminate (**Eq. 10.8**).[18] Stabilized carbanions also initially attack complexed chlorobenzene ortho and meta to the chloride, but the addition is reversible and, upon equilibration, eventually attack at the halide-bearing position, followed by loss of halide, resulting in substitution.

Although these nucleophilic aromatic substitution reactions are most extensively studied for the arenechromium tricarbonyl complexes, it is cationic areneruthenium complexes that have found the most use in complex organic synthesis, particularly in forming the aromatic ether linkage common to vancomycin[19] and ricotecin analogs (**Eq. 10.9**).[20] Cationic manganese–arene complexes have been used for similar purposes (**Eq. 10.10**).[21] Iron complexes of chloroarenes also undergo facile S_NAr reactions (**Eq. 10.11**).[22]

Eq. 10.9

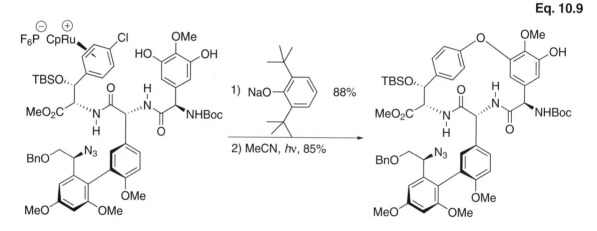

Eq. 10.10

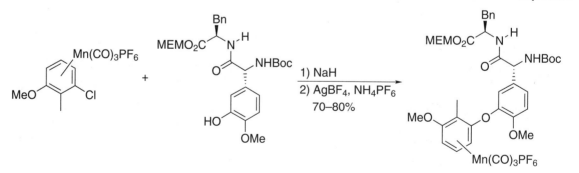

Eq. 10.11

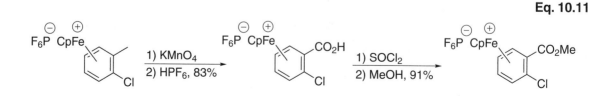

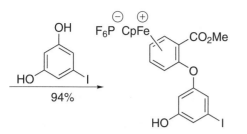

b. Addition of Carbon Nucleophiles to Metal–Arene Complexes[23]

A range of carbanions attack chromium-complexed arenes from the face opposite the metal, thus producing anionic η^5-cyclohexadienyl complexes. For all but the most nucleophilic carbanions, this process is reversible and the initial site of attack (kinetic) may not correspond to the alkylation site in the final product (thermodynamic). This η^5-cyclohexadienyl intermediate (which has been characterized by X-ray crystallography in the case of lithiodithiane as the carbanion[24]) can react further, along three different pathways (**Figure 10.3**).

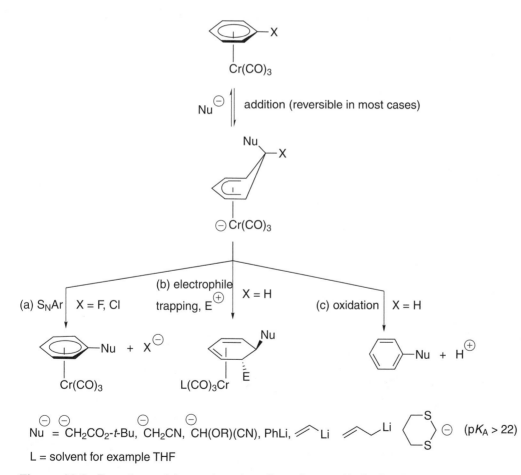

Figure 10.3. Reactions of Arenechromium Complexes with Carbanion

Path (a) in **Figure 10.3** has already been discussed above. Trapping of the η^5-cyclohexadienyl complexes with electrophiles [path (b)] is a potentially useful reaction. However, it is quite restricted by the electrophiles that undergo reaction. With carbanions that add reversibly, the only electrophile that undergoes efficient reaction is the proton (from CF_3CO_2H), leading to cyclohexadienyl complexes. All other electrophiles react preferentially with the free carbanion in equilibrium with the η^5-cyclohexadienyl complex, resulting in regeneration of the starting η^6-arene complex. It is only with carbanions that react irreversibly that general trapping by electrophiles is efficient, and this process has proven useful in several ways (see below). Oxidation of the η^5-cyclohexadienyl intermediate [path (c)] is the most extensively developed process, is efficient with all carbanions that add, and results in overall nucleophilic aromatic substitution. This is by far the most extensively utilized of the three processes and is considered first.

Because of the reversibility of alkylation with most carbanions, the regioselectivity in the product is complex. For monosubstituted arenes bearing a resonance-donor substituent (e.g., MeO, Me_2N, F, or Cl), meta attack is strongly favored, while acceptor substituents, such as TMS or CF_3, direct attack to the para position. (π-Accepting substituents such as acyl or nitrile groups undergo competitive alkylation.) The situation with methyl- and chloro-substituted arenes is more complex, and substantial amounts of ortho-alkylation can be observed, depending on the carbanion.

Regioselectivity with 1,2-disubstituted arene complexes can be high, and often is the result of attack of the sterically less-hindered position meta to the strongest donor (**Figure 10.4**). The complexity of regioselectivity in the context of indole alkylation is shown in **Figure 10.5**,[25] in which steric, thermodynamic, or kinetic factors may dominate, depending on how the reaction is run.

Despite these complexities, alkylation reactions of arenechromium tricarbonyl compounds have been used in a number of complex systems (**Eqs. 10.12**,[26] **10.13**,[27] and **10.14**[28]).

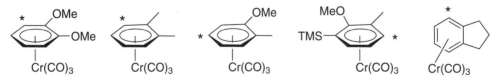

Figure 10.4. Regioselectivity of Arene Alkylation (* = major site)

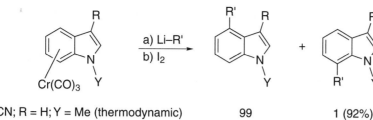

a. R' = C(Me)$_2$CN; R = H; Y = Me (thermodynamic)   99   1 (92%)
b. R' = (1,3-dithianyl); R = H; Y = Me (kinetic)   14   86 (68%)
c. R' = C(Me)$_2$CN; R = CH$_2$SiMe$_3$; Y = H (steric)   17   83 (82%)
d. R' = C(Me)$_2$CN; R = CH$_2$SiMe$_3$; Y = Si (*t*-Bu)(Me)$_2$ (steric)   95   5 (78%)

Figure 10.5. Regioselectivity of Indole Alkylation

Eq. 10.12

Eq. 10.13

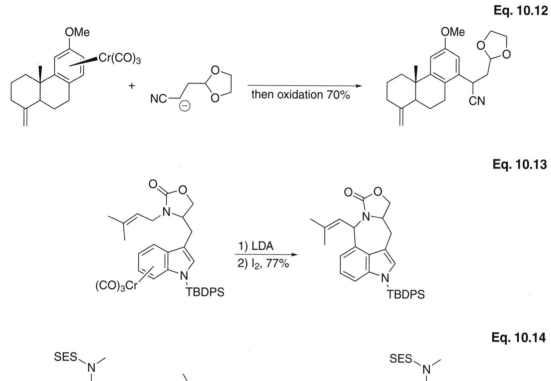

Eq. 10.14

SES = Me$_3$SiCH$_2$CH$_2$SO$_2$–
SEM = Me$_3$SiCH$_2$CH$_2$–

Although the release of the alkylated arene from the η^5-cyclohexadienyl complex is usually accomplished by oxidation, it is also possible to effect a hydride abstraction utilizing trityl cation. This results in regeneration of the (now alkylated) arene complex, permitting further use of the $Cr(CO)_3$ fragment (**Eq. 10.15**).[29] In this case, ortho-alkylation is almost surely directed by the nitrogen in the benzylic position.

Eq. 10.15

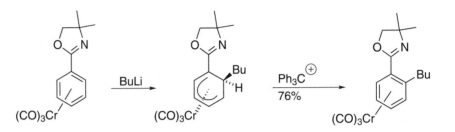

Unsymmetrically 1,2- and 1,3-disubstituted η^6-arenechromium tricarbonyl complexes are chiral, and are enantiomeric on the basis of which face the metal fragment occupies (**Figure 10.6**). They can often be resolved and, using the chemistry in **Eq. 10.15**, they can be synthesized directly with high enantiomeric excess. This has been achieved by using optically active hydrazone (SAMP) directing groups (**Eq. 10.16**),[30] and by using chiral ligands *for the organolithium reagent,* to direct it to one of the two prochiral ortho positions (**Eq. 10.17**).[31]

Reaction of η^5-cyclohexadienyl complexes with electrophiles is also an efficient process with many synthetic applications. For carbanions that add reversibly, the electrophile must be a proton. With alkoxybenzenes[32] as substrates, this is a useful procedure for the synthesis of substituted cyclohexenones (**Eq. 10.18**).[33] This ability of chromium to activate arenes towards nucleophilic attack, to direct the attack to specific positions on the arene ring, and to permit protolytic cleavage of the η^5-cyclohexadienyl complex to produce cyclohexenones has been used in a noteworthy total synthesis of acorenone B (**Eq. 10.19**).[34] Again, protons are the only electrophiles that cleave the η^5-cyclohexadienyl complex.

Figure 10.6. Chirality in Disubstituted Arene Complexes

Eq. 10.16

Eq. 10.17

Eq. 10.18

Eq. 10.19

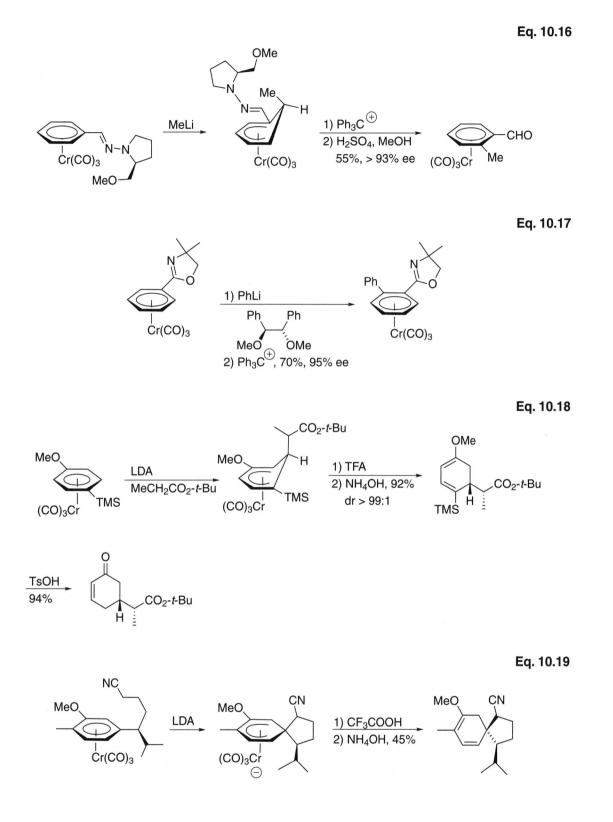

Since most alkylations (see exceptions below) are reversible, other electrophiles react preferentially with the free carbanion, regenerating the η^6-arenechromium tricarbonyl complex, and alkylating the electrophile. By incorporating a chiral alcohol by nucleophilic displacement of a fluoro group, asymmetry has been induced in the alkylation/protonation (**Eq. 10.20**).[35]

Eq. 10.20

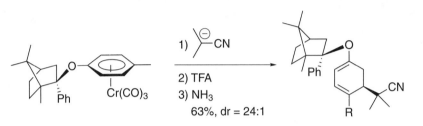

Intramolecular addition of nucleophiles to ruthenium–[36] and iridium–arene[37] complexes lead to spirocyclic or fused bicyclic η^5-coordinated complexes (**Eq. 10.21**). The spirocyclic complexes can be oxidatively decomplexed using copper(II) salts to give either free spirocyclic and/or fused compounds, depending on the type and position of the substituents.[38]

Eq. 10.21

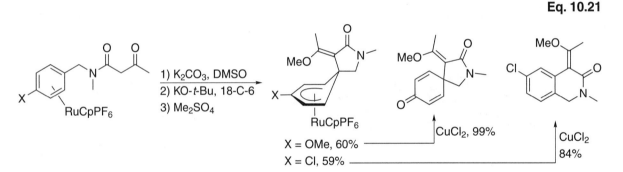

Since dithiane anion adds irreversibly to the η^6-arenechromium tricarbonyl complex, it should, in principle, be possible to trap the resulting anionic η^5-cyclohexadienyl complex with electrophiles other than a proton. Indeed, treatment of this complex with reactive alkyl halides does result in a clean reaction but, surprisingly, an *acyl* group rather than the corresponding alkyl group is introduced, and from the same face as the metal, resulting in clean trans difunctionalization of the initial arene (**Eq. 10.22**).[39]

Eq. 10.22

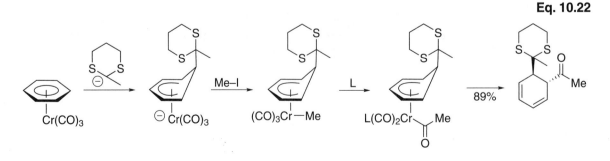

To account for this, initial nucleophilic attack by the dithiane had to occur from the face opposite the metal, as expected. This is followed by alkylation at the *metal*, then migration to an adjacent carbonyl group to produce an acyl intermediate. Reductive elimination (an acyl migration) from this complex would deliver the acyl group from the *same* face as the metal, resulting in the clean trans disubstitution observed. This chemistry works equally well with substituted naphthalenes and this one-pot addition of two carbon substituents across an arene double bond serves as the key step in a synthesis of the aklavinone AB ring (**Eq. 10.23**).[40] For benzene and electron-rich arenes, all electrophiles undergo carbon monoxide insertion. Complexes of electron-deficient arenes add some electrophiles, including allylic, benzylic, and propargylic bromides, to the ring without CO insertion. Molybdenum–arene complexes undergo related transformations, but in contrast to chromium, do not afford CO-insertion products (**Eq. 10.24**).[41]

Eq. 10.23

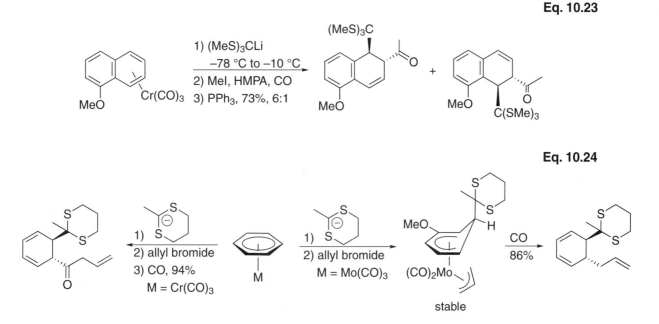

Eq. 10.24

The alkylation of complexed arenes having a chiral oxazoline directing group results in the conversion of two aromatic carbons to two new stereogenic centers with excellent de (**Eq. 10.25**).[42] In this case, asymmetric induction results from alkylation of the arene from the face *opposite* the large isopropyl group. Allyl, benzyl, and propargyl groups are transferred from the metal without CO insertion, a consequence of their lower aptitude toward migration. The second approach parallels that used in **Eq. 10.17**, and utilized a chiral ligand *for the organolithiation reagent* to direct alkylation to one of the two prochiral ortho positions (**Eq. 10.26**).[43] Utilization of both approaches in the total synthesis of (+)- and (–)-acetoxytubiprofan are shown in **Eqs. 10.27** and **10.28**.[44]

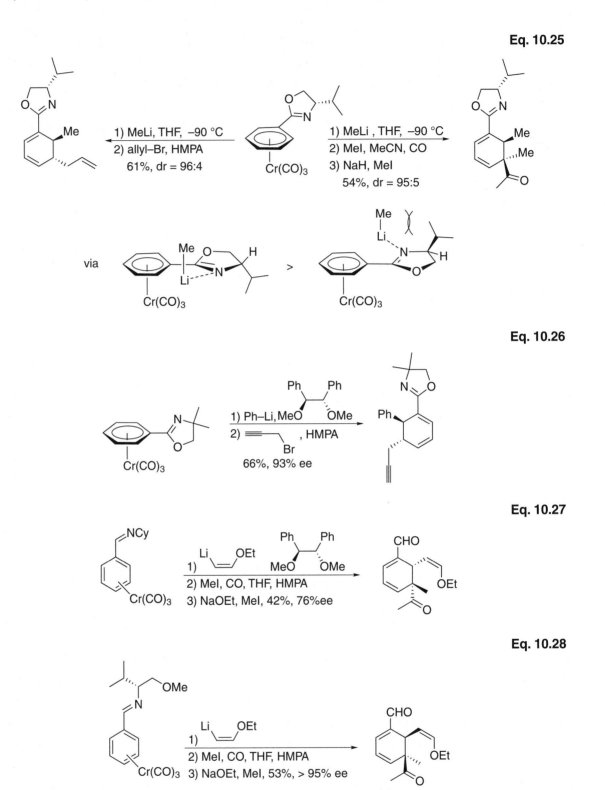

Eq. 10.25

Eq. 10.26

Eq. 10.27

Eq. 10.28

Cationic manganese arene complexes are more electrophilic compared to the neutral $Cr(CO)_3$(arene) complexes and they react with a wider range of nucleophiles, including Grignard reagents, organozinc[45] reagents, and hydride. The neutral η^5-manganese complexes obtained from nucleophilic addition do not react with carbon electrophiles.

However, a second nucleophile can be added, again from the opposite side of the metal, either directly[46] (without the insertion of carbon monoxide) or by activation of the complex by ligand exchange (**Eq. 10.29**).[47] Decomplexation of the cyclohexadiene is achieved by a simple air oxidation. A different reaction path is observed using methyl- or phenyllithium, both affording the carbon monoxide insertion products. The reason for this change in reactivity is unclear.

Eq. 10.29

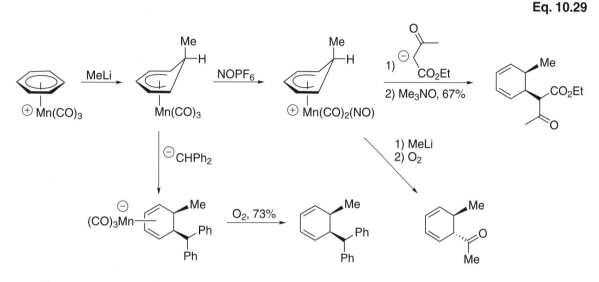

Chromium tricarbonyl arene complexes undergo substitution reactions with radicals. For disubstituted chlorotoluene complexes, meta substitution is the major product observed in all cases, regardless of the initial substitution pattern (**Eq. 10.30**).[48] Intramolecular reactions have also been described (**Eq. 10.31**).[49]

Eq. 10.30

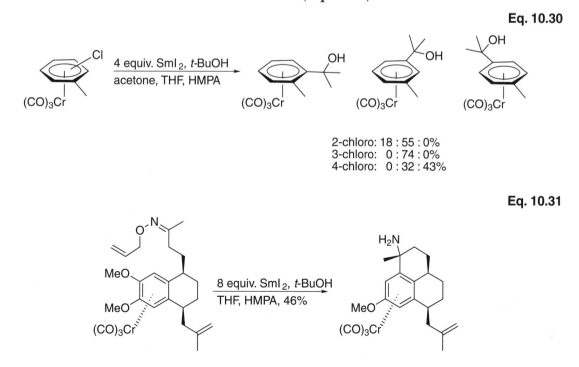

2-chloro: 18 : 55 : 0%
3-chloro: 0 : 74 : 0%
4-chloro: 0 : 32 : 43%

Eq. 10.31

c. Lithiation of Arenechromium Tricarbonyl Complexes[50]

Complexation of an arene to chromium activates the ring towards lithiation, and ring-lithiated arenechromium tricarbonyl complexes are readily prepared, and generally reactive towards electrophiles (**Eq. 10.32**).[51,52] With arenes having substituents with lone pairs of electrons, such as –OR, –NMe$_2$, –NHBoc, –CH$_2$OMe, –CH$_2$NMe$_2$, –CONEt$_2$, –F, or –Cl, lithiation always occurs ortho to the substituted position.

Eq. 10.32

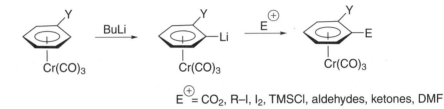

$$\overset{\oplus}{E} = CO_2, \text{ R–I, I}_2, \text{ TMSCl, aldehydes, ketones, DMF}$$

This process is efficient, even with quite complex systems, and high selectivity can be achieved. For example, treatment of dihydrocryptopine with chromium hexacarbonyl resulted in exclusive complexation of the slightly more electron-rich dimethoxybenzene ring, perhaps with some assistance from the benzylic hydroxy group. Lithiation occurred exclusively ortho to the methoxy group and peri to the benzylic hydroxy group, resulting in regiospecific alkylation (**Eq. 10.33**).[53]

Eq. 10.33

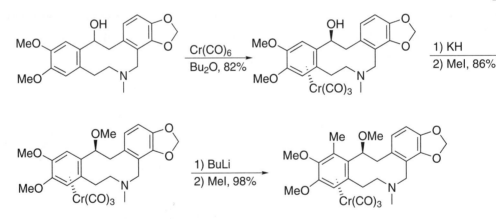

Complexed indoles are lithiated in the normally unreactive 4-position, provided a large protecting group is present on nitrogen to suppress lithiation at the 2-position (**Eq. 10.34**).[54]

Eq. 10.34

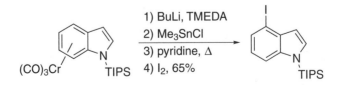

Recall that unsymmetrically disubstituted arene complexes are chiral. Ortho-lithiation/alkylation generates just such complexes and, as above, asymmetry can be induced[18] several different ways. Arene complexes having a chiral benzylic carbon bearing a donor substituent (i.e., –OMe, –OCH$_2$OCH$_3$, or –NMe$_2$) undergo clean ortho-lithiation/alkylation in excellent yield and with high diastereoselectivity. Asymmetric induction is thought to result from ligand-directed metallation to the ortho position, which allows the methyl substituent to be on the face of the complex *opposite* the metal (**Eq. 10.35**).[55] By using chiral benzaldehyde-derived ketals or aminals[56] as directing groups, optically active ortho-substituted benzaldehyde complexes were available (**Eq. 10.36**).[57] An interesting use in synthesis is shown in **Eq. 10.37**.[58,59]

Eq. 10.35

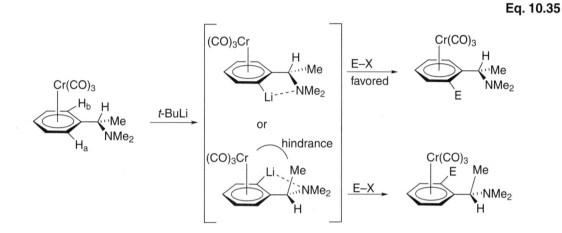

E–X = MeOSO$_2$F, Me$_3$SiCl, Ph$_2$PCl; also RCHO, R$_2$CO

Eq. 10.36

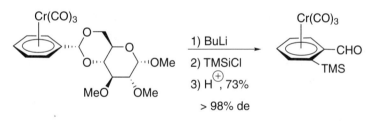

Eq. 10.37

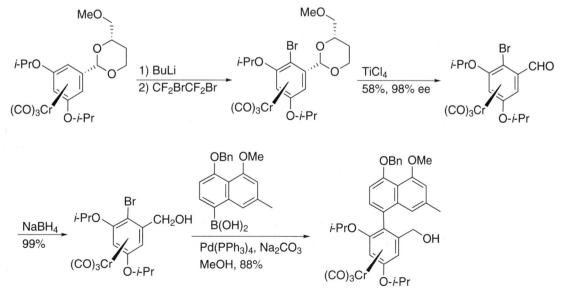

A more versatile approach to asymmetric induction in the ortho-lithiation step involves the use of chiral organolithium reagents.[60] In this instance, lithiated C-2 symmetric dibenzylamine efficiently discriminated between the prochiral ortho positions, resulting in good yields and high enantiomeric excesses[61] (**Eq. 10.38**).[62] An example of enantioselective lithiation–chlorination is shown in **Eq. 10.39**.[63]

Eq. 10.38

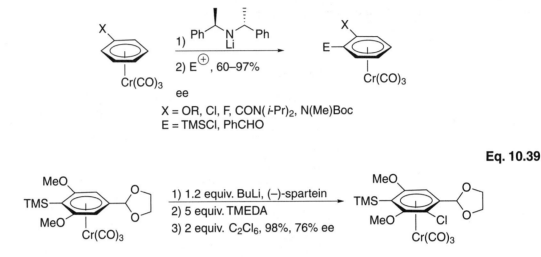

X = OR, Cl, F, CON(*i*-Pr)$_2$, N(Me)Boc
E = TMSCl, PhCHO

Eq. 10.39

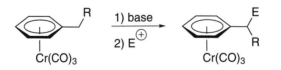

d. Side-Chain Activation and Control of Stereochemistry

Another manifestation of the "electron-withdrawing" properties of the metal complexed to arenes is its ability to stabilize negative charge at the benzylic position of alkyl side chains. This allows facile benzylic alkylation of complexed arenes with a wide range of electrophiles (**Eq. 10.40**).[64] As expected, benzylic functionalization occurs from the face *opposite* the metal with a high degree of stereo control. This is most apparent with cyclic systems, since the benzylic protons are inequivalent due to restricted rotation (**Eq. 10.41**).[65]

Eq. 10.40

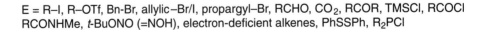

E = R–I, R–OTf, Bn-Br, allylic–Br/I, propargyl–Br, RCHO, CO$_2$, RCOR, TMSCl, RCOCl
RCONHMe, *t*-BuONO (=NOH), electron-deficient alkenes, PhSSPh, R$_2$PCl

Eq. 10.41

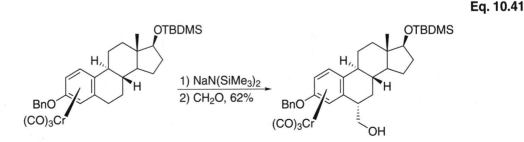

Even acyclic systems undergo benzylic alkylation with a high degree of stereo control, provided there is an ortho substituent to force the benzylic substituent to adopt a single rotomeric isomer (**Eq. 10.42**).[66]

In some cases, ring lithiation competes with benzylic lithiation. When this occurs, the ring site can be blocked by silylation, which can then be removed after successful benzylic alkylation (**Eq. 10.43**).[67] This entire process has been used in the synthesis of natural products[68] (**Eq. 10.44**).[69]

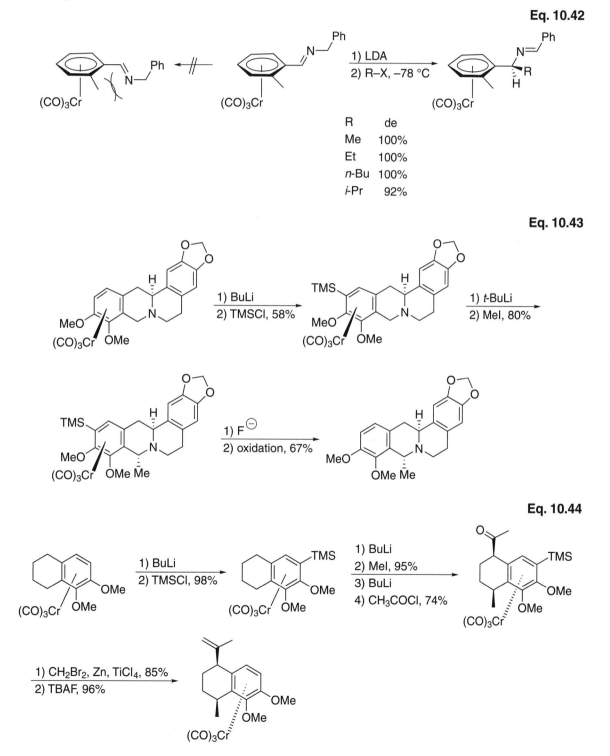

Eq. 10.42

R	de
Me	100%
Et	100%
n-Bu	100%
i-Pr	92%

Eq. 10.43

Eq. 10.44

Benzylic methylene protons of chromium–arene complexes of benzyl ethers can be deprotonated using chiral amide bases and reacted with a variety of electrophiles to give products in high yield and high enantiomeric excess. In addition to reactions with external electrophiles, the chiral anions undergo [2,3]-Wittig rearrangements (**Eq. 10.45**).[70] Deprotonation from the same face as the metal can be realized in some cases. The asymmetric silylation of chromium tricarbonyl-complexed phthalan affords the expected exo-addition product (**Eq. 10.46**).[71] However, in situ deprotonation and addition of methyl iodide give a geminally disubstituted product. This product is formed by endo deprotonation of the sterically hindered TMS-substituted carbon, even under kinetic conditions (–100°C). This unusual endo deprotonation is attributed to the higher acidity of the TMS-substituted carbon.

Eq. 10.45

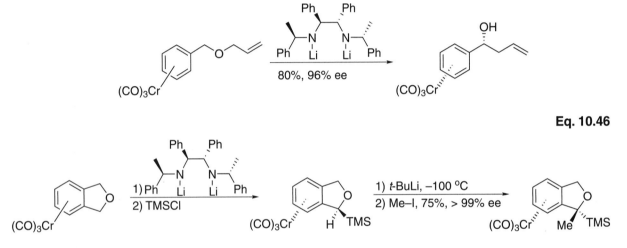

Eq. 10.46

Most remarkably, benzyl *cations* are also stabilized by chromium tricarbonyl fragments. Complexed benzylic alcohols, chlorides, acetates, and ethers and benzaldehyde acetals produce complexed benzylic cations, resulting from Lewis acid-assisted ionization from the face opposite the metal. The cation is slow to racemize and can react with a variety of nucleophiles, again from the face opposite the metal, resulting in overall retention (**Eq. 10.47**).[72]

Eq. 10.47

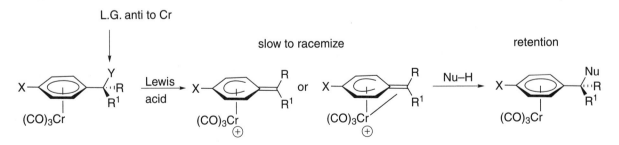

If the leaving group cannot get anti to the metal, as can be the case in cyclic systems, the replacement reaction still proceeds, albeit more slowly, and nucleophilic attack again occurs from the face opposite the metal, resulting in overall inversion. This reaction chemistry is quite efficient with complex systems (**Eqs. 10.48**[73] **and 10.49**[74]).

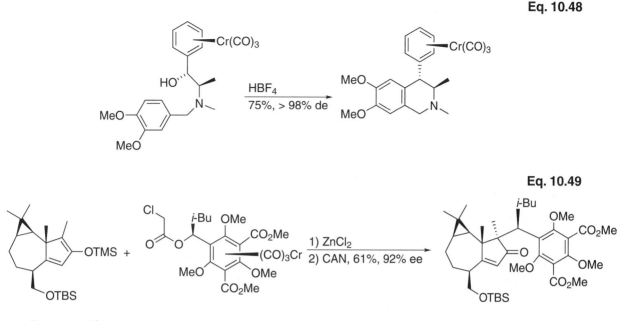

Eq. 10.48

Eq. 10.49

Propargylic cations are easily generated from arenechromium tricarbonyl-substituted propargyl acetates or alcohols using a Lewis acid. The cations are relatively stable and ^{13}C NMR spectra have been recorded.[75] The cations react with a variety of nucleophiles to give propargyl-substituted products (**Eq. 10.50**).[76] Retention of stereochemistry is observed. Addition of nucleophiles to form allenic products has been observed in some cases.[77]

Eq. 10.50

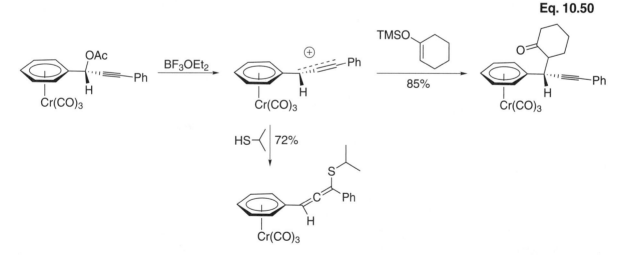

In all of the preceding η⁶-arene complexes, the metal occupies a single face of the arene, and directs reaction chemistry to the opposite face of the molecule. This results in a very high degree of stereocontrol, and has found many uses in organic synthesis. A very early example of this, using a mixture of diastereomers, is shown in **Eq. 10.51**,[78] in which α-alkylation of an enone by complexed α-methyl indanone occurred exclusively from the face opposite the metal. Removal of an activated benzylic proton permitted intramolecular alkylation of the keto group, again from the face opposite the metal, resulting in clean cis stereochemistry.

Eq. 10.51

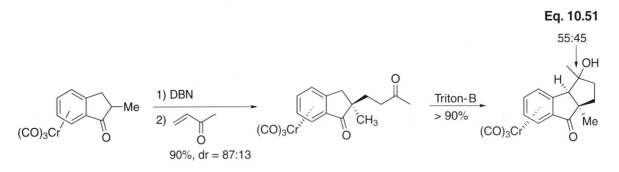

Racemic 2-indanol undergoes facially specific complexation from the same face as the OH group, presumably directed there by complexation to the oxygen. These enantiomeric complexes can be resolved, and oxidation of one of the enantiomers produces an indanone complex that is optically active solely by virtue of the face of the arene occupied by the Cr(CO)$_3$ fragment (see above). Since reactions of this optically active complex occur exclusively from the face opposite the metal, alkylation by Grignard reagents produces a single enantiomer of the indanol, α-alkylation produces a single enantiomer of the indanone, and α-alkylation followed by reduction produces a single enantiomer of the α-alkyl indanol (**Eq. 10.52**).[79] Similar transformations are feasible with tetralone.[80]

Eq. 10.52

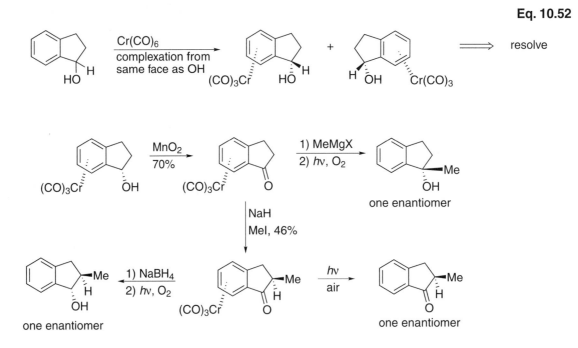

As shown above, unsymmetrically 1,2- and 1,3-disubstituted arenechromium tricarbonyl complexes are chiral and are enantiomeric on the basis of which face of the arene the chromium tricarbonyl fragment occupies (**Figure 10.6**). These can be resolved[81] or, in some cases, prepared by asymmetric synthesis.[82] Since reactions occur from the face opposite the metal, highly stereoselective reactions can be achieved.

For example, nucleophilic addition to chromium-complexed, ortho-substituted aryl aldehydes proceeds with a high degree of stereoinduction, resulting from addition to the aldehyde carbonyl group from the face opposite the metal. A single prochiral face of the aldehyde is presented to the attacking nucleophile because the ortho substituent strongly favors the less-hindered aldehyde rotamer (**Eq. 10.53**).[83] Coordination of the aldehyde oxygen and an adjacent oxygen to a Lewis acid can completely change the rotamer ratio, thus affording a diastereomeric product upon addition of a nucleophile (**Eq. 10.54**).[83] Similar factors result in high diastereoselectivity in other reactions of complexed aldehydes, such as imine alkylations,[84,85] including aza Baylis–Hilman reactions (**Eq. 10.55**),[86] Reformatsky reactions (**Eq. 10.56**),[87] Staudinger reactions (**Eq. 10.57**),[88] and aza Diels–Alder reactions[89] (**Eq. 10.58**).[90]

Eq. 10.53

Eq. 10.54

Eq. 10.55

Eq. 10.56

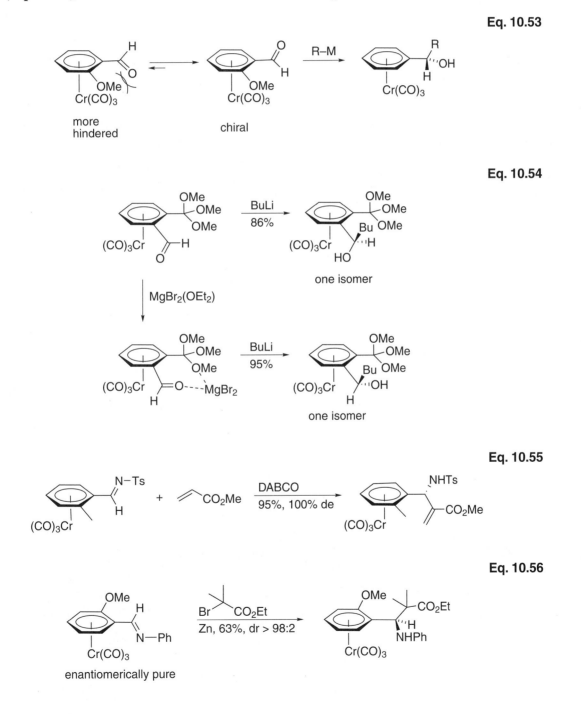

Eq. 10.57

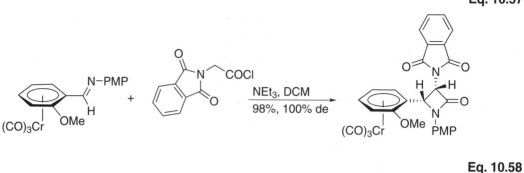

Eq. 10.58

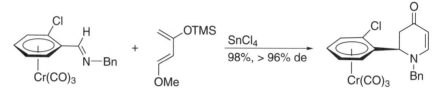

Chromium tricarbonyl-complexed benzylic radicals, derived from the reaction of an aromatic ketone or aldehyde,[91] or a benzaldehyde imine[92] and samarium diiodide, also undergo transformations with very high diastereoselectivity (**Eq. 10.59**).[93] A very nice synthetic example utilizing a majority of the features found in arenechromium chemistry is shown in **Eq. 10.60**.[94] The ketal is also cleaved in the third step.

Eq. 10.59

Eq. 10.60

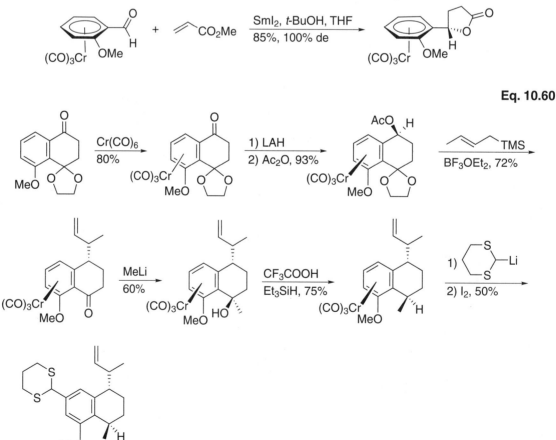

Complexation of metals to arene usually results in activation toward nucleophilic addition. However, reduction of η^6-benzene complexes of chromium or manganese carbonyl with potassium naphthalenide affords η^4-benzene–metal complexes wherein the uncomplexed alkene is activated toward electrophilic addition. Electrophilic addition of, for example, proton, benzylic halides, carbon dioxide, iminium salts, and nitrones yield η^5-cyclohexadienyl complexes that can be oxidized to η^6 complexes. As expected, addition from the face opposite the metal is observed for all electrophiles, except proton and carbon dioxide (**Eqs. 10.61**[95] and **10.62**[96]). The addition to the same face as the metal probably occur via a metal-mediated pathway. Although this is a potentially useful dearomatization reaction, it is limited to benzene and has not found use in organic synthesis to date.

Eq. 10.61

Eq. 10.62

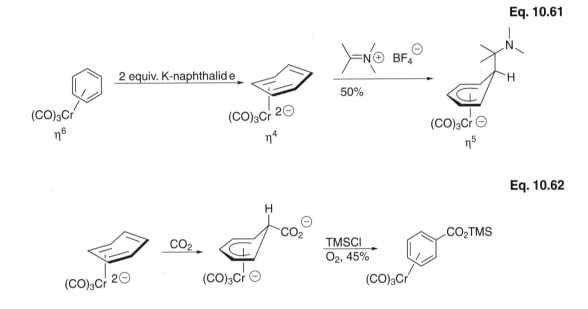

10.4 η^2-Arene–Metal Complexes[97]

The transition metals that form η^6-arene complexes are all strong π-acceptor metal–ligand systems, binding to the electron-rich π-donor cloud of the arene and acting as an electron-withdrawing group. In contrast, the metal–ligand systems that form η^2-arene complexes, particularly [Os(NH$_3$)$_5$]$^{2+}$, have powerful σ-donor ligands that are incapable of appreciable π-interaction with arenes. These complexes show a strong preference for π-*acceptor* ligands, such as alkenes, nitriles, aldehydes and alkynes, and will bind to these ligands in preference to donor ligands, such as amines, esters, ethers, alcohols, and amides. The remarkable thing about Os(NH$_3$)$_5$]$^{2+}$ is that it treats arenes as a source of an "alkene" ligand, complexing a 6π system in an η^2 fashion, essentially deconjugating the arene. Complexation to arenes occurs at the site that results in the minimum disruption of the π-system of the arene, and an external alkene (as in styrene) will complex in preference to an internal one.

These complexes are prepared by reducing $Os(NH_3)_5^{3+}$ in the presence of the arene. (This procedure requires "glove box" conditions.) Decomplexation can be achieved by oxidation with DDQ or CAN under relatively severe conditions (**Eq. 10.63**).

Eq. 10.63

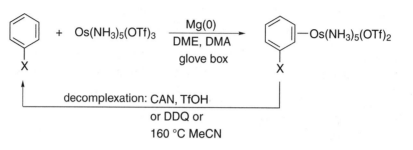

The η^2-complexation of arenes can result in some remarkable reaction chemistry. The uncomplexed alkenes can be reduced using $Pd/C/H_2$ (**Eq. 10.64**).[98] The η^2-complexation of phenols and alkoxybenzenes to $[Os(NH_3)_5]^{2+}$ [i.e., Os(II)] allows them to react like the γ-enol of enones, and to perform Michael additions to electron-deficient alkenes (**Eq. 10.65**).[99] Not surprisingly, the electrophile is added to the face opposite to the large osmium moiety. Anilines behave in a similar manner (**Eq. 10.66**).[100]

Eq. 10.64

Eq. 10.65

Eq. 10.66

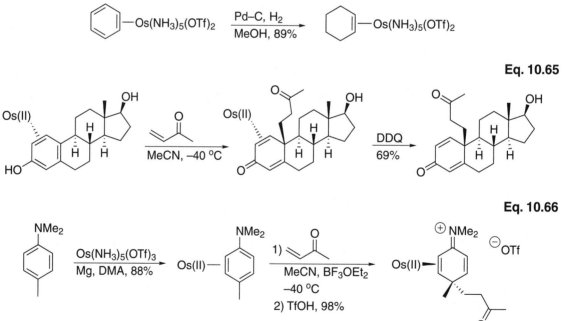

Other electrophiles, in addition to electron-deficient alkenes, can be used, including acetals, a proton, and tertiary carbocations. For alkyl benzenes and naphthalene, the sequential introduction of an electrophile and a nucleophile is possible, thus affording cis-1,4-addition products.[101] The range of nucleophiles is larger compared to the electrophile, and hydride, silyl enolates, stabilized anions, cuprates, and zinc reagents can be used (**Eq. 10.67**).

Eq. 10.67

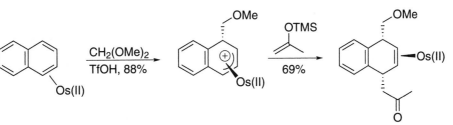

Alkoxybenzenes offer additional possibilities. The complexation of methyl (*R*)-methyl-2-phenoxypropanoate yields a chiral complex in excellent yield (**Eq. 10.68**).[102] The introduction of four different groups is possible via sequential electrophilic–nucleophilic–electrophilic–nuclephilic additions. The chiral group only directs the metal to one of the faces of the benzene ring; all other reactions are directed by the metal.

Eq. 10.68

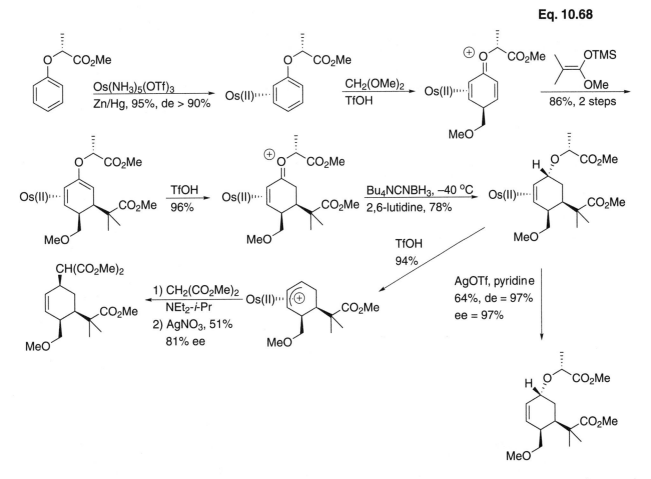

Activation of a "benzylic" position of 3-alkyl-1-methoxybenzene osmium η²-complexes can be achieved. For example, the osmium complexation of 3,4-dimethyl-1-methoxybenzene, followed by reaction with 3-buten-2-one and treatment of the adduct with a base, affords a bicyclic product (**Eq. 10.69**).[103] Further manipulations and decomplexation produce stereo-defined decalins.

Eq. 10.69

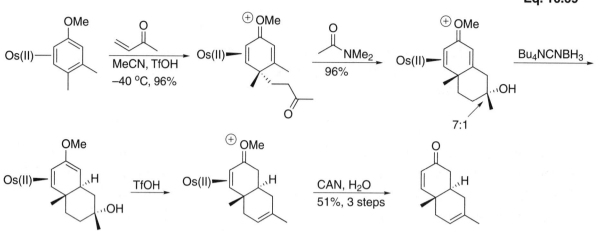

In addition to benzenes and naphthalenes, a number of heteroaromatic ring systems also undergo complexation to form osmium η^2-complexes. The 4,5-η^2-complex of pyrrole is stable and, as expected, it undergoes reactions typical of enamines, such as addition of electrophiles at the 3-position, with re-closure to the 2-position with acetylene dicarboxylates. (**Eq. 10.70**).[104] Electrophilic additions to pyrroles primarily occur in the 2-position, but complexation to osmium completely shifts the addition site to the 3-position. This is another example of the remarkable ability of transition metals to change the reactivity of a given substrate.

Although osmium preferentially complexes the 4,5-position of pyrroles, the 3,4-position is also accessible via migration. Complexation here generates a very reactive azomethine ylide, which, in the presence of 1,3-dipolarophiles, undergoes facile cycloaddition (**Eq. 10.71**).[105]

Eq. 10.70

Eq. 10.71

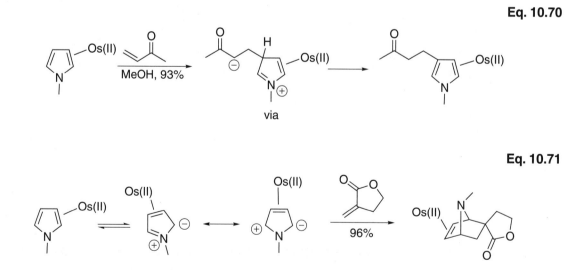

Reactions of Os complexes of furans are similar to those of pyrroles.[106] Electrophilic attack at the 4-position is favored, but because of the instability of the thus formed 4,5-η²-3H furanium species, the outcome is more complex and depends on the location of the substituents on the furan. Sequential electrophilic–nucleophilic additions furnish cis-4,5-substituted 4,5-dihydrofuran complexes (**Eq. 10.72**). Complexed 2-hydroxymethylfurans yield 2H-pyranone complexes via an intramolecular nucleophilic addition (**Eq. 10.73**).[107]

Eq. 10.72

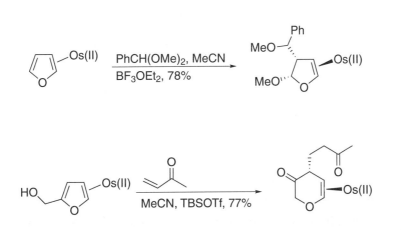

Eq. 10.73

In addition to osmium, η²-arene complexes of rhenium, molybdenum, and tungsten are also known and they exhibit very similar reactivities. Pyridines substituted in the 2-position (to "discourage" coordination to the pyridine nitrogen) form 3,4-η²-complexes from WTp(NO)(PMe₃)(η²-benzene).[108] These complexes react with electron-deficient alkenes to give [4+2] cycloaddition products (**Eq. 10.74**).

Eq. 10.74

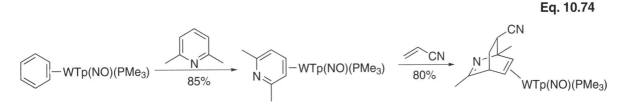

The Re, W, and Mo η²-complexes are chiral at the metal and the rhenium complexes can be resolved using (R)-α-pinene. An interesting cyclopentene formation is observed in some cases using η²-furan complexes. For example, reaction of an enantiomerically enriched complex with an electron-deficient alkene is shown in **Eq. 10.75**.[109] The metal is a stereocenter and this center was resolved prior to complexation to 2,5-dimethylfuran. The initially formed enolate-3H-furanium species undergo cleavage to give a metal-stabilized alkenyl cation. Addition of the enolate to the cation and oxidation of the product furnish the metal-free organic product.

Eq. 10.75

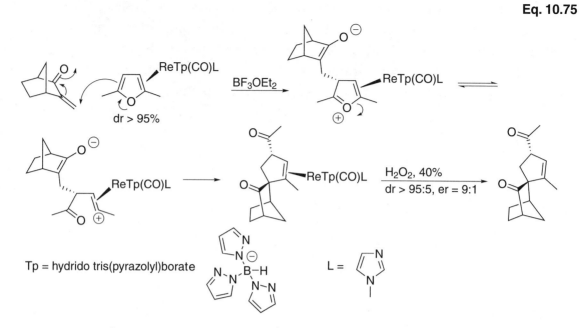

Tp = hydrido tris(pyrazolyl)borate   L =

At present, η^2-arene complexes have found little use in complex organic syntheses, since the reactions are stoichiometric in osmium, rhenium, molybdenum, or tungsten, and some of the laboratory procedures are cumbersome. However, as the methodology develops, its unique transformation chemistry should join the arsenal of the synthetic organic chemist, along with other organometallic processes, such as organopalladium chemistry, which were originally considered too acane for use in organic synthesis.

References

(1) Kündig, E. P. Transition Metal π-Arene Complexes in Organic Synthesis and Catalysis. In *Topics in Organometallic Chemistry*; Kündig, E. P., Ed.; Springer: Berlin, 2004; pp 71–94.

(2) For reviews, see a) Rosillo, M.; Dominguez, G.; Perez-Castells, J. *Chem. Soc. Rev.* **2007**, *36*, 1589. b) Semmelhack, M. F. Transition Metal Arene Complexes: Nucleophilic Addition. In *Comprehensive Organometallic Chemistry II*; Wilkinson, G., Stone, F. G. A., Abel, E. W., Eds.; Pergamon: New York, 1995; Vol. 12, pp 929–1015.

(3) a) Nicholls, B.; Whiting, M. C. *J. Chem. Soc.* **1959**, 551. b) Mahaffy, C. A. L.; Pauson, P. *Inorg. Synth.* **1979**, *19*, 154.

(4) a) Mosher, G. A.; Rausch, M. D. *Synth. React. Inorg. Met.-Org. Chem.* **1979**, *4*, 38. b) Rausch, M. D. *J. Org. Chem.* **1974**, *39*, 1787.

(5) a) Kündig, E. P.; Perret, C.; Spichiger, S.; Bernardinelli, G. *J. Organomet. Chem.* **1985**, *286*, 183. b) For preparation of the starting complex, see Desobry, V.; Kündig, E. P. *Helv. Chim. Acta* **1981**, *64*, 1288.

(6) a) Khand, I. U.; Pauson, P. L.; Watts, W. E. *J. Chem. Soc. C* **1968**, 2257. b) Astruc, D.; Dabard, R. *J. Organomet. Chem.* **1976**, *111*, 339.

(7) Abd-El-Aziz, A. S.; Bernardin, S. *Coord. Chem. Rev.* **2000**, *203*, 219.

(8) Kudinov, A. M.; Rybinskaya, M. I.; Struchkov, Y. T.; Yanovskii, A. I.; Petrovskii, P. V. *J. Organomet. Chem.* **1987**, *336*, 187.

(9) Schmidt, A.; Piotrowski, H.; Lindel, T. *Eur. J. Inorg. Chem.* **2003**, 2255.

(10) Bhasin, K. K.; Balkeen, W. G.; Pauson, P. L. *J. Organomet. Chem.* **1981**, *204*, C25.

(11) a) Sun, S.; Yeung, L. K.; Sweigart, D. A.; Lee, T. Y.; Lee, S. S.; Chung, Y. K.; Switzer, S. R.; Pike, R. D. *Organometallics* **1995**, *14*, 2613. b) Jackson, J. D.; Villa, S. J.; Bacon, D. S.; Pike, R. D.; Carpenter, G. B. *Organometallics* **1994**, *13*, 3972.

(12) This figure was adopted from an excellent (dated) review: Semmelhack, M. F. *J. Organomet. Chem.* **1976**, *1*, 361.

(13) For a review, see a) Kalinin, V. N. *Russ. Chem. Rev.* **1987**, *56*, 682. b) Uemura, M. *Adv. Met. Org. Chem.* **1991**, *2*, 195.

(14) Brown, D. A.; Raju, J. R. *J. Chem. Soc. A* **1966**, 1617.

(15) Rosca, S. J.; Rosca, S. *Rev. Chim.* **1974**, *25*, 461.

(16) Mahaffy, C. A. L.; Pauson, P. L. *J. Chem. Res.* **1979**, 128.

(17) Baldoli, C.; DelButtero, P.; Licandro, E.; Maiorana, S. *Gazz. Chim. Ital.* **1988**, *118*, 409.

(18) Semmelhack, M. F.; Hall, H. T., Jr. *J. Am. Chem. Soc.* **1974**, *96*, 7091, 7092.

(19) a) Pearson, A. J.; Chelliah, M. V. *J. Org. Chem.* **1998**, *63*, 3087 and references therein. b) Janetka, J. W.; Rich, D. H. *J. Am. Chem. Soc.* **1997**, *119*, 6488 and references therein.

(20) Pearson, A. J.; Cui, S. *Tetrahedron Lett.* **2005**, *46*, 2639.

(21) Pearson, A. J.; Shin, H. *J. Org. Chem.* **1994**, *59*, 2314.

(22) Storm, J. P.; Andersson, C.-M. *J. Org. Chem.* **2000**, *65*, 5264.

(23) For reviews, see a) Rosillo, Marta; Dominguez, Gema; Perez-Castells, Javier. *Chem. Soc. Rev.* **2007**, *36*, 1589. b) Pape, A. R.; Kaliappan, K. P.; Kündig, P. E. *Chem. Rev.* **2000**, *100*, 2917.

(24) Semmelhack, M. F.; Hall, H. T., Jr.; Farina, R.; Yoshifuji, M.; Clark, G. E.; Bargar, T.; Hirotsu, K.; Clardy, J. *J. Am. Chem. Soc.* **1979**, *101*, 3535.

(25) Semmelhack, M. F.; Wulff, W.; Garcia, J. L. *J. Organomet. Chem.* **1982**, *240*, C5.

(26) Cambie, R. C.; Rutledge, P. S.; Stevenson, R. J.; Woodgate, P. D. *J. Organomet. Chem.* **1994**, *471*, 133, 149.

(27) Semmelhack, M. F.; Knochel, P.; Singleton, T. *Tetrahedron Lett.* **1993**, *34*, 5051.

(28) In a synthesis of teleocidin A. Semmelhack, M. F.; Rhee, H. *Tetrahedron Lett.* **1993**, *34*, 1399.

(29) Fretzen, A.; Ripa, A.; Liu, R.; Bernardinelli, G.; Kündig, E.P. *Chem.—Eur. J.* **1998**, *4*, 102.

(30) Kündig, E. P.; Liu, R.; Ripa, A. *Helv. Chim. Acta* **1992**, *75*, 2657.

(31) Fretzen, A.; Kündig, E. P. *Helv. Chim. Acta* **1997**, *80*, 2023.

(32) a) Semmelhack, M. F.; Harrison, J. J.; Thebtaranonth, Y. *J. Org. Chem.* **1979**, *44*, 3275. b) Boutonnet, J. C.; Levisalles, J.; Normant, J. M.; Rose, E. *J. Organomet. Chem.* **1983**, *255*, C21.

(33) In a synthesis of *erythro*-juvabione. Pearson, A. J.; Paramahamsan, H.; Dudones, J. D. *Org. Lett.* **2004**, *6*, 2121.

(34) Semmelhack, M. F.; Yamashita, A. *J. Am. Chem. Soc.* **1980**, *102*, 5924.

(35) Pearson, A. J.; Gontcharov, A. V. *J. Org. Chem.* **1998**, *63*, 152.

(36) Pigge, F. C.; Coniglio, J. J.; Fang, S. *Organometallics* **2002**, *21*, 4505.

(37) Chae, H. S.; Burkey, D. J. *Organometallics* **2003**, *22*, 1761.

(38) Pigge, F. C.; Coniglio, J. J.; Rath, N. P. *J. Org. Chem.* **2004**, *69*, 1161.

(39) Kündig, E. P.; Cunningham, A. F., Jr.; Paglia, P.; Simmons, D. P.; Bernardinelli, G. *Helv. Chim. Acta* **1990**, *73*, 386.

(40) Kündig, E. P.; Inage, M.; Bernardinelli, G. *Organometallics* **1991**, *10*, 2921.

(41) Kündig, E. P.; Fabritius, C.-H.; Grossheimann, G.; Robvieux, F.; Romanens, P.; Bernardinelli, G. *Angew. Chem. Int. Ed.* **2002**, *41*, 4577.

(42) a) Kündig, E. P.; Bernardinelli, G.; Liu, R.; Ripa, A. *J. Am. Chem. Soc.* **1991**, *113*, 9676. b) Kündig, E. P.; Ripa, A.; Bernardinelli, G. *Angew. Chem., Int. Ed. Engl.* **1992**, *31*, 1071. c) Kündig, E. P.; Qualtropani, A.; Inage, M.; Ripa, A.; Dupre, C.; Cunningham, A. F., Jr.; Bourdin, B. *Pure Appl. Chem.* **1996**, *68*, 97.

(43) a) Amurrio, D.; Khan, K.; Kündig, E .P. *J. Org. Chem.* **1996**, *61*, 2258. b) Beruben, D.; Kündig, E. P. *Helv. Chim. Acta* **1996**, *79*, 1533. c) Kündig, E. P.; Amurrio, D.; Anderson, G.; Beruken, D.; Khan, K.; Ripa, A.; Longgang, L. *Pure Appl. Chem.* **1997**, *69*, 543.

(44) Kündig, E. P.; Cannas, R.; Laxmisha, M.; Ronggang, L.; Tchertchian, S. *J. Am. Chem. Soc.* **2003**, *125*, 5642.

(45) Yeh, M.-C. P.; Hwu, C.-C.; Lee, A.-T.; Tsai, M.-S. *Organometallics* **2001**, *20*, 4965.

(46) Roell, B. C., Jr.; McDaniel, K. F.; Vaughan, W. S.; Macy, T. S. *Organometallics* **1993**, *12*, 224.

(47) Lee, T.-Y.; Kang, Y. K.; Chung, Y. K.; Pike, R. D.; Sweigart, D. A. *Inorg. Chim. Acta* **1993**, *214*, 125.

(48) Lin, H.; Yang, L.; Li, C. *Organometallics* **2002**, *21*, 3848.

(49) Schwarz, O.; Brun, R.; Bats, J. W.; Schmalz, H.-G. *Tetrahedron Lett.* **2002**, *43*, 1009.

(50) For a review, see Semmelhack, M. F. Transition Metal Arene Complexes: Ring Lithiation In *Comprehensive Organometallic Chemistry II*; Wilkinson, G., Stone, F. G. A., Abel, E. W., Eds.; Pergamon: New York, 1995; Vol. 12, pp 1017–1037.

(51) a) Semmelhack, M. F.; Bisaha, J.; Czarny, M. *J. Am. Chem. Soc.* **1979**, *101*, 768. b) Card, R. J.; Trahanovsky, W. S. *J. Org. Chem.* **1980**, *45*, 2560. c) Gilday, J. P.; Negri, J. T.; Widdowson, D. A. *Tetrahedron* **1989**, *45*, 4605. d) Dickens, P. J.; Gilday, J. P.; Negri, J. T.; Widdowson, D. A. *Pure Appl. Chem.* **1990**, *62*, 575. e) Kündig, E. P.; Perret, C.; Rudolph, B. *Helv. Chim. Acta* **1990**, *73*, 1970.

(52) Sebhat, I. K.; Tan, Y.-L.; Widdowson, D. A.; Wilhelm, R.; White, A. J. P.; Williams, D. J. *Tetrahedron* **2000**, *56*, 6121.

(53) Davies, S. G.; Goodfellow, C. L.; Peach, J. M.; Waller, A. *J. Chem. Soc., Perkin Trans. 1* **1991**, 1019.

(54) a) Nechvatal, G.; Widdowson, D. A. *J. Chem. Soc., Chem. Commun.* **1982**, 467. b) Beswick, P. J.; Greenwood, C. S.; Mowlem, T. J.; Nechvatal, G.; Widdowson, D. A. *Tetrahedron* **1988**, *44*, 7325. c) In a synthesis of chungxinmycin methyl ester. Masters, N. F.; Mathews, N.; Nechvatal, G. Widdowson, D. A. *Tetrahedron* **1989**, *45*, 5955. d) Dickens, M. J.; Mowlen, T. J.; Widdowson, D. A.; Slawin, A. M. Z.; Williams, D. J. *J. Chem. Soc., Perkin Trans 1* **1992**, 323.

(55) a) Uemura, M.; Miyake, R.; Nakayama, K.; Shiro, M.; Hayashi, Y. *J. Org. Chem.* **1993**, *58*, 1238. b) Christian, P. W.; Gil, R.; Muñoz-Fernandez, K.; Thomas, S. E.; Wierzchleyski, A. T. *J. Chem. Soc., Chem. Commun.* **1994**, 1569.

(56) a) Alexakis, A.; Tomassini, A.; Andrey, O.; Bernardinelli, G. *Eur. J. Org. Chem.* **2005**, 1332. b) For a related reaction using a chiral oxazoline group as the director, see Overman, L. E.; Owen, C. E.; Zipp, G. G. *Angew. Chem. Int. Ed.* **2002**, *41*, 3884.

(57) a) Han, J.; Son, S. V.; Chung, Y. K. *J. Org. Chem.* **1997**, *62*, 8264. b) For related examples, see Alexakis, A.; Kanger, T.; Mangenez, P.; Rose-Munch, F.; Perrotey, A.; Rose, E. *Tetrahedron: Asymmetry* **1995**, *6*, 47; 2135.

(58) In a synthesis of korupensamines A and B. Watanabe, T.; Tanaka, Y.; Shoda, R.; Sakamoto, R.; Kamikawa, K.; Uemura, M. *J. Org. Chem.* **2004**, *69*, 4152.

(59) a) Uemura, M.; Daimon, A.; Hayashi, Y. *J. Chem. Soc., Chem. Commun.* **1995**, 1943. b) For related processes to make single atrop isomers of aryl naphthalene systems, see Watanabe, T.; Uemura, M. *J. Chem. Soc., Chem. Commun.* **1998**, 871.

(60) For a review, see Gibson, S. E.; Reddington, E. G. *Chem. Commun.* **2000**, 989.

(61) a) Ewin, R. A.; MacLeod, A. M.; Price, D. A.; Simpkins, N. S.; Watt, A. P. *J. Chem. Soc., Perkin Trans. 1* **1997**, 401. b) Price, D. A.; Simpkins, N. S.; MacLeod, A. M.; Watt, A. P. *J. Org. Chem.* **1994**, *59*, 1961. c) Kündig, E. P.; Quattrepani, A. *Tetrahedron Lett.* **1994**, *35*, 3497.

(62) Pache, S.; Botuha, C.; Franz, R.; Kündig, E. P.; Einhorn, J. *Helv. Chim. Acta* **2000**, *89*, 2436.

(63) In a synthesis of protected ent-actinoidinic acid. Wilhelm, R.; Widdowson, D. A. *Org. Lett.* **2001**, *3*, 3079.

(64) For reviews, see a) Uemura, M. *Org. React.* **2006**, *67*, 217. b) Davies, S. G.; McCarthy, T. D. Transition Metal Arene Complexes. Side Chain Activation and Control of Stereochemistry. In *Comprehensive Organometallic Chemistry II*; Wilkinson, G., Stone, F. G. A., Abel, E. W., Eds.; Pergamon: New York, 1995; Vol. 12, pp 1039–1070.

(65) Arzeno, H. B.; Barton, D. H. R.; Davies, S. G.; Lusinchi, X.; Meunier, B.; Pascard, C. *Nouv. J. Chim.* **1980**, *4*, 369.

(66) Solladie-Cavallo, A.; Farkhani, D. *Tetrahedron Lett.* **1986**, *27*, 1331.

(67) Baird, P. D.; Blagg, J.; Davies, S. G.; Sutton, K. H. *Tetrahedron* **1988**, *44*, 171.

(68) Schmalz, H. C.; Arnold, M.; Hollander, J.; Bats, J. W. *Angew. Chem., Int. Ed. Engl.* **1994**, *33*, 108.

(69) In a synthesis of 7,8-dihydroxycalamenene. Schmalz, H. G.; Hollander, J.; Arnold, M.; Duerner, G. *Tetrahedron Lett.* **1993**, *34*, 6259.

(70) Gibson, S. E.; Ham, P.; Jefferson, G. R. *Chem. Commun.* **1998**, 123.

(71) Zemolka, S.; Lex, J.; Schmalz, H.-G. *Angew. Chem. Int. Ed.* **2002**, *41*, 2525.

(72) For a review, see Davies, S. G.; Donohoe, T. J. *Synlett* **1993**, 323.

(73) a) Coote, S. J.; Davies, S. G.; Middlemiss, D.; Naylor, A. *Tetrahedron: Asymmetry* **1990**, *1*, 33. b) Coote, S. J.; Davies, S. G.; Middlemiss, D.; Naylor, A. *J. Chem. Soc., Perkin Trans. 1* **1985**, 2223. c) Coote, S. J.; Davies, S. G.; Middlemiss, D.; Naylor, A. *Tetrahedron Lett.* **1989**, *30*, 3581.

(74) In a synthesis of (–)-macrocarpal C. Tanaka, T.; Mikamiyama, H.; Maeda, K.; Iwata, C.; In, Y.; Ishida, T. *J. Org. Chem.* **1998**, *63*, 9782.

(75) Netz, A.; Drees, M.; Strassner, T.; Müller, T. J. J. *Eur. J. Org. Chem.* **2007**, 540.

(76) Netz, A.; Müller, T. J. J. *Tetrahedron* **2000**, *56*, 4149.

(77) Müller, T. J. J.; Netz, A. *Organometallics* **1998**, *17*, 3609.

(78) Jaouen, G.; Meyer, A. *Tetrahedron Lett.* **1976**, 3547.

(79) Jaouen, G.; Meyer, A. *J. Am. Chem. Soc.* **1975**, *97*, 4667.

(80) Schmalz, H. G.; Millies, B.; Bats, J.W.; Dürner, G. *Angew. Chem., Int. Ed. Engl.* **1992**, *31*, 631.

(81) a) Solladie-Cavallo, A.; Solladie, G.; Tsamo, E. *J. Org. Chem.* **1979**, *44*, 4189. b) Davies, S. G.; Goodfellow, C. L. *J. Chem. Soc., Perkin Trans. 1* **1989**, 192.

(82) a) Bromley, L. A.; Davies, S. G.; Goodfellow, C. L. *Tetrahedron: Asymmetry* **1991**, *2*, 139. b) Uemura, M.; Miyake, R.; Nakayama, K.; Shiro, M.; Hayashi, Y. *J. Org. Chem.* **1993**, *58*, 1238.

(83) Tipparaju, S. K.; Puranik, V. G.; Sarkar, A. *Org. Biomol. Chem.* **2003**, *1*, 1720.

(84) For the addition of organozinc reagents, see Ishimaru, K.; Ohe, K.; Shindo, E.; Kojima, T. *Synth. Commun.* **2003**, *33*, 4151.

(85) For reactions with silyl enol ethers, see Ishimaru, K.; Kojima, T. *Tetrahedron Lett.* **2001**, *42*, 5037.

(86) Kündig, E. P.; Xu, L. H.; Schnell, B. *Synlett* **1994**, 413.

(87) Baldoli, C.; Del Buttero, P.; Licandro, E.; Papagni, A.; Pilati, T. *Tetrahedron* **1996**, *52*, 4849.

(88) a) Del Buttero, P.; Baldoli, C.; Molteni, G.; Pilati, T. *Tetrahedron: Asymmetry* **2000**, *11*, 1927. b) Del Buttero, P.; Molteni, G. Papagni, A. *Tetrahedron: Asymmetry* **2003**, *14*, 3949.

(89) Ishimaru, K.; Koima, T. *J. Chem. Soc., Perkin Trans. 1* **2000**, 2105.

(90) Kündig, E. P.; Xu, L. H.; Romanens, P.; Benardinelli, G. *Synlett* **1996**, 270.

(91) For examples of pinacol couplings, see a) Taniguchi, N.; Kaneta, N.; Uemura, M. *J. Org. Chem.* **1996**, *61*, 6088. b) Taniguchi, N.; Hata, T.; Uemura, M. *Angew. Chem. Int. Ed.* **1999**, *38*, 1232.

(92) Taniguchi, N.; Uemura, M. *Synlett* **1997**, 51.

(93) Merlic, C. A.; Walsh, J. C. *J. Org. Chem.* **2001**, *66*, 2265.

(94) In a synthesis of dihydroxyserrulatic acid. Uemura, M.; Nishmura, H.; Minami, T.; Hayashi, Y. *J. Am. Chem. Soc.* **1991**, *113*, 5402.

(95) Bao, J.; Park, S.-H. K.; Geib, S. J.; Cooper, N. J. *Organometallics* **2003**, *22*, 3309.

(96) Corella, J. A., II; Cooper, N. J. *J. Am. Chem. Soc.* **1990**, *112*, 2832.

(97) For reviews, see a) Smith, P. L.; Chordia, M. D.; Harman, W. D. *Tetrahedron* **2001**, *57*, 8203. b) Harman, W. D. *Chem. Rev.* **1997**, *97*, 1953.

(98) Harman, W. D.; Taube, H.; *J. Am. Chem. Soc.* **1988**, *110*, 7906.

(99) a) Kopach, M. E.; Harman, W. D. *J. Am. Chem. Soc.* **1994**, *116*, 6581. b) Kopach, M. E.; Kolis, S. P.; Liu, R.; Robertson, J. W.; Chordia, M. D.; Harman, W. D. *J. Am. Chem. Soc.* **1998**, *120*, 6199. c) Kolis, S. P.; Kopach, M. E.; Liu, R.; Harman, W. D. *J. Am. Chem. Soc.* **1998**, *120*, 6205.

(100) Kolis, S. P.; Gonzalez, J.; Bright, L. M.; Harman, W. D. *Organometallics* **1996**, *15*, 245.

(101) a) Winemiller, M. D.; Harman, W. D. *J. Am. Chem. Soc.* **1998**, *120*, 7835. b) Ding, F.; Kopach, M. E.; Sabat, M.; Harman, W. D. *J. Am. Chem. Soc.* **2002**, *124*, 13080.

(102) Chordia, M. D.; Harman, W. D. *J. Am. Chem. Soc.* **2000**, *122*, 2725.

(103) Kolis, S. P.; Kopach, M. E.; Liu, R.; Harman, W. D. *J. Am. Chem. Soc.* **1998**, *120*, 6205.

(104) Hodges, L. M.; Gonzalez, J.; Koonts, J. I.; Myers, W. H.; Harman, D. *J. Org. Chem.* **1995**, *60*, 2125.

(105) Gonzalez, J.; Koontz, J. I.; Myers, W. H.; Hodges, L. M.; Sabat, M.; Nilsson, K. R.; Neely, L. K.; Harman, W. D. *J. Am. Chem. Soc.* **1995**, *117*, 3405.

(106) Chen, H.; Liw, R.; Myers, W. H.; Harman, W. D. *J. Am. Chem. Soc.* **1998**, *120*, 509.

(107) Chen, H.; Caughey, R.; Liu, R.; McMills, M.; Rupp, M.; Myers, W. H.; Harman, W. D. *Tetrahedron* **2000**, *56*, 2313.

(108) Graham, P. M.; Delafuente, D. A.; Liu, W.; Myers, W. H.; Sabat, M.; Harman, W. D. *J. Am. Chem. Soc.* **2005**, *127*, 10568.

(109) Friedman, L. A.; You, F.; Sabat, M.; Harman, D. F. *J. Am. Chem. Soc.* **2003**, *125*, 14980.

Index